Neuromimetic Semantics

Coordination, Quantification, and Collective Predicates

Neuromimetic Semantics

Coordination, Quantification, and Collective Predicates

Harry Howard

Department of Spanish and Portuguese
322-D Newcomb Hall, Tulane University
New Orleans, LA 70118-5698, USA

2004

ELSEVIER
Amsterdam – Boston – Heidelberg – London – New York – Oxford – Paris
San Diego – San Francisco – Singapore – Sydney – Tokyo

ELSEVIER B.V.
Sara Burgerhartstraat 25
P.O. Box 211, 1000 AE Amsterdam
The Netherlands

ELSEVIER Inc.
525 B Street, Suite 1900
San Diego, CA 92101-4495
USA

ELSEVIER Ltd
The Boulevard, Langford Lane
Kidlington, Oxford OX5 1GB
UK

ELSEVIER Ltd
84 Theobalds Road
London WC1X 8RR
UK

First edition 2004

Library of Congress Cataloging in Publication Data
A catalog record is available from the Library of Congress.

British Library Cataloguing in Publication Data
A catalogue record is available from the British Library.

ISBN: 0 444 50208 4

∞ The paper used in this publication meets the requirements of ANSI/NISO Z39.48-1992 (Permanence of Paper).

Transferred to Digital Printing 2009

Preface

Like many disciplines, contemporary linguistic theorization abounds in ironies that seem to be the result of historical accident. Perhaps the most striking one is the fact that practically no contemporary theorization draws inspiration from – or makes any contribution to – what is known about how the human brain works. This is striking because linguistics purports to be the study of human language, and language is perhaps the defining characteristic of the human brain at work. It would seem natural, if not unavoidable, that linguists would look to the brain for guidance. It seems just as natural that neurologists would look to language for guidance about how a complex human ability works. Yet neither is the case.

This book attempts to iron out this irony. It does so by taking up a group of constructions that have been at the core of linguistic theory for at least the last two thousand years, namely coordination, quantification, and collective predicates, and demonstrating how they share a simple design in terms of neural circuitry. Not *real* neural circuitry, but rather artificial neural circuitry that tries to mimic the real stuff as closely as possible on a serial computer. The next few pages summarize how this is done.

MODEST VS. ROBUST THEORIES OF SEMANTICS

The first chapter introduces several of the fundamental concepts of the book in an informal way. It begins by explaining how the meaning of coordinators like *and, either...or,* and *neither...nor* can be expressed by patterns of numbers and goes on to discuss simple sequential rules that can decide whether a given pattern instantiates one of these coordinators or not. It then extends this format to the quantifiers *each/every/all, some,* and *no.* It points out that for numbers greater than 100, such sequential rules cannot arrive at their answer in an amount of time that is reasonable for what is known about the speed of neurological processing.

Having followed a well-established linguistic analysis to an untenable conclusion, some alternative must be sought. Computationally, the only likely alternative is parallel processing, which is one aspect of a more general framework of neurologically-plausible or **neuromimetic** computation.

Neuromimetic computation is so alien to contemporary linguistic theorization that it cannot be taken for granted that the reader has had any exposure to it. Thus the first order of business is to accomplish some such exposure. This is ensured by a fairly detailed review of the neurology of vision, and especially the first few processing steps (retina-LGN-V1), known collectively as **early vision**. Along the way, several biological building blocks are introduced, such as pyramidal neurons, lamination, columnar organization, excitation and inhibition, feedforward and feedback interactions, and dendritic processing. More to the point of the book's goals, several computational building blocks are also introduced, such as the summation and thresholding of positive and negative inputs, redundancy reduction, Bayesian error correction, the extraction of invariances, and mereotopology.

In order to prove one computational framework to be superior to another, great care must be taken in choosing appropriate guidelines to evaluate competing proposals. The third section of the chapter reviews two main guidelines.

The first is the tri-level hypothesis of Marr (1981[1977], 1982), which examines an information processing system through the perspective of distinct though interconnected implementational, algorithmic, and computational levels. Following Golden (1996) on artificial neural networks, Marr's levels can be equated with dynamical systems theory, optimization theory, and rational decision theory, respectively. We hasten to add that our own exposition focuses on dynamical systems, leaving the other two implicit in the representational treatment. Since Marr's approach often brings up the competence/performance distinction of Chomsky (1965) and much posterior work, the usefulness of this distinction is examined, and it is concluded that competence refers to the connection matrix of the relevant grammatical phenomenon, while performance refers to the entire network in which the connection matrix is embedded.

In view of the fact that our networks are trained on data that have a linguistic interpretation, it makes sense to also review the standard guidelines for the evaluation of linguistic constructs, which, following Chomsky (1957, 1965) would be grammars or grammar fragments, as set forth in the following definitions from Chomsky, 1964, p. 29:

0.1 a) A grammar that aims for **observational adequacy** is concerned merely to give an account of the primary data (e. g. the corpus) that is the input to the acquisition device.

b) A grammar that aims for **descriptive adequacy** is concerned to give a correct account of the linguistic intuition of the native speaker; in other words, it is concerned with the output of the acquisition device.

c) A linguistic theory that aims for **explanatory adequacy** is concerned with the internal structure of the acquisition device; that is, it aims to provide a principled basis, independent of any particular language, for the selection of the descriptively adequate grammar of each language.

An appropriate rewording of these definitions around the substitution of the phrase "neuromimetic network" for "acquisition device" leads to a restatement of Chomsky's three levels of adequacy in pithy, connectionist terms:

0.2 a) An **observationally adequate** model gives the observed output for an appropriate input.
b) A **descriptively adequate** model gives an output that is consistent with other linguistic descriptions.
c) An **explanatorily adequate** model gives an output in a way that is consistent with the abilities of the human mind / brain.

These criteria provide clear guidelines for rest of the monograph.

The final bit of the chapter attempts to pull all of these considerations together into an analysis of the target phenomena of logical coordination and quantification. It is proposed that these constructions, and other logical operations listed in Chapter 3, express a signed measure of correlation between their two arguments. This statement can be seen as a definition of the function that logical coordination and quantification compute and so qualifies as their computational analysis in Marr's tri-level theory.

In view of the wide variety of topics touched on in this introductory chapter, some means of tying them all together would be extremely helpful to keeping the reader oriented towards the chapter's rhetorical goal. This 'hook' is the distinction between "modest" and "robust" semantic theories championed in Dummett, 1975, and 1991, chapter 5. A modest semantic theory is one which explains an agent's semantic competence, without explaining the abilities that underlie it, such as how it is created. In contrast, a robust semantic theory *does* explain the underlying abilities. This book defends a robust theory of the semantics of logical coordination and quantification, while looking askance at theories that only offer a modest semantics thereof.

SINGLE NEURON MODELING

The second chapter makes up a short textbook on computational neuroscience. It explains the principle neuronal signaling mechanism, the action potential, staring with the four differential equations of the Hodgkin-Huxley model and simplifying them to the single differential equation of an integrate-and-fire model, and finally the non-differential equation of a rate-based model. It spends several pages explaining how synapses pass potentials from cell to cell and how dendrites integrate synaptic potentials within a cell, both passively and actively. It winds up with a demonstration of how the biological components of a real (pyramidal) neuron correspond to the mathematical components of an artificial neuron. It is inserted in its particular place because it fleshes out many of the neurological concepts introduced in the first chapter and sets forth the neurological grounding of correlation that is the subject of the third chapter, though the reader is free to refer to it as a kind of appendix to consult when the explication of some aspect of neurological signaling is needed.

LOGICAL MEASURES

The third chapter begins to flesh out the claim that logical coordination and quantification denote a measure of correlation between two arguments. The goal is to examine the various branches of mathematics which can provide such a measure. To this end, we first review measure theory to see what it means to be a measure. Along the way, we argue that a signed measure is more appropriate than an unsigned measure, because signed measures interface with trivalent logic, and what is more important for explanatory adequacy, signed measures find a neurological grounding in the correlation of two spike trains.

With this specification of the sought-after measure, we briefly examine cardinality, statistics, probability theory, information theory, and vector algebra for candidates. Only statistics and vector algebra supply an obvious signed measure, that of Pearson's or Spearman's correlation coefficient and the cosine of the angle of two vectors. Moreover, the two can be related by the fact that the standard deviation of a variable and the magnitude of a vector are proportional to one another. Since neurological phenomena are often represented in vector spaces, this monograph prefers the angular measure of correlation.

Nevertheless, this does not mean that probability theory or information theory are of no use to us. They can provide additional tools for the analysis of the logical elements if the notions of probability and information content are extended into the signed half of the real number line. Since these are contentions that strike at the foundations of mathematics, we do not pursue them to any length, but we do show how they follow from simple considerations of symmetry.

With a sound basis for a signed measure of correlation in vector algebra, we next turn to demonstrating its topological structure, specifically in the order topology. This result paves the way for applying a family of notions related to convexity to the logical operators, in particular Voronoi tesselation, quantization, and basins of attraction. Along the way, claims to further support for the emerging formal apparatus are made through appeals to similarities to the scales of Horn (1989) and the natural properties of Gärdenfors (2000).

While the semantics of the logical operators can be drawn from the cosine of two angles in a vector space, their usage or pragmatics is better stated in terms of information theory. The asymmetry of the negative operators with respect to their positive counterparts is especially tractable as an implicature drawn from rarity or negative uninformativeness.

THE REPRESENTATION OF COORDINATOR MEANINGS

The fourth chapter finally brings us to some facts about language, and in particular about the logical coordinators AND, OR, NAND, and NOR. It is divided into two halves, the first dealing with phrasal coordination and the second with clausal coordination.

The first half attempts to prove a vector semantics for phrasal coordination by enumerating how the coordination of the various sub-clausal categories in English can be stated in terms of vector angle. It is rather programmatic and does not adduce any new data about phrasal coordination.

The second half is much more innovative. It also attempts to prove a vector semantics of clausal coordination by enumeration, but ... what should be enumerated? It argues that the relations adduced in recent work on discourse coherence provide a suitable list of testable concepts and goes on to define the thirteen coherence relations collated in Kehler (2002) in terms of the cosine of two vectors – one for each clause of the putative relation. It turns out that only three such relations license a paraphrase with AND – the three whose components are correlated, exactly as predicted by the theory of Chapter 3. Moreover, two of these three license asymmetric coordination, to wit, just the two that impose an additional constraint that that the two clauses lie in the canonical order of precedence.

The chapter winds up with a discussion of how to reduce the sixteen connectives allowed by propositional logic to the three or four coordinators observed in natural language.

NEUROMIMETIC NETWORKS FOR COORDINATOR MEANINGS

Chapter 5 continues the primer on computational neuroscience begun in Chapter 2 by testing how seven basic network architectures perform in simulating the acquisition and recognition of coordinator meanings. It divides the architectures into two groups: those that classify their input into hyperplanes and those that classify their input into clusters. Each network architecture is used to classify the same small sample of coordinator meanings in order to test its observational adequacy for modeling human behavior. The comparison is effected by simulations of network training and performance programmed in MATLAB, and full program code and data files are available from the author's website.[a] Every simulation is consequently easy for the reader to replicate down to its finest details.

In the group of hyperplane classifiers belong the single-layer perceptron networks (SLPs) and backpropagation or multilayer perceptron networks (MLPs). Such networks can easily learn the patterns made by the logical coordinators, but they are not neurologically plausible.

In the group of cluster classifiers belong the instar, competitive, and learning vector quantization (LVQ) networks. The last is the most accurate at learning patterns made by the logical coordinators and has the explanatory advantage of considerable neurological plausibility. LVQ is consequently the architecture adopted for the rest of the monograph.

---

[a] The url is: <http://www.tulane.edu/~howard/NeuroMimSem/>.

THE REPRESENTATION OF QUANTIFIER MEANINGS

The sixth chapter extends the descriptive adequacy of the correlational analysis of the logical coordinators by generalizing it to a domain with which it has long been known to share many formal similarities, namely quantification. The goal is to amplify the analysis of AND, OR, NAND, and NOR in a natural way to the four logical quantifiers, ALL, SOME, NALL, and NO. By so doing, an account can be rendered of ALL as a "big AND" and SOME as a "big OR", in the felicitous phrasing of McCawley, 1981, p. 191.

The chapter reviews the fundamental results of Generalized Quantifier (GQ) theory on the set-theoretic representations of natural language quantifiers by means of the conditions of Quantity, Extension, and Conservativity and the number-theoretic representation engendered by them. This number-theoretic representation of quantifier denotations, known as the Tree of Numbers, is shown to be descriptively inadequate by dint of not corresponding to the syntactic form of nominal quantifiers. A more adequate representation turns out to be isomorphic to the representation of coordinator meanings, allowing the correlational analysis of logical coordination to be generalized to logical quantification.

With respect to explanatory adequacy, all of the GQ conditions discussed are reduced to independently motivated principles of neurological organization.

ANNS FOR QUANTIFIER LEARNING AND RECOGNITION

The seventh chapter extends the LVQ analysis of logical coordination to logical quantification. Since most of the hard work of demonstrating the adequacy of LVQ over the other architectures is done in the fifth chapter, the seventh chapter concentrates on refining the analysis. In particular, much attention is paid to how well LVQ generalizes in the face of defective data and counterexamples. The latter consideration leads to an augmentation of the LVQ network to make it sensitive to exceptions. Finally, it is argued that the traditional universal, existential, and negation operators can be derived by allowing LVQ to optimize the number of connections from its first to its second layer, in accord with recent analyses of early vision reviewed in Chapter 1.

INFERENCES AMONG LOGICAL OPERATORS

The eighth chapter continues the exploration of the descriptive adequacy of the LVQ analysis of coordination/quantification by interpreting it as part of a larger neuromimetic grammar and demonstrating how inferences can be modeled as the spread of activation between items in this grammar. The inferences in question are organized into the Aristotelian Square of Opposition and its successors.

A grammatical design is adopted inspired on the parallel processing architecture of Jackendoff, 2002, with separate levels for concepts, semantics, syntax, and phonology. The levels are interconnected by lexical items, under the assumption that a lexical entry connects separate semantic, syntactic, and phonological stipulations. The grammar is implemented in the Interactive Activation and Competition (IAC) algorithm of McClelland and Rumelhart. This model possesses the advantages of neurological plausibility, representational transparency, and a long tradition of usage .

All of the inferences of opposition can be derived quite satisfactorily from an IAC network based on the preceding LVQ network. Each node in the IAC network is licensed directly by a lexical item or indirectly by activation spreading from a lexical item. All of the links between nodes are bidirectional, in order to simulate the ideal speaker-hearer. Thus, the speaker is simulated by turning on phonological nodes and letting activation spread up to the semantics, while the hearer is simulated by turning on semantic nodes and letting activation spread down to the phonology.

THE FAILURE OF SUBALTERNACY

The ninth chapter gives yet another push to the descriptive-adequacy envelope of LVQ/IAC coordination/quantification by taking up collective predicates in which the distributivity of coordinated and quantified subjects is blocked. Technically, the fact to be examined is why collective predicates block the subaltern implication from universals to particulars in the Square of Opposition.

The chapter begins with the collection of a number of predicates that prevent this implication from going through. Four main groups are discerned, as listed below. The pairs of sentences illustrate the failure of subalternacy between a conjunction and one of its conjuncts for these classes:

0.3 a) Centripetal constructions (motion towards a center)
The copper and the zinc united in the crucible.
≠ The copper united in the crucible.

b) Centrifugal constructions (motion away from a center)
The copper and the zinc separated under the electric charge.
≠ The copper separated under the electric charge.

c) Tandem [symmetric] constructions ('motion' in tandem)
Copper and zinc are similar. ≠ Copper is similar.

d) Reciprocals
George and Martha love each other. ≠ George loves each other.

The first two form the subgroup of center-oriented constructions. Tandem predicates are not taken up in the monograph, in order to allow space for a thorough treatment of the center-oriented classes, plus reciprocals.

The discussion goes deep into the data on center-oriented constructions in order to reach the correct level of generality. It is found that, besides an orientation towards or away from a center, they are also distinguished by the dimensionality of the center-oriented argument. If it is at least one-dimensional, e.g. "collide" or "separate", then the predicate licenses a minimum cardinality of two, but if it is at least two-dimensional, e.g. "gather" or "disperse", the predicate licenses a minimum cardinality of four or five. It follows that subalternacy implication from universals to existentials fails for those cases in which the existential falls below the minimum cardinality of the predicate.

Along the way to reaching this conclusion, LVQ networks are designed for reflexive/reciprocal and center-oriented constructions, and it is shown that they correlate in the association of one entity with its nearest complements. In this way, the LVQ/IAC approach is demonstrated to have a wide potential for descriptive adequacy.

NETWORKS OF REAL NEURONS

Chapter 10 explains why the exposition draws so much on extrapolation from vision and computational neuroscience and so little on the neurology of language. A history of neurolinguistics is sketched taking pains to demonstrate that despite more than a hundred years of investigation and even with the recent advances in neuroimaging, no technology has yet reached the degree of resolution necessary to 'see' how the brain performs coordination or quantification. This book can be taken as a guide to what to look for once the technology becomes available.

In order to not end on a note of disappointment, the chapter is rounded out with a short discursion on memory, and in particular on episodic memory as effected by the hippocampus. Shastri (1999, 2001) has proposed that the dentate gyrus binds predicates and their arguments together in a fashion reminiscent of predicate calculus. Various simulations of Shastri's system are presented using an integrate-and-fire model in order to illustrate how temporal correlation can indeed ground coordinator representations.

THREE GENERATIONS OF COGNITIVE SCIENCE

The final chapter deepens the explanatory adequacy of the LVQ/IAC analysis of coordination, quantification, and collectivization by evaluating the analysis in the perspective of three "generations" or schools of explanation in cognitive science.

The first two generations were originally identified by George Lakoff. The First Generation is characterized as the cognitive science of the "Disembodied and Unimaginative Mind", a research program pursued in classical artificial intelligence and generative linguistics which draws its descriptive apparatus from set theory and logic. The Second Generation is characterized as the the cognitive science of the "Embodied and Imaginative Mind". It rejects set theory and logic to pursue putatively non-mathematical formalisms like prototype theory, image schema, and conceptual metaphor. Its practitioners, at least in linguistics, go under the banner of the Cognitive Linguistics of George Lakoff and Ronald Langacker.

Our contribution to this debate is to point out that a Third Generation is emerging, the cognitive science of the "Imaged and Simulated Brain", that does not share the 'math phobia' of the second generation. It points to the unmistakable topological and mereological properties of prototypes and image schema and strives to derive such objects from neurologically plausible assumptions. These assumptions are embodied in, and can be tested by, artificial neural networks.

The philosophical foundations of this approach are being developed by Jens Erik Fenstad in general, by Barry Smith and others in mereotopology, and by Peter Gärdenfors in conceptual space. After a brief review of these developing frameworks, we argue that they share many common properties. Mereotopology is the more general of the two and has the potential to formalize most of the ideas that shape Gärdenfors' conceptual space. Moreover, mereotopology has an obvious realization in LVQ networks.

To bring the book full circle, the chapter ends with a reconsideration of the desiderata for knowledge representation and intelligent computation introduced in Chapter 1 in the light of LVQ and mereotopology. It is shown that these desiderata are implemented by or grounded in LVQ and mereotopology, with the result that the general principles sketched in the first chapter are fully supported by the explanatory mechanisms developed in such painstaking detail in the rest of the text.

WHO SHOULD BENEFIT FROM THIS BOOK

This book is designed primarily for the benefit of linguists, and secondarily for logicians, neurocomputationalists, and philosophers of mind and of language. It may even be of benefit to mathematicians. It will certainly be of benefit to all those that want an introduction to natural language semantics, logic, computational neuroscience, and cognitive science.

MATLAB CODE

As was mentioned in the footnote above, the book's website <http://www.tulane.edu/~howard/NeuroMimSem/> contains all of the data sets and MATLAB code for the simulations described in the text.

Acknowledgements

I have had the pleasure of the support of many people over the years that it has taken me to finish this book. Foremost are the various chairs of my department, Dan Balderston, Maureen Shea, and Nicasio Urbina. Tulane's Linguistics Program has been my home away from home, and I would like to thank Tom Klingler, Judie Maxwell, George Cummins, Ola-Nike Orie, Graeme Forbes, and Vickie Bricker for it. Tulane's program in Latin American Studies has supported my travel for research, for which I am indebted to the directors Richard Greenleaf and Tom Reese. Tulane's program in Cognitive Studies has been a constant source of encouragement, in the body of Radu Bogdan. More recently, Tulane's Neuroscience Program has accepted my collaboration with open arms, for which I am grateful to Gary Dohanich and Jeff Tasker. Finally, I have tried out some of this material on the students of my Brain and Language and Computational Neuroscience classes, and I wish to thank Aaron Nitzken, Chuck Michelson, Lisbeth Phillips, and Paulina De Santis for the insightful comments.

Outside of Tulane, I have drawn sustenance from my friendships with Steve Franks, Per-Aage Brandt, Hugh Buckingham, José Cifuentes, Clancy Clements, Bert Cornille, Rene Dirven, Bob Dewell, David Eddington, Lene Fosgaard, Joe Hilferty, Suzanne Kemmer, Ron Langacker, Robert MacLaury, Ricardo Maldonaldo, Jan Nuyts, Svend Østergaard, Enrique Palancar, Bert Peeters, Marianna Pool, Tim Rohrer, Chris Sinha, Augustin Soares, Javier Valenzuela, and Wolfgang Wildgen.

Of course, none of this would have been possible without the love of my wife, Rosa.

Table of contents

Chapter 1

Modest vs. robust theories of semantics

This chapter introduces the reader to the neuromimetic modeling of coordination and quantification by first showing how it cannot be done and then deducing from the visual system how it should be done. The review of the visual system is generalized to a series of desiderata which any cognitive ability should satisfy. These desiderata are then incorporated into specific proposals for engineering natural information-processing systems and for evaluating linguistic hypotheses based on natural computation. Finally, a particular representation of logical coordination and quantification is proposed that can serve as an input corpus for a neuromimetic dynamical system.

1.1. THE PROBLEM

Here is the problem. English has the words *and, (either)...or,* and *(neither)...nor* that can be used to link two or more syntactic categories together into a larger whole. (1.1) gives examples to show exactly what usage of these words we are focusing on:

1.1 a) Marta and Rukayyah are gourmets.
b) Either Marta or Rukayyah is a gourmet.
c) Neither Marta nor Rukayyah are gourmets.

This grammatical construction is known as **coordination**, and the words *and, (either)...or,* and *(neither)...nor* that specify the sort of coordination can be called **coordinators**. English has other words that perform a similar function, such as *but*, but they depend on information in their discourse context, which makes them more difficult to analyze. The triad *and, (either)...or,* and *(neither)...nor,* which do not depend on any contextual information for their comprehension, shall be referred to as **logical coordinators** in order to distinguish them from context-dependent coordinators like *but*. This work concentrates on the logical coordinators to the exclusion of the others.

As a final bit of nomenclature, the sort of a logical coordinator is expressed by the upper case version of the English morpheme – AND, OR, and NOR – in an attempt to refer to the coordinative procedure itself, and not the morpheme that realizes it in a given language.

1.1.1. Modest vs. robust semantic theories

To our way of thinking, there are two great families of approaches to this or any other issue in natural language semantics. The two follow from a distinction drawn in Dummett, 1975; 1991, chapter 5, between "modest" and "robust" semantic theories. A modest semantic theory is one which explains an agent's semantic competence, without explaining the abilities that underlie it, such as how it is created. In contrast, a robust semantic theory *does* explain the underlying abilities. Ludlow, 1999, p. 39 adds that it is not helpful to consign such underlying abilities to some deeper psychological theory that linguists need not traffic in, since our concern should be with investigating the knowledge that underlies semantic competence, rather than drawing pre-theoretic boundaries around the various disciplines.

The goal of this monograph is to persuade the reader that a robust perspective on the logical coordinators and quantifiers opens up new vistas of examination and explanation that are just not available under the modest perspective.

1.1.2. A modest solution: counting

How are logical coordinations like the three illustrated in (1.1) understood? A popular theory within natural language semantics equate understanding with knowing the conditions under which they are, or would be, true (see Chierchia and McConnell-Ginet 1990, Chap. 2; this approach can be traced to Tarski (1935; 1944). For **clausal coordination**, such as in (1.2), the hypothesis is that the whole coordination is true only if some combination of its clausal parts is true, as given in (1.3):

1.2 a) Marta is a gourmet, and Rukayyah is a gourmet.
b) Either Marta is a gourmet, or Rukayyah is a gourmet.
c) Marta is not a gourmet, nor is Rukayyah a gourmet.

1.3 a) 'Marta is a gourmet, and Rukayyah is a gourmet' is true if and only if both sentences 'Marta is a gourmet', and 'Rukayyah is a gourmet' are true. Otherwise, it is false.
b) 'Either Marta is a gourmet, or Rukayyah is a gourmet' is true if and only if at least one of the sentences 'Marta is a gourmet', and 'Rukayyah is a gourmet' is true. Otherwise, it is false.
c) 'Marta is not a gourmet, nor is Rukayyah a gourmet' is true if and only if none of the sentences 'Marta is a gourmet', and 'Rukayyah is a gourmet' is true. Otherwise, it is false.

The informal truth conditions of (1.2) suggest that a clausal coordination is understood by quantifying the 'truths' of its sentential parts. For AND, all sub-sentences must be true, for OR, at least one sub-sentence must be true, and for NOR, no sub-sentence can be true.

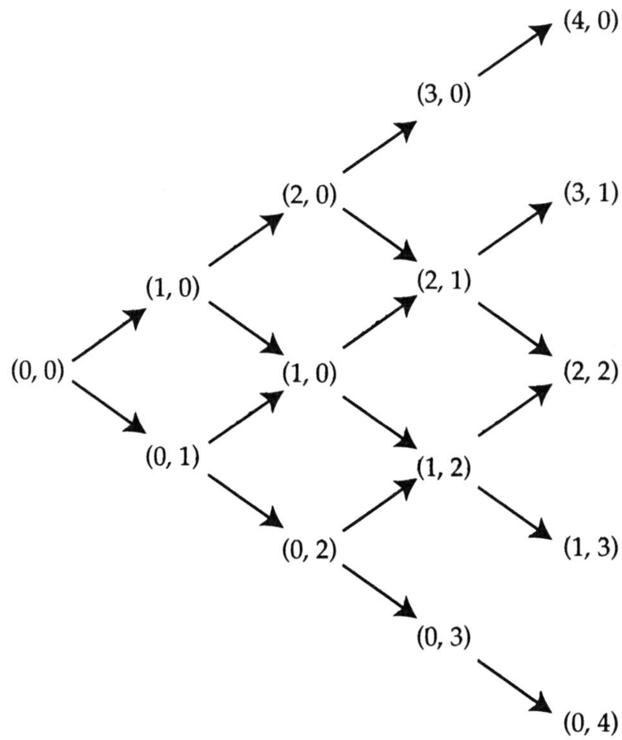

Figure 1.1. Decision tree for possible values of (true, false) for four clauses.

The drawback of truth-conditional coordination is that it only works for the coordination of units with a truth value of their own. It does not extend to the nominal coordinations seen in (1.1), which lack an overt decomposition into true or false sub-clauses. However, the generative grammar of the early 1960's elaborated a procedure for decomposing nominal coordinations into covert sub-clauses, so that the nominal coordinations of (1.1) can be evaluated semantically on a par with the clausal coordinations of (1.2). Though there are exceptions reviewed in Chapter 9 that show this procedure can not be extended to all instances of coordination, let us assume for the time being that it is indeed possible to find appropriate sub-clauses for nominal coordination and see how precise this hypothesis can be made.

One way of making it more precise is by creating a diagram that allows us to picture all of the possibilities at once. Since the procedure counts values of truth or falsity, we can abbreviate the verification process by letting the numbers stand in for the actual truth values themselves. The natural data structure for

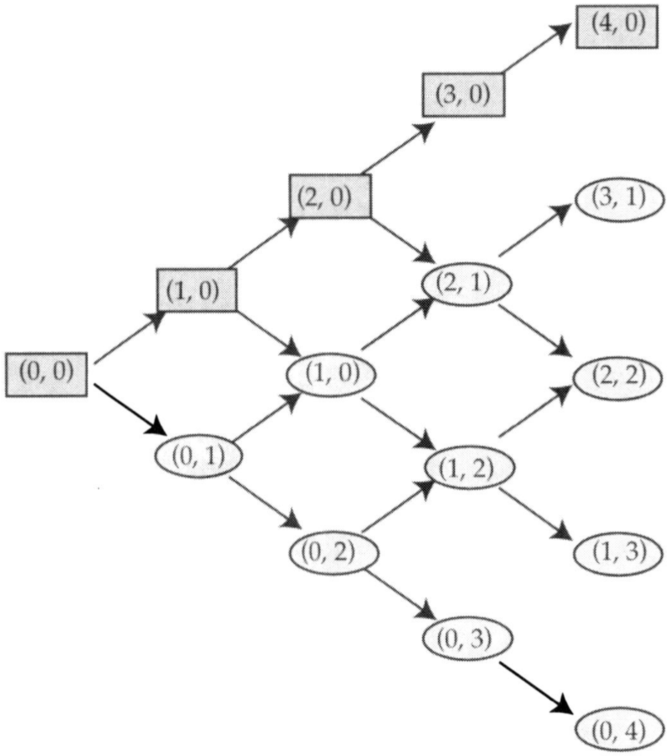

Figure 1.2. The path of AND through Figure 1.1.

these numbers holds a place for each sort of value, such as *(number true, number false)*. For four sentences, the number of possible constellations of true and false is organized into Fig. 1.1. Starting at a count of zero for both true and false, (0, 0), the arrows take the tally to the next stage according to value of the clause examined, until all four are examined.

The usefulness of this visualization of the procedure can be appreciated by highlighting that path that would be followed in the case of AND, given in Fig. 1.2, where the squares indicate the tallies that make AND true, and the circles indicate those that make it false. In prose, what happens is that the process starts at the point (0, 0) – the point at which we know nothing at all about the coordination. We ask whether the first clause is true. If it is, we add a 1 to the first or x position of (0, 0), which takes us to the new point (1, 0). If it were false, we would add a 1 to the second y position of (0, 0), which would take us to the point (0, 1), at which the verification of AND fails. Tracing the path laid out by the boxes up the top edge of the diagram demonstrates how each additional

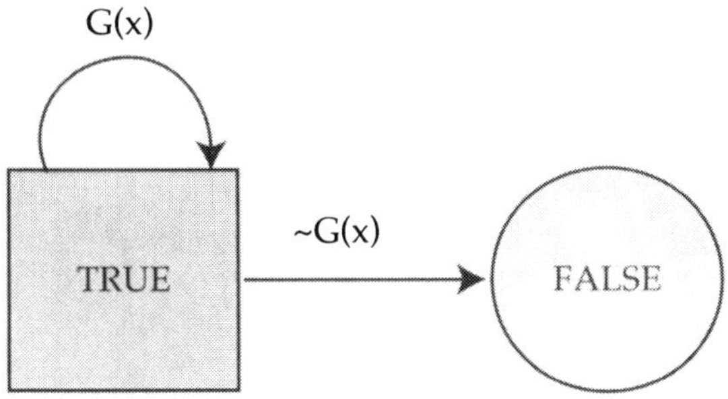

Figure 1.3. A finite state automaton for AND.

node represents the addition of another 1 to (x, 0) until all four clauses have been examined affirmatively, at node (4, 0).

1.1.3. Finite automata for the logical coordinators

This simple procedure of answering yes or no – true or false – to a list of questions is well-known in the theory of computation as a **finite automaton**. It is represented pictorially by a **finite state transition graph** or **flow diagram**. The reader may have already surmised that the potentially infinite diagram of Figures 1.1 and 1.2 can be conflated into a single square and circle as in Fig. 1.3, interconnected with the appropriate relations. The derivation begins at the square marked TRUE. There are two choices: if the first clause is true, the derivation follows the transition labeled *G(x)* – "x is a gourmet", which returns it to the **accepting state** of TRUE, and it moves down the list to the next clause. As long as the next clause is true, the derivation keeps advancing down the list, and it ends successfully when the list is exhausted. However, if at any point on the list a person is *not* a gourmet, the derivation must take the transition marked *~G(x)*, "x is not a gourmet", which leads it to the **refuting state** of FALSE, from which there is no exit. There is no reason to finish reading the list, and the derivation fails. In this case, AND cannot be used for the coordinator *X* under consideration.

Let us try this reasoning out on the opposite of AND, NOR. In contrast to AND, with NOR the procedure must be able to evaluate each clause as false, that is to say:

1.4	a)	Is Marta a gourmet?	- No (that is false).
	b)	Is Rukayyah a gourmet?	- No (that is false).

The coordinator machine for this procedure just switches the polarity of the evaluating property, as depicted in Fig. 1.4a. In other words, the evaluation of

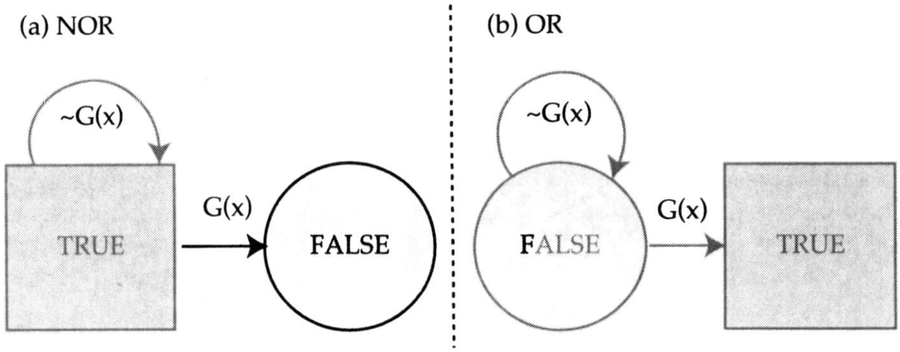

Figure 1.4. A finite state automaton for: (a) NOR, (b) OR.

the situation is true as long as no gourmets are found among the two – or any other number of – names.

The coordinator OR differs from AND and NOR in that it needs to find a single true clause to be used truthfully. Thus it does not matter that Marta is not a gourmet, as long as Rukayyah is. Such a truthful situation for OR is the one listed in (1.5):

1.5 a) Is Marta a gourmet? - No (that is false).
b) Is Rukayyah a gourmet? - Yes (that is true).

There are two others, which we leave to the reader's imagination. The procedure for OR can be constructed from that of NOR simply by switching the two states, to give Fig. 1.4b. It rejects negative answers until it finds a positive answer. If there are no positive answers, then OR is not warranted for the coordination.

There is fourth way of arranging these elements that does not respond to any single lexical item in English, or in any other natural language, as far as anyone knows. It comes from reversing the polarity of the evaluating predicates to produce a coordinator that is known as NAND, see the summary in Fig. 1.5 for its structure. The evaluation is false until it turns up a person who is not a gourmet. The sense of this coordinator can be given in English by a sentence like *It is not the case that Marta AND Rukayyah are gourmets*, where the capitalization of *and* indicates focal or contrastive stress.

The finite state representation of these four coordinators is summarized in Fig. 1.5. Automaton theory thus allows us to define four mutually exclusive yet partially related ways of coordinating a list of entities.

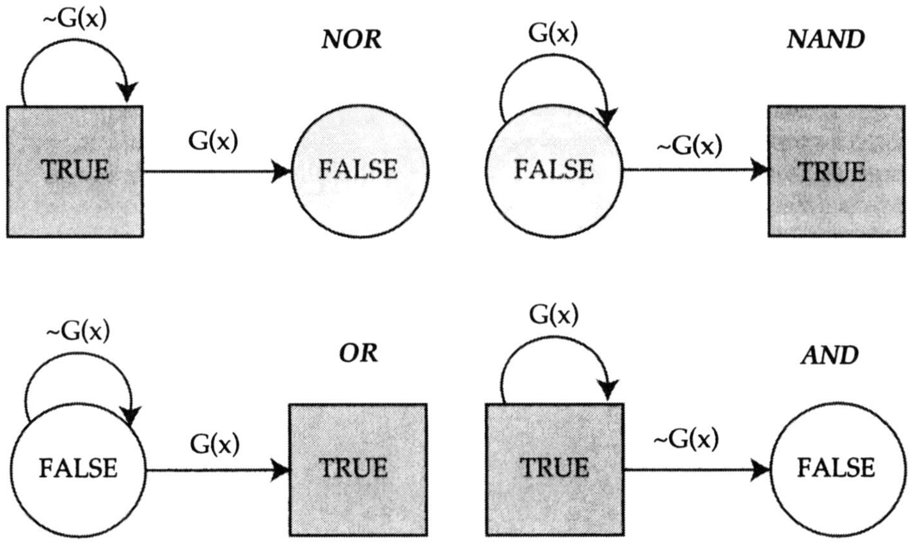

Figure 1.5. All four finite state automata for coordination.

1.1.4. A generalization to the logical quantifiers

This mechanism for deciding the truth of a coordination turns out to have a much broader range of application. Most obviously, it appears to generalize immediately to many quantified sentences. Consider a statement like *All linguists are gourmets*, under the assumption that we only know four linguists: Marta, Chao, Pierre, and Rukayyah. To verify the truth of *All linguists are gourmets* in this situation, we merely have to ask whether each individual linguist that we know is also a gourmet, as in (1.6):

1.6	a)	Is Marta a gourmet?	- Yes (that is true).
	b)	Is Chao a gourmet?	- Yes (that is true).
	c)	Is Pierre a gourmet?	- Yes (that is true).
	d)	Is Rukayyah a gourmet?	- Yes (that is true).

This of course is nothing more than the procedure for verifying AND. It follows that the coordinator automata defined above extend to the four logical quantifiers. This result recapitulates McCawley's (1981) characterization of the logical coordinators as logical quantifiers. To paraphrase McCawley, AND acts like a "big all", OR like a "big some", and NOR like a "big no". Working out the exact details of this correspondence is one of the empirical goals of this monograph.

Table 1.1. Correspondence between coordinators/quantifiers, their sets, and their cardinalities.

COOR/Q	*Sets*	*Cardinalities*
AND/ALL	$\cup P(x) = P(x)$	$\lvert \cup P(x) \rvert = \lvert P(x) \rvert$
OR/SOME	$\cup P(x) \neq \emptyset$	$\lvert \cup P(x) \rvert \neq 0$
NAND/NALL	$\cup P(x) \neq P(x)$	$\lvert \cup P(x) \rvert \neq \lvert P(x) \rvert$
NOR/NO	$\cup P(x) = \emptyset$	$\lvert \cup P(x) \rvert = 0$

1.1.5. The problem of time

The foundational problem with the computational model outlined above is time: the number of time steps or cycles needed to finish a universal computation (AND/ALL or NOR/NO) is no fewer than the number of entities in question. This is due to the fact that the automaton has to follow the list all the way down to the end to accept it. Nor are the existentials OR/SOME and NAND/NALL immune from having to traverse the entire list, since the accepting individuals may not be encountered until the very last one.

Time is the critical resource in any neurologically-plausible model of natural language, since it is known that neurons are comparatively slow, with response times measured in a few milliseconds, whereas complex behaviors are carried out in a few hundred milliseconds, see Posner (1978).[1] This does not affect the automaton analysis of coordination, since one rarely has to deal with more than four or five coordinated categories, but it is crucially relevant for the analysis of quantification, since any quantification involving more than a hundred individuals could not be computed by the automata in Fig. 1.5 in real time. Thus the number 100 should set an upper limit on what human languages can quantify – in sharp contrast to one's intuitions about such things, for which such a limit does not have any justification at all.

1.1.6. Set-theoretical alternatives

From the perspective of semantic analyses couched in model theory, the obvious solution is to draw the formal apparatus from set theory, such as the operations of union and intersection. Table 1.1 presents a first approximation to such a reanalysis, where the notation *P(x)* is read as "the x's that have the property P", $\cup$ is union, and $\lvert \bullet \rvert$ is the cardinality – the number of members – of a set. The cardinality expressions on the right are clearly parasitic on the set

[1] The importance of the 100-step limitation for the design of neurologically-plausible analyses of cognitive phenomena has been pointed out time and time again by Jerome Feldman, e.g. Feldman (1984, 1985) and Feldman and Ballard (1982).

expressions down the center, which leads one to think that counting truth values should be parasitic on some implementation of the union operation that may avoid the temporal implausibility of a serial count.

Though comforting, this thought suffers from the fact that it replaces something whose computational implementation is well understand, namely counting, with something whose computational implementation is less understand, namely set-theoretic union. Moreover, postulating set-theoretic union as the implementation of logical coordination and quantification is at best a modest solution, since we are still left wondering how it is that humans have the ability to use this operation, and whether it is learned or not.

As an alternative to the automaton and union hypotheses, let us delve into the best-understood domain of natural computation, namely vision, in order to search for inspiration for a robust theory of logical coordination and quantification. Before doing so, however, we should anticipate an objection to the entire endeavor.

1.1.7. What about modularity?

We suspect that most linguists would treat any attempt to draw linguistic insight from vision as an error in type. After all, don't vision and language deal in representations which are by definition of different types and therefore incommensurate? The answer offered in this monograph is that the neurophysiology of both domains is surprisingly similar, at least as far as is currently known. Thus the *a priori* postulation of incommensurability between visual and linguistic representations is in reality subject to empirical verification – exactly in accord with Ludlow's warning against drawing pre-theoretic boundaries between disciplines. To not interrupt the flow of our discourse, this topic can be postponed for further analysis to Sec. 1.3.4.

1.2. VISION AS AN EXAMPLE OF NATURAL COMPUTATION

Vision is perhaps the most well-studied of neural systems, and there are currently about 30 areas recognized in monkey cortex as having a visual specialization, see Felleman and Van Essen (1991). To get a glimpse of what are believed to be the global functional principles of visual processing, let us briefly survey the workings of the pathway that transduces light into neurological signals that are relayed to the brain, and then discuss how these signals are processed within the brain proper.

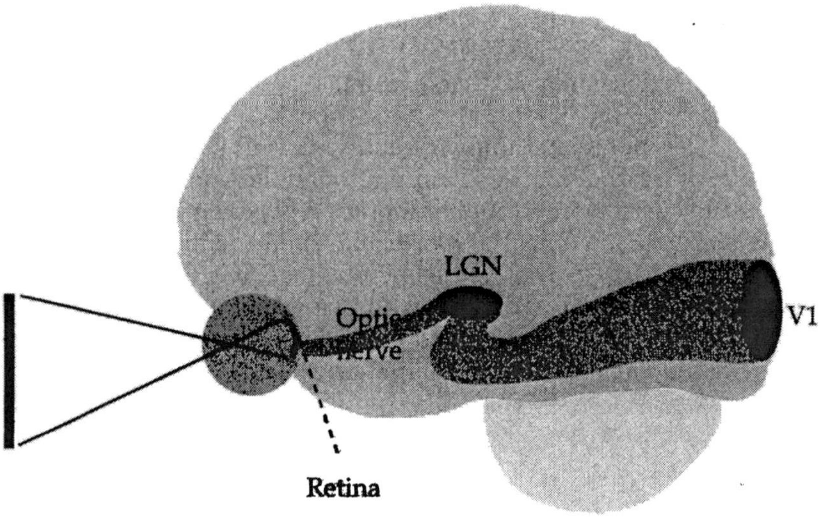

Figure 1.6. Pathway from the retina to primary visual cortex.

1.2.1. The retinogeniculate pathway

The initial pathway is laid out in Fig. 1.6 in its barest form. Light is transduced into electrical signals at the retina, and these signals are transmitted out along the optic nerve. They arrive at the lateral geniculate nucleus (LGN), part of the a subcortical structure known as the thalamus. From there, the signals are relayed to primary visual cortex (V1), part of the brain proper. Note that each of these stops comes in a left and right pair, originating at the left or right eye. In order to ensure that information from both eyes is available as early as possible in the processing chain, each optic nerve splits into two bundles, one that goes to the LGN on the same side (**ipsilateral**) of the head, and one that crosses the optic chiasm and arrives at the LGN on the opposite side (**contralateral**). Despite a point of anatomical crossover at the optic chiasm, the monocular segregation of information is maintained at the LGN, and the two streams are not combined binocularly until V1.

Even a structure as peripheral as the retina sheds light on the core functional principles of the brain. The primate retina is composed of three layers of cells and two layers of connections between them. The three layers of cells include (i)

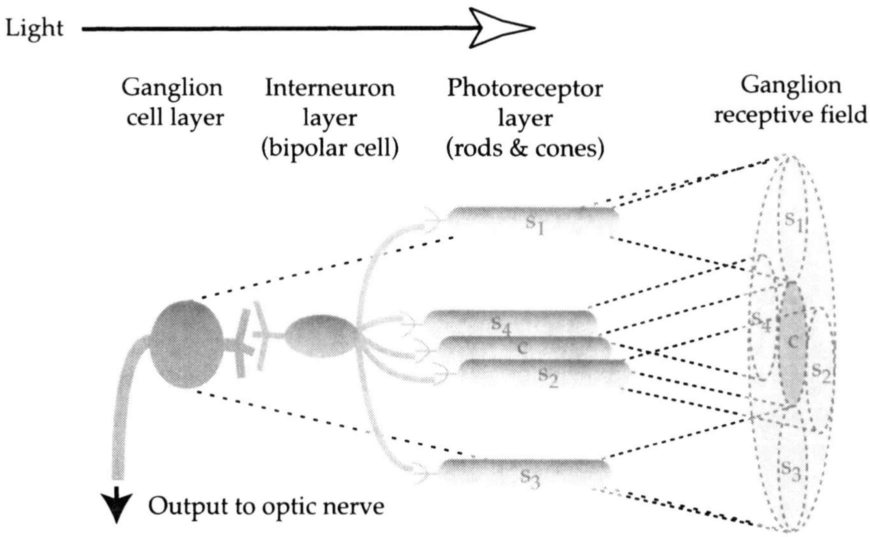

Figure 1.7. Center-surround organization of the receptive field of a ganglion cell and its downstream photoreceptors. Comparable to Delcomyn, 1998, Fig. 11-7, and Dowling and Boycott (1966).

photoreceptors (rods for luminescence and cones for color), (ii) interneurons (horizontal, bipolar, and amacrine cells), and (iii) ganglion cells.[2] The ganglion cells end in the long bundle of fibers which constitutes the optic nerve. It is a curious fact that these three layers are stacked in reverse order, so that light has to pass through the – essentially transparent – ganglion cells and interneurons to reach the photoreceptors.

This architecture is illustrated in a much simplified format in Fig. 1.7. The interneurons pool the output of the photoreceptors and pass it along to the ganglion cells. Such pooling endows each ganglion cell with a **receptive field** arranged as two concentric circles centered on a photoreceptor. By way of illustration, the receptive field of the ganglion cell in Fig. 1.7 is appended to the right end of the diagram as a dotted circle. The way in which the parts of the receptive field are projected from the physical location of the photoreceptors is established by projecting the photoreceptors onto their subfield of the overall

[2] This is the traditional classification of retinal cell types. More recent research recognizes up to fifty distinct functional elements, each carrying out a specific task. See Masland and Raviola (2000) for review.

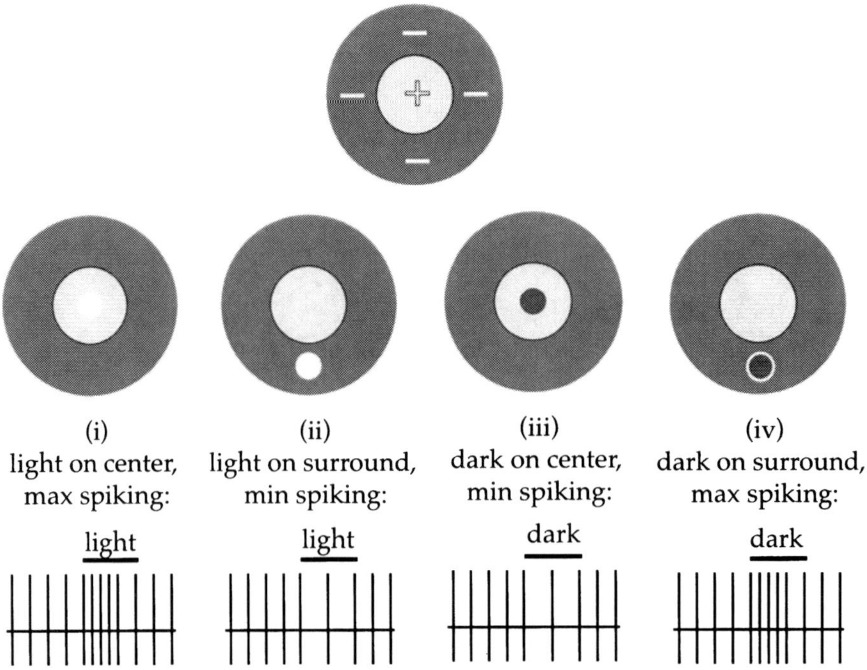

Figure 1.8. ON-center ganglionic receptive field and its response to transient illumination. Comparable to Delcomyn, 1998, Fig. 11-11, Kuffler et al. (1984), and Reid, 1999, Fig. 28.11.

field. The letter *c* labels both the center of the receptive field and the photoreceptor from which it is projected. Likewise, the letter *s* plus a subscript number labels the four areas that make up the periphery of the receptive field, plus the corresponding photoreceptors from which they are projected.

Kuffler (1953) established the fact that a ganglion cell responds most to light impinging either on the inner circle – the **center** – or the outer ring – the **surround**. The former type of ganglion cell is classified as **ON-center** and the latter, **OFF-center**. If a zone of the receptive field is not activated by light, then it is activated by darkness. Fig. 1.8 depicts four possibilities for momentarily turning on or off a spot of light within the receptive field of an ON-center cell, along with an idealized graph of the series of **spikes** or transient changes in the electrical charge of the ganglion cell's membrane undergone as a consequence of the stimulus. This distillation of Kuffler's experiments summarizes how (i) illumination of the center of an ON-center cell triggers an increase in spiking relative to the base rate, while (ii) illumination of the surround triggers a decrease. In other words, the former stimulus turns the cell on, and the latter

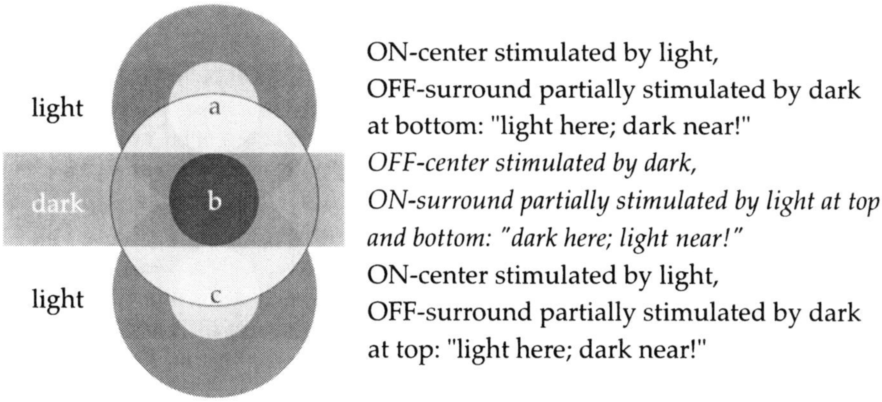

Figure 1. 9. How ganglionic receptive fields signal contrast.

turns it off. Conversely, (iii) extinguishing illumination of the cell at its center decreases its spike rate, while doing so at the surround increases its spike rate. As for an OFF-center cell, it behaves in the mirror-image fashion: illumination of its center decreases spiking relative to the base rate, while illumination of its surround increases spiking. Thus the very same stimuli have the opposite effect: central illumination turns the cell off, and peripheral illumination turns it on. It should be noted in passing that these mechanisms also create a neural correlate of the temporal dimension of the illumination, since the spike rate adapts quickly to the presence or not of the light stimulus.

The spatial function of this architecture appears to be to encode an image in terms of contrast; in Fig. 1. 8, this is the contrast between light and dark. By way of explanation, consider the kind of messages that are sent out along the optic nerve from a retinal ganglion cell. The response for an ON-center cell is a burst of spikes that can mean little more than that there is light at its point on the retina and dark nearby; conversely, the response for an OFF-center cell signals that there is dark at its point and light nearby.

Fig. 1.9 demonstrates how this dichotomy plays out in a simple image consisting of a dark band flanked by two lighter bands. Each band happens to align with the center of three overlapping ganglionic receptive fields. The top band increases the firing rate of the ON-center ganglion cell to which the top receptive field belongs, while the lower edge of its receptive field is stimulated by the dark band. This double stimulation activates the only signal that the cell is equipped to send, namely that there is light at point *a* on the retina and dark nearby. Given the shape of the pattern of illumination, the same response is evoked at point *c*, and the complementary response at point *b*. The intensity of this response – the number of spikes per second – varies with the contrast between the light and dark bands: the higher the contrast, the more spikes are emitted, until the cell reaches saturation.

A second classification of ganglion cells focuses on certain anatomical and physiological differences among them. One class is characterized by small receptive field centers, high spatial resolution, sensitivity to color, little sensitivity to motion or contrast, low frequency of spike emission, and thinner axons that conduct spikes less rapidly. All of this makes for a cell type specialized for color and fine detail at high contrast, which presumably fundamental for identifying the particular visual attributes or features of objects. Such cells are given the name **P cells**. The other class is characterized by larger receptive field centers, insensitivity to color, sensitivity to motion and contrast, high frequency of spike emission, and thicker axons that conduct spikes more rapidly. These complementary properties make for a cell type specialized for the detection of shape and motion. There are given the name **M cells**. These peculiar names owe their explanation to the way in which the two cell types are segregated in the LGN, to which we now turn.

From the retina, the series of spikes emitted by a ganglion cell, known as **spike trains**, travel through the optic nerve to the thalamus, and in particular to its lateral geniculate nucleus. This area has a distinctive V shape, whence its name *geniculate*, from the Latin "bent like a knee". Its input from the optic nerve is partitioned into separate zones so as to not mix P and M information. The two zones are known as the **parvocellular** ("small cell") and **magnocellular** ("large cell") **layers**, from whence the P and M abbreviations of the ganglion cells. Not only are these two classes maintained separate in the LGN, but also their source in the two eyes are maintained separate, so that the parvocellular and magnocellular layers themselves come in pairs, whose number depends on the species.

The LGN has traditionally been thought of as a passive relay station on the way to the cortex. For instance, simultaneous recording of neurons in the retina and in the LGN has demonstrated that the receptive-field centers of the two kinds of cells are quite similar and spatially overlapping, see Cleland et al., (1971) and Usrey et al. (1999). However, as Sherman, 2000, p. 526, points out, based on Sherman and Guillery (1998) and Van Horn et al. (2000):

> ... retinal inputs account for only 5–10% of the synapses onto geniculate cells projecting to cortex; the vast majority derive instead from local inhibitory cells, from visual cortex feedback projections and from brainstem inputs. Functionally, retinal inputs act as drivers, which carry the main information to be relayed to cortex and strongly control the firing of lateral geniculate neurons, whereas non-retinal inputs modulate the response of the geniculate cells to their driving inputs.

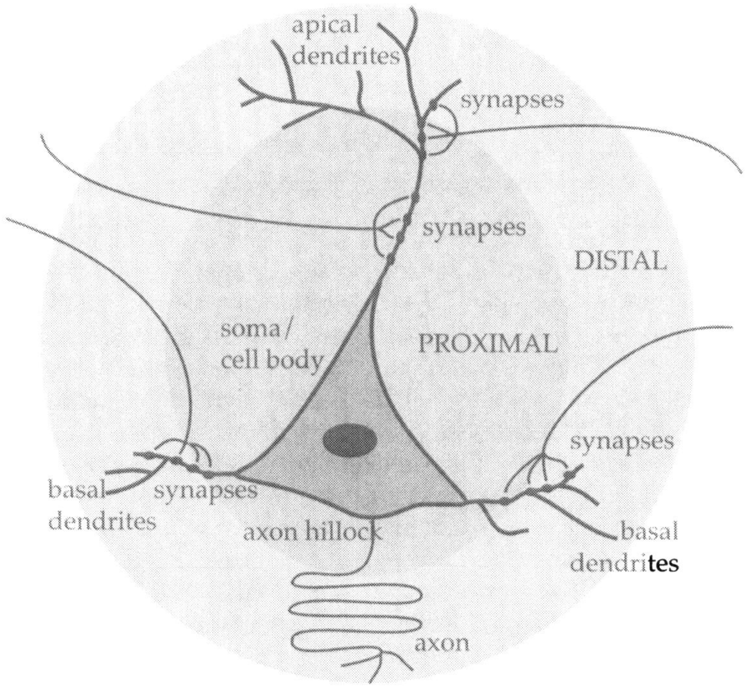

Figure 1.10. Basic anatomy of a pyramidal cell.

The exact role of these non-retinal modulatory inputs remains a mystery at present, so let us turn to something that is better understood and more central to our linguistic goals.

1.2.2. Primary visual cortex

The simple signals originating in the retina can be combined to draw a detailed description of the visual world and ultimately to categorize objects within it. This first stage of combination is known as **V1** or **primary visual cortex**. Like almost all mammalian cortex, V1 is a 2 mm thick sheet of cells which contains three morphological distinct neuron types: (i) spiny pyramidal cells, (ii) spiny stellate cells, and (iii) smooth or sparsely spinous interneurons.

The pyramidal cells, as one might guess, have cell bodies that shaped like pyramids. They comprise 75-80% of all cortical cells and mediate all long-range, and almost all short-range, excitatory influences in the cortex. Pyramidal cells can perform such mediation thanks to their extravagant design, which is depicted in an idealized form in Fig. 1.10. There are five main parts. The triangular shape in the lower middle contains the genetic and metabolic

Table 1.2. Summary of the anatomy of an idealized pyramidal cell.

Neurite	*Function*
Synapse	Site of signal transmission from one neuron to another
Dendrite	Carries a signal towards the cell body
Soma	Body of the neural cell
Axon hillock	Juncture between soma and axon where signals are initiated
Axon	Carries a signal away from the cell body

machinery of the cell and is known as the **cell body** or **soma**. The other four parts work together to channel the signals which enter the cell through **synapses** on the cell body and its **dendrites**, which are the web of fibers springing from the top and sides of the cell body. The prominent ascending dendritic shaft, called an apical dendrite, allows a pyramidal cell to form synapses in cortical layers above the layer in which its soma is located. Once in the cell, if the signals build up past a certain level, they initiate a electrical discharge at the **axon hillock** which travels down the long central channel or **axon**, and on to the next cells. Thus the fibers that synapse onto the cell in the picture come from the axons of other cells that are not shown. Given the importance of these terms for the upcoming discussion, they are summarized in Table 1.2. Fig. 1.10 also shades two zones around the soma, **proximal** for those neuron parts or **neurites** that are close to the soma, and **distal** for those neurites that are far from the soma.

The other two types of cell are considerably less numerous and are usually not a focus of computational modeling. Spiny stellate cells are generally smaller than pyramids, and their cell bodies are vaguely star-shaped. They also sport spine-covered dendrites. The interneurons have more rounded cell bodies and little or no spines on their dendrites. Their axonal and dendritic arbors branch in a bewildering variety of patterns to which anatomists have given more or less descriptive appellations over the years. In a series of papers, Jennifer Lund and colleagues (Lund, 1987; Lund et al., 1988; Lund and Yoshioka, 1991; Lund and Wu, 1997) have described over 40 such subtypes in the macaque V1 alone. Such detail escapes the introductory goal of these pages, especially since it is not at all clear whether these anatomically distinct subtypes are physiologically distinguishable from one another.

What is more concordant with our goals is the fact that these three types of neurons tend to associate with one another into a two-dimensional array of anatomically identical **minicolumns**, each a group of a few hundred neurons whose somas occupy a cylinder approximately 25-50 mm in diameter, see Hendry et al., 1999, p. 666 or Kandel and Mason, 1995, p. 432 for an overview and Mountcastle (1997) for a more detailed review. Minicolumns are the functional units of sensory cortex, and also appear to be the functional units of associative cortex, see Mountcastle (1997).

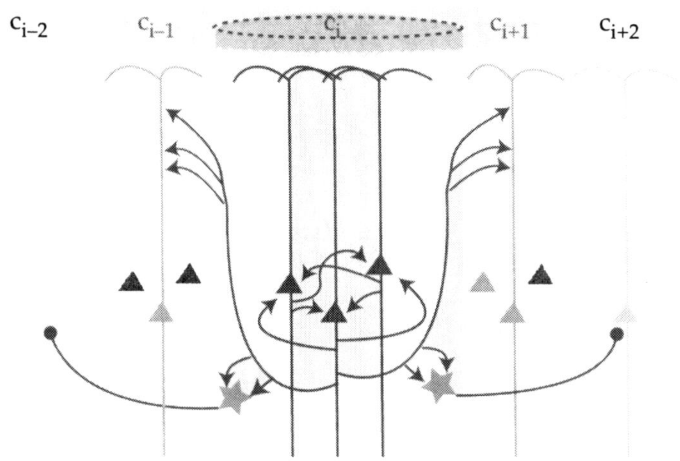

Figure 1.11. Columnar organization of neocortex. Comparable to Spitzer, 1999, Fig. 5.3.

They fulfill this role through a circuit design like that of Fig. 1.11. Each minicolumn is composed of three pyramidal neurons and two smooth stellate interneurons, the latter of which are omitted from all but the central minicolumn of Fig. 1.11 for the sake of clarity. The diagram focuses on the central minicolumn, labeled *i*, showing how pyramids within it excite one another and the co-columnar stellate cells. Moreover, since a pyramidal cell's basal dendrites radiate laterally from its soma for up to 300 mm, it can sample input from axons in minicolumns other than its own. In this way, the pyramidal cells in minicolumn *i* can also excite neurons in contiguous minicolumns, as depicted by the arrows from *i* to its two flankers *i*±*1*. The flankers are not excited enough to become active all by themselves, but they are excited enough to respond more readily to similar signals. Finally, the stellate interneurons are inhibitory, so their excitation inhibits the pyramidal cells at the further remove of minicolumns *i*±*2*. In this way, an excitatory 'halo' extends around the active minicolumn *i* that dissipates with increasing distance from *i*. This halo is represented directly at the top of Fig. 1.11 and also indirectly by the shading of the pyramidal somas: the darker the color, the more active the cell is.

Fig. 1.11 suggests a certain layering of V1, and this is indeed the case. Mammalian neocortex is conventionally divided into six layers or **laminae**, according to where the soma and dendrites of pyramidal neurons bunch together. The traditional convention is to label each layer with a Roman numeral from I to VI, where layer I lies under the **pia**, the membrane just under the skull that protects the brain, and layer VI lies on top of the **white matter**, the bundles

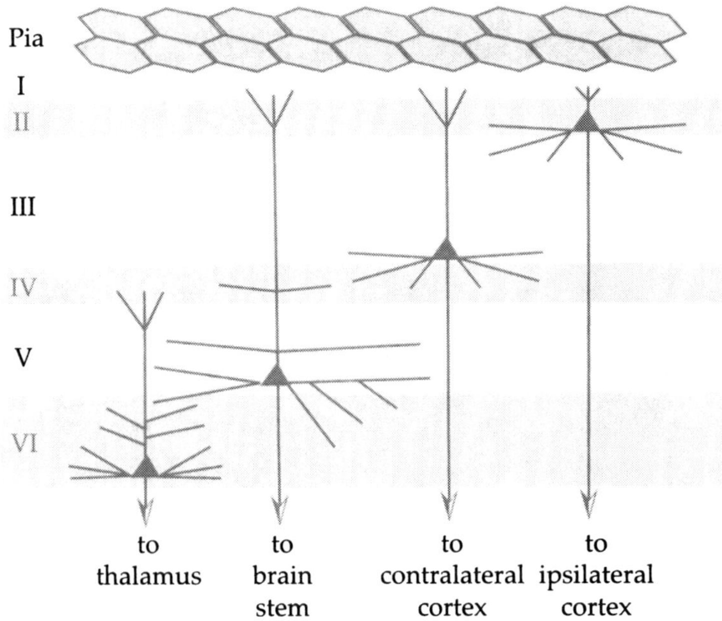

Figure 1.12. Layering of neocortex, schematized by outputs of pyramidal cells. Comparable to Hendry, Hsiao and Brown, 1998, Fig. 23.7, Jones (1985), and Douglas and Martin, 1998, Fig. 12.8.

of fiber that connect individual areas of the cortex. Fig 1.12 gives a very schematic diagram of what this looks like. For instance, the soma of neurons whose output propagates to cortex nearby in the same hemisphere of the brain tend to bunch together in layer II, whereas the soma of neurons whose output propagates to the corresponding cortical area in the other hemisphere of the brain tend to bunch together in layer III. Recent usage prefers to conflate these two layers and further subdivide layer IV – and to use Arabic numbers for the Roman ones.

There is a fairly strict hierarchical pathway through these laminae, which follows the feedforward path of 4 → 2+3 → 5 → 6 and the feedback path of 5 → 2+3 and 6 → 4. Primary visual cortex in addition evidences the need to subdivide layer 4 into separate paths for parvocellular and magnocellular input from the LGN, which merge at layer 2+3. Finally, the next paragraphs will show that in primary visual cortex it is crucial to distinguish between simple cells, which receive direct input from the LGN, and complex cells, which take in the output of the simple cells. Fig. 1.13 attempts to conflate all of this complexity into a single diagram.

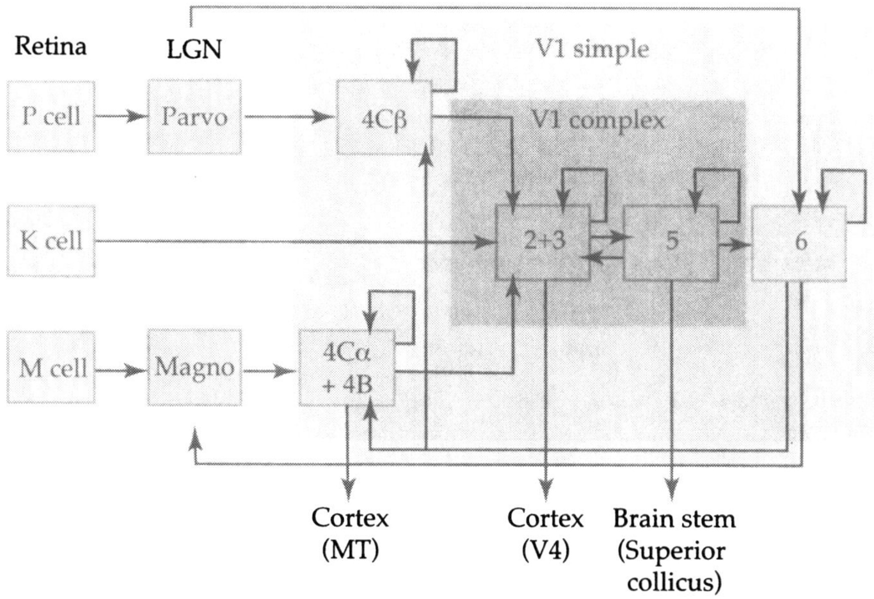

Figure 1.13. Retinogeniculocortical excitatory connectivity. Comparable to Reid, 1999, Fig. 28.11, Gilbert and Wiesel (1985), Nicholls et al., 2001, Fig. 21.7, and Merigan and Maunsell (1993).

1.2.2.1. Simple V1 cells

Primary visual cortex was the subject of ground-breaking studies by David Hubel and Torsten Wiesel in the late 1950's, see especially Hubel and Wiesel (1962). Hubel and Wiesel discovered that V1 neurons are selective for the orientation of elongated visual stimuli, even though their inputs from the LGN have no such selectivity. The mechanism for orientation selectivity turned out to be embodied in two types of V1 cells, simple and complex. A simple cell responds strongest to a bar of light (or darkness) that has a particular orientation and position in visual space. Fig. 1.14 illustrates this phenomenon with a bar of darkness oriented vertically, diagonally, and horizontally within the simple cell's receptive field, where the field is most sensitive to the vertical orientation.

Hubel and Wiesel reasoned that such a percept could be formed from LGN receptive fields if they were aligned and if similarly aligned receptive fields projected to a single simple cell. Fig. 1.15 illustrates this organization, with receptive fields standing in for the corresponding geniculate input cells. An arrow in Fig. 1.15 represents an excitatory connection from the LGN cell to the V1 simple cell. By this is meant that an active input cell tends to activate the cell

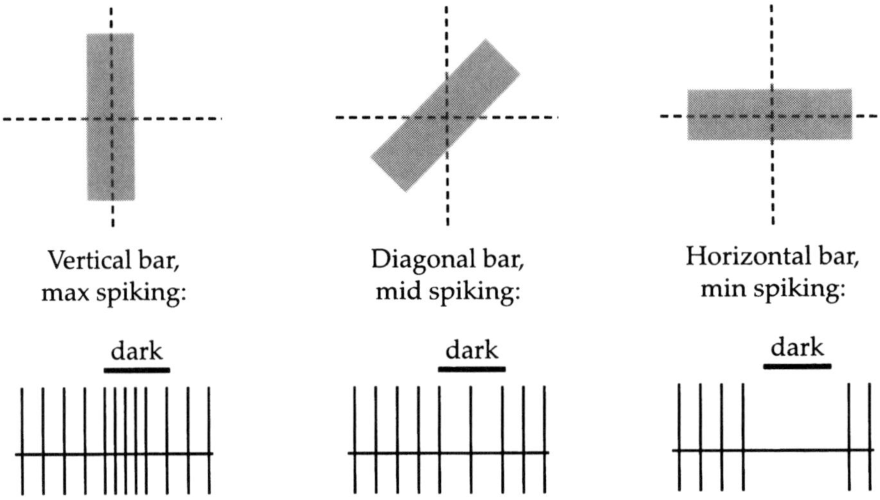

Figure 1.14. The response of a simple cell to a bar of darkness at different orientations to its receptive field. Comparable to Delcomyn, 1998, Fig. 11-14.

to which it is connected. A group of such connections is understood as (linear) summation of their activity within the receiving cell. For example, if each LGN cell of Fig. 1.15 sends a spike train that measures '1' on some appropriate scale, then the simple cell receives an input of 1+1+1+1+1 or 5.

This reliance on linear summation creates a conundrum for the summating cell. Since it also receives input from LGN cells with similar but slightly different orientations, the simple cell could also respond to a non-preferred orientation, in contradiction of Hubel and Wiesel's observations. Fortunately, the response mechanism of the simple cell, namely the generation of a spike, acts in such as way as to resolve this conundrum. It turns out that the input to any neuron must cross a certain threshold before it triggers a spike. Given that the number of LGN cells exciting a simple cell at a non-preferred orientation will be less than that of the preferred orientation, the threshold of the spike-generation mechanism will act to filter out such small, 'spurious' signals and ensure that the maximal spike generation capacity of the cell corresponds to optimally-oriented LGN input. For instance, in Fig. 1.15, a horizontal line would activate one of the five LGN cells that connect to the simple cell – the central one that is shared with the vertical line. Thus the input to the simple cell would be 1. However, if the threshold for firing a spike is, say, 3, then the horizontal line will fail to do so. The vertical line, in contrast, supplies more than enough input to cross the threshold and generate a spike, and the system consequently

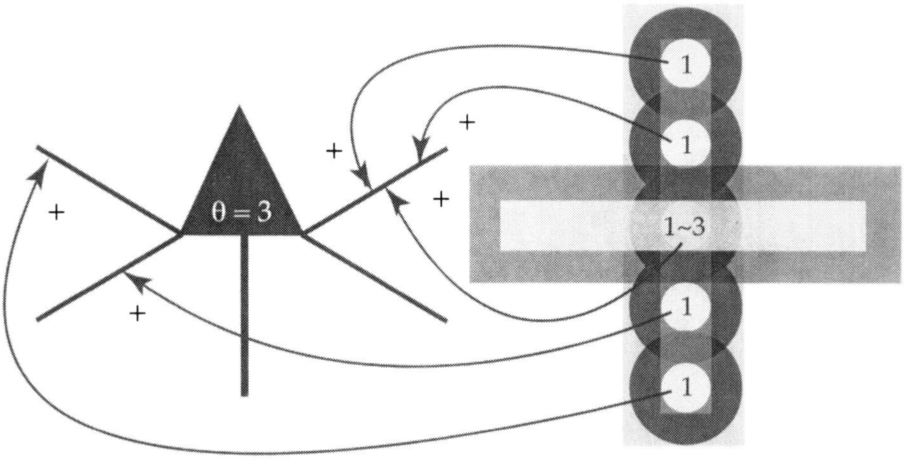

Figure 1.15. Similarly aligned ON-center receptive fields of LGN cells (not shown) extend excitatory (+) connections to a simple V1 cell with a threshold = 3. The cell can recognize a vertical line, indicated by the pale vertical bar superimposed on the centers of the LGN receptive fields, with input = 1. A high-contrast orthogonal stimulus, with input = 3 could also activate this cell.

reproduces quite accurately Hubel and Wiesel's observations. The overall picture of the simple cell, therefore, is that it sums together its LGN input linearly and then rectifies this sum non-linearly by not responding to inputs evoked by improperly oriented stimuli.

The sketch of orientation selectivity in simple V1 cells abstracts away from considerable complexity, but it is detailed enough to expose a glaring empirical problem with the model, whose solution leads to a clearer understanding of the logic of natural computation. The problem is this: one of the complications that we have ignored is the fact that higher contrast in a visual stimulus is encoded directly by greater activity in the retinogeniculate pathway. The model outlined in the previous paragraph therefore predicts that a simple V1 cell could receive as much supra-threshold input from a high-contrast bar at an orthogonal orientation as it does from a low-contrast bar at the preferred orientation. Such a confounding stimulus is also depicted in Fig. 1.15 by means of the high-contrast horizontal bar, for which the activity of the central LGN cell soars to 3. Thus the high contrast stimulus alone would pass the simple cell's threshold and cause it to emit a spike. The simple cell would wind up responding equally well to both sorts of stimuli. This predicted response does not take place, however. The

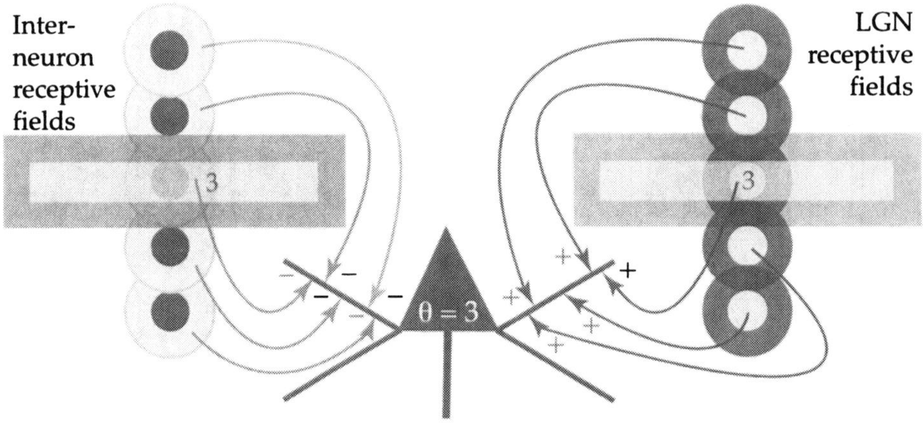

Figure 1.16. Push-pull inhibition of a simple V1 cell from inhibitory interneurons that pass through an LGN orientation at the opposite phase of the preferred orientation of the simple cell. High-contrast input = 3, simple cell threshold = 3.

obvious conclusion is that our model lacks some means by which to make the simple cell orientation tuning invariant to differences in contrast.

A promising source for this invariance is already implicit in Hubel and Wiesel (1962), though the mechanism that they proposed was not addressed specifically to contrast invariance. What they proposed, and others followed up on, see Troyer et al. (1998), is called **antiphase** or **push-pull inhibition**. By inhibition, we mean the opposite of excitation, that is, the activity of the an inhibitory cell serves to decrease the activity of a cell that it is connected to. In the typology of cortical neurons presented above, inhibition is supplied by the interneurons. More exactly, the idea of push-pull inhibition is that simple cells receive strong OFF inhibition in their ON sub-fields, and strong ON inhibition in their OFF sub-fields. This inhibition comes from the inhibitory interneurons that all pyramidal neurons are associated with, under the assumption that these interneurons receive LGN input at the exact same spatial organization as the simple cell does, but with the opposite phase. Fig. 1.16 adds a column of such interneurons to Fig. 1.15.

The stimulus depicted here is that mentioned above: a horizontal line at a higher contrast than, and orthogonal to, the preferred vertical line. If we assume that the higher contrast line is activating the relevant LGN cells and interneurons at the value of 3, then the simple cell receives +3 from the central cell of the LGN group and –3 from the surround of the central interneuron, to give an input of 0. This falls short of the simple cell's threshold of 3, ensuring

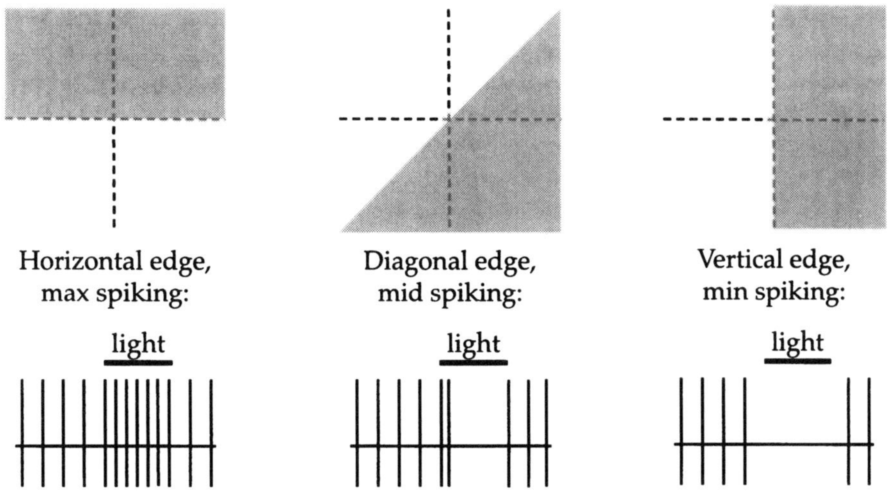

Figure 1.17. The response of a complex cell to an edge at different orientations to its receptive field. Comparable to Delcomyn, 1998, Fig. 11-16.

that it does not recognize the horizontal line, despite its high contrast. The cell is thereby rendered invariant to contrast, in accord with Hubel and Wiesel. It is crucial to emphasize that the antiphase assumption of interneuron orientation is what maintains the simple cell's orientation tuning: if the interneurons contributed inhibition that was in phase with the preferred orientation of the simple cell, they would effectively suppress all positive input to the simple cell and prevent it from recognizing anything – a rather pointless function for a circuit.

1.2.2.2. Complex V1 cells

The other type of neuron that Hubert and Wiesel discovered in V1, the complex cell, responds strongest to an edge with a particular orientation, but without regard to position in visual space. Fig. 1.17 illustrates this phenomenon with an edge oriented horizontally, diagonally, and vertically within the receptive field of a complex cell that is attuned to the horizontal orientation. Hubel and Wiesel reasoned much as before: an edge could be recognized from a combination of the receptive fields of simple cells if they were aligned and if similarly aligned receptive fields projected to a single complex cell. Fig. 1.18 illustrates this organization.

An alternative account of complex-cell computation has emerged since the mid-1990's, somewhat ironically through work on simple cells. Ben-Yishai et al. (1995), Douglas et al. (1995), Somers et al. (1995), and Sompolinsky and Shapley (1997) take as their starting point the fact that the feedforward input from the

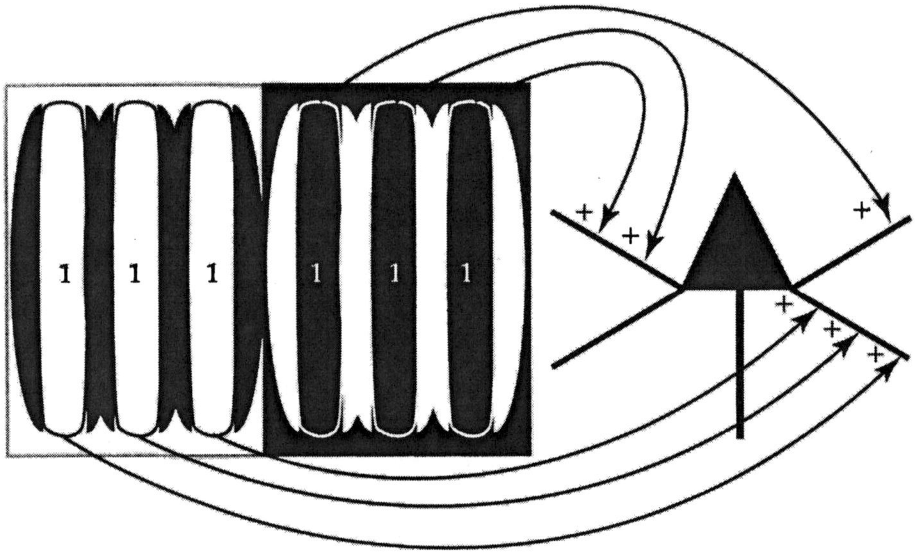

Figure 1.18. Similarly aligned receptive fields of simple V1 cells (not shown) project to a complex V1 cell to recognize a vertical edge. Comparable to Reid(1999) Fig. 28.10.

LGN that drives simple cells in the Hubel and Wiesel model is relatively weak and perhaps overshadowed by excitation coming from nearby simple cells, see Toyama et al. (1981) for evidence thereof. In a nutshell, what these authors argue is that a group of simple cells receiving similar input from the LGN will tend to reinforce or amplify one another if they are tied together by mutually excitatory connections. Thus their selectivity for a given orientation increases.

Chance et al. (1999) turn this argument on its head by demonstrating that such recurrent excitation can serve to *decrease* the selectivity of a group of cells if their feedforward input is drawn from a heterogeneous set of patterns. With heterogeneous inputs, the group generalizes across the variety in its input patterns to recognize a more abstract version thereof, in which the individual patterns are 'smeared' together. And this is exactly what a complex V1 cell appears to do: it receives relatively weak feedforward input with a restricted range of orientational or spatial-phase preferences and extracts a spatial-phase invariance. Chance et al. would say that the complex V1 cell becomes insensitive to spatial phase through the amplification of a consensual pattern that emerges through repeated exposure of all of the input patterns to all of the fellow complex cells. As a consequence of such cortical amplification, the phase selectivity of a complex cell decreases. As is our custom, we endeavor to summarize this tricky bit of argumentation with a picture that emphasizes the

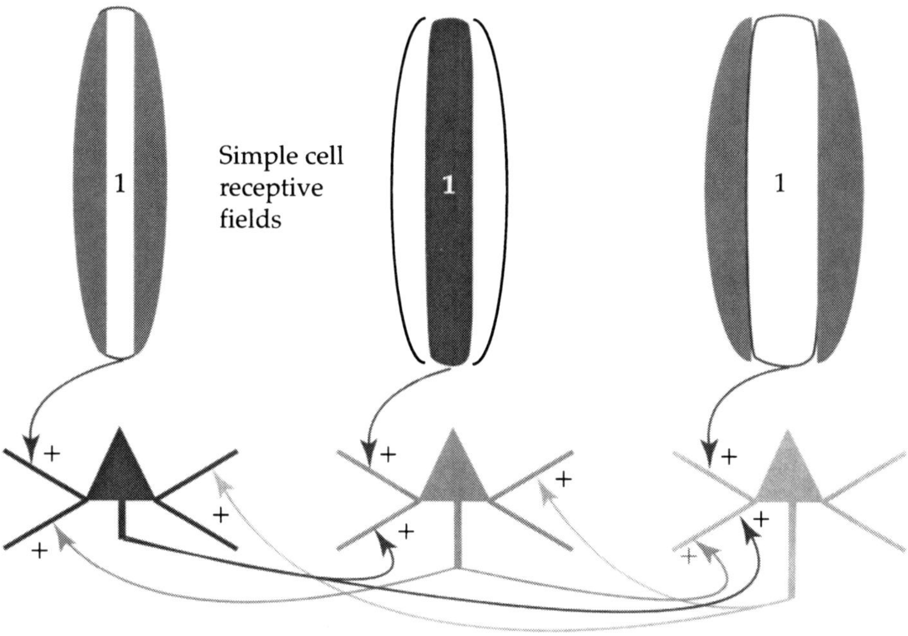

Figure 1.19. Similarly aligned receptive fields of simple V1 cells (not shown) project to complex V1 cells connected via recurrent excitation. Simple cell input = 1.

integration of the new information with the old, Fig. 1.19. In accord with the text, in Fig. 1.19 each complex V1 cell sends an excitatory connection to its fellow complex cells.

In closing, let us point out an additional advantage of this analysis in the elegant way in which it extends to the simple-cell circuit. Chance, Nelson and Abbott show that with the same blueprint of recurrent excitatory connections, weak coupling among them facilitates the emergence of simple-cell selectivity, whereas strong coupling among them facilitates the emergence of complex-cell invariance.

While bringing an accurate and parsimonious solution to the enigma of computation in complex V1 cells, this solution engenders its own drawbacks: small changes in coupling strength can dramatically modify the degree of amplification, and as the firing rate in a complex V1 network increases, the level of excitation rises to a point at which the circuit looses it stability and begins to respond too slowly to rapidly changing stimuli.

Much as in the case of the initial, purely excitatory analysis of the simple V1 circuit, a correction comes from the application of inhibition to moderate the

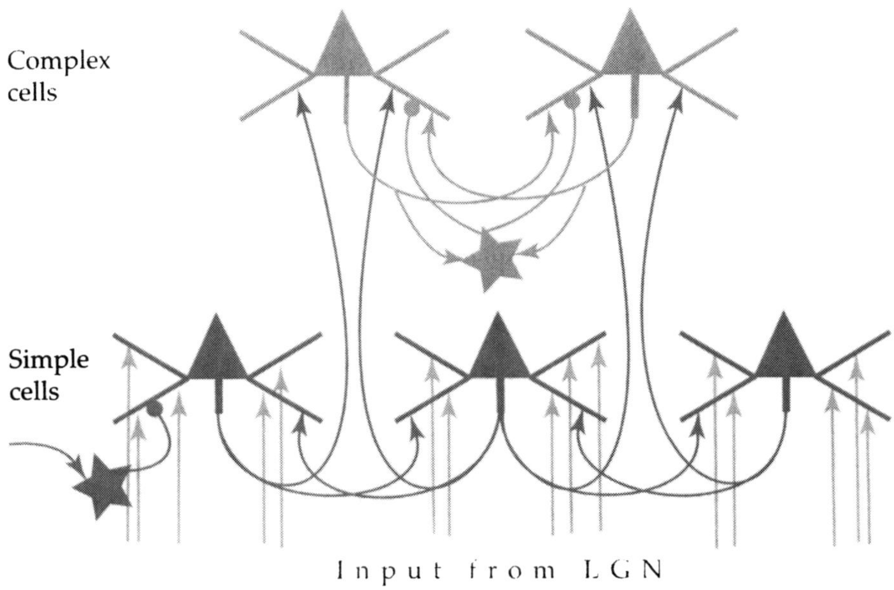

Figure 1.20. The composite V1 circuit.

growth of excitation. Chance and Abbott (2000) introduce the notion of divisive inhibition, in contrast to the subtractive inhibition that was pressed into service above. The idea of divisive inhibition, as its name indicates, is that inhibition acts so as to divide excitation. As long the divisor is positive and does not fall to zero, there will always be some recurrent excitation left over to perform the mixing of the feed-forward patterns, in contrast to subtractive inhibition, which can subtract it all away. Neurophysiologically, Chance and Abbott describe how the relevant connections can be organized on the dendrites to bring about the mathematical effect. Recurrent synaptic inputs would be located at the ends of a dendritic tree along which inhibitory inputs shunt current flow, whereas feedforward inputs are located nearer to the soma and are unaffected by the shunting. By **shunting** is meant a lessening of the cell membrane's ability to carry a positive charge through a loss of the positive to the exterior of the cell. In this way, the complex cells become self-regulating and so avoids runaway recurrent excitation.

1.2.2.3. The essential V1 circuit: selection and generalization

The point of our review of primary visual cortex, the best understood area of the brain, is to illustrate the building blocks of natural computation. And we have seen practically all of them: feed-forward excitation, recurrent excitation, and two regimes of inhibition: subtractive and divisive. We have also seen how changing the pattern of connectivity among these components can alter the computation that the whole performs. A happy by-product of this review is that we have accumulated enough knowledge to draw a simple circuit for V1, that of Fig. 1.20. Excitatory input from the LGN drives the layer of simple cells in layer 4, which are connected by reciprocal excitation. This coupling is weak, however, so Fig. 1.20 does not include any divisive inhibition to regulate it. The diagram does include subtractive inhibition from an antiphase simple cell (beyond the left edge) to prevent spurious activation of the left-most simple cell by partially-matching patterns. This is but a sample of how all the simple cells should be regulated. The axons of the simple cells ascend to layers 2/3 to drive the complex cells, also connected by reciprocal excitation. This coupling is strong, and an inhibitory cell is introduced into the circuit to keep it in an acceptable range.

But what does this circuit do? Following Ferster and Miller (2000), we may call the input driving a cell pattern *A*. For our single simple cell in Fig. 1.15 and 1.16, *A* is the set of signals transmitted from the five vertical LGN cells. Continuing in this vein, push-pull inhibition would naturally be called the complement of *A*, namely pattern *A′*, since it consists of all the input to the simple cell at the same orientation, but opposite polarity. In terms of activation, it is the pattern that is least coactive, or most anticorrelated, with *A*. All inputs from orthogonal orientations can be grouped together as the set of patterns *B*. These are the patterns that share some co-activation with both *A* and *A′*, but this co-activation is uncorrelated or random.

These definitions allow Ferster and Miller (2000) to lay bare the logical basis of contrast-invariant orientation selectivity:

> In simple cells receiving the input *A* alone ..., we have seen that orientation selectivity becomes contrast dependent, because input pattern *B* of sufficiently large amplitude (an orthogonal stimulus of high contrast) can activate the cell. Adding strong push-pull inhibition translates into making the cell selective for the pattern "A AND NOT A". As a result, *B* of any strength, since it activates both *A* and *A′* to some degree, can no longer activate the cell when push-pull inhibition is present. The cell becomes selective for pattern *A*, independent of stimulus magnitude.

Ferster and Miller's overall conclusion is that layer 4 of V1 divides its inputs into opposing pairs of correlated input structures in such a way that a cell responds only when one is present without the other.

Ferster and Miller (2000) interpret this hypothesis of the complex V1 cell as claiming that the complex cell, or the layer 2+3 of V1 in which most complex cells are embedded, recognizes an oriented stimulus independent of its polarity. This observation could be assimilated to Troyer et al.'s (1998) model of antiphase inhibition by supposing that a complex cell responds to a pattern of the form "A OR NOT A", which is to say that it extracts the information that *A* and *A'* have in common, namely orientation, while discarding the information that distinguishes them, namely polarity. The relevance of this kind of OR to the linguistic OR is discussed below.

While we find the elegant circuit of AND feeding OR as attractive as Ferster and Miller do, we are far less convinced that it is accurate. The principal drawback is that the complex cell is just as prone to a high-contrast orthogonal pattern evoking a false positive response as the simple cell is. This parallelism suggests that the complex cell should also be subject to antiphase inhibition, so that both layers of V1 compute "A AND NOT A", and "A OR NOT A" is not computed at all. Such a suggestion could be supported by the existence of a parallel set of interneurons supplying push-pull inhibition, but neither Troyer et al. (1998) nor Ferster and Miller delve into the physiology of complex cells, and the issue is not crucial enough to out concerns to pursue at length here. What is sufficient is the existence of push-pull inhibition for simple cells.

Nevertheless, a way to save the complex cell OR-computation comes readily from Chance et al.'s (1999) understanding of recurrent excitation. Such excitation makes each complex cell sensitive to its group's input, thereby performing a logical OR across the grouped cells. In Ferster and Miller's terms, we could symbolize this computation as "A1 OR A2". In simple English terms, we could say that simple cells select from among the broad variety of inputs, and complex cells generalize across this selection.

1.2.2.4. Recoding to eliminate redundancy

We have come far enough to pause and reflect on what has been learned along the way. We have reviewed the classical model of visual receptive fields, in which receptive fields at one level combine the receptive fields of cells that feed into them from a previous level. Nicholls et al. (2001) draw a rather clever picture to summarize this progressive expansion of the receptive fields of the three types of neurons that participate in early visual processing, to which we have appended a fourth type, that of the photoreceptors that get the whole thing started in the first place, to produce Fig. 1.21 below. The four different RF types are arrayed from left to right.

At the initial level, the small photoreceptor receptive fields tile the top left corner of the white rectangle. The four inside the rectangle are being illuminated by it and so are active, which is indicated by darkening the perimeter of the circles. At the next level, the ganglion and LGN center-surround receptive fields intervene. The small four bars radiating from them like the cardinal points of the compass symbolize the activity of each RF according to its location along the edge of the rectangle. The top one is only slightly active, since illumination is both exciting it through the ON-center and inhibiting it through the OFF-surround. The RF beneath is the most active of the four, since its OFF-surround is partially in the dark, while its ON-center is still illuminated. The OFF-center

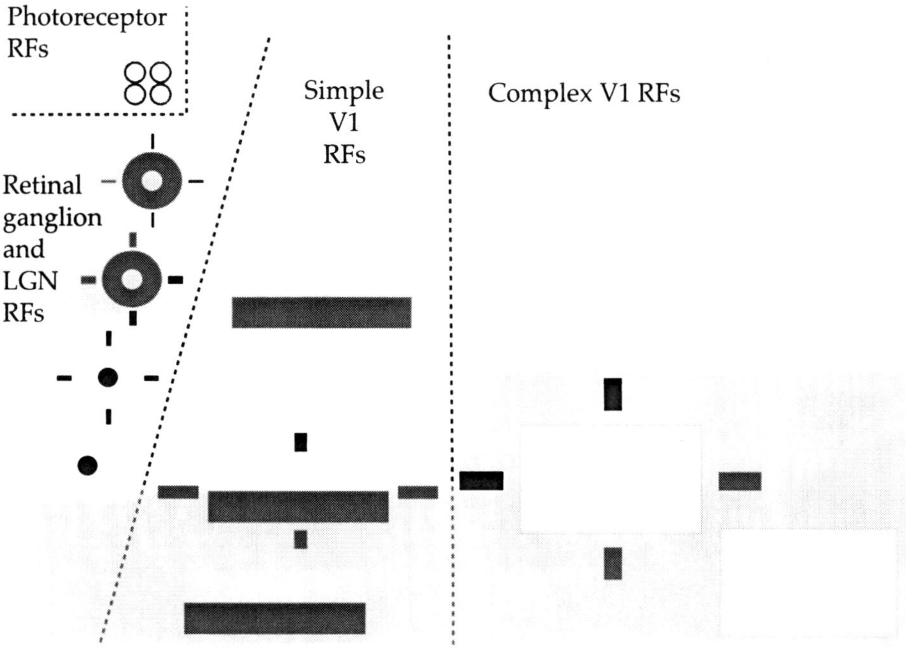

Figure 1.21. Responses of receptive fields of early visual cells to a rectangular patch of light, where a white sub-field represents ON and a black sub-field represents OFF. Comparable to Nicholls et al., 2001, Fig. 20.16.

cell below it is somewhat less active, through the illumination of the right side of its ON-surround. Finally, the bottom RF is inactive, due to its lack of illumination.

The simple V1 cells take in the LGN output and transform it into a selectivity for oriented lines. In Fig. 1.21, only the RF that aligns correctly with an edge of the rectangle becomes active; the other two are inactive. For the one that is entirely illuminated, inhibition cancels excitation; the other is in the dark. Finally, the complex V1 cells select for oriented edges, a criterion which, for the particular RF shown in the diagram, is only satisfied for the third RF from the top.

It is hoped that seeing all four RF types side-by-side has helped to fix their behavior more firmly in the reader's mind. Yet there is a more profound and rewarding reason for spending a few moments perusing this diagram (and its somewhat prolix explication). It has to do with the puzzling fact that the early visual system chooses not to respond to areas of constant illumination; the RFs

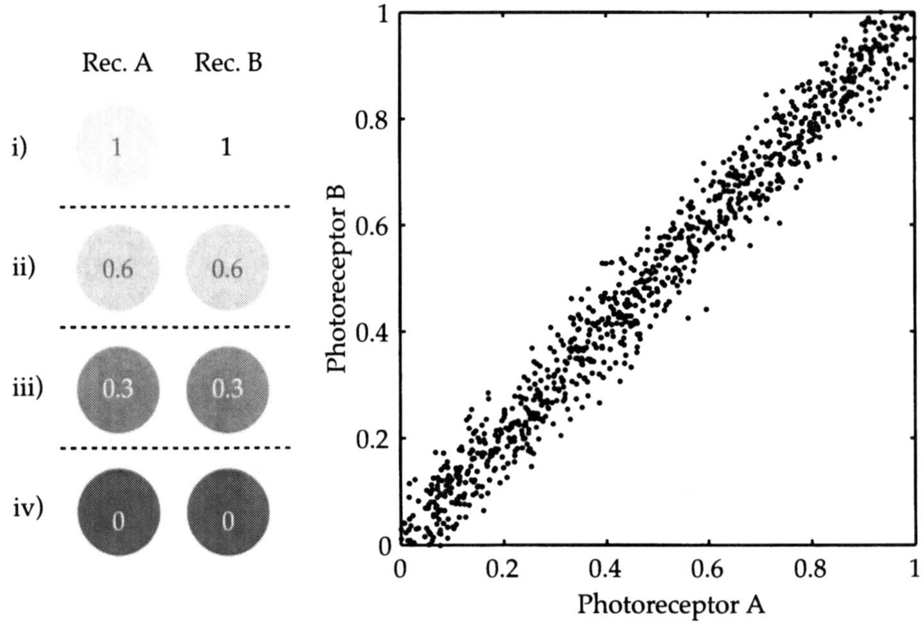

Figure 1.22. Two photoreceptors illuminated to the same degree, and a plot of many such pairs. Comparable to Field, 1994, Fig. 2, and Olshausen and Field, 2000, Fig. 4.

in Fig. 1.21 that respond most strongly are exactly those that overlay a change or discontinuity in illumination.

To see the utility of this choice, we can begin by imagining two photoreceptors side-by-side and ask simply what it would look like for them to frequently be illuminated to the same degree. It would like somewhat like Fig. 1.22. Along the left edge are stacked the receptive fields of two photoreceptors, pictured simply as circles with the same degree of darkening in each case, and varying from 0 (no illumination) to 1 (the maximum to which they can respond). The graph on the right plots a large number of such pairs, with a small amount of noise added to mimic more closely a real set of observations. Both representations are meant to drive home the same point: if nearby photoreceptors receive the same illumination, then their input is highly correlated. If such correlation is a property of visual scenes in general, then they are highly redundant. The visual system could economize its expenditure of metabolic resources by stripping out this redundancy – which is equivalent to compressing the input image. In fact, such correlation is a robust property of natural images, see for instance Fig. 4 in Olshausen and Field (2000), which plots the brightness values in photographs for points that are adjacent, two, and four pixels apart. Each plot reproduces the pattern of correlation seen in our Fig. 1.22, though with decreasing clarity.

This simple observation points the way towards a conceptual framework for early visual processing, and perhaps other sensory and perceptual processes, namely of "recoding to reduce redundancy" (Phillips and Singer, 1997, p. 659). This is in fact a long-standing hypothesis within computational neuroscience, dating at least to Attneave (1954); see Phillips and Singer, 1997, p. 659, for additional references, plus more recent work that shall be mentioned shortly. As Phillips and Singer, 1997, p. 659, put it in their review,

> The underlying idea is that the flood of data to be processed can be reduced to more manageable amounts by using the statistical structure in the data to recode the information it contains, with frequent input patterns being translated into codes that contain much less data than the patterns themselves.

In the example of Fig. 1.21, the frequent input pattern of diffuse or constant illumination is translated into the code of a low firing rate, while the less frequent pattern of a change in illumination is translated into the code of a high firing rate.

This hypothesis suggests in turn that the proper theoretical framework in which to develop it is information theory. One of the more fruitful principles forthcoming from this field of mathematics is that of **maximum information transfer** or **infomax**, see Linsker (1988), Atick and Redlich (1990), and Bell and Sejnowski (1995). As Friston, 2002, p. 229, puts it,

> This principle represents a formal statement of the common-sense notion that neuronal dynamics in sensory systems should reflect, efficiently, what is going on in the environment. In the present context [i.e., neuromimetic models], the principle of maximum information suggests that a model's parameters should be configured to maximize the mutual information between the representations they engender and the causes of sensory input.

It should be underscored that "information" is used here in the technical sense of Shannon's information theory, see Shannon (1948) and voluminous posterior work, as well as Sec. 3.2.4 of the current work. **Information** in Shannon's sense really means **entropy**, which is a measure of the predictability or surprise value of a response. Frequent responses come to be expected and so are more predictable or less surprising. Infrequent responses are not expected and so are less predictable or more surprising.

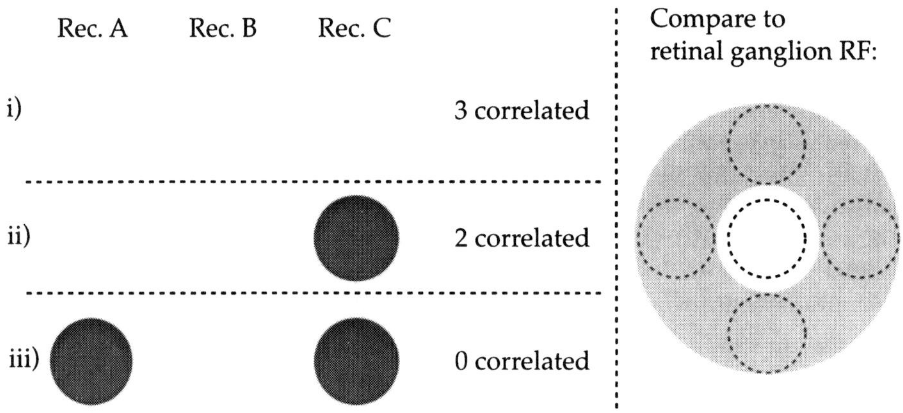

Figure 1.23. On the left, a neighborhood of three photoreceptors pictured under three patterns of illumination, marked with the number of photoreceptors correlated by the same degree of illumination. On the right, the input structure of an ON-center retinal ganglion cell, for comparison.

Of course, there is a trading relation among these two extremes. Very infrequent responses may be very unpredictable and therefore deserve a large surprise value, but the fact of their infrequency excludes them from making a large contribution to the overall entropy measure. It follows that an implementation of the entropy measure will be most efficient if it draws from the middle of the distribution, where responses are both surprising enough, but also frequent enough to be encountered in a limited sample.

We can construct a thought experiment from these observations by trying to imagine what physical organization the retinal ganglion cells would have to assume in order to reduce the redundancy of the photoreceptor output. Building on Fig. 1.22, let us add one more photoreceptor to the array of two and illuminate them differentially. There are three main patterns, laid forth on the left of Fig. 1.23. Starting at row (i), each photoreceptor is (equally) illuminated, a pattern answering to the description of the worst case for efficiency: all three photoreceptors are correlated. In row (ii), two photoreceptors are illuminated, giving an intermediate case of redundancy. In row (iii), only one photoreceptor is illuminated, which results in the most efficient pattern, with no redundancy whatsoever.

Now compare these three patterns to the entire array of photoreceptors postulated to constitute an ON-center retinal ganglion receptive field, reproduced for the reader's convenience on the right side of Fig. 1.23. In particular, take each row on the right to correspond to the horizontal subfield of the RF on the left. Case (i) on the right side would turn off the subfield on the left, since the inhibitory input from the two photoreceptors in the OFF-surround would overwhelm the excitation of the ON-center photoreceptor. Case (ii) on the right would lead to a weak signal from the RF, since the single inhibitory OFF-surround photoreceptor would not counterbalance the retinal ganglion cell's heightened sensitivity to the ON-center photoreceptor. Finally, case (iii) on the right would produce the highest output from the RF, due to the absence of an active photoreceptor in the surround to attenuate the excitation coming from the center photoreceptor. The conclusion is that we have proved informally that the retinal ganglion center-surround receptive field is an excellent, if not optimal, mechanism for the reduction of redundancy in a small patch of photoreceptor array.

From the information-theoretic perspective, the peculiar structure of the retinal ganglion RF can be interpreted as a means for finding almost any discontinuity in illumination among the photoreceptors to be surprising and transmitting it to V1 via the LGN by means of a high rate of spiking. The retinal ganglion layer takes the variation in the probability of differing visual stimuli impinging on the photoreceptors and discards the most predictable stimuli and passes on a representation of the rest as spike trains. The discarded residue tends to be the information with the least statistical structure, namely the most and least predictable stimuli. The retained information tends to be **decorrelated**, which is to say that there is little statistical correlation among the various items retained, such as overlapping RFs. The mutual information of this recoding process is the information that is represented at both layers. The recoding itself acts to maximize the amount of information sent to the higher layer, and thus is said to maximize information transfer.[3]

[3] The process described for the LGN coding – presumably equivalent to the ganglion coding – is not idle theoretical speculation. Yang et al. (1996) report on recordings taken from the cat LGN cells that were found to be decorrelated with respect to natural, time-varying images, i.e. movies.

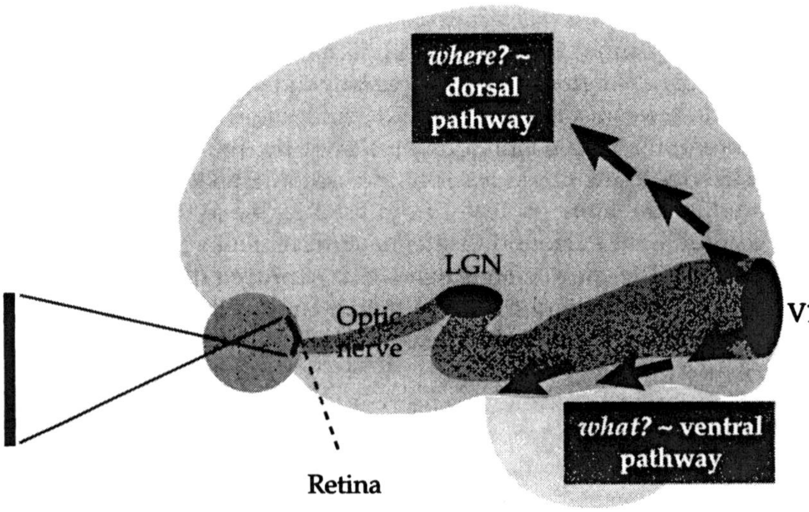

Figure 1.24. The two principle visual pathways superimposed on the human brain.

Turning to the next stage in the pathway, at V1, the simple cell RF can be interpreted as the exact mechanism needed to be surprised by the presence of lines within the LGN output and to pass this information on to the complex cells. The complex cell RF can likewise be interpreted as the exact mechanism needed to be surprised by the presence of edges within the simple cell output and to pass this information on to higher cortical areas.0

To round out this argument, let us estimate how much of a metabolic savings can be achieved under the decorrelation regimen. We can begin by counting how many receptive fields of each type beyond the photoreceptors it would take to tile the rectangular stimulus in Fig. 1.21 without overlapping. It would take about 88 ganglion and LGN cells, about 25 simple V1 cells; and about 13 complex V1 cells. However, the only cells that respond strongly are those that cover an edge at the proper orientation. Let us say that the edges are covered by about 34 ganglion and LGN cells, about 11 simple V1 cells; and about 11 complex V1 cells. By dividing the number of neurons firing strongly by the total number needed to tile the image, we can calculate a metabolic 'effort' of only 39% for the ganglion/LGN layer, 44% for the V1 layer, and 85% for the V2 layer.

The pathway consequently becomes metabolically more efficient at each step; on the scale of a real brain, the savings would be vast.[4]

In the next few subsections we review what some higher visual areas make of the information that they receive from the early visual pathway, but before doing so, let us summarize what has been said about the physical implementation of the recoding process at V1.

1.2.3. Beyond primary visual cortex

V1 is but the first of thirty or so areas of the cortex that work together to endow primates with sight. If these areas did not cooperate in some way, it would be almost impossible to make sense of them. Yet, fortunately, they do. From V1, visual information is routed into two major streams through the brain, a **dorsal** stream along the top, and a **ventral** stream along the bottom. Fig. 1.24 superimposes these two pathways on the surface of the brain, along with the retinocortical pathway introduced in the previous section.

1.2.3.1. Feedforward along the dorsal and ventral streams

The dorsal stream continues the magnocellular pathway of the LGN and V1, while the ventral stream continues the parvocellular pathway. Thus what begins as a small physiological difference between retinal ganglion cells grows into one of the major functional structures of the entire brain. The dorsal stream is often referred to more mnemonically as the ***where?*** pathway, since its main function is to localize a visual stimulus in space. By analogy, the ventral stream is often referred to as the ***what?*** pathway, since it is chiefly concerned with identifying what the stimulus is. This split was first conceptualized in *where?/what?* terms by Ungerleider and Mishkin (1982) and then in dorsal / ventral terms by De Yoe and Van Essen (1988). It has received considerable elaboration since these initial statements.

---

4 One can even do real experiments to arrive at the same conclusion. Barlow (1972) estimates that there are 10^8 cells in the retina, but only 10^6 cells in the optic nerve. Thus some compression of the retinal image must be achieved even before it passes down the optic nerve. See Laughlin et al. (2000) for a more exact calculation of the metabolic cost of sensory and neural information in the retina of the blowfly.

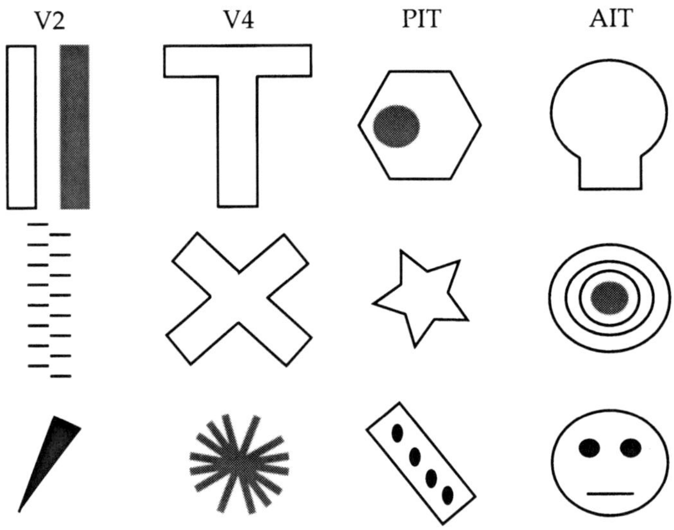

Figure 1.25. Some patterns recognized by areas along the ventral pathway. Comparable to Oram and Perrett, 1994, Figure 4, and Kobatake and Tanaka (1994).

In the classical view, the main computational principal of both streams extends the classical feed-forward analysis of V1 by Hubel and Wiesel to the rest of the visual system: features or patterns recognized at one area are combined to form a more complex feature or pattern at the next. This generalization is most easily described for the ventral/what?/parvocellular stream, to which we dedicate the next few paragraphs.

The ventral stream is best known from recordings of neurons in monkeys, see Oram and Perrett (1994) and Kobatake and Tanaka (1994), plus the overview in Farah et al. (1999). The sequence is roughly V2 → V4 → posterior inferior temporal cortex (PIT) → anterior inferior temporal cortex (AIT) → superior temporal polysensory area (SPTa). At each stage, neurons are activated by a larger area of the visual image and more complex patterns. Fig. 1.25 attempts to convey the increasing complexity of pattern recognition. Similar to V1, cells in V2 select for bars of light of a fixed length, width, or orientation. Some also respond to oriented illusory contours, which are apparent lines formed from the juxtaposition of other elements. The middle pattern in the V2 column of Fig. 1.25 depicts an illusory horizontal line. V4 cells can select for both the length and the width of a bar, as well as the junctions of lines and bars. V4 is also the first area that is sensitive to color, though our diagram makes no attempt to represent such a sensitivity. Moving into posterior inferior temporal cortex, the patterns

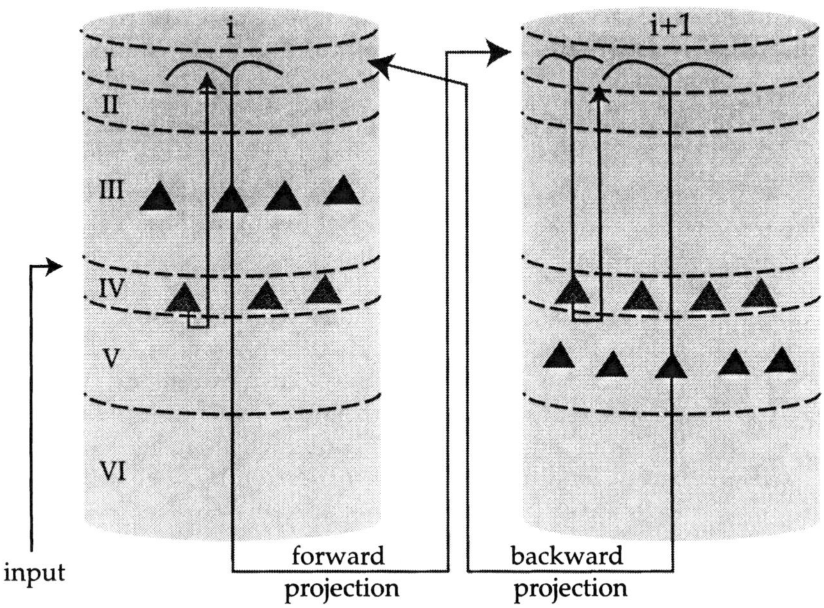

Figure 1.26. Bidirectional processing in neocortex. Comparable to Cotterill (1998), Fig. 7.8, and Rolls and Treves (1998), Fig. 10.7.

cells respond to become more elaborate and less easy to describe in words. This is even more true of anterior inferior temporal cortex, though it is here that the first cells that are attuned to faces become evident. Moreover, it is in AIT that neurons first begin to loss their responsiveness to a pattern if it is presented repeatedly – presumably a kind of memory signal – and to maintain their activation after the pattern has been withdrawn, if it is needed for a posterior task. Finally, the superior temporal polysensory area is not represented in the diagram because our sources only discuss it with respect to its ability to recognize faces, see Farah et al., p. 1349, for an overview.

There is considerably more that could be said about these areas, but it would not add much more insight to what we have already learned from the retinocortical pathway. The computations that these areas perform are poorly understood, though it is presumed that each one combines the output of its downstream source to build a more complex pattern detector. The existence of a corresponding dorsal pathway has been confirmed in humans through functional neuroimaging techniques, but these techniques do not at present have the resolution to attain the level of detail that can be achieved from single-cell recordings of monkeys.

Having completed in a cursory way our overview of feedforward visual processing, or at least of the object-recognition component of feedforward visual processing, let us change direction and take up visual feedback, which turns out to constitute an entirely different kind of visual processing.

1.2.3.2. Feedback

The sketch of the visual system given so far has ignored two aspects of vision that undoubtedly have considerable import. One is anatomical: there are extensive connections from higher areas of visual cortex back to lower areas, and even to the LGN; see Fig. 1.26 for a generic illustration. In this figure, not only does minicolumn *i* activate upstream minicolumn *i+1*, but *i+1* also sends its activation downstream back to *i*, as first observed by Rockland and Pandya (1979), see Cotterill, 1998, pp. 199-204, 227ff, and Rolls and Treves, 1998, p. 240ff, for overview. Zeki, 2001, p. 61 goes so far as to raise this observation to the rank of a "universal rule of cortical connectivity, with no known exceptions: An area A that projects to B also has return projections from B", evidence for which he remits the reader to Rockland and Pandya (1979) and Felleman and van Essen (1991). By definition, any feed-forward description omits such feed-back connections, but what is more disturbing is that the feed-forward analysis seems to account for the data on visual perception adequately all by itself, so why have backward connections at all?

1.2.3.2.1. Generative models and Bayesian inference

It has only been in recent years that researchers have come to appreciate and try to account for the contribution of these top-down or feedback connections. One framework emerging from such work incorporates both directions of processing, but assigns them different roles. The feedforward processing that has just been reviewed is known as a **recognition model**. It learns a probability distribution of the underlying causes from the sensory input. This recognition model is modulated by the feedback connections of a **generative model**, see Zemel (1994), Hinton and Zemel (1994), and Hinton et al. (1995) for initial formulations. A generative model tries to reconstruct the input accurately by drawing an inference about its cause and testing this prediction against the observed input. It relies on backward connections to compare the prediction made at a higher level to an input at a lower level. Such a process consequently inverse the feedforward challenge, which is to find a function of an input that predicts its cause.

Friston (2002) finds one of the most compelling aspects of generative modeling to be the fact that it emphasizes the brain's role as an inferential machine, an emphasis that Dayan, Hinton, Neal and Zemel (1995) attribute ultimately to Helmholtz, presumably Helmholtz (1925). In particular, the brain is characterized as a Bayesian inference machine.

The original statement of Bayesian inference is credited to an essay by the English clergyman Revd. Thomas Bayes published in (1764). According to Earman, 1992, p. 7, up to the time of Bayes' essay, most of the work on what we now call probability theory involved what Earman calls "direct inferences":

> given the basic probabilities for some chance setup, calculate the probability that some specified outcome will occur in some specified number of trials, or calculate the number of repetitions needed to achieve some desired level of probability for some outcome: e.g. given that a pair of dice is fair, how many throws are needed to assured that the chance of throwing doubles sixes is at least 50/50?

The innovation of Bayes was to grapple with the inverse inference:

> given the observed outcomes of running the chance experiment, infer the probability that this chance setup will produce in a given trial an event of the specified type.

This is not the most enlightening explanation, though it does put us in the right context.

A more perspicuous inductive introduction concerns the 'canonical example' of a precocious newborn who observes his first sunset, and wonders whether the sun will rise again or not. An article in the *Economist* of 9/30/00 tells it this way:

> He [the precocious newborn] assigns equal prior probabilities to both possible outcomes, and represents this by placing one white and one black marble into a bag. The following day, when the sun rises, the child places another white marble in the bag. The probability that a marble plucked randomly from the bag will be white (ie, the child's degree of belief in future sunrises) has thus gone from a half to two-thirds. After sunrise the next day, the child adds another white marble, and the probability (and thus the degree of belief) goes from two-thirds to three-quarters. And so on. Gradually, the initial belief that the sun is just as likely as not to rise each morning is modified to become a near-certainty that the sun will always rise.

The means by which the precocious newborn arrives at this solution is known as **Bayes' rule** or **Bayes' theorem**.

To illustrate how Bayesian inference works, we adapt the tutorial example of Knill et al., 1996, p. 9ff, which is "crudely" analogous to geometric projection in vision.[5] It begins with a message set of four three-dimensional solid objects, a cube, a pyramid, a prism, and a tetrahedron. They can be seated into a flat transmitter that has a square and a triangular slot in it, each of which is large

[5] The example of pattern recognition (the classification of fish) of Duda et al., 2000, p. 20ff is also helpful, though more cumbersome

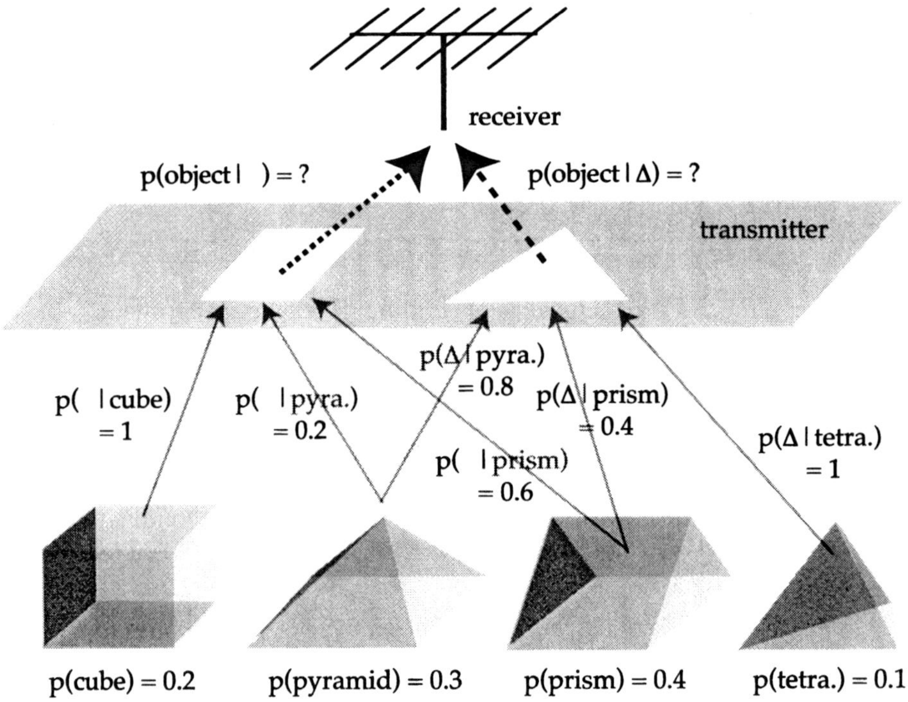

Figure 1.27. Bayesian shape categorization system. Comparable to Knill et al. 1996, Fig. 0.2.

enough to fit the object that has a side with that particular shape. This device has the effect of reducing a 3D object to a 2D silhouette, in analogy to how the visual system reduces 3D objects in the visual environment to 2D images on the retina. An object that is seated successfully in a slot triggers the emission of a signal that is feed to a receiver. Knill et al. use color-coded signals, but this is a rather confusing choice in the context of shape. We prefer to use a spike-train-like signal that can be simply called 'short' or 'long' – short for the square slot and long for the triangular slot. Fig. 1.27 depicts the set-up that we have in mind, along with the relevant probabilities, which are explained below. This system classifies the message set in the following manner.

The challenge for the receiver in detecting the silhouettes is they do not uniquely determine the shape of the objects selected by the transmitter, since two of the objects, the pyramid and the prism, can be seated in either slot. The information provided by a silhouette is therefore ambiguous and can only be described probabilistically. The probability of encountering one of these objects in the environment is listed across the top of the diagram (and they are arrived

Table 1.3. Posterior probability distribution, *p(object | silhouette)*. Bold face highlights the highest one per object.

	Square	*Triangle*
cube	(1 x 0.2)/0.5 = **0.4**	(0 x 0.2)/0.5 = 0
pyramid	(0.2 x 0.3))/0.5 = 0.12	(0.8 x 0.3)/0.5 = **0.48**
prism	(0.6 x 0.4))/0.5 = **0.48**	(0.4 x 0.4)/0.5 = 0.32
tetrahedron	(0 x 0.1))/0.5 = 0	(1 x 0.1)/0.5 = **0.2**

at after some observation that is independent of the model). For instance, one is more apt to run across a prism in the environment than a tetrahedron. In Bayesian systems, these are known as the ***a priori*** or **prior probabilities**, and the entire list is known as the **prior distribution**. We assume that each side of an object, when dropped onto one of the transmitter's slots, has an equal probability of facing down, so the probability that an object will be coded as a given silhouette, *p(silhouette | object)*, is simply the proportion of the sides of the object which have the silhouette's shape. This probability, the **conditional likelihood**, is appended to the arrows leading from the objects to the silhouettes in the figure. Finally, we need to know the probability of occurrence of a given silhouette, *p(silhouette)*. In our simple, noise-free example, there are only two shapes, so the probability of picking one, *p(silhouette)*, is obviously 1/2. Knill et al. do not label this probability with a specific name, while Duda et al., 2001, p. 22 refer to it as the less than perspicuous "evidence". We can call it simply the **likelihood**.

Given that all of these probabilities can be measured in our toy experiment, we now have enough facts to ascertain the probability that an object is categorized as a given silhouette by the receiver, *p(object | silhouette)*, known as the (all-important) ***a posteriori*** or **posterior probability**. This is where Bayes supplies the crucial insight, for what he says in essence is:

1.7. $$\text{posterior} = \frac{\text{conditional likelihood} \times \text{prior}}{\text{likelihood}}$$

Instantiating (1.7) with the particulars of our example results in (1.8):

1.8. $$p(\text{object} \mid \text{silhouette}) = \frac{p(\text{silhouette} \mid \text{object}) \times p(\text{object})}{p(\text{silhouette})}$$

Solving (1.8) for each case produces the distribution of posterior probabilities tabulated in Table 1.3. This distribution has the effect of ranking each classification by its posterior probability. Such a ranking points to an easy way of deciding whether an object is correctly classified or not: just pick the classification with the highest posterior. These are highlighted in bold face in the

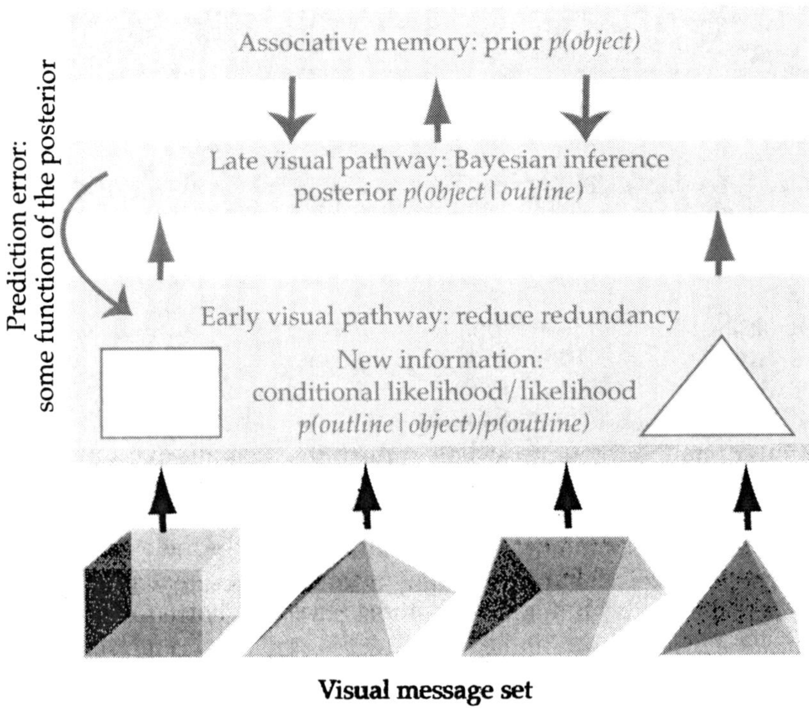

Figure 1.28. Two-way visual processing for the message set of Fig. 1.27. Comparable to figures in Knill et al. (1996), Barlow (1996), and Friston (2002).

table. In effect, what we have done is infer a cause (an object in the environment) from an effect (a silhouette).

To summarize, Bayesian probability provides a normative model for how prior knowledge should be combined with sensory data to make inferences about the world. Human vision is assimilated into this framework under the assumption that it draws visual inferences from an implicit model of the posterior probability distribution. This model incorporates assumptions about image formation and the structure of scenes in the environment – it "characterizes the world to which the human visual system is 'tuned' and in which humans would be the ideal perceptual inference makers" (Knill et al., 1996, pp. 15-6).

Much debate and disagreement has surrounded the role of the priors. Knill et al. take them to be "low-level, automatic perceptual processes that do not have access to our cognitive database of explicit knowledge about the world." (p. 16) In the context of our previous discussion, this characterization practically defines the redundancy-reducing properties of the feedforward pathway. However, Friston, 2002, p. 240 takes the opposite perspective, identifying the likelihood term with the bottom-up (i.e. feedforward, redundancy-reducing) direction and the prior term with the top-down (i.e. feedback) direction. We are inclined to agree with Friston, for the overall logic of the system favors the priors as part of general cognitive knowledge, see for instance Fig. 12.2 of Barlow (1996).

We consequently offer Fig. 1.28 as an example of the overall function of the system upon the combination of the feedforward and feedback hypotheses. The gaze takes in the image of an object in the environment, and visual processing commences. The early visual pathway reduces the redundancy among correlated cells, producing, say, an outline of an image. This 'leftover' is what Barlow calls "new information", the information that the system must classify. It can be coded by a firing rate in which the outlines that fire the most are the ones most activated by the input. This collection of firing rates qualifies as the conditional likelihood probability distribution.

It seems reasonable to assume that the likelihood is also estimated during this process. If it is used as a divisor for the conditional likelihood, the conditional likelihood is scaled or normalized, and half of Bayes' rule is executed during early visual processing. These results propagate to the late visual pathway where they are multiplied with the priors in order to calculate the posterior probability in accord with Bayes' rule. Parenthetically, to our way of thinking, the priors are not visual objects, but rather arise from one's knowledge of how the environment is structured. Fig. 128 therefore draws them from associative memory, beyond the bounds of the visual system. Returning to the main flow of the diagram, the predictive value of the posteriors can then be feed back to earlier levels as an error signal, exciting the subpaths that lead to the highest posterior and inhibiting the others; see Friston, 2002, p. 233ff for a mathematically more complex statement of a similar idea.[6]

---

[6] Again, generative modeling is conceptually compatible with predictive coding, by adding a feedback loop which turns off the output of the previous layer if it contradicts the prediction of the current layer. Rao and Ballard (1999) devise an intriguing application of predictive coding to the end-stopped cells of V1, which are omitted from the previous discussion of V1 for the sake of simplicity. An end-stopped cell is sensitive to an oriented line in its receptive field – unless the line extends off the edge of the RF, in which case the cell stops responding. Rao and Ballard argue attribute this to the statistical fact that short lines tend to be part of longer lines. Thus if a line runs off the end of a (short) V1 RF, it is predicted by a higher area to be part of a longer line and consequently a better stimulus for the higher area. The higher area tells the V1 cell to not bother representing the line by turning it off – thus saving the metabolic cost of the redundant,

1.2.3.2.2.Context

The ideas reviewed above put the notion of visual context on a much firmer footing, and in fact show it to be fundamental for the understanding of a scene. Friston, 2002, p. 240 states this new understanding of the visual system quite elegantly:

> The Bayesian perspective suggest something quite profound for the classical view of receptive fields. If neuronal responses encompass both a bottom-up likelihood term and top-down priors, then responses evoked by bottom-up input should change with the context established by prior expectations from higher levels of processing. In other words, when a neuron or population is predicted by top-down inputs, it will be much easier to drive than when it is not.

Others have attributed an empirical deficit to the feed-forward account attendant on ignoring context. Albright and Stoner, 2002, p. 342 describe it as so:

> As a simple illustration, consider an orientation-selective neuron in primary visual cortex (Hubel and Wiesel, 1968). The RF of the neuron illustrated in Figure [1.16] was characterized without contextual manipulations, and the data clearly reveal how the neuron represents the proximal (retinal) stimulus elements of orientation and direction of motion. From such data, however, it is frankly impossible to know whether this neuron would respond differentially to locally identical image features viewed in different contexts.... In other words, the full meaning of the pattern of responses in Figure [1.16] is unclear, as that pattern is not sufficient to reveal what the neuron conveys about the visual scene.

A few pages on, Albright and Stoner, 2002, p. 344 state the drawbacks of trying to understand vision by means of limited and uncontextualized stimuli in terms that a linguist can appreciate. Please allow us another long quote:

> To illustrate this assertion, imagine attempting to understand the function of language areas of the human brain armed only with non-language stimuli such as frequency-modulated audible noise. Imagine further that, using these stimuli, you make the remarkable discovery that the response of neurons within Wernicke's area (for example) are correlated with behavioral judgments (of, for instance, whether the frequency modulation was high-to-low or low-to-high). Although this

---

representation. See Koch and Poggio (1999) for comment and Schultz and Dickinson (2000) for additional examples.

> finding would offer an interesting (and perhaps satisfyingly parametric) link between single neurons and perceptual decisions, it seems clear that stimuli constructed of words and sentences would yield results more likely to illuminate language processing. Just as we can progress so far using nonsense sounds to explore language function, so are we constrained by visual stimuli lacking the rich cue interdependencies that permit the perceptual interpretation of natural scenes.

The next few subsections discuss some of the ramifications of taking context seriously.

Albright and Stoner (2002) review several experimental paradigms that have demonstrated an effect of context on visual perception, especially during early visual processing. Given that there are a potentially vast number of ways in which context interactions could reveal themselves, Albright and Stoner focus on spatial interactions, in which information in one region of an image influences the interpretation of another. An especially rich source of such effects is constituted by images in which 'missing' information must be recovered. In the next paragraphs, a particularly clear example is reviewed.

Sugita (1999) reports on an experiment in which V1 neurons were demonstrated to respond to lines outside their classical receptive field. The control trials of the experiment tested the effect of moving a line with respect to the classical V1 receptive field. Projected at the preferred orientation of a cell, a line was moved in the preferred direction of the cell, say left to right, and in the contrary direction, see case (i) of Fig. 1.29. The cell responded strongly to the former motion, but barely at all to the latter, which is indicated in the figure by the relative thickness of the two arrows underneath the diagram. The experiment was repeated, changing the single line to two segments separated by the receptive field. The cell barely responded to movement of the line in either direction, as summarized by case (ii). These two procedures mainly replicate previous work and set the stage for the core contribution of the experiment.

In the next series of trials, the two line segments were made to appear stereoscopically separated by an occluding patch in such a way that the inference could be made that the they were actually one and the same line, part of which was merely hidden from view by the patch. To support this inference or not, the planes of the image were arrayed so that the occluder would appear to be at the same depth as the line, on top of it, or underneath it, see cases (iii) to (v) of Fig. 1.29. As the pair of arrows underneath each stimulus indicate, only the positioning of the planes that supported the inference that the patch was occluding a single line underneath elicited a significant response from V1, case (iv). The result could not be more striking: V1 responds to a line in a classical RF *which it does not actually see*, as if it had 'X-ray vision', to use Albright and Stoner's colorful term.

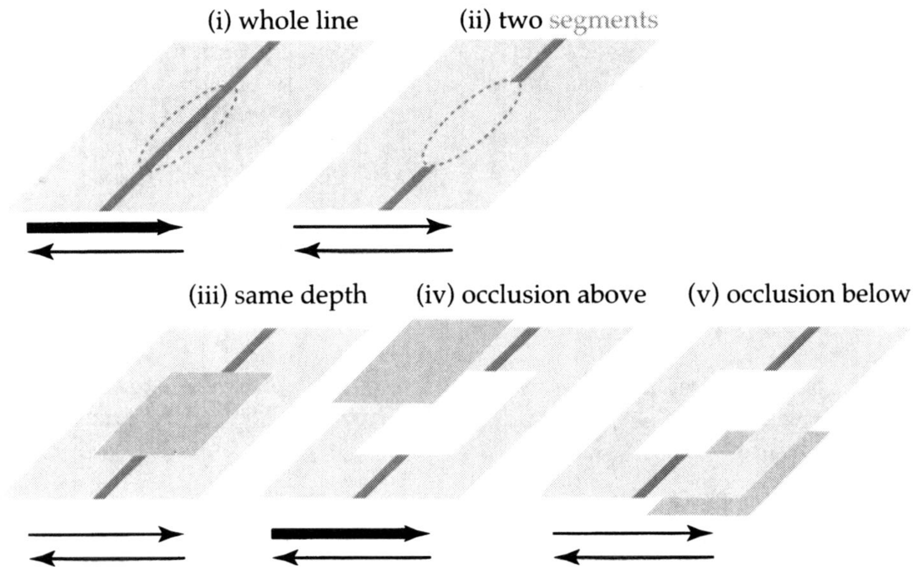

Figure 1.29. Five ways to see a line. The dotted oval in the top two diagrams outlines the receptive field of the V1 neuron in question; see text for an explanation of the rest. Comparable to Sugita, 1999, Fig. 2.

Sugita found that the response latencies of case (iv) were on a par with case (i) and therefore concluded that the 'X-ray' response was likely to be carried over horizontal connections in V1 or fast feedback from V2. Albright and Stoner, 2002, p. 357 put this into a broader perspective by adding that whatever the exact neuronal mechanism is, it depends on contextual cues – cues which are ignored in the classical framework.[7]

[7] It is instructive to consider Rao and Ballard's (1999) analysis of the end-stopped cells of V1 mentioned in the previous footnotes in the context of Sugita's results on occlusion at V1. Rao and Ballard proposed that end-stopping is a signal from a higher level to ignore a redundant line at the lower level. Sugita demonstrates the converse: the higher level (or perhaps nearby cells at the same level) signals the lower level to postulate a redundant line. What emerges is a pattern of the higher level overriding the lower level's concern for redundancy reduction.

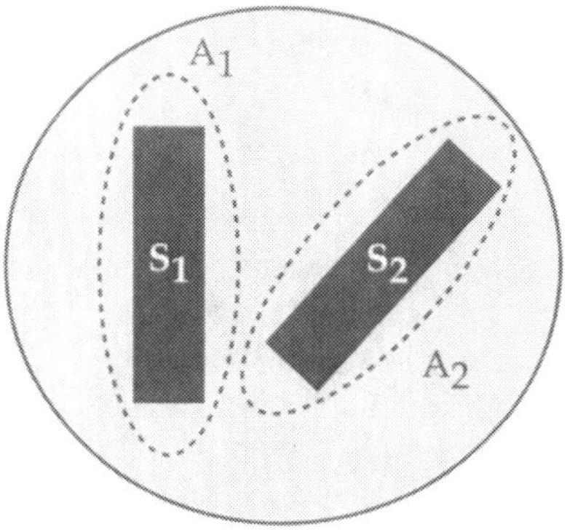

Figure 1.30. A large receptive field containing a compound stimulus S_1+S_2, and its decomposition by attention into units A_1 and A_2.

1.2.3.2.3. Selective attention and dendritic processing

One of the areas of potential contextual interaction that Albright and Stoner exclude from their review is that of attention. Since Posner and Boies (1971), psychologists distinguish three types of attention: arousal, vigilance, and selective attention. Arousal describes the capacity of an organism to receive, process, or generate information. Vigilance refers to the capacity of an organism to maintain focus over time. Selective attention labels the means by which an organism chooses which of the multiple internal and external stimuli to subject to additional processing. As Coslett, 2000, p. 257 explains, "this notion is perhaps best captured by the concept of 'gating'; at any moment, the nervous system is processing, at least to 'pre-conscious' level, a wide range of external and internally generated stimuli, yet we are typically conscious of only a limited number of stimuli at any given time." It is selective attention that has been of most concern to researchers in vision.

By way of introduction, consider the fact that neurons in higher visual cortex have relatively large receptive fields. For example, neurons representing the central visual field in macaque area V4 have receptive fields up to 5° across, see Desimone and Schein (1987). Such large receptive fields often contain many potentially significant features in a single image, as illustrated for an invented example of just two such features in Fig. 1.30. A natural question to ask about the 'cluttered' organization of such large receptive fields is, how can information about individual items be extracted from them?

In their recent review of this topic, Kastner and Ungerleider (2000) point out that multiple stimuli compete for representation in visual cortex. That is, they are not processed independently but rather interact with one another in a mutually suppressive way. Evidence for this conclusion comes from recordings

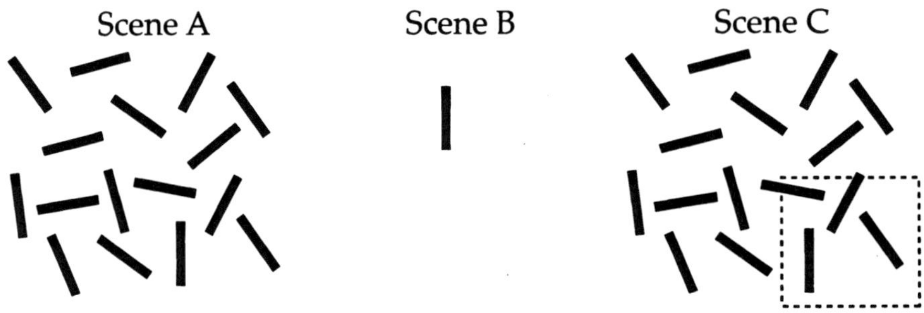

Figure 1.31. Find the vertical bar in three cluttered scenes. Comparable to Kastner and Ungerleider, 2000, Fig. 1.

from individual V4 neurons for which simultaneously presented stimuli compete to set the output firing rate, see Luck, Chelazzi, Hillyard, and Desimone (1997) and Reynolds, Chelazzi, and Desimone (1999). These experiments found cells in which one stimulus, presented by itself, produces a strong response and another stimulus produces a weak response. Presenting the two stimuli together generally produces a response less than the combined response of the two taken separately, but more than either one of them alone. It would appear that the "weak" stimulus is facilitative for the cell when presented alone, since it increases the cell's response, but suppressive when presented in tandem with the "strong" stimulus.

This suppression can be biased by both bottom-up and top-down mechanisms. An example of a bottom-up, sensory-driven mechanism would be an item with high contrast in an array being perceived as more salient and thereby 'popping out' from the lower contrast background. More interesting for our current concerns are the top-down mechanisms. One of the best-known examples comes from the seminal work of Moran and Desimone (1985), which showed that when multiple stimuli are present within the RF of a V4 neuron, attention effectively reduces the RF extent of the cell, so that only the attended feature contributes to its output. Fig. 1.31 provides a specific illustration of both of these mechanisms. Scene A is the worst case, in which the absence of any bias forces one to do a serial search of the entire image. Scene B demonstrates the efficacy of a sensory-driven bias in contrast enabling the high-contrast vertical bar to 'pop out' from (what is interpreted as) the low contrast background. Finally, in Scene C the dotted box corresponds to a verbal instruction to attend to the bottom-right quadrant. Such selective attention enables the vertical bar to be localized almost effortlessly.

Kastner and Ungerleider, 2000, p. 332 list four general effects of selective attention:

1.9. a) the enhancement of neural responses to attended stimuli,
b) the filtering of unwanted information by counteracting the suppression induced by nearby distractors,
c) the biasing of signals in favor of an attended location by increases in baseline activity in the absence of visual stimulation, and
d) the increase of stimulus salience by enhancing the neuron's sensitivity to stimulus contrast.

These effects can bias competition among simultaneously-presented stimuli to the point of overriding sensory-driven inputs. Their source lies outside of visual cortex, though we defer to Kastner and Ungerleider for references. Winning the competition channels a stimulus to memory for encoding and retrieval and to motor systems for the guidance of action and behavior.

From this short synopsis, it would seem that selective attention fits quite snugly into the theory of the visual system outlined in the previous sections. Whereas Bayesian inference is a process that accounts for statistical regularities beyond the realm of the purely visual, selective attention is a feedback process that engages if statistical regularities fail to reduce the visual input to a manageable size. VanRullen and Thorpe, 1999, p. 912, state the challenge with particular clarity:

> Consider a neural network performing object recognition. With one neuron selective to a particular object for each spatial location, such a system does not need any attentional mechanism to perform accurately. ... The problem arises in real networks such as the human visual system, where the amount of resources, namely the number of neurons, is limited.
>
> Clearly, the human visual system cannot afford one 'object detector' for each object and each retinotopic location. It is well known that neurons in the visual system have increasing receptive fields sizes, and many neurons in the latest stages, such as the inferotemporal cortex, have receptive fields covering the entire visual field. They can respond to an object independently of its spatial location. Such a system needs far fewer neurons. But how can it deal with more than one object at the same time? With no attentional mechanism, if you present to that network an image containing many objects to detect, it is impossible to decide which one it will choose. Furthermore, there is a risk that features from different objects will be mixed, causing problems for accurate identification. This is an aspect of the well-known 'binding problem' (Triesman, 1996).

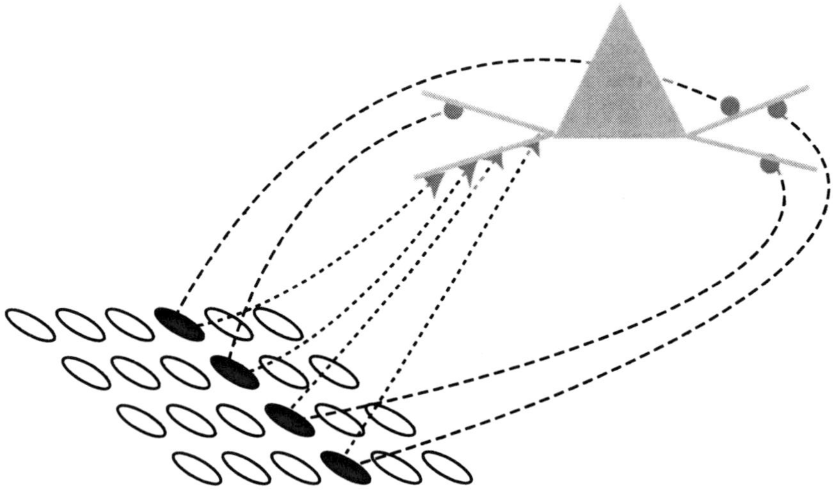

Figure 1.32. A stimulus decomposed into simpler units and their connections to separate dendrites.

In support of these ideas, VanRullen and Thorpe design a neural network simulation of the way in which selective attention improves the accuracy of a simple object-recognition system.

However, our real reason in bringing up the topic of selective attention is that there are now quite explicit hypotheses about how it works. Desimone (1992) noted that one way this attentional modulation could be performed is to assign input from each RF sub-region of an image like that of Fig. 1.31 to a single dendritic branch of the V4 neuron; modulatory inhibition could then "turn off" branches, so that subregions of the RF could be independently gated. Fig. 1.32 illustrates this notion. The excitatory afferents of a given sub-stimulus all connect to the same branch of the complex neuron, while the inhibitory afferents connect randomly to other branches.

Archie and Mel (2001) elaborate this hypothesis by moving even more of the processing onto the dendritic architecture and its connections. They test the following hypotheses: (i) segregation of input onto different branches of an excitable dendritic tree could produce competitive interactions between simultaneously presented stimuli, and (ii) modulatory synapses on active dendrites could constitute a general mechanism for multiplicative modulation of inputs. Fig. 1.33 demonstrates the full connectivity of the sub-stimuli to the complex neuron, using the labels of Fig. 1.30. The input from each stimulus (not shown) excites both the pyramidal neuron and an inhibitory interneuron. In a parallel fashion, the input from selective attention (not shown) excites the pyramidal neuron, but a different inhibitory interneuron. Each inhibitory

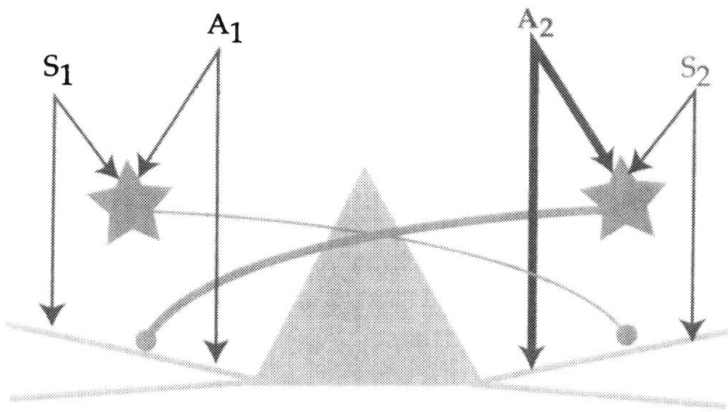

Figure 1.33. Compound stimulus and its segregation onto separate dendrites and inhibitory interneurons. The labels are drawn from Fig. 1.30, in which *S* stands for stimulus and *A* for selective attention. The heavier lines emanating from A_2 indicate that attention is being directed through its sub-path.

interneuron connects to the opposite dendritic branch of its inputs. Without any input from selective attention, the two stimuli inhibit each other, which accounts for the fact that their occurrence together in the RF results in a lower response than their occurrence separately. Whichever one has the greater intrinsic activation will tend to be the 'strong' stimulus.

The computational power of the system comes from the order of the other connections onto the dendritic branch. Attention to one stimulus excites its own dendritic branch while inhibiting the end of the opposite branch where the other stimulus connects. This is indicated in the figure for A_2, which enhances the excitation emanating from S_2 on its own branch while blocking the excitation emanating from S_1 on the opposite branch. This accounts for the ability of selective attention to override the sensory input. This is a simple yet powerful hypothesis of dendritic processing that will put to good use in the analysis of the semantic phenomena with which this monograph is concerned.

1.2.4. Overview of the visual system

If our introduction tot he visual system could be condensed into a single idea, it would have to be Albright and Stoner's, 2002, p. 339, assertion that "the challenge facing the visual system is to extract the 'meaning' of an image by decomposing it into its environmental causes". We have sketched the classical results of Kuffler's and Hubel and Wiesel's recordings of individual neurons in the lateral geniculate nucleus and primary visual cortex and located these results within a broader theory of visual function. The initial function is to reduce as much redundancy as possible. Yet, given that such reduced representations are often ambiguous, a parallel function is to assign them to their most probable environmental cause. Along the way, it will often be advantageous to single out one of the many stimuli for further processing by means of selective attention.

There has also been an implementational facet to our review. We have introduced almost all of the principle building blocks of the neocortex: pyramidal neurons, their internal organs (soma, dendrites, axon, etc), and their external organization(lamina and minicolumns). We have also introduced the principles by which they communicate among themselves (excitation, inhibition, and variable connectivity). All of this has been accomplished with a minimum of implementational detail, which will be the goal of the next chapter. For now, let us see what light the neurophysiology of vision can throw on language. But before doing so, there are two more organizational principles to introduce.

1.2.4.1. Preprocessing to extract invariances

The preceding discussion has on several occasions drawn a distinction between early and late visual processing, without putting much effort into specifying precisely what the distinction is, at least beyond a basic anatomical differentiation. It turns out that the theory of statistical pattern recognition distinguishes two phases in the pattern recognition process that can be pressed into service to help elucidate the early / late dichotomy.

In the parlance of pattern recognition, any 'massaging' of the original data into some more tractable form is known as **preprocessing**, see for instance the overview in Chapter 8 of Bishop (1995). Such preprocessing is justified by the necessity to build an invariant characterization of the data. Once the relevant invariances are found, then the second stage of classification applies to actually extract some grouping of the data into useful classes.

A simple example comes from the classification of objects in two-dimensional images. A particular object should be assigned the same classification even if it is rotated, translated, or scaled within the image, see Bishop 1995, p. 6. Imagine what would happen if the visual system did not extract a representation that was invariant to these three transformations: the viewer would effectively have to memorize every tiny geometric variation in the presentation of an object. For complete coverage of the entire visual field available to a human, a vast number of units would be needed – a number which rises exponentially with the number of cells in the retina (Körding and König, 2001, p. 2824). Preprocessing to extract invariances reduces this combinatorial explosion to a manageable size.

With this background, we can hazard the hypothesis that the 'early' part of the visual system extracts invariances which the 'later' part groups into classes that are useful for the human (or monkey) cognitive system.

1.2.4.2. Mereotopological organization

The preceding sections have concentrated as much on *why* the visual system is organized it is as they have on the details of *how* it is organized. This is due to

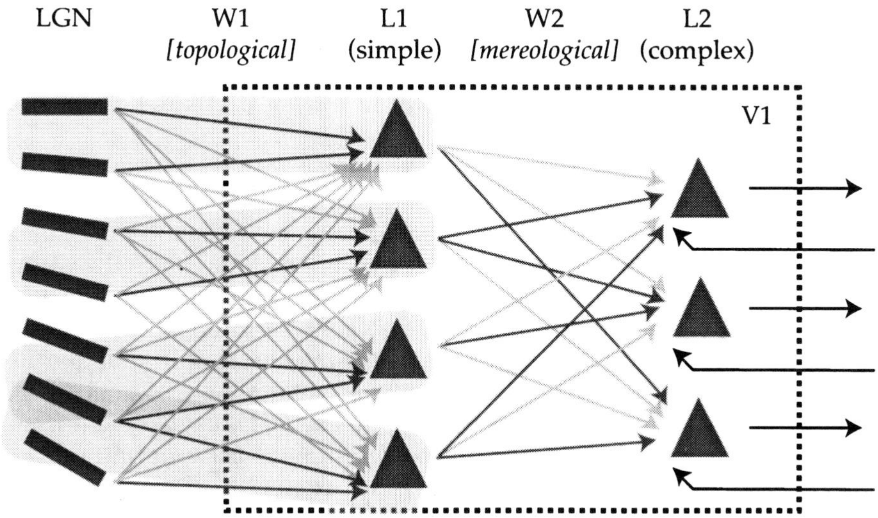

Figure 1.34. Mereotopological organization of V1. Dark arrows have greater strength than light arrows.

the fact that, even though we do not know enough to make point-for-for-point physical comparisons from vision to language, we may know enough to make functional comparisons. And one of the salient functional properties of the visual system is its directional organization.

We have reviewed two main directions, feedforward and feedback. The feedforward path combines parts into ever larger parts or wholes. It is crucial to emphasize the fact that at every level of organization on this route, a given feature or part detector is surrounded by detectors of similar features or parts. Thus a given level is organized into a space in which elements are located in terms of similarity to their neighbors, where similarity can be treated as a measure of spatial distance. The feedback route either evaluates the goodness of fit of a part into the whole (Bayesian inference), or it focuses on one part in preference to another (selective attention).

In mathematics, the study of parts and wholes is known as mereology, while the study of spaces is known as topology. The confluence of the two is known as mereotopology, which is discussed in further detail in Chapter 11. The visual system has at least a mereotopological physical structure, and it does not seem too farfetched to postulate a mereotopological functional structure for it, either.

Fig. 1.34 attempts to convey the essence of this claim. Along the left edge are arrayed a series of bars whose orientation changes from horizontal in 5° increments, which represent a sample of LGN output into V1. Arrows connect these groups of LGN cells to triangles representing the simple cells of V1, labeled "L1" for "layer 1". The standard practice in neuromimetic modeling is to encode a layer of neurons as a vector, where each element of the vector represents the level of activation of a neuron. The next chapter goes into considerable detail on how such activations are calculated.

Each simple cell is host to every LGN output, which makes up the **connection matrix** for a simple cell. However, two contiguous outputs have stronger connections than the others. In neuromimetic modeling, it is standard practice to call this variation in connection strength a **weight**. Hence the entire connection matrix is labeled "W1" and is encoded as a two-dimensional matrix in which each number represents the strength of a given input to a given cell. Thus the weight matrix W1 has the effect of sorting the simple cells by the similarity of their input angles, so that as one follows the column of cells down, each simple cell is sensitive to a progressively greater LGN displacement from horizontal. This answers to the description of a topological ordering on the simple cells, in which the small range of LGN angles that they are sensitive to locates them in a space demarcated by measures of angle.

The simple cells are in turn connected to a smaller number of complex cells, L2, by weight matrix W2. In the real V1, the complex cells are also topologically ordered, but the W2 connections have been scrambled in Fig. 1.34 in order to illustrate what a connection matrix looks like without such an ordering. Thus nearby cells in L1 do not consistently connect to nearby cells in L2. What is depicted instead for L2 is an arbitrary composition from parts found in L1. This isolates the contribution of mereology, which is simply the combination of potentially arbitrary parts into a whole.

The result is a canonical circuit for V1 whose pattern of connectivity can be formalized in terms of mereotopology. We will come to know this circuit very well in the upcoming chapter as the model for the logical coordinators and quantifiers, but let us first review some guidelines that have been drawn form the neurophysiology of the visual system that will steer us towards a neurologically realistic computational regime for language.

1.3. SOME DESIDERATA OF NATURAL COMPUTATION

Artificial intelligence, cognitive science, and generative grammar all grew up in the 1960's in the shadow of the computational paradigm of the day, serial processing and declarative organization. This paradigm is severely limited in its capacity to account for complex natural phenomena, such as that of vision as sketched in the previous sections. This failure has given impetus to a new paradigm founded on parallel processing and self-organization. In the next few subsections we review some of the desiderata that this new paradigm satisfies.

1.3.1. Amit on biological plausibility

One of the first statements that we know of, of the criteria for elaborating a realistic cognitive model comes from Amit, 1989, p. 6. The central one is:

1.10. **Biological plausibility**: the elements composing the model should not be outlandish from a physiological point of view.

Physiological outlandishness comes from ignoring the other five criteria:

1.11 a) **Parallel processing**: the basic time-cycle of neuronal processes is slow, but the model as a whole should react quickly to tasks that are prohibitive to high-speed serial computers.

b) **Associativity**: the model should collapse similar inputs into a prototype, e.g. a picture viewed from many angles and in different light and shading still represents the same individual.

c) **Emergent behavior**: the model can produce input-output relations that are rather unlikely (non-generic) given the model's elements.

d) **Freedom from homunculi**: the model should be free from little external observers that assign ultimate meaning to outputs.

e) **Potential for abstraction**: the model should operate similarly on a variety of inputs that are not simply associated in form but are classified together only for the purposes of the particular operation.

The slowness of neuronal processes mentioned in (1.11a) has already been met in Sec. 1.1.5 as the mention of Posner's 100-cycle limit on basic computation. Given that the five criteria of (1.11) are rather general, they are refined gradually in the following pages by examination from further viewpoints.

1.3.2. Shastri on the logical problem of intelligent computation

Shastri (1991) puts many of these criteria in a global framework by means of a thought experiment to deduce the architecture of an intelligent computational system from simple considerations of how such a system must work. The simplest consideration is that intelligent behavior requires dense interactions between many pieces of information:

> ...the mundane task of language understanding – a task that we perform effortlessly and in real time – requires access to a large body of knowledge and is the result of interactions between a variety of knowledge pertaining to phonetics, prosodics, syntax, semantics, pragmatics, discourse structure, facial expression of the speaker and that nebulous variety conveniently characterized as commonsense knowledge. (ibid., p. 260)

A von Neumann computer, that is, one with a central processing unit (CPU) that sequentially processes elements drawn from an inert repository of knowledge, is not adequate to performing this task, because during each processing step, the CPU can only access an insignificant portion of the knowledge base. It follows that a more cognitively plausible architecture would be for each memory unit to itself act as a processing unit, so that numerous interactions between various pieces of information can occur simultaneously. This architecture answers to the description of a massively parallel computer.

From this initial postulation of parallelism, Shastri deduces several other desiderata. The most obvious one is that the serialism of a CPU not be reintroduced surreptitiously in the guise of a single central controller. More subtle is the desideratum for making the best usage of parallelism by reducing communication costs among the processing units.

Communication costs are divided into the costs of encoding and decoding a message and the cost of routing a message correctly to its destination:

> The sender of a message must encode information in a form that is acceptable to the receiver who in turn must decode the message in order to extract the relevant information. This constitutes encoding/decoding costs. Sending a message also involves decoding the receiver's address and establishing a path between the sender and the receiver. This constitutes routing costs.

Routing costs can be reduced to zero by connecting each processing element to all processors with which it needs to communicate. Thus there is no need to set up a path or decode an address, because connections between processors are fixed in advance and do not change. Encoding/decoding costs can be reduced by removing as much content as possible from the message: if there is no content, there is nothing to encode or decode. Though this may at first sight appear counterintuitive – after all, the point of communication among processors is to communicate *something* – it can be approximated by reducing content to a simple indication of magnitude. Thus a message bears no more information than its source and intensity, or in Shastri's more colorful terms, "who is saying it and how loudly it is being said".

To summarize, Shastri argues that the computational architecture of an intelligent system should obey the following design specifications, whose technical terms are given in square brackets:

1.12 a) Large number of active processors [**massive parallelism**].
b) No central controller.
c) Connections among processors are given in advance [**hard wired**] and are numerous [**high degree of connectivity**].
d) Messages communicate a magnitude without internal structure [**scalars**].
e) Each processor computes an output message [**output level of activation**] based on the magnitudes of its input messages [**input levels of activation**] and transmits it to all of the processors to which it is connected.

These are the principal features of a neuromimetic system, as is explained in the next chapter on basic neural functioning.

1.3.3. Touretzky and Eliasmith on knowledge representation

Touretzky (1995) and Eliasmith (1997) review several challenges for any theory of knowledge representation, which we have melded into the five in (1.13):

1.13 a) neural plausibility
b) cross-domain generality
c) flexibility of reasoning
d) self-organization and statistical sensitivity
e) coarse coding and structured relationships

Touretzky and Eliasmith discuss how symbolic and non-symbolic representational systems fare against these challenges in a way that complements Amit's and Shastri's concentration on processing. We examine each one in turn, except for neural plausibility, which has already been mentioned and will be discussed in more detail below.

The human cognitive system must represent a variety of non-language-based domains, such as sight, taste, sound, touch, and smell. The propositionality of symbolic representations prevents them from explaining psychological findings drawn from these domains, see Kosslyn (1994). Of course, if language is a module within the human the cognitive system, it may be plausible to advocate a special symbolic subsystem for it. However, positing a special propositional symbolic module for language makes it difficult to explain how it would have evolved from a non-propositional neurological substrate. Any paradigm that handles both linguistic and non-linguistic domains with the same primitives would be favored by Occam's razor over a bipartite organization. More is said about modularity in Sec. 1.3.4.

The third desiderata of flexibility of reasoning encompasses two observations: partial pattern-matching and resistance to degradation. There is abundant evidence in humans for partial retrieval of representations or complete retrieval from partial descriptions. There is also abundant evidence for accurate retrieval from a human conceptual network in the face of minor damage or **lesioning**, see Churchland and Sejnowski (1992) and Chapter 10. In symbolic reasoning, in contrast, an input must match exactly what the rules of inference expect or the system cannot function, and even minor damage to a symbolic conceptual network causes the loss of entire concepts. As Eliasmith says, "[a symbol] is either there (whole), or it is not there (broken)", a property often known as **brittleness**. Such categorical brittleness is not characteristic of human cognitive behavior.

The fourth desideratum of self-organization is the notion that a human cognitive system must construct appropriate representations based on its experience. In particular, low-level perception in all animals appears to depend on the extraction of statistical regularities from the environment, a notion that we have gone to great length to illustrate in the first half of this chapter. The extraction of statistical regularities also appears to play an important role in human development, see Smolensky (1995). Given their propositional, all-or-nothing nature, symbolic representations would not appear to be constructable from such regularities, which leaves their acquisition as an open, and rather vexing, question.

Perhaps more than just vexing. Bickhard and Terveen (1995) argue that artificial intelligence and cognitive science are at a foundational impasse due to the circularity and incoherence of standard approaches to representation. Both fields are conceptualized within a framework that assumes that cognitive processes can be modeled in terms of manipulations of encoded symbols. These symbols get their meaning from their correspondence to things in the world, and it is through such correspondences that things in the world are represented. This leads to circularity in that an agent that only has access to symbols could never know what they correspond to. The agent could only know that the symbols must correspond to something. In contrast, an agent in a self-organizing cognitive system presumably has access to the statistical regularities in the environment that the system is sensitive to and so has no difficulty in recognizing how its concepts are **grounded** in these regularities.

Finally, the one desideratum in which symbolicism has the upper hand is that of the representation of structured relationships, such as the one between Mary and John in *Mary loves John*. First order predicate logic was specifically designed to represent this sentence as the proposition *love(Mary, John)*. The challenge is to derive this advantage from what is known about how the brain encodes concepts. We will review one such proposal at the end of Chapter 10, where we take up an encoding of predicate logic in the hippocampus.

1.3.4. Strong. vs. weak modularity

The discussion of the desiderata for natural computation opens the door to treating language as just one instantiation of general neurological principles of perception and cognition, and not as a unique ability. Under this conception, the study of language could indeed be informed by the study of vision, since the two partake of similar computational principles. This monograph does not attempt verification of this conception directly, but rather must be satisfied by an indirect approximation, in which principles from one domain demonstrate a fruitful application to another. Let us call this approach **weak modularity**, and define it thus: an ability is weakly modular if its neurophysiological architecture consists of dedicated areas and pathways connecting them.

There is considerable consensus that vision and language are weakly modular, given that they are localized to separate cerebral areas: the parvo- and magnocellular pathways discussed in Sec. 1.2 bear visual information and the left perisylvan cortex discussed in Chapter 10 houses linguistic information.

Thus visual and linguistic representations could be different because there is some advantage to routing them through dedicated pathways, but they could still be processed in a comparable manner that would allow insight from one domain to apply to the other. This is not as far-fetched as it may seem, in view of the homogenous six-layer lamination depicted in Fig 1.12 that is found throughout the cerebral cortex and which is the ultimate substrate for both visual and linguistic cognition.

The sort of modularity that holds vision and language to be incommensurate, however, is built of sterner stuff. Let us say that an ability is **strongly modular** if it is weakly modular, and in addition the processing algorithms of a dedicated area are unique to that area and are not instantiated in any other area, or at least not in an area to which comparison is directed. It is convenient to reserve the term **proprietary** for such algorithms. If vision and language are strongly modular, their representations are processed by proprietary algorithms and are therefore computationally incommensurate. Understanding of one ability will not shed light on the other.

The reader undoubtedly realizes the debt that this invocation of modularity owes to Jerry Fodor's influential book *Modularity of Mind*, Fodor (1983); see also Fodor (1985), and more recent developments in Fodor (2000) and the discussion that this work engendered, e.g. Bates (1994), among others. To refresh the reader's memory of Fodorian modularity, we reproduce the distillation of Bates (1994):

> ...Fodor defines modules as cognitive systems (especially perceptual systems) that meet nine specific criteria. Five of these criteria describe the way that modules process information. These include ***encapsulation*** (it is impossible to interfere with the inner workings of a module), ***unconsciousness*** (it is difficult or impossible to think about or reflect upon the operations of a module), ***speed*** (modules are very fast), ***shallow outputs*** (modules provide limited output, without information about the intervening steps that led to that output), and ***obligatory firing*** (modules operate reflexively, providing pre-determined outputs for pre-determined inputs regardless of the context). ... Another three criteria pertain to the biological status of modules, to distinguish these behavioral systems from learned habits. These include ***ontogenetic universals*** (i.e. modules develop in a characteristic sequence), ***localization*** (i.e. modules are mediated by dedicated neural systems), and ***pathological universals*** (i.e. modules break down in a characteristic fashion following some insult to the system). ... The ninth and most important criterion is ***domain specificity,*** i.e. the requirement that modules deal exclusively with a single information type, albeit one of enormous relevance to the species.

This characterization of modularity is enormously relevant to the proper understanding of language, since Fodor follows Chomsky (1965, 1980) in postulating language as the archetypal example of a cognitive module. Considerable controversy surrounds the accuracy of this postulation, however.

We do not wish to enter the fray, for Fodor's nine criteria have the effect of fleshing out our notion of weak modularity, and we have already conceded that vision and language are weakly modular. What is at stake is whether vision and language are processed by proprietary algorithms within their separate pathways which do not shed light on one other. This is an empirical issue which goes beyond the limits of Fodor's philosophical methodology to address. The properties of natural computation introduced above and in upcoming paragraphs suggest that intramodular processing is not proprietary, at least not to the best of current knowledge, or ignorance, in the case of the neurology of language.[8]

1.4. HOW TO EVALUATE COMPETING PROPOSALS

We are fast approaching the point where we will need specific criteria for comparing theories of language and computation. This section is dedicated to reviewing the major proposal for either topic. In order to compare theories of computation we appeal to David Marr's well-known proposal to analyze information-processing systems into three levels, which is considered one of the cornerstones of cognitive science. In order to compare theories of language we appeal to Noam Chomsky's less well-known proposal of three levels of adequacy of a grammar. Given that we need to organize our previous observations into a more cohesive computational framework, we start with Marr's work.

---

[8] Neuroscience's current degree of ignorance about the neurophysiology of language means that we are sympathetic to Chomsky when he says, "...When people say the mental is the neurophysiological at a higher level, they're being radically unscientific. ... The belief that neurophysiology is implicated in these things *could* be true, but we have very little evidence for it. So, it's just a kind of hope; look around and you see neurons; maybe they're implicated." (Chomsky, 1993, p. 85) However, if there is one conclusion to be drawn from the history of scientific progress since the Renaissance, it is that betting that our ignorance will last very long into the future is a fool's bet. There is certainly enough known already about neurocomputation to begin formulating hypotheses that are explicit enough to be tested when we have the technology to do so. And of course, it goes without saying that the field will not advance until someone takes the effort to formulate the most explicit hypotheses that are compatible with our current knowledge.

1.4.1. Levels of analysis

Marr (1981[1977], 1982) argues that an information processing system can only be understood completely if is described at different levels of analysis. As he says in Marr, 1982, 19-20:

> Almost never can a complex system of any kind be understood as a simple extrapolation from the properties of its elementary components ... If one hopes to achieve a full understanding of a system ... then one must be prepared to contemplate different levels of description that are linked, at least in principle, into a cohesive whole, even if linking the levels in complete detail is impractical.

Marr advocates three levels of description.

1.4.1.1. Marr's three levels of analysis

The **computational level** specifies the task that is being solved by the system. It states the input data, the results they produce, and the overall aim or goal of the computation. In other words, it defines the type of function or input-output behavior that the system can compute. The **algorithmic level** specifies the steps are being carried out to solve the task. In computer parlance, these steps are known as an algorithm, that is, a system of mathematical formulas that can be instantiated as an executable computer program. It is concerned with how the input and output of the system are represented, and how input is transformed into output. One approach is to identify the overall problem and then break this into subgoals, which in turn can be broken into subgoals, and so forth, a process which Cummins (1983) has termed functional analysis. Thus the relation between the computational and the algorithmic levels can be equated with the relation between a function that is computable and a specific algorithm for calculating its values. The **implementational level** specifies the physical characteristics of the information-processing system.[9]

There is a one-to-many mapping from the computational to the algorithmic level, and a one-to-many mapping from the algorithmic to the implementational level. More simply put, there is one computational description of a particular information processing problem, many different algorithms for solving that problem, and many different ways in which a particular algorithm can be implemented physically. We can summarize these mappings graphically as in Fig. 1.35, where the arrows point out the mapping between levels.

---

[9] Marr's framework has been extended to several other domains beyond its original application to vision, as summarized in the introductory section of Frixione (2001), and it has been subject to various refinements which do not concern us here, for which the reader is again referred to Frixione (2001), as well as Patterson (1998).

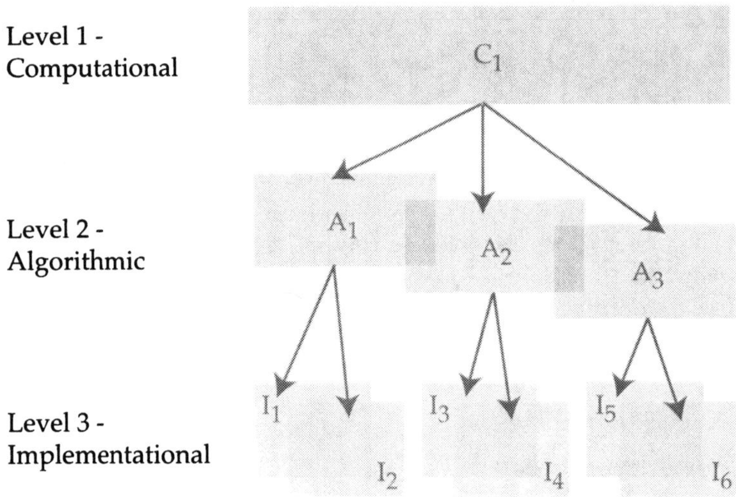

Figure 1.35. Marr's three levels of description of an information-processing system. Note the top-down numbering of levels.

The flow from top to bottom in Fig. 1.35 reflects Marr's conviction that understanding at lower levels will be achieved through understanding at higher levels. To Patterson, 1998, p. 626, what Marr means is that as one moves down through the levels, explanation become progressively more detailed. First we understand what function a system computes, then the procedure by which it computes it, and finally how that procedure is implemented physically in the system. Dennett, 1987, p. 227, conveys this growing particularization of understanding as the image of a 'triumphant cascade through Marr's levels'.

1.4.1.2. Tri-level analysis in the light of computational neuroscience

Franks, 1995, p. 478, points out that a successful cascade of this sort requires 'inheritance of the superordinate': "Given a particular Level 1 starting point, any algorithm must compute the same function, and any implementation must implement the same algorithm and compute the same function." That is to say, the Level 1 function must map on to the Level 2 algorithm, which must map on to the Level 3 implementation.

The significance of superordinate inheritance is that a mismatch between any two levels will block the cascade of description. In the words of Patterson, 1998, p. 626:

> If a system S is physically unable to implement the algorithm specified at Level 2, we cannot explain S's ability to compute the Level 1 function in terms of its executing that Level 2 algorithm.

> If the Level 2 algorithm does not compute the function specified at Level 1, we cannot explain S's ability to compute that function in terms of that algorithm.

More concisely, constraints at a lower level percolate up to a higher level. Given that any elaboration of a higher-level hypothesis involves some amount of idealization of lower-level detail, the hypothesizer runs the risk of overlooking relevant lower-level specification. By way of illustration, we continue to quote Patterson, 1998, p. 627:

> ... if we are working with an idealized conception of system S's hardware, we may think that S can implement an algorithm which in fact it cannot. Then our Level 2 algorithm will not map on to S's actual physical structure. Or if we are working with an idealized conception of S's abilities, so that the function specified at Level 1 is not one which S in fact computes, we will be unable to complete the cascade by mapping that function on to an algorithm which S actually implements.

Inheritance of the superordinate thus binds an algorithm to its implementation, a result which violates Marr's presupposition of independence of levels.

Churchland and Sejnowski, 1990, p. 248, also criticize the presupposition implicit in "Marr's Dream" that the three levels can be formulated independently of one another. Churchland and Sejnowski point out that the potential computational space is so vast – "too vast for us to be lucky enough to light on the correct theory simply from the engineering bench" – that we have no choice in practice but to let ourselves be guided by what Nature has already accomplished. Moreover, Nature's solutions may be better than our own. They therefore advocate a bottom-up approach, grounded in the implementational level of neurophysiology.

Churchland and Sejnowski also fault tri-level analysis for the 'tri' part, since they fail to find three such levels of organization in the nervous system. As they put it,

> Depending on the fineness of grain, research techniques reveal structural organization at many strata: the biochemical level; then the levels of the membrane, the single cell, and the circuit; and perhaps yet other levels, such as brain subsystems, brain systems, brain maps, and the whole central nervous system. But notice that at each structurally specified stratum we can raise the functional question: What does it contribute to the wider, functional business of the brain? (ibid., p. 249)

They go on to guess that each level should be adjudicated a distinct task description, along with a distinct algorithm to perform it. Thus the uniqueness of 'the' algorithmic level dissolves into a multiplicity of local algorithms, one for each task. There can be no global algorithm for human cognition.

The empirical situation is even bleaker than Churchland and Sejnowski's philosophical brush paints it. Arbib, 1995, p. 15, adduces studies which show there to be no uniqueness even between computation and implementation. Several distinct functions may share the same neural circuitry, and the same function may be distributed among several distinct circuits.[10] The inescapable conclusion is that computational neuroscience has been jarred awake from the peaceful slumber of induced by tri-level analysis.

The interdependence of levels, and especially inheritance of the superordinate, forms the philosophical backbone of our interest in neurologically-plausible models of language, under the assumption that neurology constitutes language's implementational level. It also is of particular aid in fleshing out Dummett's distinction between modest and robust semantics: a modest (computational- or algorithmic-level) semantics runs the risk of overlooking significant neurological limitations and so blocking the descriptive cascade. Thus we wind up appropriating tri-level analysis for our own purposes, though we apply it in a bottom-up, or even bidirectional fashion that Marr would have rejected.

Moreover, we endorse a local application of tri-level analysis that is implicit in the long quote from Churchland and Sejnowski above. Table 1.4 illustrates what we have in mind, by explicating the levels at which the simple V1 neuron can be analyzed. This table also attempts to further justify the utility of tri-level theory by locating the various desiderata of natural computation at the most reasonable level. Of course, a moment's inspection reveals that the 'tripartiteness' of the scheme could not be maintained, as we were forced to preface Marr's three with a fourth or 'zeroth' level for evolutionary analysis, but let us take up one thing at a time.

From Hubel and Wiesel's perspective, the function that a simple V1 cell computes is to recognize a line laying at a certain orientation in its LGN input. and signal this recognition to certain V1 complex cells. From the more recent information-theoretic perspective, the function recognizes a specific sort of reduction in the redundancy of its LGN input, namely an oriented line, and signals this recognition to certain V1 complex cells. The algorithm used to accomplish either type of recognition is to simply add up the signals coming in from the LGN and fire a spike train if the sum exceeds a threshold. The actual addition can be implemented by a family of equations that are discussed in more detail in the next chapter. The result is a tidy tri-level decomposition of the target phenomena.

---

[10] Arbib, 1993, p. 278, makes a similar remark with respect to Newell's (1990) version of tri-level analysis.

Table 1.4. Tri-level theory, the simple V1 cell, and natural computation.

Level	*Simple V1 cell*	*Natural computation*
Environmental (L0)	Best vision for lowest commitment of neurons	Freedom from homunculi, self-organization, emergent behavior
Computational (L1)	LINE(LGN, V1c), where both arguments are scalars	Scalar message, input/output activation, hard wiring; statistical sensitivity, associativity
Algorithmic (L2)	If the sum of the input is above a threshold, emit an output; otherwise, don't	Parallelism
Implementational (L3)	Action-potential equations, e.g. Hodgkin-Huxley, FitzHugh-Nagumo, integrate-and-fire, rate	[slow speed > 100-step problem]

1.4.1.3. The computational environment

This tidy decomposition is notably silent on why the simple V1 neuron should exist in the first place, and indeed, such a question lies well beyond what such a decomposition can accomplish. To look for answers to this ultimate *why* question is the reason why we located an environmental level at the head of the other three. The simple V1 cell presumably increases the adaptive fitness of the seeing organism by increasing its sensory range at an optimal commitment of costly neurons, which is to say that any fewer simple V1 neurons would impair vision, and any more would not provide a significant enhancement. While such an optimization of cost/benefit trade-offs is fundamental to the understanding of any biological process, we assume that Marr did not include it in his original formulation because, as a computer scientist, the role of evolution is played implicitly by the scientist judging the value of a computer program. The only programs that thrive are those whose expected cost of development is low compared to their expected utility in solving some problem. In other words, Marr himself was such an integral part of the analytic process that he overlooked his own intrinsic participation. Biologists, on the other hand, do not have the luxury of overlooking the ultimate arbiter of adaptiveness.

To help to flesh out this notion of an environmental level of analysis of natural computation, we would like to adopt a notion from ecological psychology, see Gibson (1972) and much posterior work. The notion we have in mind is **affordance**, a term which Gibson (1977) coins to refer to the actionable properties that exist between the world and an actor (read, a human or other animal). Less technically, an affordance is a perceived property of something, particularly a property that suggests an action that can be taken with the thing. The prototypical example is the way in which the shape of a door handle, by matching the shape of the human hand, suggests to a human that the handle should be grasped and turned.

Gibson introduced the term as part of his theory of visual perception, but it can be kneaded into a broader application. In particular, we wish to claim that affordances are the units of environmental analysis. They are the properties of the environment that an organism can act on, presumably to solve some problem. However, such actions always come at a cost to the acting organism, so the organism must decide how to balance the benefit it derives from taking advantage of an affordance with the metabolic price it must pay for doing so. Optimization falls out as the language of environmental analysis. Normally, such optimization is performed on an evolutionary scale.

1.4.1.4. Accounting for the desiderata of natural computation

Returning to the global picture, the 'quad-level analysis' that results from adding a superordinate environmental level is expressive enough to organize the various desiderata of intelligent computation into a coherent whole. Table 1.4 inserts the relevant criteria into one of the four levels of analysis in its left column. It is instructive to walk through the four levels, starting at the bottom.

The only desideratum of natural computation which resides at the implementational level is the speed of the neural response, which gives rise to the 100-step problem. Yet this is a tremendous constraint, and one which superordinate levels inherit and must abide by. Following Amit's and Shastri's line of argumentation, any algorithm must execute in parallel in order to make up for the slowness of the neurons which implement it. Addition is an operation that can be performed in parallel, especially when it is the addition of ionic currents, discussed in the next chapter. In a similar vein, to reduce the time devoted to decoding an input and encoding an output, the inputs and outputs of the algorithm must be scalar quantities, which limits the representation of the function at the computational level. The membrane potential of a neuron supplies the requisite scalars, as we will also see in the next chapter. This conclusion also assigns the desideratum of input/output levels of activation to the computational level. Moreover, to eliminate any time being spent on routing messages among neurons, their pathways are hard-wired in advance. This is the reason why the input and output arguments of the LINE function name specific neural populations.

Environmental considerations impose a selective pressure on the functional specification of the system. Having external criteria for selecting some functions over others free the system from reliance on homunculi, the "little external observers that assign ultimate meaning to outputs" in Amit's words. To examine the problem from the positive side, if there are no external observers that shape the system, then the system must shape itself. This pushes natural computation towards self-organization and emergent behavior. Either perspective confirms our decision at the beginning of the chapter to view the simple V1 sub-system as self-contained and understandable by opening it up and looking at its parts.

Of course, the effect on a given system of the environmental pressure towards self-organization may not be immediately obvious, but the information-theoretic approach to early vision argues that what is highly prized is a redundancy-free representation of the visual image. Thus the function stated at the computational level must learn to extract statistical regularities from the environment with no supervision. As we will see in Chapter 5, this tends to favor functions that are associative.

There still remain a few desiderata to be accounted for, but we have not yet seen the neurophysiology that will do so. We retake this topic at the end of the next chapter, after a more detailed analysis of neural signaling. In the meantime, the notion of environmental analysis stands in need of further refinement.

1.4.2. Levels of adequacy

Our on-going efforts to convince the reader of the importance of robust semantics could be made less strenuous if linguistic theory had some means to evaluate competing hypotheses. Actually, generative theory used to debate the question of how to tell whether one grammar is superior to another, in the guise of Chomsky's delineation of three levels of adequacy of a grammar. Unfortunately, such debate faded to the background as generative grammar gained converts, leaving many of the initial assumptions unquestioned (as reflected in Table 1.6 below). It is our contention that this debate has been absent from linguistics for far too long. The next few paragraphs bring Chomsky's original formulation up to date with the challenges to generative grammar posed by natural computation.

1.4.2.1. Chomsky's levels of adequacy of a grammar

Given the importance of a grammar within generative grammar, it was natural for Chomsky to define some means of evaluating competing grammars. Chomsky, 1964, p. 29, stakes out three levels of evaluation:

1.14 a) A grammar that aims for **observational adequacy** is concerned merely to give an account of the primary data (e.g. the corpus) that is the input to the acquisition device.

b) A grammar that aims for **descriptive adequacy** is concerned to give a correct account of the linguistic intuition of the native speaker; in other words, it is concerned with the output of the acquisition device.

c) A linguistic theory that aims for **explanatory adequacy** is concerned with the internal structure of the acquisition device; that is, it aims to provide a principled basis, independent of any particular language, for the selection of the descriptively adequate grammar of each language.

These definitions are so thoroughly assimilated into contemporary linguistic practice that they are rarely mentioned, much less examined critically, nowadays.

1.4.2.2. Adequacy of natural (linguistic) computation

The need for a reevaluation of these guidelines springs from the way in which natural computation calls into question the serial and declarative algorithm implicit in them. In particular, natural computation organizes a network in parallel only on the basis of the input data, without the guidance of an external program. Thus there is no physical differentiation between a language acquisition device and a grammar; all there is, is one and the same network that structures itself in accord with its input. It follows that the terms "acquisition device" and "grammar" must be collapsed into a single term, for which we shall simply use the word **model**. (1.15) details a first draft of the necessary editorial changes to Chomsky's formulation in (1.14):

1.15 a) A model that aims for **observational adequacy** is concerned merely to give an account of the primary data (e.g. the corpus) that is its input.

b) A model that aims for **descriptive adequacy** is concerned to give a correct account of the linguistic intuition of the native speaker; in other words, it is concerned with its output.

c) A model that aims for **explanatory adequacy** is concerned with its internal structure; that is, it aims to provide a principled basis for the descriptively adequate self-organization of a linguistic network independent of any particular language.

However, the conflation of "acquisition device" and "grammar" into a single component brings a certain contiguity to the input and output that is absent from generative grammar and has at least one ramification that goes beyond the merely editorial. Such contiguity makes it implausible to consider the processing of the input separately from the processing of the output, as Chomsky's distinction between observational and descriptive adequacy does. The next few paragraphs examine this and other ramifications in greater detail.

As will become clear from the discussion of neuromimetic learning in the upcoming chapters, such a system learns a representation for its input corpus. The final state of the network after training on the corpus establishes a function from the input to the output, but the two are tied so closely together that it is not practical to try to evaluate one in the absence of the other. Changing the input corpus can potentially change the output produced by the network, and the output cannot be changed without changing the input or the network itself. Thus it is not realistic to separate the two as Chomsky does, a conclusion that spurs us to rewrite observational adequacy as in (1.16):

1.16. An **observationally adequate** model gives the observed output for an appropriate input.

In contrast, to Chomsky's formulation, (1.16) is concerned with the internal functioning of the input-output function – it is concerned to reject those functions that do not produce the right output.

The definition of observational adequacy in (1.16) knocks the feet out from under descriptive adequacy, leaving it with nothing to do. However, it is still important to rule out linguistically *ad hoc* analyses, so let us hypothesize this as the domain of our refurbished notion of descriptive adequacy:

1.17. A **descriptively adequate** model gives an output that is consistent with other linguistic descriptions.

This definition enforces a criterion of coherency in the model's output which leads to a convergence of representational formats. As with observational adequacy, (1.17) addresses the internal functioning of the input-output function, in particular by rejecting those functions that produce descriptions that are not linguistically plausible. It would perhaps be more accurate to refer to it as "representational adequacy", but "descriptive adequacy" is so well-entrenched that we would rather not encumber the reader with any more terminology than is strictly necessary.

If both observational and descriptive adequacy deal with the internal functioning of the input-output function, what is there left to do? Well, there still remains one area of coherency that has not been touched on – the computational structure that embeds the linguistic input-output function as a whole, namely, the brain. That is to say, we would like to exclude those input-output functions that are not biologically possible. (1.18) makes this the domain of explanatory adequacy:

1.18. An **explanatorily adequate** model gives an output in a way that is consistent with the abilities of the human mind/brain.

This criterion for choosing input-output functions is quite different from that of generative grammar, for it is founded on extra-linguistic considerations. Among them are: (i) neurological inspiration, which is to say that the model in question has already been found in the brain or is likely to be found there; (ii) componential fit, which is to say that the model in question can reasonably be expected to support input from or output to allied nonlinguistic components such as speech production or amodal memory, and (iii) computational efficiency, which is to say that the model in question is just powerful enough to accomplish its task, and no more.

Before making our closing comments, let us pull all three definitions together into a single group in order to appreciate their overall consistency and effect:

1.19 a) An **observationally adequate** model gives the observed output for an appropriate input.

b) A **descriptively adequate** model gives an output that is consistent with other linguistic descriptions.

c) An **explanatorily adequate** model gives an output in a way that is consistent with the abilities of the human mind / brain.

(1.18 / 1.19c) provides the premise to draw a conclusion that vertebrates this monograph:

1.20. The corollary of optimal explanation: the most explanatory model approximates actual brain function, i.e. it is neuromimetic.

Once again, our reasoning has lead us back to one of the premises of natural computation. We may add parenthetically that a modest semantics at best reaches descriptive adequacy, while a robust semantics takes the further step to explanatory adequacy.

1.4.3. Levels of adequacy as levels of analysis

Did the coincidence between Marr laying out *three* levels of analysis and Chomsky laying out *three* levels of adequacy pique the reader's curiosity? It certainly piqued our own, but then we added a *fourth* layer of analysis, which erases the parity with the three levels of adequacy. Fortunately, nothing but an exercise in imagination prevents us from positing a fourth type of adequacy to reestablish parity with the four levels of analysis. The following claim states our best first guess:

1.21. An environmentally adequate model gives the environmental niche that a behavior fills.

Environmental adequacy rates a model to the extent that it provides an account for the way in which language in general or a grammatical construction in particular fits into its environment. That is to say, the extent to which the target phenomenon solves an environmental problem at a reasonable cost. With respect to language as a whole, environmental adequacy should account for the evolution of human language from the cognitive endowment of our pre-human ancestors, and the kind of linguistic ability that could be expected to be found in a contemporary non-human species given their cognitive makeup. At various points through-out this book we will address the issue of the environmental niche filled by the specific constructions of coordination and quantification. In this way, we attempt to convince the reader that environmental adequacy has enough empirical support and plays an important enough roll in the overall system to merit inclusion on an equal footing with the others.

Table 1.5. Summary of five-level theory.

Level of analysis	*Unit*	*Kind of adequacy*
Environmental (L0)	affordance	environmental
Computational (L1)	(mathematical) function	observational
Algorithmic (L2)	algorithm	descriptive
Implementational (L3)	cell assembly	explanatory
Genetic (L4)	gene(s)	?

1.4.4. Summary of five-level theory

Table 1.5 brings together all of the structural claims of four-level theory into a single package. Starting at the top, we have argued for a new layer of environmental analysis that locates an organism's action in its ecological context, where it can be refined by evolution. Without putting a great deal of thought into it, we take evolutionary refinement to consist of the optimization of the benefit derived from an action against its metabolic cost. The unit of environmental analysis is Gibson's affordance, and models can be evaluated on their environmental adequacy.

It should be added parenthetically that in linguistic theory, an affordance is commonly called the function of a linguistic construction. We prefer affordance, because it is more precise and suggests that linguistic analysands can be reduced to more general cognitive or psychological analysands. Moreover, the word *function* is already taken (see the next paragraph), and we prefer to banish as much ambiguity as possible from our terminology.

The next stratum down analyzes an affordance as a mathematical function. We take a function to be an operation that assigns an output to a given input. The unit of analysis is obviously such a function, and a model can be evaluated on the degree to which its input-output mapping matches that of the target data. Such an evaluation describes the observation adequacy of the model.

The next stratum down reifies a function as a procedure for deriving the output from the input. Its unit of analysis is obviously the algorithm, and a model can be evaluated on the degree to which its algorithm produces an input-output mappings that are consistent with other input-output mappings produced by the organism. Such an evaluation specifies the model's descriptive adequacy. It would also seem desirable to evaluate the algorithm in terms of the accuracy with which its input-output mapping matches the target data, but we are assuming, perhaps incorrectly, that this evaluation percolates down from the superordinate computational level.

The next stratum down executes the algorithm on the organism's biology. For our cognitive concerns, the relevant biology in its most general form is taken to be a **cell assembly**. Hebb (1949) introduced this term to mean a group of cortical neurons that function to sustain some memory trace. Our cell assemblies will be computationally simulated idealizations of the real thing. A cell assembly model can be evaluated on the extent to which it is consistent with what is known about the actual biology. This degree of conformity denotes its explanatory adequacy.

Finally, we felt it to be an inexcusable intellectual oversight to not include a bottom layer of genetic analysis, at which the genesis of the cell assembly in question could be investigated. Unfortunately, we know practically nothing about genetics, and so will not pursue such investigation in this volume.

Before moving on, we should briefly recapitulate the claims of four-level analysis that are most disconcerting with respect to its origins in Marr's and Chomsky's work.

From Marr's perspective, the most disconcerting aspect of four-level analysis would be the fact that the computational level can be constrained by levels beneath it, as well as by the new level above it. Gibson was quite clear about this, arguing that an affordance is jointly determined by the environment and the make-up of an organism. That is to say, an affordance is a possibility allowed by physiological constraints internal to the organism and environmental constraints external to it. This internal/external dichotomy projects onto the bottom/top axis of four-level theory. The computational level finds itself sandwiched in the middle, and so topologically subject to influences climbing up and down the hierarchy.

From Chomsky's perspective, the most disconcerting aspect of four-level analysis is undoubtedly the way in which it dethrones the computational level as the locus of explanatory adequacy. One can perform a simple thought experiment to support the necessity for this outcome. Given that the computational level stipulates an input-output mapping without any regard to how the mapping is effected, as far as it is concerned, the mapping could consist of an arbitrary list of input-output correspondences $\{<i_1, o_1>, <i_2, o_2>, \ldots, <i_n, o_n>\}$. Of course, such a list cannot be reduced to any more compact form, which means that there is no generalization that it expresses. There is consequently nothing to explain; a list is just a list. The fact that the computational level cannot distinguish between functions implemented by listing and functions that assimilate the input-output mapping to more general processes dashes any hope that it can be the locus of explanation.

Having sketched a theory of cognitive analysis conditioned by natural computation in order to undergird our neuromimetic analysis of coordination and quantification, we can now take up the objection that neuromimetics is nothing but a theory of performance.

1.5. THE COMPETENCE/PERFORMANCE DISTINCTION

The early work of Noam Chomsky on generative grammar, e.g. Chomsky, 1965, p. 4, lays out a dichotomy between competence and performance in language:

> Linguistic theory is concerned primarily with an ideal speaker-listener, in a completely homogeneous speech-community, who

> knows its language perfectly and is unaffected by such grammatically irrelevant conditions as memory limitations, distractions, shifts of attention and interest, and errors (random or characteristic) in applying his knowledge of language in actual performance [...]. To study the actual linguistic performance, we must consider the interaction of a variety of factors, of which the underlying competence of the speaker-hearer is only one. [...] We thus make a fundamental distinction between competence (the speaker-hearer's knowledge of his language) and performance (the actual use of language in concrete situations).

By way of clarification, linguistic **competence** refers to a speaker's/hearer's knowledge of a language, and is conceptualized as a **grammar**: a set of rules and/or principles which specify the legal strings of the language. Linguistic **performance** refers to how a speaker/hearer uses her linguistic competence to produce and understand utterances in the language. In posterior work, e.g. Chomsky (1986), Chomsky replaces competence with internalized language or **I-language.** Performance is dropped in favor of externalized language or **E-language**, but the initial terminology is still widely employed.[11]

1.5.1. Competence and tri-level theory

The reason for bringing up Chomsky's competence/performance distinction in the vicinity of five-level theory is that Marr ascribes linguistic competence to his Level 1. Marr (1981[1977]) writes "Chomsky's (1965) notion of a 'competence' theory for English syntax is precisely what I mean for a computational theory for that problem." Others have extended the ascription by defining linguistic performance as an instance of Level 2 or algorithmic description.[12]

---

[11] See Jackendoff, 2002, p. 29ff. for historical background on Chomsky's initial statement and some help in relating it to the I-/E-languagedichotomy.

[12] It is curious to note that Marr's interest in Chomsky's work is not reciprocated. Stemmer, 1999, p. 397, quotes Chomsky as denying that language is an input-output system. Given that tri-level theory is a cornerstone of the computational practice of cognitive science, one can only wonder what Chomsky's alternative is, and whether generative grammar really offers the "new understanding of the computational systems of the mind/brain" that Chomsky, 1993, p. 52, claims it does. See also Botha, 1989, 159-164, for criticism of the psychological reality of generative grammar and Edelman and Christiansen, 2003, for criticism of the neurological plausibility of Chomsky's more recent work.

In the light of the preceding discussion, and especially of Franks' notion of superordinate inheritance, the marriage of Chomsky's and Marr's ideas exposes the linguistic theorizer to implementational error. Indeed, this is the point of Franks' paper. Yet Patterson decries the Chomsky-Marr union, arguing that Chomsky and his followers have always maintained a notion of competence/performance that does not map onto the top two floors of tri-level theory.

Patterson does so by emphasizing that competence and performance are different theoretical objects, whereas each cognitive level describes the same theoretical object with a different degree of detail. One of the most enlightening passages of her exposition centers on a quote from Chomsky, 1991, p. 19, that compares I-language (competence) to a parser:

> The parser associates structural descriptions with expressions; the I-language generates structural descriptions for each expression. But the association provided by the parser and the I-language will not in general be the same, since the parser is assigned other structure, apart from the incorporated I-language. There are many familiar examples of such divergence: garden path sentences, multiple self-embedding, and so on.

This divergence between the mappings provided by I-language and a parser follows from the fact that a I-language is not a parser – it is a grammar, a non-idealized description of linguistic knowledge, which interacts with other factors, such as the memory resources of the parser, when language is actually used.

Thus a grammar is not an idealized description of the mapping computed by a parser when language is put to use. The result is that a grammar cannot be assimilated to Marr's computational level, nor can a parser or the representation that it computes be assimilated to Marr's algorithmic level. Franks' conception of Chomsky's theory therefore misrepresents Chomsky's intention, and the competence/performance distinction is not felled by any implementational shortcomings.

What is felled is any hope of assimilating the competence/performance distinction to any more general practice of cognitive science. But perhaps a rapprochement can be found by redirecting the spotlight of five-level analysis onto performance itself. Since Chomsky, 1986, p. 24, conceptualizes a grammar as a "system of knowledge of language attained and internally represented in the mind/brain", perhaps we should be applying five-level description to this theoretical object – which is in any event an information-processing system in Marr's sense and so susceptible to five-level analysis. It follows that the desideratum of neurological plausibility that underlies robust semantics falls out from the unavoidable need for implementational description in five-level theory. Or in more provocative terms, neurological plausibility is not an aspect of performance which ignores or obscures the actual competence of a speaker/hearer – it is competence itself, at its deepest level of the elucidation!

However, before taking to heart this new marriage of Marr and Chomsky, we should mention the evidence that calls into question distinguishing competence from performance in the first place.

1.5.2. Problems with the competence/performance distinction

Allen and Seidenberg, 1999, p. 2ff, review three problems with the competence/performance distinction, which we have organized into the list of (1.22):

1.22 a) demarcation of performance from competence in primary data
b) demarcation of performance from competence in analysis
c) potential for exclusion of informative data as 'performance'

We take them up in turn.

The primary data of linguistic theorization are judgments of the well-formedness of utterances made by native speakers. Such judgments are affected by limitations of memory, changes in attention and interest, mistakes, false starts and hesitations, and the plausibility or familiarity of the utterance, as well as by the internalized grammar. Thus for the naive informant, competence is just one factor in the judgment process. It is the job of the specialist to abstract away from such 'grammatically irrelevant' distractions in order to infer the properties of the underlying grammar. Yet this task is encumbered by the absence of a general theory of how grammaticality judgments are made, a weakness which ultimately calls all inferences drawn from grammaticality judgments into question. Allen and Seidenberg, 1999, p. 3, summarize the conundrum as so:

> Considering the enormous number of performance factors that have been identified as potentially influencing the judgment process, and how poorly they are understood, it is not surprising that a careful review of the evidence leads Schütze (1996) to conclude that "it is hard to dispute the general conclusion that metalinguistic behavior is not a direct reflection of linguistic competence".

In other words, if there is no systematic means of distinguishing performance from competence in grammaticality judgments, there can be no assurance that the specialist's inferences from these judgments are valid.

If it is problematic to separate performance from competence in the collection of data, then it will likewise be problematic to separate performance from competence in any analysis based on the data collected. And indeed, Allen and Seidenberg speak of a "systematic ambiguity in the field regarding the extent to which competence grammar should figure in accounts of performance". Despite Chomsky's insistence that the ordering of operations in grammatical theory is an abstraction of neurological properties that does not imply any temporal realization, see Chomsky (1995), any analysis that tries to go beyond the abstract

Table 1.6. The generative vision of linguistics and an alternative.

Characteristic	*Generative*	*Non-generative/Experiential*
theory of cognitive processes	none? (linguistic representations are shaped by a repertoire of innate ideas)	cognitive processes involve the manipulation of representations that allow the organism to interact successfully with its environment
modularity	linguistic representations are unique w.r.t. other cognitive domains	linguistic representations are not unique w.r.t. other cognitive domains
the goal of linguistic theory	... is to devise primitives that describe the set of sentences an idealized speaker/hearer would accept	... is to make explicit the experiential and constitutional factors that account for the development of the knowledge structures underlying linguistic performance
a child learns a language	... by learning the rule set that characterizes it	... by learning how to produce and comprehend utterances
judgments of grammaticality	... reflect the rule set	... are just one aspect of knowing how to produce and comprehend utterances

specification of a grammar and explain how language is acquired, used, or impaired by injury must make very specific assumptions about the temporal ordering of operations and thus wrestle with the implementation of grammatical knowledge in real systems.

An obvious corollary of the difficulty of correctly parceling out performance from competence in the primary data is that too narrow a view may unwittingly exclude crucial information from consideration.

1.5.3. A nongenerative/experiential alternative

A family of alternatives to the generative paradigm has been unfolding over recent years. Since the place of linguistic theory within cognitive science is the topic of Chapter 11, we will use this space to briefly sketch how an alternative vision can be counterpoised to the outline of generative linguistics set out above. We again draw on Allen and Seidenberg, by distilling their own summary, Allen and Seidenberg, 1999, p. 119ff, into the opposed characteristics organized into Table 1.6. Note that these authors do not name their alternative; the labels *non-generative* and *experiential* are our own suggestions and will be explained in Chapter 11. The thrust of the non-generative/experiential alternative is to strip language of its gaudy generative trappings as a faculty independent of the rest of human cognition and drape it in the perhaps somewhat drabber uniform worn by other aspects of human intellect.

By way of explanation of the effect that this repackaging has on the pursuit of linguistic analysis, Allen and Seidenberg, 1999, pp. 120-1, draw an analogy between learning to read and learning to speak:

> The beginning reader's problem is to learn how to read words. There are various models of how the knowledge relevant to this task is acquired [reference omitted]. Once acquired this knowledge can be used to perform many other tasks, including the many tasks that psychologists have used in studying language and cognition. One such task is lexical decision: judging whether a stimulus is a word or not. Even young readers can reliably determine that *book* is a word but *nust* is not. Note, however, that the task confronting the beginning reader is not learning to make lexical decisions. By the same token, the task confronting the language learner is not learning to distinguish well- and ill-formed utterances. In both cases, knowledge that is acquired for other purposes can eventually be used to perform these secondary (metalinguistic) tasks. Such tasks may provide a useful way of assessing people's knowledge but should not be construed as the goal of acquisition.

From this analogy, it emerges that Allen and Seidenberg's non-generative alternative relates a competence grammar only indirectly to the knowledge that underlies language use. What is more directly engaged in language use is some neurological structure:

> Grammars represent high-level, idealized descriptions of the behavior of these networks that abstract away from the computational principles that actually govern their behavior. Grammatical theory has enormous utility as a framework for discovering and framing descriptive generalizations about languages and performing comparisons across languages, but it does not provide an accurate representation of the way knowledge of language is represented in the mind of the language-user. (ibid., p. 121)

This book, like Allen and Seidenberg's own work, offers a neuromimetic framework that 'does' language, rather than a grammar fragment that does not.

However, we believe that Allen and Seidenberg err much as Franks (1995) does in rejecting competence just because it is not implemented neurophysiologically in generative grammar. Competence *can* be modeled neurophysiologically, as we will take pains to demonstrate in this monograph – it can be formulated minimally as the connection matrix of an artificial neural network introduced in Sec. 1.2.4.2. The broader experiential vision of language espoused by Allen and Seidenberg embraces other mechanisms beyond the connection matrix. Moreover, we can tie all of this together with semantics by identifying Dummett's modest semantics with a connection matrix, and

Dummett's robust semantics with the entire neuromimetic network. It will take us the rest of this book to substantiate this claim.

1.6. OUR STORY OF COORDINATION AND QUANTIFICATION

Returning to the main thread of our story, we left off the analysis of the logical coordinators and quantifiers upon realizing that the serial approach of automaton theory quickly runs up against real-time processing counterevidence. More abstract alternatives such as those of set theory leave one in the dark as to how (or whether) they are implemented neurophysiologically. The only option left is to look for a neurologically-realistic parallel-processing approach, and the outline of the visual system has supplied us with the requisite background and a few clues. In the next subsections, we introduce two neurologically-realistic parallel-processing approaches, whose further elaboration and defense will form most of the rest of the book.

1.6.1. The environmental causes of linguistic meaning

The environmental level of five-level analysis claims that natural computation takes place in an environment in which certain adaptations are favored and others are suppressed. It goes without saying that we take language to be a specific instantiation of natural computation, so it too should play out in an environment of selective adaptation. Exactly what the selective forces may be is rather obscure, but let us once again look to other human (and primate) faculties, such as vision, for inspiration. What we see is that an appropriate rewording of Albright and Stoner's assertion quoted at the beginning of Sec. 1.2.4 about vision could also characterize language. The requisite rewording is the following: *the challenge facing the linguistic system is to extract the 'meaning' of an utterance by decomposing it into its environmental causes.* Is this a plausible thing to claim? And, to bring us back to the overriding concern of the chapter, does this challenge lead us to a robust theory of semantics?

The word "meaning" is set off in scare quotes in the linguistic version of Albright and Stoner's assertion because our digression into vision enables us to entertain the hypothesis that a linguistic utterance displays two kinds of meaning, the expected cognitive or semantic kind, but also a physical or phonological kind. Let us dispatch the latter briefly in order to concentrate on the former.

Just as the visual system is conjectured to extract the meaning of an image by decomposing it into its environmental causes, so too can it be conjectured that *the challenge facing the phonological system is to extract the 'meaning' of an utterance by decomposing it into its environmental causes – and to create new causes by producing an utterance.* Some theories take up this challenge more directly than others. For instance, the Motor Theory of Speech Perception (Liberman, Cooper, Shankweiler, and Studdert-Kennedy, 1967; Liberman and Mattingly, 1985; 1989) and Articulatory Phonology (Browman and Goldstein, 1986; 1989; 1990a,b; 1992) would readily agree with our conjecture, and even add the proviso that the "environmental causes" of an utterance are the articulatory gestures that create its acoustic form. Unfortunately, the Motor Theory of Speech Perception and Articulatory Phonology are not mainstream theories of phonology, so it would take us too far afield to locate them within more popular approaches, such as Optimality Theory, see for instance Archangeli (1997) for an introduction and further references. We regretfully must leave this fascinating perspective on phonology for another venue and return our attention to semantics proper.

Restating our conjecture for semantics produces *the challenge facing the semantic system is to extract the 'meaning' of an utterance by decomposing it into its environmental causes – and to create new causes by producing an utterance in response.* All semantic theories agree on the decompositional proviso of this statement; where they disagree is on the nature of the environmental causes. Presumably, the irreducible environmental cause of a semantic object is a human's intention to convey some message. Where theories part company is on the extent to which 'humanness' colors the message to be conveyed. In the terms of Chapter 11, objectivist theories such as truth-conditional semantics do not make allowance for any particular 'human' component to semantic causes, whereas experiential theories go to the opposite extreme, taking 'humanness' to inform all semantic causes.

This monograph finds evidence for both positions. By analogy to vision, 'humanness' should make itself manifest in the semantic system through Bayesian and attentional mechanisms. The priors of the semantic system will be concepts that enhance humans' adaptiveness in their environment. In a similar vein, the aspects of complex situations that selective attention will be drawn to will be those that humans find salient or compelling. Conversely, yet also by analogy to vision, there are general mechanisms that the human semantic system is subject to that have nothing to do with any particular human concern. Redundancy reduction is one such mechanism– the *sine qua non* without which humans and any other complex organism would be overwhelmed by the detail and variation in their environment. It is only by keeping in mind both types of processing that one can properly elucidate the environmental causes of the meaning of an utterance. And to do so constitutes the foundation of a truly robust semantics.

1.6.2. Preprocessing to extract correlational invariances

The first step is to decide on a data structure for logical coordination and quantification. We have already introduced the traditional data type of truth-conditional semantics, namely the truth evaluations *true* and *false*. In Chapter 3 it is demonstrated that these evaluations are structured in a particular fashion. Without getting too far ahead of ourselves, let us simply assert a structure here, for the sake of argument. The structure we have in mind is to compare the number of possible truth evaluations to the number that are true for a given logical operator. For instance, if there are two truth values and the operator is

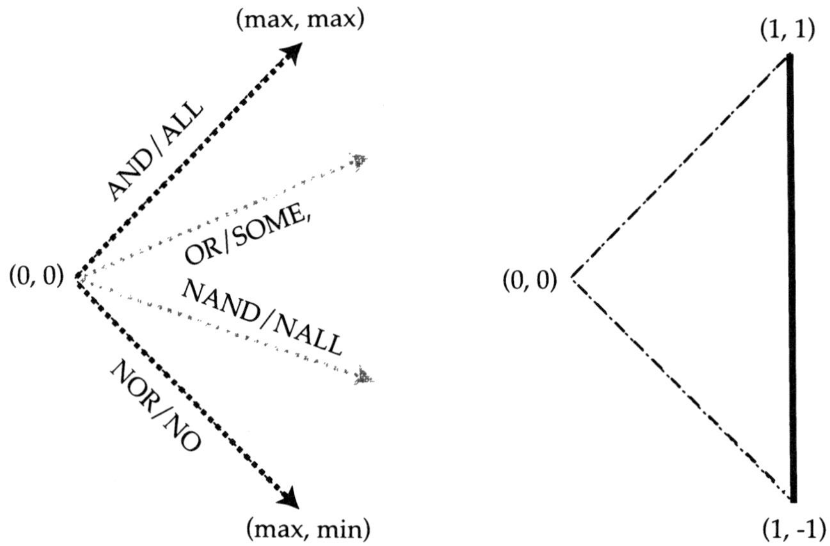

Figure 1.36. Structured COOR/Q truth values (left), and their normalization (right).

AND/ALL, then we expect the two values to be true. This is not very controversial.

The major change is to augment the two-valued logic used so far with a three-valued logic, true (1), false (-1), and undetermined (0). Arraying the number of truth values along the x axis and the number that are actually true along the y axis produces a graph like the left side of Fig. 1.36. The arrows pick out the patterns that specific operators take: (i) AND/ALL as a ray at 45°, since every evaluation is true (ii), NOR/NO as a ray at -45°, since all values are false, and (iii) the other operations in between these two extremes, with (exclusive) OR/SOME in the positive quarter and (exclusive) NAND/NALL in the negative quarter.

This representation has two fascinating properties. One is that AND/ALL reminds one of the graph of receptor correlations in Fig. 1.22, which also traces a diagonal line across the northeast quadrant. The interpretation of this observation is different in the linguistic context, however, for it suggests that a function for the logical operators, namely the expression of correlation. AND/ALL marks maximally correlated truth values, NOR/NO marks maximally anticorrelated truth values, and the others mark the space in between, with a discontinuity at (n, 0) for uncorrelated truth values.

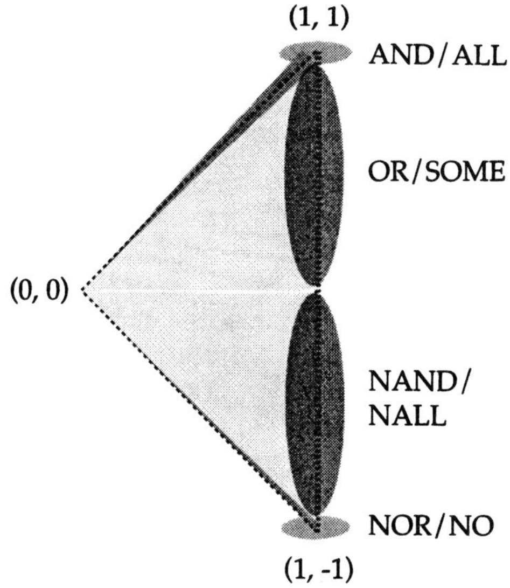

Figure 1.37. Receptive fields for the logical operators on the normalized scale.

The second fascinating property of Fig. 1. 36 is that the logical operators do not care what the actual quantity of truth evaluation is; they can be manifested by any point along their corresponding ray(s). This observation cannot help but bring to mind the functional property ascribed to the early visual system by information theory, namely the task of redundancy reduction. We can consequently postulate an 'early semantic system' analogous to the early visual system that removes a certain type of redundancy, namely numerical redundancy, in order to reveal the invariant of correlation.

The exact method by which this redundancy reduction is achieved is of considerable interest, because it must be neurologically plausible. The simplest method is normalization, which is a generic term for reducing the numerical complexity of a data set to a standard highest value, say 1, and a standard lowest value, say 0. Performing this reduction on the left side of Fig. 1.36 produces the scale on the right side of the same figure.

Once this invariant has been extracted, the patterns of the logical operators can be located along it. The result will be the four elliptical receptive fields depicted in Fig. 1.37. For instance, the OR/SOME receptive field will respond to any normalized truth value that falls within its darkened area; NOR/NO responds to all the others (and vice versa). Several formal interpretations of these fields are elaborated in Chapters 3.

By means of this type of preprocessing, the absolute number of truth values, which presents such a problem to the automaton approach, is converted into a relative number. A solution computable in real time is thereby within our grasp.

1.6.3. Back to natural computation and experiential linguistics

We now have a statement of the problem that is precise enough to weave together the several strands of knowledge spun in this chapter. Points drawn from the representation asserted in Fig. 1.36 will become the input corpus on which a dynamical system is trained in Chapters 5 and 7. These same points will be built into a slightly different dynamical system in Chapter 8 in order to illustrate how inferences can be drawn from them. Given that these dynamical systems will be as neurologically accurate as they can be for our rather humble expository purposes, the resulting simulations constitute a theory of natural computation for natural language semantics.

By assimilating logical coordination and quantification to the general perceptuo-cognitive categories of invariant extraction, correlation and attention to exceptions, our dynamical system satisfies the second desideratum of Table 1.6, namely that linguistic representations are not unique with respect to other cognitive domains, and it implies the first desideratum, that cognitive processes involve the manipulation of representations that allow the organism to interact successfully with its environment. The human environment is rampant with signals that the human brain (probably) represents and manipulates in terms of invariant correlations and exceptions. By providing appropriate data and a dynamical system to process them, the simulations that we run in Chapters 5, 7, and 8 will also meet the third and fourth desiderata of Table 1.6: they make explicit the experiential factors (the input) and the constitutional factors (the learning algorithms) that account for the development of the knowledge structures underlying linguistic performance, and show directly how a child learns how to produce and comprehend coordinative and quantificational utterances. It is crucial to add that the knowledge acquired in this way – for instance, the four receptive fields of Fig. 1.37 – do not map in any obvious way to a set of grammatical rules. Finally, the system can be used to produce grammaticality judgments – submit a set of truth values to it and see which receptive field responds the strongest – but this is clearly not its goal or reason for being. Its reason for being is to extract regularities from the learner's environment and package them into a signal of recognition for posterior processing. A grammaticality judgment is just a side effect of this broader function.

1.7. WHERE TO GO NEXT

The rest of this book substantiates the theory of logical coordination and quantification sketched above, and adds some investigation of collectivity. It thus answers to Dummett's description of a robust semantic theory – so robust, in fact, that it opens the door to understanding coordination, quantification, and collectivity as instances of a general form of neural organization. But our first step is to clarify what we mean by a neuron and how an artificial neural network can be built from neuromimetic computer programs.

Chapter 2

Single neuron modeling

The main function of the nervous system is to process information. Incoming sensory information is coded into biophysical signals – electrical and chemical – and then processed so as to determine whether a response should be made: the movement of an arm, the recognition of a face, the pleasure of a piece of music. This process typically leaves a trace, a memory, within the system which can be used to improve its performance the next time it receives the same or similar sensory information. This chapter details some of the ways in which this happens.

2.1. BASIC ELECTRICAL PROPERTIES OF THE CELL MEMBRANE

Keyes (1985) raises the question of what makes a good computational device, and answers with three desiderata. A "good" computational system is one that survives in the real world, for which it (i) must operate at high speeds, in order to anticipate and react to a fast-changing environment, (ii) must have a rich repertoire of computational primitives, in order to have a wide range of responses, and (iii) must interface with the physical world, in order to represent sensory input accurately and decide on appropriate motor output.

As Koch , 1999, p. 5, adds, the membrane potential of an excitable cell such as a neuron is "the one physical variable that fulfills these three requirements". It can change its state quickly and over neurologically large distances; it results from the confluence of a vast number of nonlinear sub-states, the various ionic channels; and it is the common currency of the nervous system: visual, tactile, auditory, and olfactory stimuli are transduced into membrane potentials, and action potentials in turn stimulate the release of neurotransmitters or the contraction of muscles. Drawing on Keener and Sneyd (1998), this section introduces the basic components of natural or biological computation.

2.1.1. The structure of the cell membrane

A cell is awash in a fluid that approximates seawater – more technically, a dilute aqueaous solution of dissolved salts, mainly sodium chloride, NaCl, and potassium chloride, KCl. A cell's internal environment consists of a similar aqueaous solution, which is separated form the external solution by a two-level or bilayer of phospholipids known as the **cell membrane**.

The cell membrane regulates the exchange of molecules between the internal and external environments. Some molecules can diffuse right through the membrane, such as oxygen and carbon dioxide, because they dissolve in lipids.

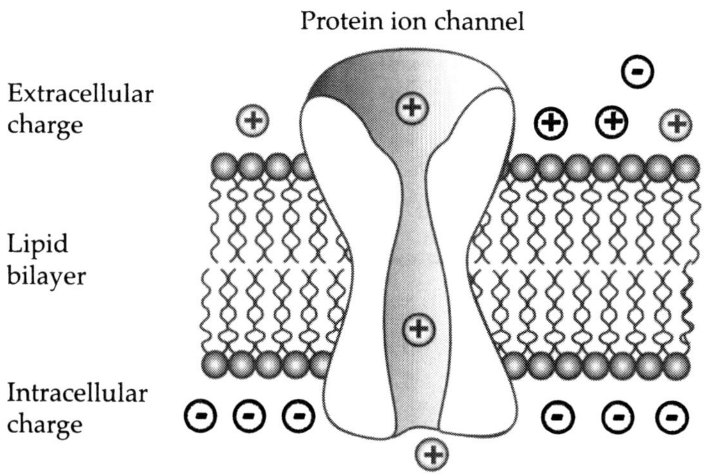

Figure 2.1. Schematic representation of a patch of cell membrane showing an accumulation of charge across the insulating lipid bilayer and ion passage through protein channels. Comparable to Hille (1992).

All others must have a specific means of transport. The cell membrane offers three different avenues. It is punctured here and there by small pores, as well as by larger, protein-lined channels, and it has embedded in it large globular proteins. The pores permit the diffusion of small molecules, such as water and urea, while the globular proteins attach to larger macromolecules such as sugars to pivot them across the membrane. Fig. 2.1 illustrates the lipid bilayer and a protein-lined channel, and also anticipates the chemical and electrical behavior of this structure.

2.1.2. Ion channels and chemical and electrical gradients

What is of most interest to us are the protein-lined channels, for they permit the passage of the small ions that ultimately account for the electrical activity of neurons. There are two main kinds, sealable or **gated** channels, which can be open or closed, and **passive** or **resting** channels, which are always open. When a channel is open, any disparity between the ionic concentrations inside and outside the neuron will decrease.

The reason for this variety of **selective ion channels** is that a cell's metabolism is constantly changing the concentration of ions and large molecules within it. If the concentration of these products were to become too large, osmotic pressure would force water into the cell, causing it to swell and burst. Thus for a cell to survive, it must have some means of regulating the

Table 2.1. Concentration of major ions and electrical potentials for squid giant axon, adapted from Keener and Sneyd, 1998, Table 2.1.

Ion	*Intracellular concentration*	*Extracellular concentration*	*Nernst potential* (E_{ion}, V_{ion})	*Membrane potential* (V_m, V_{rest}, E_m)
Na^+	50 mM	437 mM	+56 mV	-65 mV
K^+	397 mM	20 mM	-77 mV	
Cl^-	40 mM	556 mM	-68 mV	

concentration of the chemical species within it, and this is achieved mainly through the selective ion channels.

As was mentioned above, both the extra- and intracellular environments contain dissolved sodium and potassium chloride. These two salts disassociate into the ions Na^+, K^+, and Cl^-. As a sample of how these ions can vary in concentration inside and outside a cell, the left side of Table 2.1, relates their concentrations with respect to the squid giant axon. For all three ions, there is a marked asymmetry in concentration on either side of the axon membrane that produces a **diffusion gradient** from areas of high concentration to areas of low concentration. Under the influence of this gradient, the small potassium cation K^+ readily diffuses through the passive channels and out of the cell, where it is much less abundant. In contrast, the low intracellular concentration of Na^+ is maintained through a mechanism known as the sodium-potassium exchange pump, which uses energy to remove three atoms of Na^+ against the sodium diffusion gradient of the extracellular fluid, while bringing in two atoms of K^+ against the potassium diffusion gradient of the intracellular fluid.

The discussion of the movement of these ions under the influence of the various diffusion gradients ignores a crucial fact: their electrical charge. Given the differential concentration of ions between the interior and the exterior of a cell, an electrical charge accumulates on the interior surface of the cell membrane that exerts an **electrical gradient** across the cell membrane. Every potassium cation that leaves the cell makes its interior more negatively charged. The accumulation of negative charges starts attracting K^+ back *into* the cell. Eventually, an equilibrium is reached at which the outflow and the inflow balance each other, and no more net change in accumulation takes place. The same holds for the other two ions, though the cell membrane has fewer channels through which they can pass, preventing them from playing a larger roll in the overall charge that builds up within the cell.

The difference in electrical potential across the membrane necessary to counterbalance the concentration gradient for a given ion is calculated by the Nernst equation, and the result is called the ion's **Nernst** or **equilibrium potential**, E_{ion} or V_{ion}. However, the Nernst equation is an idealization in that it assumes that only a single ionic species moves through an open channel. Given that there is a small probability that another ion of a similar size and charge will also pass through the same channel, it becomes necessary to use the Goldman or Goldman-Hodgkin-Katz equation, to find the actual potential in this mixed environment, called the **reversal potential**. If the probability of 'contamination' by different ions is small enough, the two equations produce the same results. The Nernst potentials for the squid giant axon are reproduced in the third column of Table 2.1.

The global charge that accumulates on the interior of the membrane from the mix of intracellular ions can be found by the Goldman-Hodgkin-Katz equation, and is called the **membrane potential**, V_m. In a neuron, the membrane potential is also known as the **resting state** or **resting potential** of the neuron, V_{rest}, though it should be borne in mind that the membrane is not actually at rest; it is constantly expending energy to run the sodium-potassium pumps and so maintain the equilibrium between the influx and efflux of ions. In mammals, this expenditure accounts for half of the metabolic consumption of the brain, see Ames (1997). The membrane potential for the squid giant axon are reproduced in the fourth column of Table 2.1.

2.2. MODELS OF THE SOMATIC MEMBRANE

The reader may have been surprised to see references to the squid giant axon in the preceding paragraphs – for, after all, the goal of all this neurophysiology is to understand that most human of abilities, language, and squid are not known for their linguistic proficiency. The reason for an unavoidable mention of cephalopods in an introduction to human neuroscience lies in the fact that the first measurements and models of the membrane signaling event were made on squid giant axons in the 1940's and early 1950's. Given the technique of inserting a glass micropipette electrode into the neurite to be studied, the giant axon of the North Atlantic squid *Loligo pealei* was a convenient target, because it is several centimeters long and one millimeter in diameter.

The size of microelectrodes has for decades limited the neurites whose electrical behavior can be studied to the axon and soma, and they continue to be the best known, and simplest, cases. For this reason, our initial models ignore dendrites and concentrate on the widest part of the neuron.

2.2.1. The four-equation, Hodgkin-Huxley model

Hodgkin and Huxley (1952) devised one major equation and three supporting equations to model the signaling event in squid giant axons known as the action potential. This mathematical model describes the initiation and propagation of action potentials so well that, not only has it not been replaced in the intervening four decades, but rather has become the standard used for simulations of the squid giant axon, as well as the usual form in which equations for other cell membranes are cast – not to mention winning a Nobel prize for Hodgkin and Huxley in 1962.

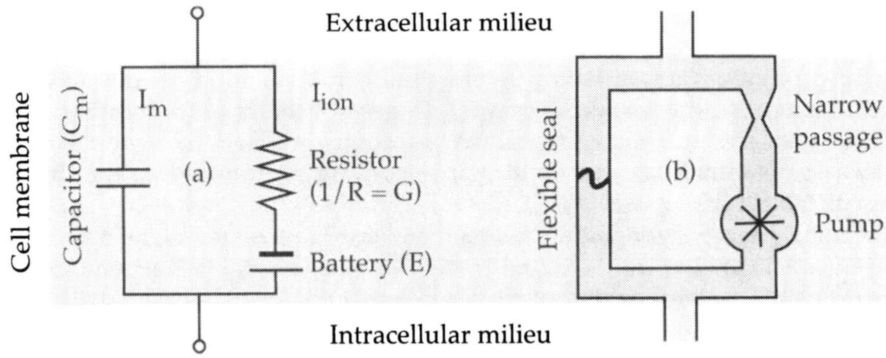

Figure 2.2. Models of the cell membrane. (a) Equivalent electrical circuit. The left branch represents the displacement current I_m, and the right branch represents the conduction current I_{ion}; (b) analogous hydraulic circuit.

2.2.2. Electrical and hydraulic models of the cell membrane

The Hodgkin-Huxley model springs from an earlier insight that the electrical behavior of a neuron cell membrane can be modeled by three electrical components, a capacitor, a battery, and a resistor. The insulating lipid bilayer acts like a capacitor in that a charge tends to build up on the inside wall of the cell membrane that is opposite in polarity to the charge outside the cell. The equilibrium potential of the cell acts like a battery that supplies current if some load on the circuit takes it out of equilibrium. The flow of ions through a protein channel acts like a resistor in the sense that the narrow protein channel restricts the flow of ions greatly. The standard circuit diagram for a capacitor and a resistor acting in parallel is given in Fig. 2.2a, which labels each component with the corresponding mathematical expression.

Since it is often difficult to grasp exactly what is happening in an electrical circuit without some formal training, Fig. 2.2b sketches a hydraulic analog to the direct current diagram of Fig. 2.2a which the reader may find more intuitively understandable. The flexible seal on the left branch bulges in response to current flow, thereby dividing its pipe into a half in which the fluid is compressed – building up positive pressure – and a half in which the fluid is rarefied – building up negative pressure. This is the hydraulic analog of a capacitor. The hydraulic pump corresponds to the electric battery as a source of current. The constriction in the pipe imposes a drag on current flow that reproduces a resistor's impedance. Note that neither circuit has a direction of current flow imposed on it, since direction varies according to the ion.

2.2.2.1. The main voltage equation (at equilibrium)

Returning to the electrical circuit, the fact that its components have well-understood mathematical properties can be used to construct a mathematical idealization of the membrane potential, and thus of the neural signaling mechanism.

We are initially interested in the resting state of the circuit – the equilibrium point at which the charge accumulating at the capacitor is balanced by the charge escaping through the resistor – so the currents running through the two branches must counterbalance each other. By Kirchhoff's current law – the sum of all currents flowing into or out of a node must be zero – this means that the two expressions must sum to zero, as in Eq. 2.1:

2.1. $$I_m + I_{ion} = 0$$

Let us explain this equation in more detail, since it is the foundation on which the rest of the model is built.

Starting on the left, when there is no change in charge, the capacitance of an insulator such as the cell membrane, C_m, is defined as how much charge Q needs to be distributed across the membrane in order for a certain potential V_m to build up, as expressed in Eq. 2.2:

2.2. $$C_m = Q/V_m$$

When the voltage across the capacitance changes, a current will flow. We want to know how much current is flowing, and since current is defined as the change in charge over time, we first solve Eq. 2.2 for the electrical charge Q, giving Eq. 2.3:

2.3. $$Q = C_m V_m$$

We now restate this result in terms of change of some quantity X over time, dX/dt, i.e. differentiate it, to give Eq. 2.4, where the changing quantity is the charge Q:

2.4. $$I_m = dQ/dt = C_m(dV_m/dt)$$

I_m is equated to the terms derived from Eq. 2.3 to state that the right-hand side of Eq. 2.4 measures the displacement current moving on or off the capacitance. Recall that no charge actually crosses a capacitor from one side to the other; instead it redistributes itself across both sides of the capacitor by way of the rest of the circuit. Thus the membrane capacitance imposes a temporal constraint on how quickly the membrane potential can change in response to a current – the larger the capacitance, the slower V_m can change. Imagine the hydraulic

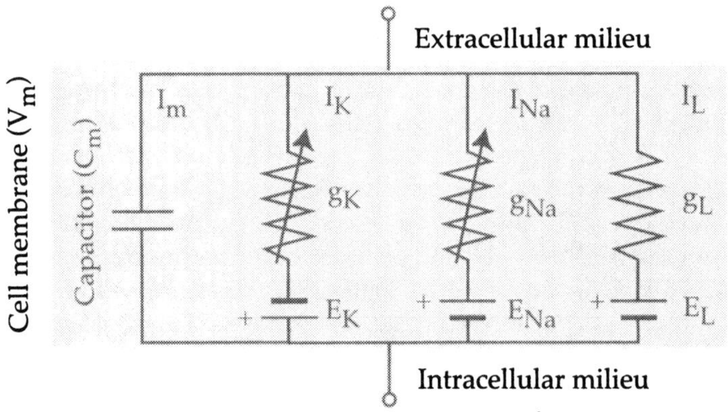

Figure 2.3. Equivalent electrical circuit for the cell membrane that includes the three Hodgkin-Huxley ionic conductances. The arrows across the potassium and sodium resistors indicate that they are active, i.e. triggered by voltage. The leak resistor is passive.

analogy: current flowing into the cell will make the flexible seal bulge outward, but this displacement depends on how thick the seal is and so limits the speed at which current can enter the cell.

As for the resistance current I of a given ion, I_{ion}, the simplest assumption is that it is can be derived from the membrane potential. As was mentioned above, the membrane supports two kinds of ionic flows, those that follow a diffusion gradient and those that follow an electrical gradient. For a given ion, the former is calculated by the Nernst equation and the latter by multiplying the transmembrane current I_{ion} by the channel's resistance r. Summing these two together gives the membrane potential, which is the import of Eq. 2.5:

2.5. $$V_m = E_{ion} + rI_{ion}$$

Solving Eq. 2.5 for the current I_{ion} produces Eq. 2.6:

2.6. $$I_{ion} = (V_m - E_{ion})/r$$

In order to not have to worry about keeping track of the division by r, it is convenient to transform it into a constant $1/r = g$. This alteration changes the quantity to be measured from (specific membrane) resistance in ohms per cm^2 of the membrane to (specific leak) conductance in Siemens/cm^2. The mathematical change follows the steps in Eq. 2.7:

2.7. $$I_{ion} = (V_m - E_{ion})(1/r) = (V_m - E_{ion})g_{ion} = g_{ion}(V_m - E_{ion})$$

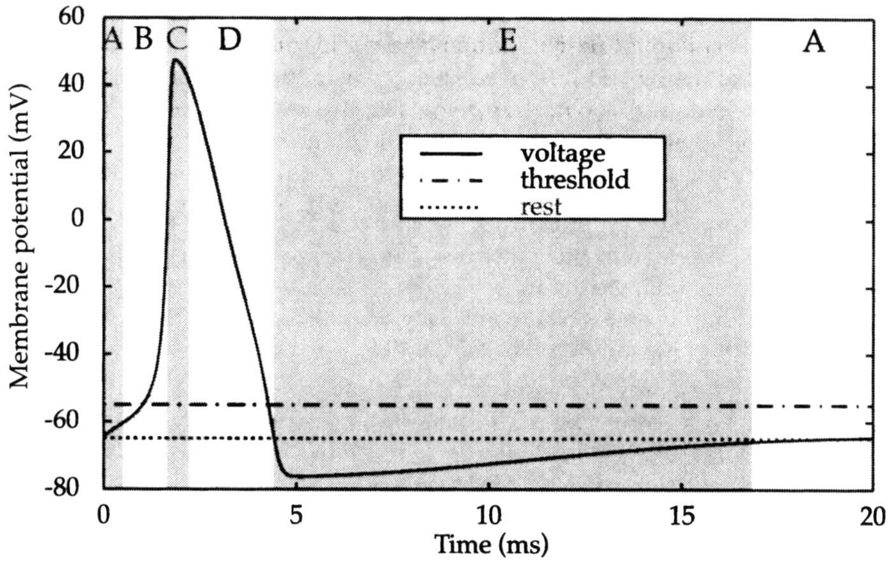

Figure 2.4. Reaction of the cell membrane of an excitable cell to supra-threshold current input. (**p2.01_hodghux_single_spike.m**[13])

The conductance *g* will ultimately depend on the number of channels found in a unit area of membrane.

Substituting the right side of Eq. 2.4 and the right side of Eq. 2.7 into the corresponding currents of Eq. 2.1 give the complete version of Eq. 2.8:

2.8. $$C_m(dV_m/dt) + g_{ion}(V_m - E_{ion}) = 0$$

The final step is to specify the ion variables.

The insight of Hodgkin and Huxley is that the Na^+ and K^+ conductance currents cross the membrane in separate but parallel pathways that are controlled by voltage, along with the passive diffusion of K^+ that maintains the resting potential. Thus the overall conductance should sum together these three ionic currents, as set forth in Eq. 2.9, where the subscript *L* indexes the terms for miscellaneous ionic 'leakage':

[13] The expression "**p2..01_hodghux_single_spike.m**" names the MATLAB program that produces this graph.

2.9. $$C_m(dV_m/dt) + g_K(V_m - E_K) + g_{Na}(V_m - E_{Na}) + g_L(V_m - E_L) = 0$$

Fig. 2.3 augments the electric circuit of Fig. 2.2a to reflect the contribution of each ionic term. The next step is to find out what happens when this system is perturbed from its equilibrium.

2.2.2.2. The action potential and the main voltage equation

If voltage is applied momentarily to the cell membrane of a 'normal', unexcitable cell, the membrane potential quickly returns to its resting state. However, in an excitable cell such as a neuron, the return to the initial state happens only if the applied voltage is below a certain value. If it is above this threshold value, the membrane potential shoots up to a maximum level, before falling precipitously back down to – and then below – its resting state. For instance, graphing the total membrane potential *V* over time reveals a single spike, which is modeled in Fig. 2.4. In prose, the entire sequence consists of: (A) an initial resting state of the membrane potential of –65mV, (B) an upstroke (depolarization) up to (C) the excited state near 50 mV, (D) repolarization as the membrane potential returns to the resting state and then (E) a refractory period during which the potential overshoots the resting state and falls to –75mV, and (A) recovery to the resting state.

Since the action potential constitutes a change in membrane potential, Eq. 2.9 can be solved for the capacitance term in order to calculate Fig. 2.4:

2.10. $$C_m(dV_m/dt) = -g_K(V_m - E_K) - g_{Na}(V_m - E_{Na}) - g_L(V_m - E_L)$$

This quantity can be computed by an ordinary first-order differential equation, if all the constants are known.

2.2.2.3. The three conductance equations

In view of the fact that the three ionic terms $(V_m - E_{ion})$ in Eq. 2.10 are practically equivalent, the complex dynamics shown in Fig. 2.4 must be localized to the single expression that differs for each, the three conductance terms, g_{ion}. We first consider that of the simpler potassium conductance, g_K.

Experimental measurements show that the potassium conductance rises in an S-shaped manner, and then falls precipitously. The potassium conductance curve calculated for an action potential in Fig. 2.5 approximates this shape. To a mathematician, this looks like a sigmoidal function followed by an exponential function. Hodgkin and Huxley modeled the exponential part by introducing a new term, the rate constant *n*, raised to the fourth power and multiplied by the actual potassium conductance, $\bar{g}_K$, to give Eq 2.11:

2.11. $$gK = n4\bar{g}_K$$

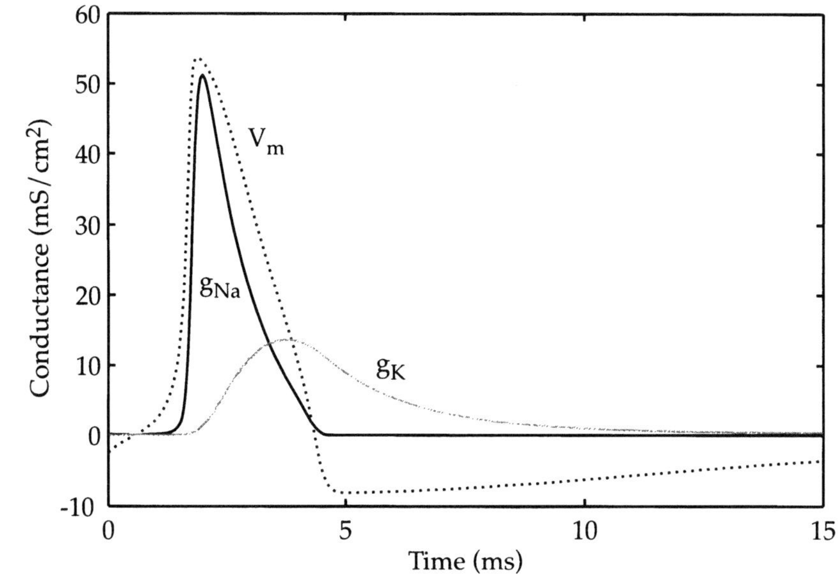

Figure 2.5. Time × conductances of Na^+ and K^+, from the Hodgkin-Huxley equations. An action potential is overlaid as a temporal reference by scaling it so as to fit onto the *y* axis. Comparable to Delcomyn, 1998, Fig. 5-6; Fain, 1999, Fig. 5.19; Weiss 1996, Fig. 4.34; and Cooley and Dodge 1966, Fig. 2. (**p2.02_hodghux_k_na_V.m**)

Hodgkin and Huxley did not know whether *n* captured any actual physiological phenomenon, seeing it more as a mathematical idealization that makes the model work. Their reasoning was roughly that *n* measures the probability that a potassium channel is open, so that its raising to the fourth power can be understood as the assumption that there are four charged "particles" per channel, all of which must move for potassium to flow. For instance, if '1' means that a particle has moved, the entire channel would be open only if all four particles multiply to '1', $n * n * n * n = 1$, which in turn only happens if all four particles take on the value of '1': $1 * 1 * 1 * 1 = 1$. Any value of '0' effectively closes the channel.

Posterior investigators have shown Hodgkin and Huxley to have been largely correct. Their "particles" have been identified as part of a protein that undergoes conformational changes in orientation under the influence of a change in voltage. This protein weaves in and out of the cell membrane in the same way that a stitch weaves in and out of a piece of fabric. It is composed of six domains at which the protein crosses the membrane. These six domains, named as S1 through S6, are considered the functional subunits of the protein. One of them, S4, has seven positively charged amino acids which are thought to respond to a change in voltage by tilting S4 towards or away from the channel pore and thus opening or closing one quarter of the channel. The pore through which ions pass is found between S5 and S6, see Fain (1999), Chapter 6, and Koester (1995) for detailed review and Doyle et al. (1998) and Jiang et al. (2002) for more recent results. As a potassium channel is made up of four copies of the six-domain protein arranged into a circle, it takes all four S4's to be tilted into the open position for the entire channel to be open to current flow. This is the physical mechanism that undergirds Hodgkin and Huxley's postulation of a power of four for *n*.

If the rate constant *n* expresses the probability of all four S4 domains being open, the relationship between the open and closed states of a single domain is given by the first-order kinetic equation 2.12:

2.12. $$n \underset{\alpha_n(V)}{\overset{\beta_n(V)}{\rightleftarrows}} 1-n$$

β is a voltage-dependent rate constant that expresses how many transitions occur per second from the open to the closed state. The probability of being in the closed state is found by subtracting *n* from unity, a trick based on the fact that probabilities must sum to 1. α expresses the converse number of transitions from closed to open. The product of a rate constant and the corresponding probability creates a probabilistic rate of change in one direction. The overall rate of change for a S4 domain is the difference between both directions, given by the differential equation 2.13:

2.13. $$dn/dt = \alpha_n(V_m)(1-n) - \beta_n(V_m)n$$

Eq. 2.13 effectively implements Hodgkin and Huxley's insight that the opening or closing of an ion channel depends on the membrane potential.

Turning to the more complex change in sodium conductance, its abrupt rise and fall illustrated in Fig. 2.5 lead Hodgkin and Huxley to surmise that it originates in two processes, one that turns sodium channels on, formalized by the rate constant *m*, and another that turns them off, *h*. Eq. 2.14 puts this hypothesis into the format of Eq. 2.11, where $\bar{g}_{Na}$ is the conductance of sodium:

2.14. $$g_{Na} = m^3 h \bar{g}_{Na}$$

In Hodgkin and Huxley's terms, m^3 states the probability that three sodium particles are in their open state, whereas *h* states the probability that one additional particle is *not* in its closed state. The outcome still recapitulates that of potassium: the sodium channel is only open if multiplying the constants together reaches unity, $m * m * m * h = 1$.

The two rate constants m and h are themselves described by the two differential equations of in Eq. 2.15, which have the same form as that of Eq. 2.13:

2.15 a) $$dm/dt = \alpha_m(V_m)(1 - m) - \beta_m(V_m)m$$

b) $$dh/dt = \alpha_h(V_m)(1 - h) - \beta_h(V_m)h$$

In parallel to Eq. 2.13, these equations are derived from first-order kinetic equations isomorphic to that of Eq. 2.12.

The sodium channel is thought to have a physiological structure similar to that of the potassium channel, with two significant differences. One is that the four proteins of the potassium channel are linked together to form a single large molecule that makes up the sodium channel. Nevertheless, the sodium channel still has the same four mobile S4 domains, and we have not been able to find any explanation for why this similarity in internal structure does not require m to be raised to the fourth, rather than the third, power. The other difference lies in the inactivation mechanism responsible for h, for which Armstrong and Bezanilla (1977) proposed that a part of the protein on the intracellular side has a chain of amino acids dangling from it that ends in a ball. Opening of the channel at the pore – the effect of m – allows the ball to swing up and seal it, either by electrostatic or hydrophobic forces. This accounts for the independence of the m and h gating properties, though it is more accurate to say that they are *coupled*: first the pore opens – the effect of m – and then it is being sealed by the ball – the effect of h.

Finally, leakage of ions through ungated or passive channels occurs at a small, steady rate, so the membrane conductance which it creates can be represented by a single term, $\bar{g}_L$: The current passing through passive channels is found by multiplying this constant by same terms as before:

2.16. $$I_L = \bar{g}_L(V_m - E_L)$$

Such leakage is responsible for a neuron's resting potential, as was discussed above.

We now have enough information to calculate the total current passing across the membrane. Substituting the complete versions of the three conductance constants gives the full form of the Hodgkin-Huxley equation:

2.17. $$C_m(dV_m/dt) = -n^4\bar{g}_K(V_m - E_K) - m^3h\bar{g}_{Na}(V_m - E_{Na}) - \bar{g}_L(V_m - E_L) + I_{app}$$

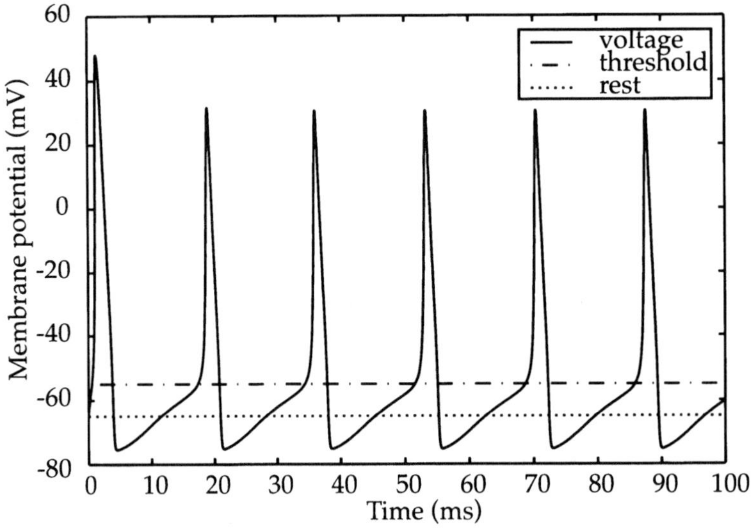

Figure 2.6. Hodgkin-Huxley action potentials or spike train, $I_{app} = 7$. (**p2.03_hodghux_train.m**)

Note that a new term, I_{app}, has been included at the end to represent the current that is applied from the exterior. It is this equation from which most of computational neuroscience has sprung.

2.2.2.4. Hodgkin-Huxley oscillations

One of the most fascinating properties of the Hodgkin-Huxley model appears when an above-threshold stimulus is applied to it without interruption. The result is depicted in Fig. 2.6. What the graph shows is that, after some initial settling-in, the action potential repeats itself periodically and with no variation. This sequence of repeated firings of a neuron is known as a **spike train**. The fact that the Hodgkin-Huxley model produces spike trains is considered to constitute another confirmation of its empirical validity, given that trains such as those of Fig. 2.6 have been observed repeatedly in living neural tissue.

The periodic rise and fall of the membrane potential in Fig. 2.6 describes an **oscillation** between minimum and maximum values. A few technical terms will help to make this notion more precise. The **trajectory** $\vec{X}(t)$ of a dynamical system $X(t)$ is the time course of the system from some initial conditions. For instance, Fig. 2.6 plots the trajectory $\vec{V}(t)$ of the solution to the voltage equation of the Hodgkin-Huxley model from $t = 0$ to $t = 100ms$, with the initial conditions $y0$ set forth in the MatLab script **hodghux_train_plane.m**. A trajectory $\vec{X}(t)$ is an **oscillation** if adding some supplemental time T to $\vec{X}(t)$ does not change $\vec{X}(t)$, a condition whose mathematical formulation is Eq. 2.18:

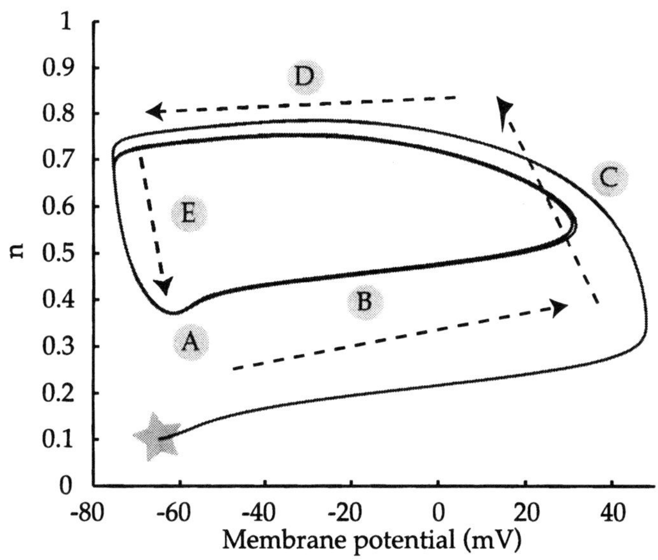

Figure 2.7. Labeled phase-plane portrait of Hodgkin-Huxley system: membrane potential x n. (**p2.04_hodghux_plane.m**)

2.18. $\vec{X}(T + t) = \vec{X}(t)$, for some unique $T > 0$ and all t.

The import of this condition is that the system always returns to the same state after *T*. The **period** of an oscillation is the smallest *T* for which Eq. 2.18 holds. The **frequency** of an oscillation is the reciprocal of the period, *1/T*.

Not only does the spike train of Fig. 2.6 depict an oscillation, but it is also depicts a particular sort of oscillation, one whose shape does not change after the initial settling-in of the first spike. However, the pictorial format of Fig. 2.6 does not necessarily let us make this determination with confidence, since it could suffer from variations that are too small to be revealed by the resolution of the image. A more robust representation is to plot the variables of the dynamical system against one another. Such a graph is known as the **state space** or **phase space** of the system, since it depicts the various states or phases the system can undergo. One such space is illustrated for the Hodgkin-Huxley system in Fig. 2.7 by plotting *V(t)* by *n(t)*, the probability of the potassium gate being open. Starting from the initial conditions marked by the star, the two variables follow an oval trajectory until they enter a closed curve. This closed curve is known as a **limit cycle**. Each spike of the action potential describes one circuit around the limit cycle, which can be deduced from the labeling of the limit cycle with the five states of the action potential introduced in Fig. 2.4. Even at the rather low

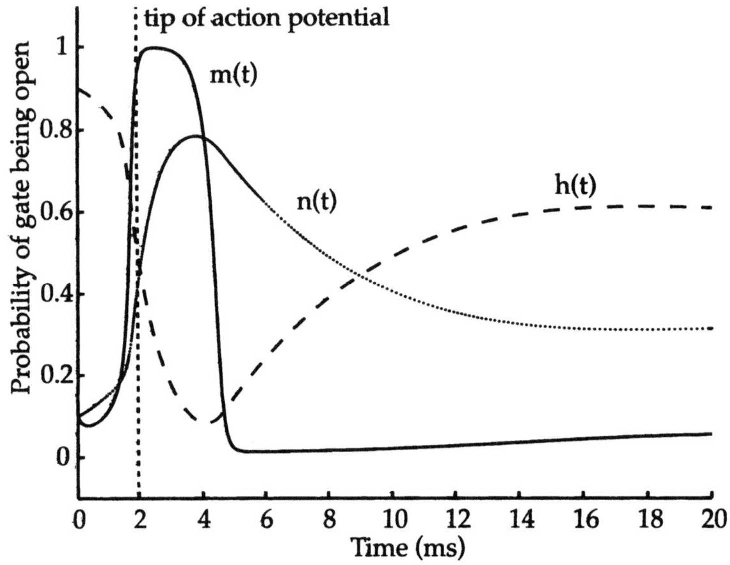

Figure 2.8. Time × probability of gates being open. **(p2.05_hodghux_all_gates.m)**

resolution of Fig 2.7, it seems clear that the six circuits – the six spikes of Fig. 2.6 – have the same shape.

2.2.2.5. Simplifications and approximations

The representational system of phase space has a rich mathematical structure that can be exploited to shine additional light on the genesis of the action potential. Unfortunately, the Hodgkin-Huxley system of four differential equations plus several ancillary equations is so complex as to resist even the basic attempts at analysis that interest us here. It would seem to be an inescapable conclusion that some way must be found to simplify or approximate the level of detail embodied in the Hodgkin-Huxley model. FitzHugh, 1969, p. 14 puts it quite clearly:

> For some purposes it is useful to have a model of an excitable membrane that is mathematically as simple as possible, even if experimental results are reproduced less accurately. Such a model is useful in explaining the general properties of membranes, and as a pilot model for performing preliminary calculations.

To put it more succinctly, sometimes it more helpful to have an approximate qualitative model than a dead-on-the-mark quantitative one. This is certainly the

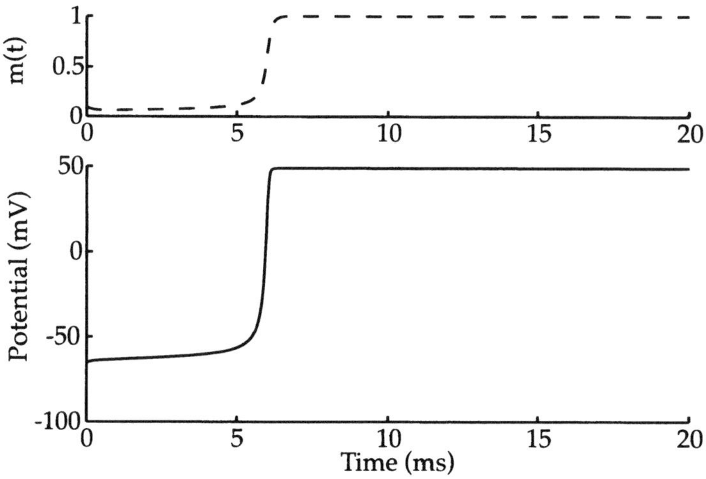

Figure 2.9. Action potential for Hodgkin-Huxley fast system. **(p2.06_hodghuxfast_script.m)**

case of the linguistic phenomena studied in this book, for which any quantitative knowledge of the neural substrate is sorely lacking.

2.2.3. From four to two

2.2.3.1. Rate-constant interactions and the elimination of two variables

The tightly orchestrated opening and closing of the ionic gates that produces an action potential can perhaps be explained more perspicuously by comparing them in a diagram. Fig. 2.8 plots all three against time. At their extremes, h and n cancel each other out. As the graph shows, both n and h are near their extremes most of the time during the action potential, so the sodium conductance must be inactive most of the time. It is only in the vicinity of the spot where h and n assume their medial values – and cross each other – that the sodium conductance becomes active. And indeed, it is there that the action potential reaches its highest point.

Moreover, Fig 2.8 clearly shows the large contribution that m makes to the action potential, since it reproduces the shape of the action potential rather well. This observation can be illustrated even more clearly by plotting the sodium and potassium conductances against time, as in Fig. 2.5. It is easy to appreciate the large spike in the sodium conductance, whose tip overlaps almost exactly the maximum of the action potential, marked by the vertical line at 1.9 ms.

The small contribution that the rate constants n and h make to the generation of an action potential in the Hodgkin-Huxley model holds out the promise that they can be held constant without too much distortion of the qualitative behavior of the model. The reason why is that they change much more slowly than m (and V), and so for much of the action potential they stay near a constant value. Holding h and n constant would enable us to eliminate their differential equations from the four-equation system, thereby achieving a significant simplification to just two equations.

Unfortunately, a simulation and plot of the resulting 'fast system' uncovers a grave deficiency in it. In Fig. 2.9, the graph on the bottom shows that the membrane potential rises up to its maximum – but then stays there. The graph on the top shows why: nothing turns m off, so the sodium channel stays open indefinitely. Such an uninterrupted efflux of sodium would eventually drain the neuron and end its ability to signal, if not kill it outright.

2.2.3.2. The fast-slow system

Clearly, a simplified version of the Hodgkin-Huxley model must include one of the slow rate constants, h or n. This observation was developed in similar ways by three independent researchers. FitzHugh (1960, 1961, 1969) noticed that h and n sum up to about 0.8 during the course of an action potential, so that one can be reduced 0.8 minus the other. H would seem to be the better candidate for elimination, given the redundancy of the other sodium variable, m, with V – consider in this respect the similarity of the membrane potential and the curve of $m(t)$ in Fig. 2.5, which we have already remarked upon. Rinzel (1985) drew the same conclusion, but made the subtraction from unity, $h = n - 1$, for the sake of greater simplicity. Moreover, both researchers recognized the redundancy of m and developed a means of reducing it to V.

The result can be called a **fast-slow** version of the Hodgkin-Huxley system. The slow variable based on n is called rather un-mnemonically W for the degree of accommodation or refractoriness of the system or – our mnemonic preference – R for recovery, since it resets V. A train of action potentials produced by a fast-slow system is plotted in Fig. 2.10. The action potentials of Fig. 2.10a have a similar overall shape as the unreduced ones of Fig. 2.6, and they can be divided by visual inspection into the same five phases. It therefore appears that little is lost in the way of precision while great gains are made in simplicity.

The greatest gain in simplifying the four-equation model to two equations is that the entire system can be visualized in a single two-dimensional phase-plane portrait, as in Fig. 2.10b. What we see, after an initial settling-in cycle, is a rhomboidal trajectory traced by the two variables. The trajectory starts – and comes close to finishing – at the resting state, (A). The bottom leg (B) shows the upstroke of the action potential, during which the membrane potential rises from -65 to 50 while the gating recovery variable R barely budges from its minimum value at 0.35. The right leg (C) shows the excited phase at which the membrane potential stays at its peak while R begins to rise. In this model, the rise is so gradual that (C) merges seamlessly with (D), the decrease of the membrane potential as it is turned off by rising R. The membrane potential

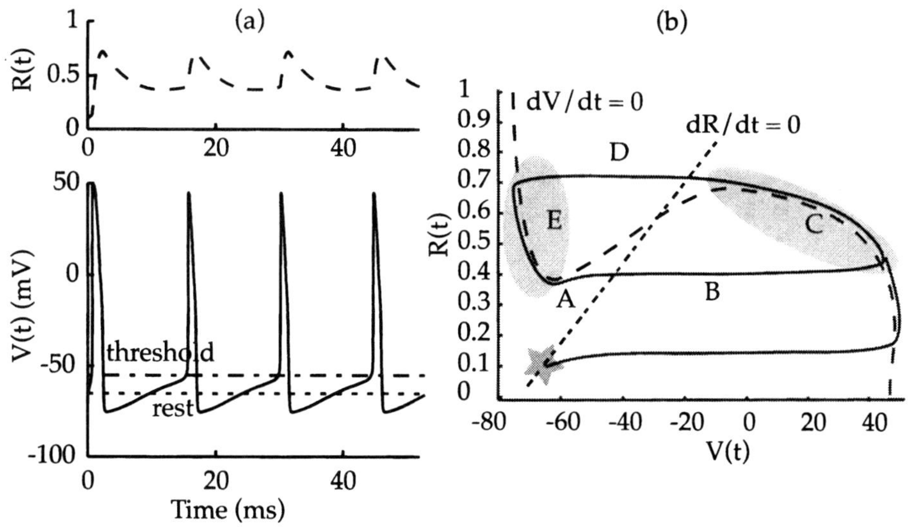

Figure 2.10 (a) Action potentials for Hodgkin-Huxley fast-slow system, I = 7; (b) phase-plane portrait of (a), labeled with phases of action potential and nullclines, and shading of branches of the cubic nullcline. (**p2.07_hodghuxfast_script.m**)

overshoots its resting potential, at which point, (E), *R* resets to its minimal value, which brings the membrane potential up to its resting value, and the whole cycle can repeat itself once again. Thus the two-dimensional system brings out the essential mechanism of the Hodgkin-Huxley action potential, which is the opening and closing of the voltage 'spigot' at (C) and (E) by means of *R*

A second advantage is that the phase-plane portrait itself can be reduced to the concurrent satisfaction of the constraints imposed by the two variables. This line of research has revealed that the membrane potential instantiates a cubic equation, while *R* is monotonically increasing. The cubic and monotonically increasing curves are superimposed on the phase-plane portrait of Fig. 2.10b as null isoclines or nullclines. An **isocline** is a curve in the (*V*, *R*) plane along which one of the derivatives is constant, i.e. not changing. The two most interesting ones are the ones at which the curve along either *V* or *R* is not changing, or null. Such **nullclines** are the steady-state values of a differential equation; its solution in the absence of the changing input from an external variable.

The cubic nullcline, labeled $dV/dt = 0$, is the more interesting of the two. It can be divided into three branches at the two points where it changes direction. These are named, from top to bottom, the **left**, **middle**, and **right branches**. The relevance of this naming convention to our concerns is that two of the four legs

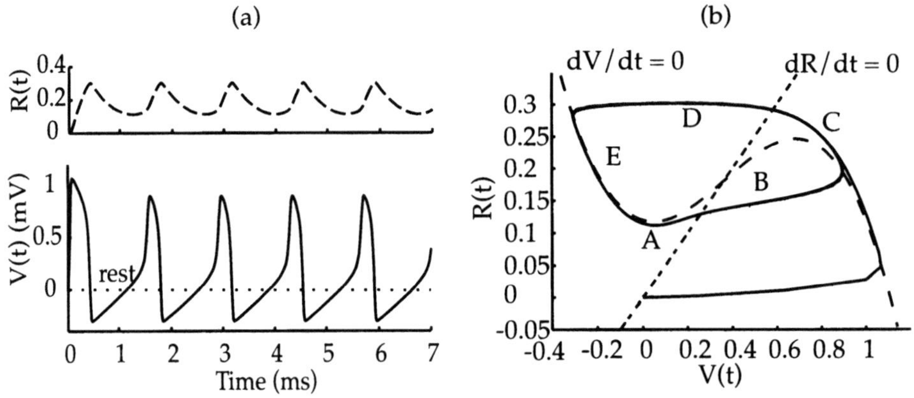

Figure 2.11 (a) FitzHugh-Nagumo action potentials, I = 0.12; (b) phase portrait of (a), labeled with phases of the action potential and nullclines. **(p2.08_fitzhughnagumo_script.m)**

of the phase-plane trajectory follow the two external branches of the cubic: (E) follows the left branch, and (C) the right branch – the two that are shaded in the figure. In view of this correlation, a hypothesis about the shape of the limit cycle suggests itself: the left and right branches are stable in some sense, and once the solution trajectory reaches the end of one of these zones of stability, it follows the slow variable to the other zone. The modeler would say that the whole point of the complex neurophysiology resumed in the Hodgkin-Huxley model is just to ensure this simple behavior.

This dependence of the phase plane trajectory on the two nullclines suggests a next level of simplification: having suppressed two variables of the original Hodgkin-Huxley system, the remaining two can be pared down to their mathematical bare essentials, namely, a cubic and a monotonically increasing equation that interact in the requisite ways.

2.2.3.3. The FitzHugh-Nagumo model

This is the purpose of the FitzHugh-Nagumo model, based on the mathematical analysis of FitzHugh cited above, plus the electrical circuit that implements it, built by Nagumo and colleagues in the 1960s and described in Nagumo, Arimoto and Yoshizawa (1962). It replaces all of the biological constants so painstakingly worked out by Hodgkin and Huxley with a handful of artificial constants whose only motivation is to make the two variables interact in a manner that is analogous to the fast-slow system just described.

The actual equations used for the simulations undertaken here are those of the cubic polynomial in Eq. 2.19a and the monotonically increasing equation in Eq. 2.19b:

2.19 a) $\varepsilon * dV/dt = V(V - 0.1)(1 - V) - R + Iapp$
b) $dR/dt = v - 0.5R$

It is also important to draw the reader's attention to the artificial nature of the parameters of these equations. For the sake of perspicuity, they are designed to keep the trajectory of *V* and *R* within a small amplitude, close to the bounds of zero and one, as illustrated in Fig. 2.11a. Multiplying the output of the system by some constant, such as 30, will more closely approximate the 'real' Hodgkin-Huxley spike.

What is more revealing of the actual behavior of the system is the phase-plane portrait. The reader should be able to discern once again the rhomboidal trajectory that the two variables take on when plotted against each other, as done in Fig. 2.11b, in which the legs of the rhombus are labeled with the phases of the action potential. This system is so simple mathematically that it is straightforward to calculate the nullclines of the two underlying equations. The resulting curves are superimposed on the phase-plane portrait of Fig. 2.11b. The nullcline $dV/dt = 0$ has the tell-tale inverted humps of a cubic polynomial, while the nullcline $dR/dt = 0$ has the tell-tale linear shape – it is a straight line.

One fundamental question about Fig. 2.11b is how the FitzHugh-Nagumo system can trace a path from the initial conditions of the simulation, marked by the star, to the limit cycle. The paths from a given initial state to the limit cycle are not traced at random but rather follow a definite direction, which the reader can test by running **p2.08_fitzhughnagumo_script.m** with different values for the initial conditions, *y0*. This observation suggests that there is some underlying 'terrain', or prevailing 'wind' on the phase plane which is not apparent in a portrait such as Fig. 2.11b.

Fortunately, we already have the tools to uncover this additional patterning. All that needs to be done is choose a representative set of points from the phase plane and use them to solve the two equations. This procedure produces a vector $[x, y]^T$ for each sample point which indicates the amount of change associated with each point. Anticipating the geometric interpretation of vectors discussed in the next chapter as directed line segments, that is, lines with a direction and magnitude, the output vectors can be superimposed on the phase-plane portrait, pointing in the direction of the vector and scaled in proportion to their magnitude. Such a manipulation is performed on Fig. 2.11b to convert it into Fig. 2.12. One global trait jumps out immediately: the 'flow' of the vector field is mainly horizontal. The reason for the triumph of the horizontal axis – the *x* or *V(t)* component of the vectors – can be understood by comparing the magnitudes of *V(t)* and *R(t)* in Fig. 2.11a. They vary over an amplitude of about 1 and 0.2, respectively, which makes the rate of change of *V(t)* about five times larger than that of *R(t)*. Such an unequal ratio is the motive for hedging above by

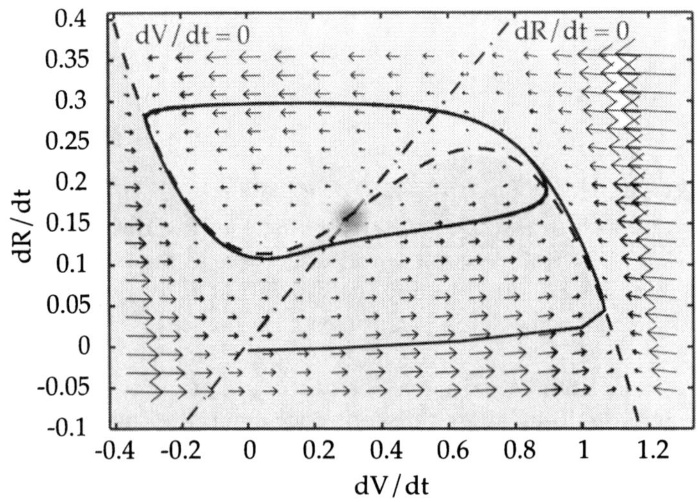

Figure 2.12. Direction and magnitude of change in the FitzHugh-Nagumo phase plane. (**p2.09_fitzhughnagumo_quiver.m**)

saying "*mainly* horizontal": there is a vertical component to the field, but at a fifth the size of the horizontal component, the skewing that it imparts is so slight as to be nearly imperceptible at Fig. 2.12's level of resolution.[14]

In fact, this conclusion can be taken a step further and elevated to a defining property of the FitzHugh-Nagumo model. Recall that we distilled the FitzHugh-Nagumo model from a reduction of the Hodgkin-Huxley model to the "fast" V equation reset by the "slow" n equation. The horizontal direction of flux pointed out in the previous paragraph follows from this difference in speed between the two FitzHugh-Nagumo equations. Under the interpretation that R is slow with respect to V, as reflected by their difference in amplitude, then R is effectively stationary while V changes, and instantaneous otherwise. This description characterizes a large set of dynamical systems, under the rubric of **singularly perturbed systems**, as systems that evolve principally due to the perturbational effects of a 'fast' component, see Cronin (1987) and Koch, 1999, p. 175.

A second large-scale property of Fig. 2.12 is that the direction of the vector field is oriented with respect to the nullclines: the vectors point to the right if

---

[14] The reader can examine the vertical contribution of R by increasing the *arrow_scale* constant at the end of **p2.09_fitzhughnagumo_quiver.m** to some larger number, such as 5, and enlarging the graph window as much as possible.

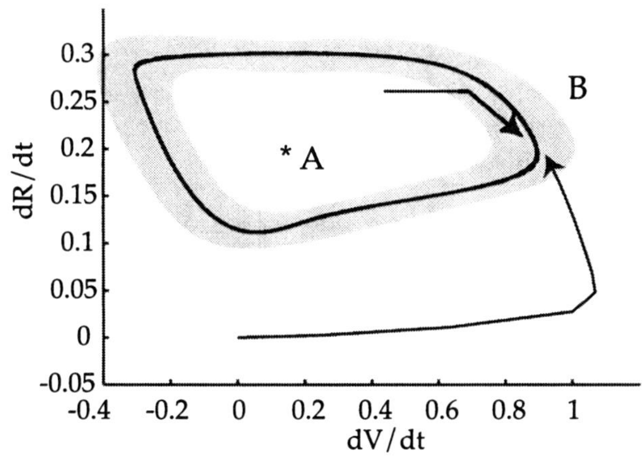

Figure 2.13. Poincaré/Bendixon theorem applied to the FitzHugh-Nagumo limit cycle. Comparable to Wilson, 1999, Fig. 8.4.

they are under the *V* nullcline, and to the left if they are above it. The half of the graph under $dV/dt = 0$ is shaded to highlight this cleavage. This directionality makes sense intuitively by considering the different functions of the top and bottom halves of the limit cycle. The top half, embracing the legs (C) and (D), constitutes the upstroke of the action potential and its reset by *R*; the bottom half, embracing the legs (E), (A) and (B), constitutes the downstroke, overshoot, and recovery. Thus the right-to-left direction of the top half of the field merely reflects the fact that the limit cycle is increasing above the *V* nullcline, while the opposite direction of the bottom half reflects the fact that the limit cycle is decreasing beneath the *V* nullcline.

These two properties contrive to create a sense of laminar or rectilinear rotation both around and within the limit cycle. Yet within this flow there are points of little or no change – those that have the smallest arrows. In particular, there is one such point that is distinguished by its behavior, the **critical** or **equilibrium point**. This is the point where nullclines cross, which is near $[0.355, 0.175]^T$ in Fig. 2.12, at the center of the outwardly diffusing circle which highlights it. Since it is at this point that the two nullclines cross, it is at this point where neither variable changes, and the system stands at equilibrium. In theory, if the system were to start at this point, it would not evolve – no action potential would be fired. In practice, the dense nature of the real number line makes it extremely difficult to fix a location exactly 'at' a given point, and even if one were to achieve it, noise in the system would nudge the location to somewhere in the immediate vicinity of the critical point.[15]

The notion of an equilibrium point provides the final building block on which to raise the mathematical analysis of a limit cycle. The first step is to define more precisely what we mean by a limit cycle:

2.20. An oscillatory trajectory $\vec{X}(t)$ in the state space of a nonlinear system is a **limit cycle** if all trajectories in a sufficiently small region enclosing $\vec{X}(t)$ are spirals. If these neighboring trajectories spiral towards $\vec{X}(t)$ as $t \rightarrow \infty$, then the limit cycle is said to be **asymptotically stable**. If, however, neighboring trajectories spiral away from $\vec{X}(t)$ as $t \rightarrow \infty$, the limit cycle is said to be **unstable**. (Wilson, 1999, p. 117)

By way of illustration, the vector field of Fig. 2.12 shows that any trajectory entering the small window enclosing the FitzHugh-Nagumo trajectory will be forced by the vector field into spirals, just as the trajectory starting at the initial conditions of $[0, 0]^T$ is. Therefore the FitzHugh-Nagumo trajectory counts as a limit cycle. Moreover, the vector field forces any trajectory to spiral in towards the limit cycle, just as the trajectory starting at $[0, 0]^T$ does. Thus the FitzHugh-Nagumo limit cycle qualifies as asymptotically stable.

These two concepts permit the statement of a theorem that describes the functional essence of a limit cycle, attributed to Poincaré and Bendixon:

2.21. Suppose that there is an annular region in an autonomous two-dimensional system that satisfies two conditions: (a) the annulus contains no equilibrium points; and (b) all trajectories that cross the boundaries of the annulus enter it. Then the annulus must contain at least one asymptotically stable limit cycle. (Wilson, 1999, p. 119)

Following Wilson, 1999, p. 119, we sketch an intuitive proof of (2.21) using a diagram, that of Fig. 2.13, which is based on the previous graphs of the FitzHugh-Nagumo phase plane. The annulus in question is the gray ring around the FitzHugh-Nagumo limit cycle, which divides the phase plane into an internal region *A* and an external region *B*. The arrows depict representative trajectories that enter the annulus across both its inner and outer boundaries. Once they enter the annulus, the conditions of the theorem guarantee that they can neither leave, (2.21b), nor come to rest, (2.21a). Moreover, because the system is **autonomous** – the time variable *t* is not a parameter of (is not found on

---

[15] Again, the reader can try this for him or herself by changing the initial conditions for **p2.09_fitzhughnagumo_quiver.m**, given on or near line 22, to the equilibrium point,

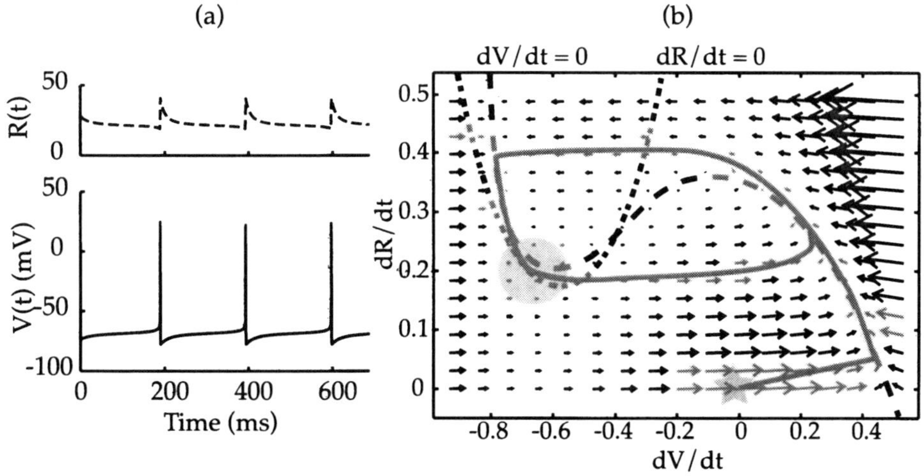

Figure 2.14 (a) Spike train of FitzHugh-Nagumo model of Type I neuron, I = 0.22; (b) phase-plane portrait and vector field of (a), along with nullclines. The second zone of nullcline proximity is highlighted. **(p2.10_type1_script.m)**

the right side of) any of the equations – the two trajectories can never cross one another. Thus as trajectories enter from *A* and *B*, they must approach each other asymptotically, which implies that they are separated by an asymptotically stable limit within the annulus. This is the claim of the proof.

This is also an accurate if not too long-winded characterization of a limit cycle. And, as we have been endeavoring to demonstrate in the last several subsections, it is the limit cycle that provides the best means of understanding the oscillatory nature of an action potential.

2.2.3.4. FitzHugh-Nagumo models of Type I neurons

As fate would have it, the squid giant axon turns out to be atypical in having only one Na^+ and one K^+ current, undoubtedly due to the limited dynamic range of such a simple system: it cannot fire at rates of less than 175 spikes/s and only increases its firing rate modestly with increasing input current. The typical neuron expands its dynamic range with the addition of a second K^+ current, faster than the first, which permits the cell to fire at lower spike rates and with a longer delay to firing when the input is low. This second K^+ current, and third overall, was first characterized and added to the Hodgkin-Huxley model by Connor, Walter and McKown (1977), with the label of I_A. Because it illustrates an alternative way of turning off the membrane potential – and because of its ubiquity, especially in the human neocortical neurons that interest us here, let us examine it briefly.

Though Connor, Walter and McKown's original model augmented the Hodgkin-Huxley model with additional equations, Rose and Hindmarsh (1989) demonstrated that many of the effects of I_A could be approximated by a FitzHugh-Nagumo model in which the monotonically increasing equation for the recovery variable is made quadratic, i.e. its highest power is two. Eq. 2.22 reproduces the equations used in Wilson, 1999, p. 147:

2.22. a) $dV/dt = 1/\tau(-(17.81 + 47.58V + 32.8V^2)(V - 0.48) - 26R(V + 0.95) + I)$

b) $dR/dt = 1/\tau_R(-R + 1.29V + 0.79 + 2.3(V + 0.38)^2)$

A sample spike train is graphed in Fig. 2.14a. At about five per second, the Type I spike rate is much lower than the Hodgkin-Huxley spike train of Fig. 2.6, and at about 200ms, the time to spiking is much longer.

Just by looking at the spike train of Fig. 2.14a, there is no way of knowing why the Type I system is so different quantitatively from the Type II system. It is only by examining the phase-plane portrait of the spike train, which is plotted in Fig. 2.14b, that we can begin to understand the qualitative difference between the two systems. The difference is that the quadratic equation for *R(t)* produces a U-shaped nullcline that crosses, or nearly crosses the *dV/dt* nullcline in two places. This second zone of nullcline proximity mimics an equilibrium point in that the rate of change for both variables is extremely slow, a fact that is corroborated by consideration of the vector field. Compared to the vector field of Fig. 2.12, the vector field of Fig. 2.14a is greatly depressed in magnitude throughout its entire lower left quadrant. Such a decrease in magnitude means that the change in *R(t)* will be small in this region, so that it will inactivate *V(t)* for a longer period, making the time between spikes much longer. This is the essence of Type I spiking behavior.

2.2.3.5. Neuron typology

A classification of human/mammalian cortical neurons in terms of their dynamical properties uncovers four major sorts: (i) fast-spiking cells, (ii) regular-spiking cells, (iii) intrinsic-bursting cells, and (iv) bursting or chattering cells, see Connors and Gutnick (1990), Gutnick and Crill (1995), and Wilson, 1999, p. 169. These four classes are dynamically similar in that their models all contain the Type I voltage and recovery equations of Eq. 2.22. They differ in the number of additional equations, describing additional ionic currents, that their models contain. Fast-spiking cells lack any additional currents; regular-spiking cells have one additional current, and intrinsic-bursting and bursting/chattering cells have two. Sample action potentials for the three dynamical systems are plotted in Fig. 2.15. Thus, although neocortical neurons possess about twelve different ionic currents, see Gutnick and Crill (1995) and

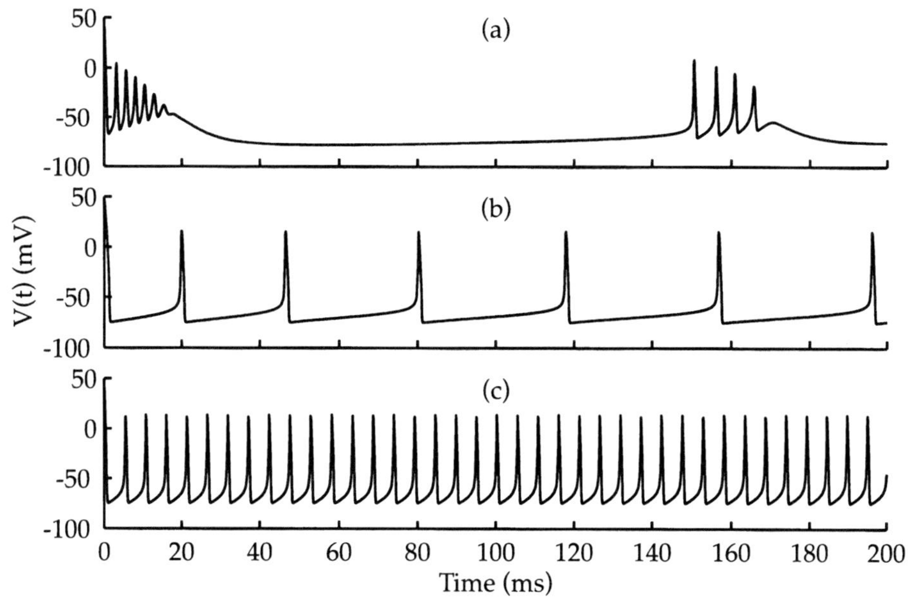

Figure 2.15. Dynamical taxonomy of cortical neurons, I = 0.85. (a) Bursting; (b) regular-spiking; (c) fast-spiking. (**p2.11_taxonomy_script.m**)

McCormick (1998), the entire gamut of cortical action potentials fits snugly in the confines of a four-dimensional dynamical system.

Fast-spiking neurons are distinguished by the rapid rise and fall of their action potentials and by the fact that their spike rate does not gradually decrease or **adapt** during continued stimulation. Fast-spiking neurons are almost always in an excitable state, since their action potentials terminate quickly and do not adapt, i.e. they run down over time. This maintenance of a constantly excitable state makes them ideal for inhibitory neurons, in order to quickly dampen any runaway excitation that would lead to seizures, if not death. The model of Wilson (1999) used in (2.22) can approximate fast-spiking action potentials by reducing the recovery time constant τ_R to 2.1 ms., which produces the fast-spiking plot of Fig. 2.15.

Regular-spiking characterizes those excitatory neurons whose action potential that has a rapid rise but a much slower decay and whose spike rate adapts during continued stimulation. The model of Wilson (1999) used in (2.22) is already optimized to reproduce the size and shape of regular-spiking action potentials, but it must be extended with an additional differential equation that represents an after-hyperpolarizing current with a very slow time constant, 99 ms., that has no effect on the shape of the action potential but rather slowly

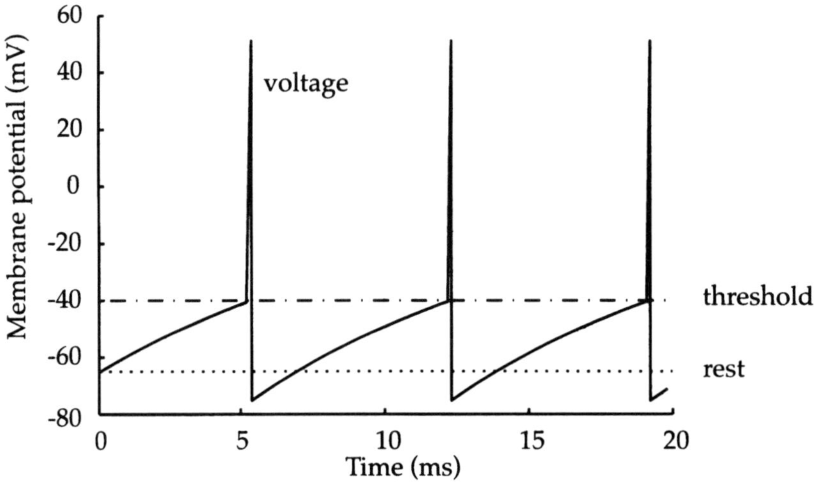

Figure 2.16. Integrate-and-fire action potentials. (**p2.12_if_one_script.m**)

counteracts the input current, thereby slowly reducing the spike rate. The resulting three-equation system is implemented in **tax_regular_ode.m** and produces the spike train of the middle graph of Fig. 2.15.

Bursting characterizes a variety of spike rates in which a quick succession of spikes is followed by a period of inactivity. As mentioned above, it comes in two sorts and is mediated by two additional currents that interact with the voltage and recovery variables. A through discussion of bursting goes beyond the bounds of this book, and the reader is referred to chapter 10 of Wilson (1999), from whence the system in **tax_bursting_ode.m** is taken and used to generate the spike train in the top graph of Fig. 2.15.

2.2.4. From two to one: The integrate-and-fire model

Can we take the logical next step and reduce the two differential equations to one? This indeed can be done, but by this ultimate act of simplification we leave behind almost all of the neurophysiological landmarks that have guided our tour so far and enter the realm where the neuron is not studied for its own sake, but rather for what it does, and in particular for how its simple signals are integrated to perform complex computations.

The one-equation model dates back at least to Lapicque (1907), and was revived in Knight (1972). The general idea is that a cell membrane gradually accumulates a charge until its threshold is crossed, at which point a spike is emitted and the membrane potential is instantaneously reset to its resting state. In Hodgkin-Huxley terms, the active membrane properties – the sodium and potassium conductances – are lumped together into the computational mechanism of "spike emission and instantaneous reset", leaving the membrane potential to be expressed via the 'passive' property of the leakage conductance. Thus the Hodgkin-Huxley equation reduces to the leakage portion, reproduced in Eq. 2.23 from Eq. 2.9:

2.23. $$C_m \, dV/dt = -\bar{g}_L(V - V_L) + I_{app}$$

A train of action potentials calculated by a choice for the parameters of Eq. 2.23 taken form the Hodgkin-Huxley simulation is graphed in Fig. 2.16. This is one version of what is known as the **integrate-and-fire** model.

Note that the voltage falls beyond the resting level after a spike and then the membrane immediately begins to recharge itself. This is how our particular integrate-and-fire implementation models the refractory period of real neurons, though other versions may implement it by not having the membrane respond at all for a few milliseconds after a spike.

2.2.4.1. Temporal or correlational coding

The fact that an integrate-and-fire neuron produces spikes makes it the simplest model that retains a sensitivity to the internal structure of a spike train. By "internal structure of a spike train", we mean the fact that action potentials can occur in particular sequences. Consider the two spike trains in Fig. 2.17. As the thick lines highlight, the top train consists of a sequence of one slow spike followed by three fast ones, while the bottom train runs in reverse: three fast spikes followed by one slow one. This difference could be the means by which the system encodes different patterns or pattern features in the environment. Such a contrast in sequencing could be used in at least two ways by the central nervous system, to respond as quickly as possible and to detect coincidences.

If a neuron is to respond as quickly as possible, it must be sensitive enough to be activated by the first or at most the second spike from some input source. Under this constraint, a neuron receiving the two trains in Fig. 2.17 would be activated by the bottom train after about 15 ms, effectively distinguishing between the two. That some components of the central nervous system do indeed behave in this manner has been argued for extensively by Simon Thorpe and colleagues, see Thorpe et al. (2001) for recent review and references.

Just as useful to other components of the central nervous system is the fact that the two trains in Fig. 2.17 overlap. The shaded boxes draw attention to this fact by enclosing instants at which a spike occurs simultaneously in both sequences. There are at least two: the slow spike of the top train coincides with the last fast spike of the bottom train, while the slow spike of the bottom train correlates to the first fast spike of the top train. If these two trains were being received by a third integrate-and-fire neuron, it would receive double the input at the two moments of temporal overlap or correlation in the separate sequences. This would endow the receiving neuron with the ability to respond

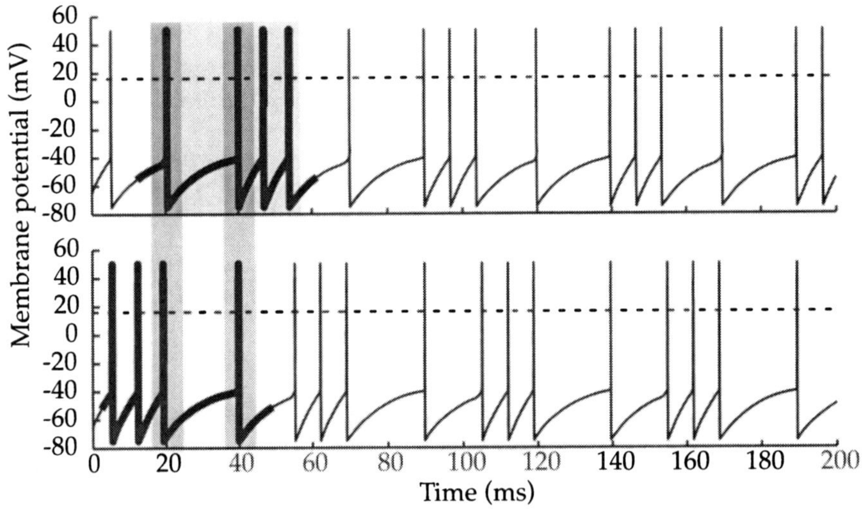

Figure 2.17. Illustrative temporal structure of spike trains. Dotted horizontal lines measure average firing rate. (**p2.13_if_two_script.m**)

to coincidences in the two input trains. In fact, if the excitation that the third neuron receives from the temporal correlation is large enough, it could become synchronized with the correlated spikes and only fire an action potential upon receiving coincident input. Such synchronization has been observed in several neural systems and is the object of considerable current research, see for instance Singer (2000) for review.

2.2.5. From one to zero: Firing-rate models

So far, we have seen that the response of a neuron to a supra-threshold input is one or more action potentials or spikes. Such a train of spikes is what another neuron receives as input and must use to calculate a response. Despite the aforementioned evidence for a sensitivity to particular spikes or patterns of spikes within a spike train, there is contravening evidence that neurons are sometimes only sensitive to an average number of spikes over some relatively long period. By way of illustration, consider again Fig. 2.17. The two dotted lines trace the average potential over 50 ms, which, at 16mV, is the same for both spike trains. Thus, from the perspective of the *rate* of spiking, the two trains are identical. The only way to distinguish them is to change the rate of one, as is done in Fig. 2.18. The dotted lines show an average voltage found by multiplying the number of spikes per 50 ms by the peak potential of 50 mV. This calculation converts the discontinuous spike train into a smooth firing rate.

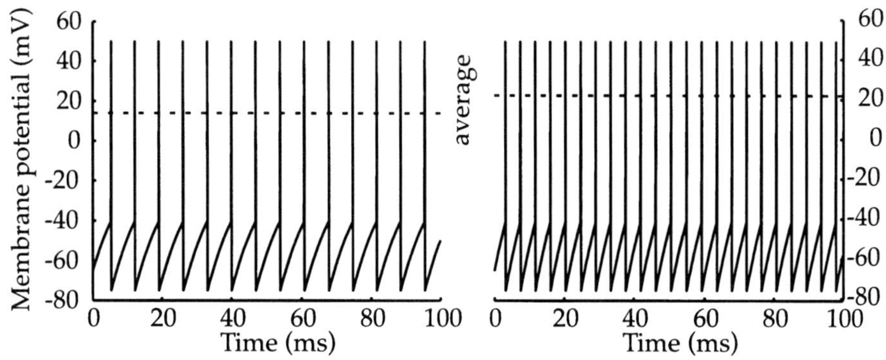

Figure 2.18. Average potential over 50 ms for two different firing rates. **(p2.14_if_diff_rates.m)**

Empirical support for such a simplification of spike-train dynamics comes from experiments in which an animal's behavior can be predicted by counting the spikes emitted over a relatively long period of time by a single neuron, as reviewed by Koch, 1999, p. 331.

This has spurred the development of the ultimate simplification of the Hodgkin-Huxley model, one which contains no differential equation at all. An enormous variety of such **firing-rate models** have been developed, but most have the very simple form of Eq. 2.24:

2.24. $$f = g(V)$$

That is, the output of the neuron is some function g of its voltage, where g is known as the **transfer** or **activation function**, a mathematical object which is discussed more fully in the final section of this chapter.

2.2.6. Summary and transition

This long section introduces the reader to the fundamental signal of the central nervous system, the spike or action potential, by tracing a progressive simplification from the Hodgkin-Huxley model to the firing-rate model. The reader may have gotten the impression that computational neuroscience allows one to examine a neurological phenomenon at almost any level of detail, from a one-to-one representation of the relevant physiological events to an ethereal level of functional abstraction. It is left to the reader's judgment to decide whether this latitude of choice is good or bad for a field. From the author's perspective, it is good, for there is practically nothing known about the single-cell behavior of the neuronal assemblies responsible for logical coordination and quantification. In this state of ignorance, one has no choice but to assume a high level of functional abstraction and hope that one's results will be precise enough to be tested when the tools for examining human linguistic function at the single-cell level are finally invented.

Having devoted so much space to the genesis of the action potential – and, implicitly, the axon hillock where it is generated, let us turn our attention to the parts of the neuron that collect the input from other neurons that will ultimately trigger the firing of an action potential or not.

2.3. THE INTEGRATION OF SIGNALS WITHIN A CELL AND DENDRITES

Up to now, our description of the neuron has not attributed much in the way of internal structure to it, as if it were an indivisible mathematical point. This implicit **point model** is of course the simplest description, and the one on which most neuromimetic modeling is based. Unfortunately, real neurons have a highly complex internal structure due to their dendritic appendages, to which we now turn.

2.3.1. Dendrites

A dendrite is an extension of the neuron cell body with a complex branching shape that is specialized for receiving excitatory synaptic inputs. Their branching structure dwarfs the cell body to such an extent that dendrites make up the largest structure in the brain in terms of both volume and surface area. These branching structures are known as trees or **arbors** and can take on a diversity of forms. This varied morphology can be classified in terms of the degree to which the arbor fills the space it projects into and the shape of the projection. The degree of filling varies between a minimum at which the arbor connects to a single neighboring cell ("selective arborization") to a maximum at which the arbor appears to fill an entire region ("space-filling arborization"). These two extremes are illustrated at the left and right edges of Fig. 2.19, respectively, with the center occupied by an example of an intermediate density ("sampling arborization"). This figure also illustrates the various patterns of radiation that the dendritic arbor can assume within the various degrees of density, of which the biconical and fan patterns are depicted. The reader is referred to Fiala and Harris, 1999, pp. 4-6 for a more complete typology.

One of the first things to realize about the dramatic structural differences mentioned in the preceding paragraph is that it is unlikely that the membrane potential is the same at every point. It is much more likely that such intricate ramifications create spatial gradients in the membrane potential which can be taken advantage of for some functional specialization. Yet an understanding of such potential specialization has remained out of reach until recently, for two reasons. On the one hand, dendrites are too thin to bear the glass micropipette electrodes used by Hodgkin and Huxley to measure current flow in the axon. On the other hand, their branching structures are so complex as to preclude any obvious mathematical simplification that would elucidate their functional role.

(a) (b) (c)

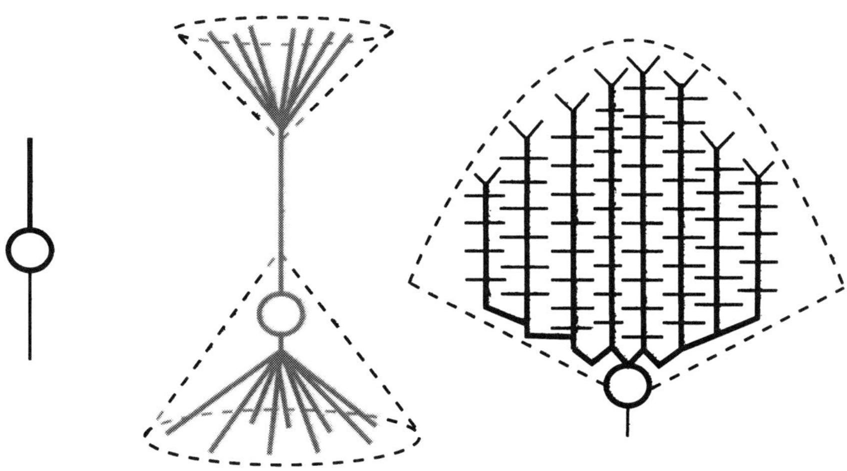

Figure 2.19. Dendrite densities and arborization patterns. (a) selective arborization; (b) sampling arborization (biconical radiation); space-filling arborization (fan radiation). Comparable to diagrams in Fiala and Harris, 1999, Table 1.2.

Burdened by these impediments to empirical and theoretical tractability, it is understandable that dendrites received little attention up until the late 1950's, and have been excluded from the most popular artificial neural network algorithms. The next subsections review some of what has been learned since then, following the gross outlines of the explication of Wilson, 1999, chapter 15, with contributions from Koch, 1999, chapter 11, and Keener and Sneyd, 1998, chapter 8.

2.3.2. Passive cable models of dendritic electrical function

Due to their shape, dendrites have a ready interpretation as electrical cables, which enabled Rall (1959) to adapt the apparatus of **cable theory,** developed by Lord Kelvin in 1855 to describe the transmission of electricity across the first transatlantic telegraph cable, to their analysis. Rall's adaptation relies on the insight that each dendritic filament can be modeled as a cylindrical cable, which interconnects with other filaments to form ever larger, or smaller, branches. At any point along such a cable, current is either flowing along its length or into its walls, which in neurological terms means that current is either charging the membrane capacitance or crossing the membrane resistance and leaking out of the cell. If there is no other variation in current flow, especially if there is no

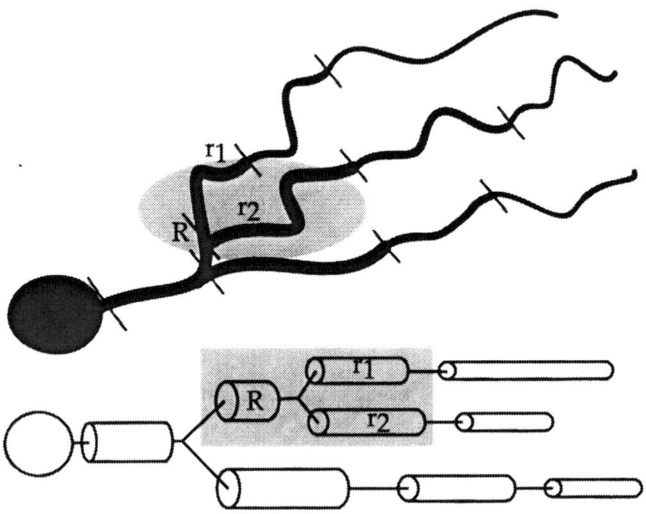

Figure 2.20. Partition of a dendritic arbor along diagonal cut points (top) into a set of membrane cables (bottom). Comparable to Segev and London, 1999, Fig. 9.1.

voltage-dependent flow across the membrane as in an action potential, then the dendrite is said to **passive**, and its behavior can be described by the **(passive) cable equation** of Eq. 2.25:

2.25. $$\tau \frac{\partial V(x,t)}{\partial t} = \lambda \frac{\partial^2 V(x,t)}{\partial t^2} - V(x,t), \text{ where } \tau = r_m c_m \text{ and } \lambda = \frac{r_m}{r_i}$$

Rall showed that the flow of current in a dendritic tree could be calculated by connecting cables of different lengths and widths, as illustrated at the bottom of Fig. 2.20. By matching the predictions made by the mathematical model to physiological measurements, considerable progress was made in elucidating the contribution of dendrites to neural processing. In fact, perhaps the principal discovery was the way in which a dendritic arbor filters postsynaptic input on its way to the spike-initiation zone of the axon hillock.

2.3.2.1. Equivalent cables/cylinders

A next step was taken in Rall (1962), where it was observed that current flow through a membrane cable is proportional to dendritic cross-sectional area, which in turn depends on the cable radius *R* raised to the 3/2 power. Similar considerations may be applied to the smaller **daughter** cables to show that current flow at their junction with a larger **parent** cable will also be proportionate to $R^{3/2}$. Rall's crucial insight was that, under the assumption that electrical constants are identical for both daughters and parent, if the sum of the currents entering the daughters equals the current leaving the parent, then the daughters are mathematically equivalent to an extension of the parent. Since this sum itself depends on the respective radii, the parent and daughter equality is guaranteed by the sum of the daughter radii at the junction with the parent, i.e. by Eq. 2.26:

2.26. $$R^{3/2} = r_1^{3/2} + r_2^{3/2}$$

For illustration, see the three cables labeled in Fig. 2.20. Since Eq. 2.26 easily generalizes to equality across a dendritic junction with N daughters of radii r_n, Eq. 2.27, an entire tree can be collapsed into a single cylinder by summing the N daughters:[16]

2.27. $$R^{3/2} = \sum_{n=1}^{N} r_n^{3/2}$$

If either of these equations does not hold, then there will be an accumulation of ionic concentration on one side of the junction, and it is inaccurate to simplify the tree to a single cylinder. However, the painstaking anatomical measurements of Bloomfield, Hamso and Sherman (1987) show Eqs. 2.26/27 to be a good approximation to actual dendritic junctions in some areas of the brain.

2.3.2.2. Passive cable properties and neurite typology

The cable equation posses two constants that constrain the way in which action potentials accumulate or not within a neural component. The **length** or **space constant λ** constrains the extent of combination of two or more inputs from different locations that occur at about the same time. The **time constant τ**, in contrast, constrains the combination of two or more inputs from the same or different locations that occur at different times. These constants interact to

[16] Summation notation enables us to reduce a series of sums of the form $x_1 + x_2 + \ldots + x_n$ to a capital sigma, for "sum", augmented with three variables: $\sum_{i=1}^{n} x_i$. Following the sigma is the variable over which summation is performed, here x_i, with i a variable for the numerical indices of x. Appended underneath the sigma an indication of the beginning of the series, here $i = 1$. Appended above the sigma is an indication of the end of the series, here n, which abbreviates $i = n$.

Table 2.2. Typology of space constants λ.

	Wide neurite, r_i *is low*	*Narrow neurite,* r_i *is high*
Leaky membrane, r_m *is low*	$r_m / r_i \approx$ medium λ (soma)	$r_m / r_i \approx$ min λ (dendrite)
Tight membrane, r_m *is high*	$r_m / r_i \approx$ max λ (axon)	$r_m / r_i \approx$ medium λ

Table 2.3. Typology of time constants τ.

	Narrow neurite, c_m *is low*	*Wide neurite,* c_m *is high*
Leaky membrane, r_m *is low*	$r_m * c_m \approx$ min τ	$r_m * c_m \approx$ medium τ (soma)
Tight membrane, r_m *is high*	$r_m * c_m \approx$ medium τ	$r_m * c_m \approx$ max τ

determine whether a neuron sums together its postsynaptic potentials slowly and from synapses that are far from the axon hillock or quickly and from synapses that are close to the axon hillock – or even differentially: slowing down some and quickening others in order to perform calculations much more complex than simple addition. The next paragraphs sketch how the two constants classify neural components, following the lead of Spruston, Stuart and Häusser (1999).

The length or space constant λ is defined as the resistance of a unit length of membrane divided by the resistance of a unit length of intracellular fluid or cytoplasm. A large space constant means that a postsynaptic potential can spread a relatively long way from its origin, while a small space constant means that it cannot.

The resistance of the membrane r_m depends on how leaky it is: if it is leaky, a potential will escape the confines of the membrane before it can travel very far. Conversely, if it is tight, none of the potential will leak out, so the distance it can travel is limited by other effects. The resistance of the cytoplasm r_i depends on the diameter of the neurite: if it is wide, it is easier to go around any impediments, and the distance an action potential travels is limited by other effects. Conversely, if it narrow, impediments cannot be side-stepped, and an action potential will not spread very far. The two sorts of resistance cross-classify give the typology in Table 2.2. By far the tightest membranes are those of the axon, wrapped in insulating layers of myelin, which ensure long-distance propagation of action potentials at a small metabolic cost. In contrast, the narrow, unmyelinated dendrites make for a minimal space constant, implying that postsynaptic potentials will not necessarily propagate to the soma.

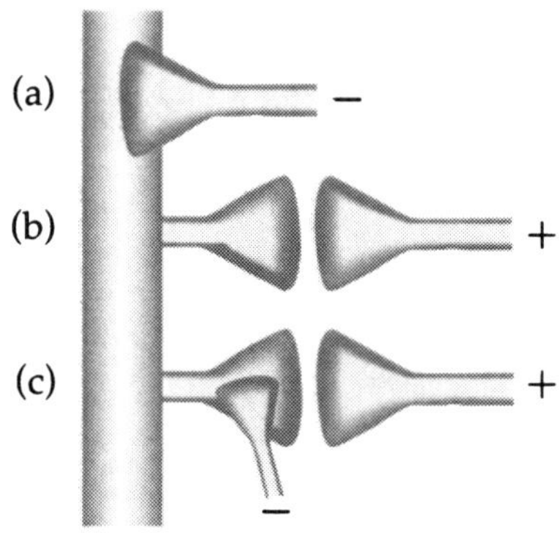

Figure 2.21. Three main types of axodendritic synapses.

Once the postsynaptic ion channels have closed, the amount of time that a potential will last at a given location is expressed by the time constant τ of the membrane. It is defined mathematically as the product of the membrane resistance and its capacitance, while it is determined experimentally as the time it takes for a constant voltage to build up to about 63% of its final value. A large time constant means that a postsynaptic potential will last relatively long, while a small time constant means that it will not. The longer a potential lasts, the longer it is available for interaction with other potentials.

As a product of two terms, a deeper understanding of the time constant depends on a clarification of its two multiplicands. Given that the effect of membrane resistance was discussed in the previous subsection, let us take up capacitance here. To remind the reader, capacitance measures the ability of the membrane to retain an electric charge. A high capacitance permits the membrane to store ions that would otherwise seep across its boundary and out of the cell. Such 'extra' ions are available to prolong any potential that passes through the region, slowing its decay. Conversely, a low capacitance permits only a small charge reservoir to accumulate at the membrane and so hastens the decay of a local potential. Though the capacitance per unit area of membrane varies little, between 0.7 and 1 $\mu F/cm^2$, the capacitance reservoir available to a given potential depends on the membrane area acting as substrate: the larger the area, the more capacity available. Under this interpretation, membrane capacitance does vary morphologically enough to cross-classify with membrane resistance, producing the typology in Table 2.3.

2.4. TRANSMISSION OF SIGNALS FROM CELL TO CELL: THE SYNAPSE

Signals are transmitted from one neuron to another at a junction point known as the **synapse**. Synapses are formed where an axon comes into close

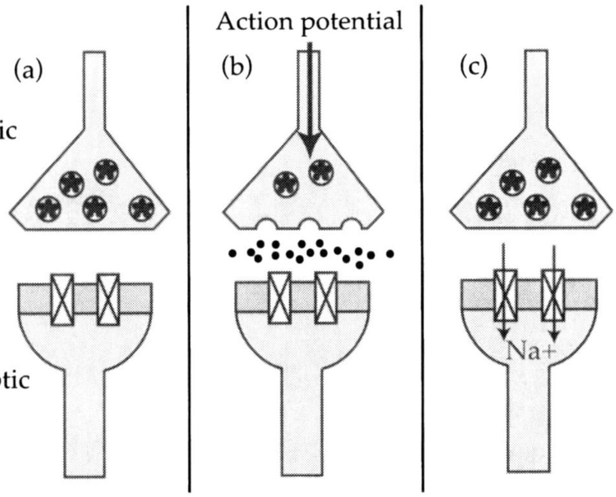

Figure 2.22. Activation of a chemical synapse between an axonal button and a dendritic spine by an action potential. Comparable to Jessell and Kandel, 1995, Fig. 11-8.

contact with a neuron at the soma or on the dendrites. The axonal side of a synapse consists of a small bulbous head known as a **terminal button**. A terminal button can synapse onto a dendrite in one of the three ways illustrated in Fig. 2.21. It can synapse directly onto the smooth surface of a dendrite, Fig. 2.21a, or indirectly, onto an outgrowth of a dendrite called a **spine**, Fig. 2.21b. Finally, one button can synapse directly onto another, as in Fig. 2.21c. Given that dendritic spines receive the majority of excitatory synapses, intense interest has swirled around them as improvements in experimental procedures have permitted an understanding of their behavior. Some of this research is cited in the upcoming sections, as well as the import of the plus and minus signs.

2.4.1. Chemical modulation of synaptic transmission

Since the neurological signal itself is electrical in kind – the membrane action potential – it is natural to suppose that it is an electrical event that is propagated at the synapse. Natural, but not what nature has chosen to do. Only a minority of synapses are electrical, and they are found in pathways where quick or accurate transmission of signals is necessary, such as in cardiac muscle or between the rods of the retina. For the cortical areas that interest us here, the existence of electrical synapses is unknown.

What is known is that there is a huge number of chemical synapses in cortical areas, mediated by a variety of neurotransmitters. A chemical synapse consists of a collection of pockets or **vesicles** of neurotransmitter, see Fig 2.22a, which

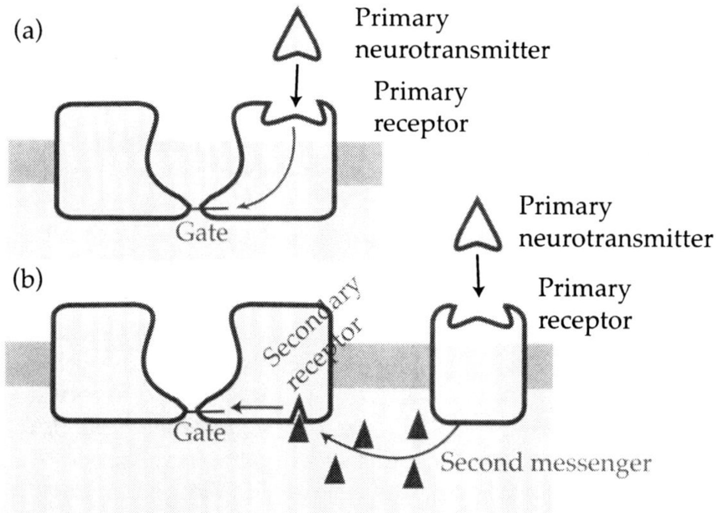

Figure 2.23. Postsynaptic receptors. (a) ionotropic. (b) metabotropic. Comparable to figures in Hille (1992).

under the stimulation of an action potential rise to the surface of an axonal button and release their contents into the gap between the button and its postsynaptic receptor, Fig 2.22b. The neurotransmitter causes gated channels of the postsynaptic neuron to open and suck in Na^+, while at the same time being reabsorbed into the presynaptic neuron, Fig 2.22c. If enough channels open, the postsynaptic neuron can undergo a depolarization of its own.

The chemical mediation of message transmission between the presynaptic and postsynaptic neurons paves the way for a dizzying variety of transmission schemes. They can be classified broadly into the release of various neurotransmitters and the response of various receptors.

Neurotransmitters are organized into three main classes according to their chemical composition, amino acids, biogenic amines, and neuromodulators. Amino acids are responsible for synaptic transmission in the central nervous system of vertebrates that is fast, acting in less than 1 ms, and brief, lasting about 20 ms. Biogenic amines are the next fastest group, with a slower onset and lasting from hundreds of milliseconds to seconds. Neuromodulators comprise a catch-all group composed of neuropeptides and hormones. Neuropeptides modulate the response of postsynaptic neurons over the course of minutes. Hormones are transported in the bloodstream and act over the same if not longer intervals. To use Koch's, 1999, p. 93 felicitous phrase, the release of a long-lasting neuromodulatory substance will affect all of the neurons in the vicinity and so act like a global variable within a computer program, by being efficacious for the entire program rather than for just a particular procedure.

Postsynaptic receptors are classified into ionotropic and metabotropic families. Ionotropic receptors are directly coupled to ionic channels, making for the transient and almost instantaneous opening and closing of channels that is characteristic of rapid perception and motor control. Metabotropic receptors, in contrast, are coupled to ionic channels only indirectly, by means of a cascade of biochemical reactions that send "second messengers" to the channel within the cell. These multiple intracellular steps can greatly amplify or squelch the incoming signal by acting in multiplicative chains: a single occupied receptor may activate many proteins in the first link, which in turn activate many proteins in the second link, and so on. Fig 2.23 describes their general mode of action in pictures Not unexpectedly, the dependency of metabotropic reception on intermediate reactions greatly decreases the speed of signal transmission, to the order of seconds or more. What the nervous system gains in exchange is a tremendous flexibility of response. One cannot help but quote one of Koch's most lyrical passages:

> It is difficult to overemphasize the importance of modulatory effects involving complex intracellular pathways. The sound of stealthy footsteps at night can set our heart to pound, sweat to be released, and all of our senses to be at a maximum level of alertness, all actions that are caused by second messengers. They underlie the difference in sleep-wake behavior, in affective moods, and in arousal, and they mediate the induction of long-term memories. It is difficult to conceptualize what this amazing adaptability of neuronal hardware implies in terms of the dominant Turing machine paradigm of computation. (Koch, 1998, pp. 95-6)

For a linguist, it is particularly difficult to overemphasize the importance of the modulatory effects that create long-term memories.

2.4.2. Synaptic efficacy

The biophysical sketch of the synapse offered in the previous subsection suggests a mathematical model with at least two variables, to wit: the number n of neurotransmitter release sites, and a measure q of the postsynaptic effect of the release of a single vesicle of neurotransmitter. It is natural to assume that the larger the presynaptic release, the greater the postsynaptic response, and likewise, the larger the postsynaptic effect, the greater the postsynaptic response. Arithmetically, this means that the two variables should be related multiplicatively to produce a postsynaptic response R, or $R = n * q$. However, one fundamental factor has been overlooked.

Given the pivotal intermediary role of the synapse in the transmission of neurological signals, it comes as a considerable surprise to learn that the synapse makes for a rather unreliable intermediary. Experimental investigation has shown that the probability of a postsynaptic current following the injection of a current into the presynaptic area can be as low as 0.1, which is to say that only one out of ten input spikes provoke an output spike. Thus some allowance must be made for a new variable, the probability p of release of a vesicle of neurotransmitter following a presynaptic action potential. Its effect should also be multiplicative of the other two: a lower probability of release lowers the response proportionally, while a higher probability raises it. The final version of the equation becomes that of Eq. 2.28:

2.28. $$R = n * p * q$$

Of course, this begs the question of how the nervous system can function so well with such undependable components. And Eq. 2.28 only allows us three variables in to play with.

An obvious solution is to inflate the number of neurotransmitter release sites n. And this is indeed what happens at the junctures where a motor neuron synapses onto a muscle, for which a single axon project a thousand release sites onto the muscle, see Katz (1966). This ensures that when you intend to touch the tip of your nose with your index finger, you actually do so, instead of doing nothing or maybe even sticking your finger in your eye instead. Yet for the cortical and hippocampal neurons that interest us here, the number of contacts can be quite small, from one to a dozen. How does the brain ensure reliable signal transmission for such small n, where the failure of even a single site would seriously degrade the signal to be passed?

2.4.3. Synaptic plasticity, long-term potentiation, and learning

The answer lies in a family of processes that allow the efficacy of a synapse to vary according to its past history by altering p and/or q. This ability of synapses to modulate their response is known as **plasticity**. The various processes that create synaptic plasticity can be classified according to whether they involve only a change in p, the probability of neurotransmitter release, or both a change in p and in q, the postsynaptic effect of release. This latter class is the one that interests us the most, since its results can last from thirty minutes to an entire life-time. It is labeled **long-term potentiation**, **LTP**, if it leads to an enduring increase in the efficacy of a synapse, or **long-term depression**, **LTD**, if it leads to an enduring decrease.

Long-term potentiation is by far the better understood process, since it has been studied extensively since first being described in the mammalian hippocampus by Bliss and Lomo (1973). Nevertheless, it is also still highly controversial. Koch, 1998, pp. 318-9, culls three generally agreed-upon observations from the large and unwieldy literature reproduced in (2.29):

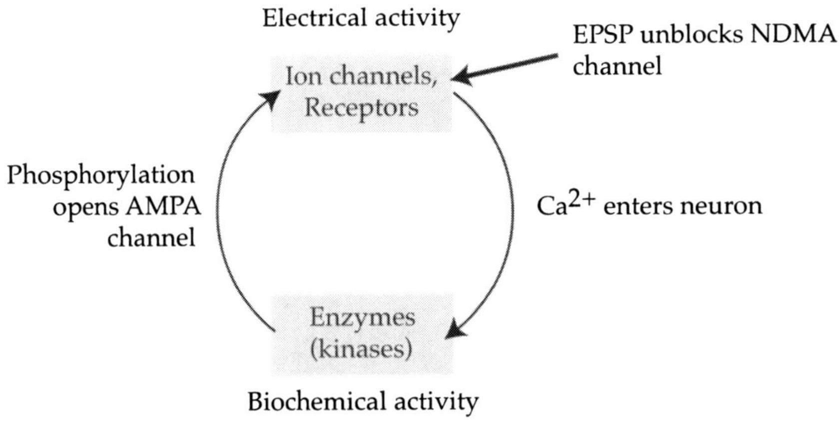

Figure 2.24. Coupling of electrical and biochemical activity in a neuron. Comparable to Helmchen, 1999, Fig. 7.1a.

2.29 a) LTP is induced by nearly simultaneous presynaptic neurotransmitter release and postsynaptic depolarization.

b) LTP is induced through activation of N-meythl-O-aspartic acid (NMDA) receptors, which are unique among receptors in opening only when both the presynaptic and postsynaptic neurons are activated.

c) LTP is induced by localized increase in postsynaptic calcium, Ca^{2+}.

The simplest story that ties these three observations together hinges on the peculiar properties of the NMDA receptor.

The NMDA receptor requires both a presynaptic neurotransmitter and a postsynaptic membrane depolarization to open. The presynaptic neurotransmitter is glutamate, which binds to the NMDA receptor for a relatively long time and so represents a reliable indicator of presynaptic activity. Glutamate triggers the opening of the NMDA gate, but the channel remains blocked by stray Mg^{2+} within it. It is only under the electrical influence of a postsynaptic potential that the magnesium ions are flushed out of the channel, unblocking it to the extracellular fluid. With all obstacles removed, Ca^{2+} ions rush into the postsynaptic neuron and trigger the changes that eventually lead to potentiation by activating enzymes that modulate q, the postsynaptic effect. These enzymes, known as kinases, allow a phosphate group to be removed from an ATP molecule and added to a target protein, a process known as phosphorylation. For the particular case of LTP, Ca^{2+} activates Ca^{2+}/calmodulin kinase II, which phosphorylates the receptor α-amino-3-hydroxy-5-methyl-4-

isoxazolepropionic acid (AMPA for short), making it more sensitive to glutamate.

But here the story branches off into many directions, since potentiation may also require a modulation of the presynaptic variables n and p, and maybe even a new one such as the amount of glutamate in each vesicle. All of these would require a signal to be propagated *backwards* across the synapse, presumably by the diffusion of a novel class of retrograde messengers; see Nimchinsky et al. (2002) and especially the review article of Yuste and Bonhoeffer (2001) for more detailed discussion.

Be that as it may, the fundamental conclusion for our modeling efforts in the upcoming chapters is that there is a plausible physical substrate for long-term changes in synaptic efficacy that not only explains how the brain trusts its most fundamental process of signal transmission to the highly variable structure of the synapse, but also, almost as an epiphenomenon, explains how the brain can learn from experience.

2.4.4. Models of diffusion

The sketch of LTP highlights a close coupling between electrical and chemical activity in a neuron, which the simple cycle in Fig. 2.24 attempts to communicate. This subsection gives a bird's-eye view of chemical diffusion, with the aid of Koch, 1999, chapter 11, as well as contributions from Helmchen (1999), Segev and London (1999), Wilson, 1999, chapter 15, and Keener and Sneyd, 1998, chapter 8.

Typically, any substance diffuses from a zone of higher concentration to a zone of lower concentration, due to the probabilistic thermal agitation of molecules. This simple observation about the entropy of the world supplies a first approximation to the movement of ions within a dendrite by equating it to the diffusion of a substance in some appropriate space. A long, thin cylinder with no internal obstacles is the simplest choice of space, because its radius is much shorter than its length, making the time for radial diffusion very short, if not negligible, with respect to the time for longitudinal diffusion. In this way, a potentially three-dimensional problem can be pared down to a single dimension, the length of the cylindrical compartment. And by a happy coincidence, dendrites tend to be shaped like long, thin cylinders.

In such a cylinder, we are interested in the temporal change of a concentration C of a diffusible substance located at position x at time t, abbreviated as $C(x, t)$. This concentration varies according to the influx and efflux from both sides of $C(x, t)$, which are standardly identified as $C(x+\Delta x, t)$ and $C(x-\Delta x, t)$, where Δx is the distance over which the substance has diffused. These abbreviations obey the general convention that relationships are stated from left to right, so that the measurement in this canonical direction is positive or additive, while in the opposite direction it is negative or subtractive. The entire prose description is given visual form in Fig. 2.25. To recapitulate the prose description, the target concentration $C(x, t)$ finds itself enclosed in the

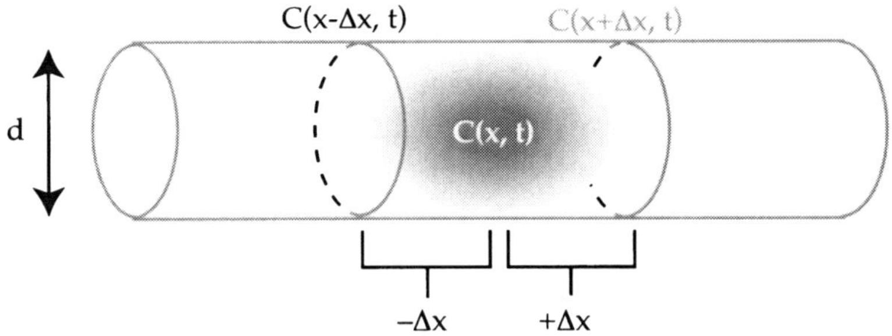

Figure 2.25. Diffusion of a substance in a cylinder. Comparable to Koch, 1999, Fig. 11.2.

compartment delimited by the boundaries $C(x{+}\Delta x,\ t)$ on the right and $C(x{-}\Delta x,\ t)$ on the left.

To make a long story short, the **flux** into and out of this location is found by means of the **diffusion equation** 2.30, where D is referred to as the **diffusion coefficient**:

2.30. $$\frac{\partial\, C(x,\ t)}{\partial t} = D\frac{\partial^2\, C(x,\ t)}{\partial t^2}$$

The reader may notice that Eq. 2.30 is practically isomorphic to the cable equation 2.25, with concentration C taking the place of voltage V and the diffusion constant D taking the place of the space constant λ. The only structural differences are the subtraction of a leakage term from the right side of the cable equation and the multiplication of its left side by the membrane time constant τ. The leakage term of the cable equation is necessary to account for the loss of ions across the membrane resistance, as mentioned above. A parallel term could have been included in the diffusion equation to model chemical seepage through the cylinder walls, but was not for the sake of simplicity.

Differentiation of Eq. 2.30 produces Eq. 2.31, where S_0 is the initial amount of the substance:

2.31. $$C_\delta(x,\ t) = \frac{S_0}{\sqrt{2\pi}}\ \frac{1}{\sqrt{2Dt}}\ e^{\pm\frac{x^2}{4Dt}}$$

The effect of these equations can be appreciated by simulation of the injection of calcium into the cylinder of Fig. 2.25 at position $x = 0$ and time $t = 0$. From this point onward, the calcium diffuses in space as depicted in Fig. 2.26. What we see

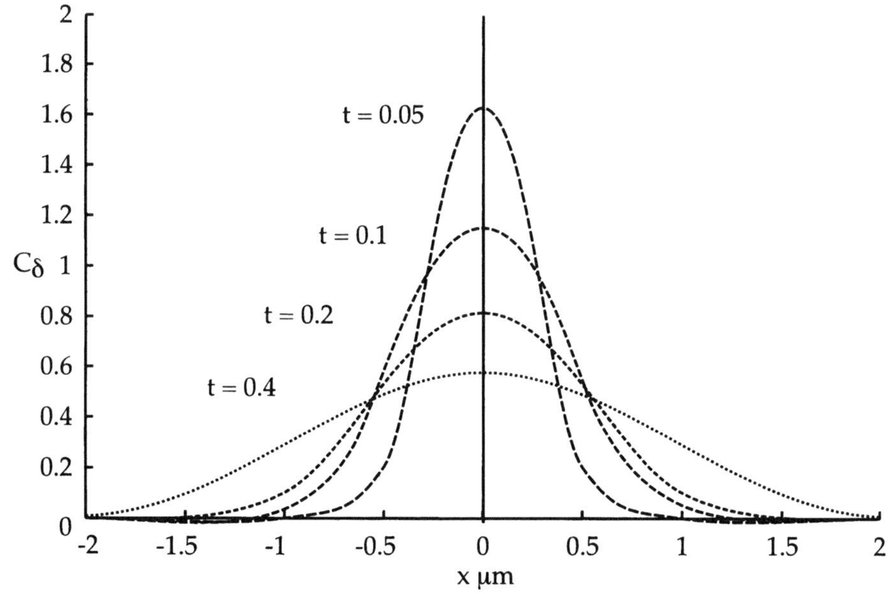

Figure 2.26. Spread of concentration $C_\delta(x, t)$ of calcium from $C_\delta(0, 0)$ as a function of space, with $S_0 = 1$ and $D_{Ca} = 0.6$ mμ²/ms. Times are in ms. Comparable to Koch, 1999, Fig. 11.4a. **(p2.15_diff_space.m)**

is that the concentration of calcium assumes the familiar shape of a bell-shaped curve, known in mathematics as a Gaussian function, whose peak gets steadily lower and wider. This steady flattening out means that the concentration is trying to fill the entire cylinder to a constant proportion, but what is fundamental is the speed at which it does so.

Fig. 2.27 graphs the diffusion of calcium with the same initial conditions as in Fig. 2.26, but as a function of time, not space. At the initial point of injection, the concentration falls off steeply as time elapses, while at successive points, the concentration first rises with the arrival of the calcium wave and then falls off gradually. All four curves will eventually converge at the concentration at which calcium is evenly dispersed throughout the cylinder.

All of this is just background for the observation that is crucial to understanding the role of diffusion in constraining dendritic function, namely the rate at which an injected concentration decreases. It is reflected in the graph of the initial point at the top, which suggests that the decrease in concentration from the point of injection is inversely proportional to the square root of time. Additional mathematical analysis show this intuition to be on the right track,

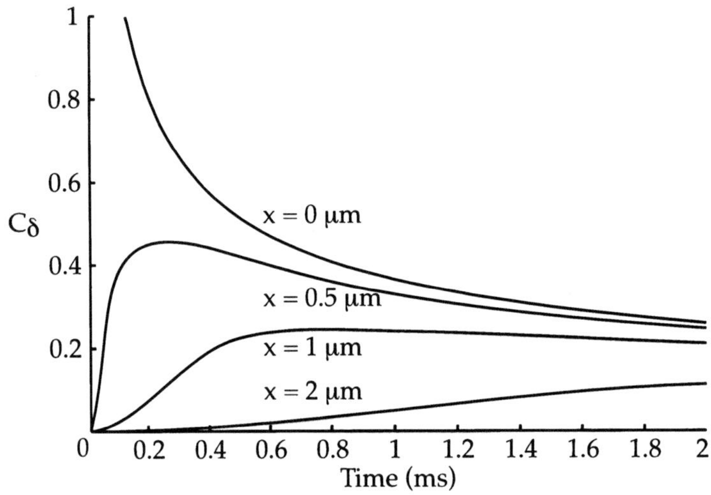

Figure 2.27. Spread of concentration $C_\delta(x, t)$ of calcium from $C_\delta(0, 0)$ as a function of time, with $S_0 = 1$ and $D_{Ca} = 0.6$ mμ^2/ms. Comparable to Koch, 1999, Fig. 11.4b. (**p2.16_diffusion_time.m**)

though holding for a different term: the displacement *x'* from the point of injection is proportional to the square root of time and the diffusion coefficient, a relationship known as the square-root law of diffusion:

2.32. $$\bar{x} = \sqrt{Dt}$$

This relationship is also implicit in the Gaussian spreading of Fig. 2.26, for which it can be shown that the standard deviation increases with the square root of time.

The upshot of this rapid rate of diffusion is that any chemical messenger released into a long, thin cylinder such as a dendrite and especially one of its spines will not be able to travel very far before it dissipates to an ineffectual level of concentration. The functional result is that large dendritic arbors such as those of neocortical neurons should act to compartmentalize their input, presumably to perform a multitude of separate calculations on them in parallel. And it may be that dendritic spines are the ultimate compartments in this process, as the following short review culled from Holmes and Rall (1995), Yuste, Majewska and Holthoff (2000), and Yuste and Majewska (2001) explains.

2.4.5. Calcium accumulation and diffusion in spines

When a neuron is first formed it does not yet have dendrites, and consequently also lacks spines. The precursors to spines are thought to be filopodia, long, thin protrusions from a dendrite with a dense actin matrix and few internal organelles, and lacking a bulbous head. During maturation of the brain, filopodia are replaced by spines, and there is a peak growth time during post-natal development. Yet part of the maturation process appears to involve the pruning of spines. Fewer spines are present in adults than children – up to 50% fewer. However, over-pruning can have disastrous consequences. The deformation or absence of spines on certain neurons have been associated with brain disorders, such as stroke, epilepsy, and fragile X syndrome – see Segal (2001).

Dendritic spines were first described by Santiago Ramón y Cajal, who discovered that certain cells in the cerebellum had small "thorns" (Sp. *espina*) that projected outward from their dendrites like leaves from a tree. Shortly afterward, he put spines in the spotlight by proposing that they serve to connect axons and dendrites and that they might be involved in learning, see Ramón y Cajal (1888, 1891, 1893), respectively. Half a century later, the landmark study of Gray (1959) using electron microscopy confirmed Cajal's prediction that spines were a site of synaptic contact.

Because synapses can be made onto dendritic shafts directly, see Fig. 2.21, it is natural to suppose that spines must have some function in addition to receiving synaptic inputs. Speculations about this function have covered dozens of possibilities. The peculiar morphology of spines, which bear a small (less than 1 μm diameter) head connected to the dendrite by a thin ($\approx$ 0.2 μm diameter) neck, has fueled speculation about their function as biochemical, rather than electrical, compartments and, specifically, as a means for compartmentalizing calcium.

Diffusion theory tells us that any change in a spine's shape will have dramatic effects on its chemical (and electrical) behavior. A short spine is more closely linked to its parent dendrite and so reflects changes in the parent's Ca^{2+}, whereas a long, thin spine regulates its Ca^{2+} transients independently of the parent dendrite, see Segal and Anderson (2000). The overall computational picture is of a regime in which spines restrict calcium diffusion in order to isolate different inputs and so modulate local synaptic plasticity.

In the first study of spines using the high-resolution technique of two-photon microscopy, Yuste and Denk (1995), three different functional patterns of calcium accumulation in the spines of hippocampal pyramidal neurons were described. Postsynaptic action potentials propagated through the dendritic tree and triggered generalized Ca^{2+} accumulation in spines and dendrites, while subthreshold synaptic stimulation produced Ca^{2+} increases restricted to individual spines. Finally, the co-occurrence in a spine of an action potential and synaptic stimulation produced Ca^{2+} accumulation that exceeded the sum of the two taken separately. These three patterns of calcium accumulation have a direct computational correspondence: postsynaptic action potentials are the output of a cell, synaptic stimulation is its input, and the temporal pairing of the two represents the detection of output/input coincidence at the synapse that is hypothesized to underlie long-term potentiation.

More recently, a triumvirate of papers that report the observation of the formation or growth of spines in association with LTP or strong electrical synaptic stimulation have garnered considerable attention (Engert and Bonhoeffer (1999), Maletic-Savatic et al. (1999), and Toni et al. (1999)); see Anderson (1999) and Pray (2001), as well as prominent mention in review articles, such as Segal and Anderson (2000). Yet Segal and Anderson counsel caution, since it is not clear that the spinal growth observed in these experiments with cultured cells would result in functional synapses in the real thing.

Not only does a spine grow *in vitro*, but with the recent advent of high resolution imaging methods for living cells in culture, it has been discovered that a spine can continually change its shape, elongating its neck to stretch away from its dendrite or retracting it to huddle down closer, see Fischer et al. (1998) and Dunaevsky et al. (1999). For instance, using video imaging, Fischer et al. observed that within two seconds, the edge of a spine could move as far as 100 nm, while over two minutes it could move by more than 300 nm – up to 30% of the total width or length of the spine.[17]

It is natural to link this **motility** to the apparent function of spines in compartmentalizing calcium. Segal (2001) proposes that calcium controls the change in spine shape in a bell-shaped manner: (i) lack of Ca^{2+} due to lack of synaptic activity causes transient outgrowth of filopodia but eventual elimination of spines; (ii) a moderate rise in Ca^{2+} causes elongation of existing spines and formation of new ones, while (iii) a massive increase in Ca^{2+}, such as that seen in seizure activity, causes fast shrinkage and eventual collapse of spines.

2.5. SUMMARY: THE CLASSICAL NEUROMIMETIC MODEL

A neuron can be thought of as a cell that can manipulate the electrical properties of its cell membrane to transfer a signal to other cells. This potential for communication is the insight that drives neuroscience in general, and neuromimetic modeling in particular.

The generic form of the signal-inducing mechanism is that channels specifically sized for sodium ions open in some region of a neuron's membrane, sucking sodium ions into it under the combined force of the electrical gradient (the interior surface of the cell membrane is negatively charged, while the sodium ions are positively charged) and the diffusion gradient (there is

---

[17] See the book's web site for viewing these and other videos over the Internet.

normally much more sodium outside a neuron than inside it). Due to the influx of extracellular ions, the membrane loses its negative polarization, or depolarizes, and then momentarily reverses to a positive charge. In so doing, it creates an impulse that constitutes a striking departure from its normal state and so can be used to carry a message – though it is a very simple one, not much more than the cry of "I am on!".

In functional terms, excitatory inputs sum together to depolarize the membrane, and if the resulting depolarization reaches threshold, an action potential is generated. Inhibition takes on the role of opposing depolarization and so increasing the number of excitatory inputs required to reach threshold. An inhibitory neuron therefore works in the opposite way of the Hodgkin-Huxley neuron analyzed above, which is to say that instead of producing a depolarization of the cell membrane, it produce a hyperpolarization – the action potential is even more negative than the resting state, usually on the order of –75mV. This comes about by a net reduction of the positive charge within the neuron, triggered by an influx of Cl^- or an efflux of K^+ through the appropriate open channels.

Inhibition can oppose excitatory depolarization in one of two ways, as was already anticipated in Sec. 1.2.2.2. An inhibitory neuron can synapse onto a dendrite that is host to synapses from other neurons and so mute all of its upstream inputs, which describes the **postsynaptic inhibition** of the synapse depicted in Fig. 2.21a with respect to the synapse of Fig. 2.21b. This silent or shunting inhibition answers arithmetically to division of the on-going sum of excitatory currents. The alternative is for an inhibitory neuron to synapse onto a single excitatory spine and so mute its output, which describes the **presynaptic inhibition** depicted in Fig. 2.21c. This characterization of inhibition answers arithmetically to subtraction from the on-going sum of excitatory currents.

Passive cable theory elaborates this general electrotonic theory by explicating the contribution of the dendritic arbor. Condensed from Segev and London (1999), the major insights from passive cable theory are listed in (2.33):

2.33 a) Due to their complex arbors, dendrites are electrically distributed in such a way that voltage flow attenuates from synapse to soma.
b) Voltage attenuation is asymmetrical, being greater in the dendrite-to-soma direction than in soma-to-dendrite direction.
c) Nevertheless, a significant proportion of the synaptic charge flowing in the dendrite-to-soma direction reaches the soma.
d) Dendrites slow down the transmission of action potentials to the soma, and also slow down the action potentials themselves.
e) The time window for summation of synaptic inputs is much shorter at dendrites than at the soma.
f) Dendrites are favorable sites for synaptic plasticity.

For our concerns, the two key properties are (2.33e, f), since they control the ability of dendrites to implement correlation and to learn.

The putative mechanism for learning, long-term potentiation embodied in synaptic efficacy induced by NMDA, puts the finishing touches on a story about information processing in the nervous system, a story which can be told as the answer to a query. It goes something like this:

The fascinating question is why the nervous system opts for the rather Byzantine mechanism of chemical transmission when it could make do with faster and more accurate electrical transmission. The answer seems to be that it is only the peripheral nervous system that is concerned with quick and accurate transmission of signals, in order for the central nervous system to have timely and reliable information. The central nervous system, and especially the linguistic components that we are interested in, is much more concerned with computation, such as the extraction of features from incoming sensory data. The multifarious chemical events that take place at both boundaries of the chemical synapse make it the mechanism of choice for dealing with the extraction of ever-changing features from an ever-changing environment. Or to put it somewhat more prosaically, if the synapse had evolved a high fixed value for *R*, say by refining the probability of failure of vesicle release to less than 10^{-14} – that is, *p* would be $1 - 10^{-14}$, which is the probability of failure per switching event in a contemporary digital computer – then it would have been robbed of the dynamic range it needs to adapt to changing conditions, such as those of a child learning a language.

In this story, there are three implicit assumptions about synapses that make them the "mechanism of choice" for the modeling of cortical information storage and transmission. The assumptions are that synapses: (i) are stable on both short and long time scales, (ii) have a high resolution, and (iii) are sufficient unto themselves, i.e. they can be represented by a single real number which does not depend on anything else.

2.5.1. The classical model

Fig. 2.28 brings all of these notions together as they are understood in the firing-rate model. The reader should recognize the generic pyramidal neuron from Chapter 1 on the left of Fig. 2.28. To the right of it is a mathematical idealization of its major functional components. The axonal inputs from other neurons are symbolized by the row of *V*'s across the top of the artificial neuron. Each such input is multiplied by a weight *w*, which represents the efficacy of the synapse for that particular axonal connection. Within the circle, whose biological analog is the soma, the summation notation indicates that the weighted inputs are summed together to get an intermediate output V_i. Thus, in the spirit of passive cable theory, the model ignores the soma-dendrite contrast altogether and treats all synapses as being uniformly arranged on the soma. V_i is then passed through a **transfer** or **activation function** $g(V_i)$, to produce a quantity

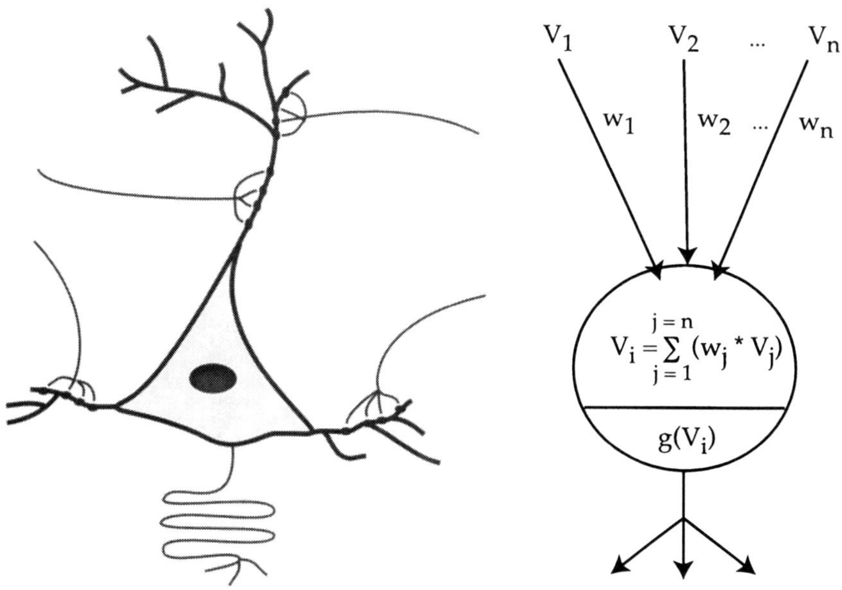

Figure 2.28. From real to artificial neurons.

that represents the firing rate of the neuron. This number is then broadcast to any other neuron that it is connected to in the network.

The models that are tested in Chapter 5 all instantiate this classical prototype. The only aspect of this model that has not been discussed in sufficient detail is the activation function, to which we now turn.

2.5.2. Activation functions

An activation function delimits how the voltage of the neuron is integrated to produce its output, a firing rate. Fig. 2.29 illustrates five of the most commonly-used activation functions, which we will spend the next few paragraphs reviewing.

The touchstone for any of these functions is the linear or identity function at the top. For a linear function, the voltage input equals the output. Thus if you find the 0.5 position along the x axis and follow it up to the graph of the function, and look to the left side where the output is indicated on the y axis, you see that it is also 0.5. This is the mathematical way of indicating compositionality: the sum of any two inputs is the same as the sum of their outputs. Such compositionality is good for transmitting information unchanged, but as we have had the occasion to mention several times, the human cerebral cortex is less interested in the mere transmission of information and more

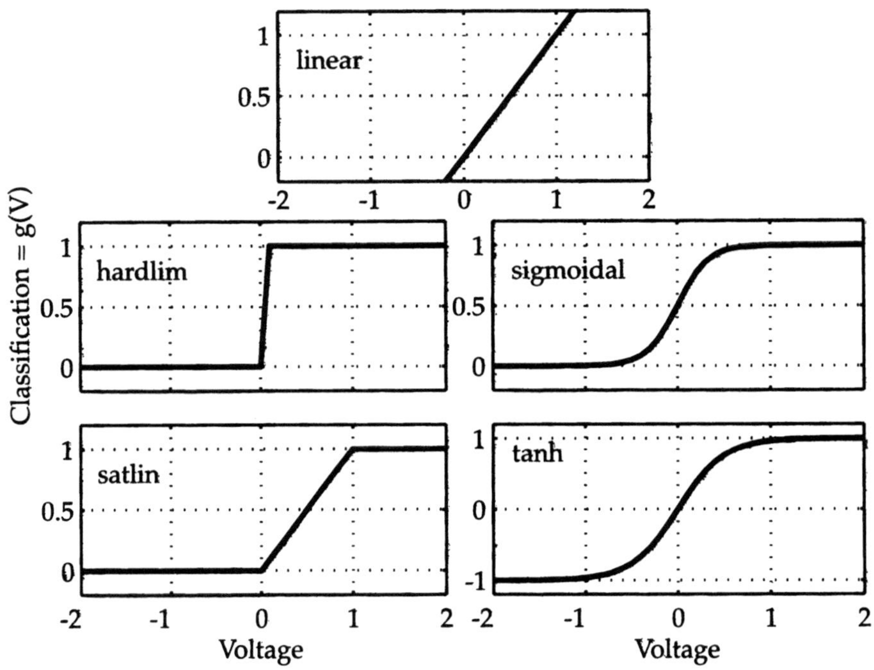

Figure 2.29. Plot $f = g(V)$ for five transfer or activation functions g. Comparable to Kartalopoulos, 1996, Fig. 2-4. (**p2.17_act_fun.m**)

interested in the creation of new information. Thus linear activation functions are only used to achieve very special effects, such as the linear layer of learning vector quantization reviewed in Chapter 5.

It is much more common for the activation function to be drawn from the class of nonlinear functions, especially those that are continuous, saturating, and positive monotonically increasing, such as the bottom four in Fig. 2.29. These properties endow the relevant nonlinear functions with an incipient ability to classify their input – to reject it or accept it – which will become the mathematical basis of many of the pattern-classification techniques introduced in Chapter 5. It thus behooves us to devote a few words to how such seemingly abstract notions can lay the foundations for the fundamental ability of humans to classify sensory stimuli into the relevant cognitive categories.

A function is continuous if it has no gaps in its graph, so that it produces an output for any input. Thus a continuous function always makes a decision about its input; it is never at a loss to emit a signal of acceptance or rejection, whether it is the correct one or not. A function is saturating if its extremes both evaluate to a single minimum and a single maximum value. For instance, the lower inputs of the nonlinear functions of Fig. 2.29, from $-\infty$ to about 0, all produce the same output of 0, while their upper inputs, from about 1 up to ∞, all produce the same output of 1. This saturation of extremes provides the function with a window of attention: it produces the most varied output for just a small range of input values, while ignoring everything else. Given that such a function ignores every input beyond the confines of its window of attention by compressing them into the same output, such functions are often known as "squashing functions". Moreover, a continuous saturating function tends to divide its input space into two halves, one where the response is approaching 1 and another where it is approaching 0 or -1. Hence it excels at making sweeping cuts through its input space, which is the essence of classification.

A function is positive if its slope goes uphill from left to right; it is negative if its slope goes downhill. Finally, a function is monotonically increasing if any two values in its input for which one is less than or equal to the other are mapped to outputs which preserve the relationship of one being less than or equal to the other. Such functions tend to preserve a correlation between input and output. For the linear function, this is perfect correlation.

Each of the four non-linear functions in Fig. 2.29 are specialized for different effects. For instance, the hardlim function rejects any input below 1 and so can represent decisions in two-valued logic, which evaluate either to true (1) or false, (0). The sigmoidal (S-shaped) function, on the other hand, is more accurate at representing neuronal dynamics. It can be understood as reproducing the gradual charging of a capacitor as a gradual increase in output from zero to its maximal level. A common equation for this function is Eq. 2.34:

2.34. $$g = 1 / 1 \pm e^{-2 * b * V}$$

The *b* variable controls the steepness of the slope from 0 to 1: the higher the value for *b*, the steeper the slope becomes. Choosing *b* as 3 and plotting *V* in the range [-2 2] produces the corresponding graph of Fig. 2.29.

2.6. EXPANDED MODELS

The classical model of the generic pyramidal neuron is the simplest and most conservative one. Indeed, it was familiar to Ramón y Cajal in the late 19th century, who formulated the **law of dynamic polarization**, which in the translation of Shepherd, 1999b, p. 364, states:

> The transmission of neuronal electrical activity takes place from the dendrites and cell body toward the axon. Therefore every neuron has a *receptive component*, the cell body and dendrites; a *transmission component*, the axon; and an *effector component*, the varicose terminal arborization of the axon. (italics added)

That is to say, as even as far back as the late 19th century, Ramón y Cajal appreciated the major functional units of the rate model depicted in Fig. 28. It is on the strength of this longevity that it has attained the status of "classic" in computational neuroscience.

It is only in recent years that enough evidence has accumulated to call into question the simplicity and elegance of the classical doctrine. This new evidence does not shake the foundations of the classical pyramidal neuron, but rather disputes the claim that the computational unit is the entire neuron. What has been found is that various subparts of the neuron can perform computations – subcomputations, as it were – on their own. The final section of this chapter review some of these more recent findings, and especially those that reveal sub-unit computation.

2.6.1. Excitable dendrites

Perhaps the most important insight of passive cable theory is that dendrites are not passive. On one hand, the dendritic membrane conductance increases with distance from the soma, making distal dendrites leakier than proximal dendrites, see Segev and London, 1999, p. 214. On the other hand, dendrites contain voltage-dependent ion channels that actively propagate the synaptic action potential towards the soma, much as the axon propagates the somatic action potential towards other neurons. The effects of dendritic ion channels were first recorded intracellularly in the late 1950's and early 1960's (Eccles, Libet and Young (1958); Spencer and Kandel (1961)). Since then, it has become evident that the dendrites of pyramidal cells contain a large number and variety of voltage-gated Na^+ and Ca^{2+} channels; see Nusser (1999) and Magee (1999) for review, as well as and Poirazi and Mel, 2001, p. 779 for a slightly more recent list of references. These channels open in response to membrane depolarization and in turn cause further depolarization, which results in a regeneration of the dendritic current. In some cases, they are efficient enough to produce their own responses, including full-blown spikes; again see Poirazi and Mel, 2001, pp. 779-80, for extensive references. It is these dendritic channels that are of more interest in the current context, for they lay the groundwork for sub-unit computation.

2.6.1.1. Voltage-gated channels and compartmental models

The initial observations of dendritic ion channels motivated Rall (1964) to elaborate a **compartmental model** of dendritic function, in which the continuous cable equation is 'discretized' into a finite set of electrical compartments, each of which lumps a section of dendritic membrane into a resistance-capacitance (RC) element such as that of Fig 2.2. The current flowing through compartment *j* in such a model is given by Eq. 2.35:

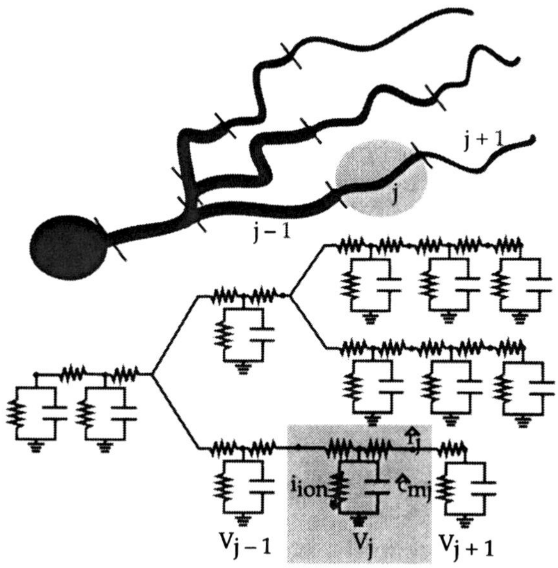

Figure 2.30. Partition of a dendritic arbor along diagonal cut points (top) into RC compartments (bottom). Comparable to Segev and London, 1999, Fig. 9.1.

2.35. $$c_{mj}\frac{dV_j}{dt} = \frac{d}{4r_j}\frac{V_{j-1} - 2V_j + V_{j+1}}{\Delta x^2} - i_{ion_j}$$

This **compartment equation** is isomorphic to the cable equation 2.25, with the following changes in constants. The new constants are the membrane capacitance of the jth compartment c_{mj}, d, the axial resistance r_j, and the ionic current that leaks through the compartment membrane i_{ion_j}. Fig. 2.30 is intended to aid the reader to grasp the effect of compartmentalization. Rall (1964) showed that if the length of the dendritic section is sufficiently small, the solution for the compartmental model converges to that of the corresponding cable model.

The fundamental question to be asked of any compartmental model is, how many compartments are necessary? For instance, Mainen et al. (1995) use about 275 compartments to simulate the dendritic tree of a particular rat layer 5 pyramidal cell, which requires a thousand coupled nonlinear differential equations to be solved. Solving a thousand coupled nonlinear differential equations is not for the computationally faint of heart, with the aggravation that there is little physiological data for many of the parameters in need of specification. The results may therefore not have the precision that at first glance they would be expected to have.

Such considerations have lead to several attempts to collapse compartments comparable in spirit to Rall's collapsing of equivalent membrane cylinders. For instance, Bush and Sejnowski (1993, 1995) develop a technique for collapsing 400 compartments into eight or nine, while Destexhe et al. (1996) demonstrate how three compartments can reproduce the results of 230. The greatest level of reduction is clearly that at which the entire dendritic arbor is collapsed into a single compartment and connected to a somatic compartment in an appropriate fashion. Rinzel and colleagues, e.g. Pinsky and Rinzel (1994), have developed such a two-compartment model.

2.6.1.2. Retrograde impulse spread

In keeping with Ramón y Cajal's law of dynamic polarization, the flow of electrical activity has so far been characterized as unidirectional, from dendrites into the soma and out through the axon. Yet it has been known since intracellular recordings performed in the 1950's that an action potential can spread backwards from the axon hillock into the soma and dendrites, see Spruston et al, 1999, p. 248ff, and Wilson, 1999, p. 268.

The utility of this phenomenon is only now receiving scrutiny, but this scrutiny is rather intense, given that retrograde impulse spread could contribute to a variety of fundamental processes. The mini-review in Shepherd, 1999b, p. 383, lists four possibilities. Let us only mention one here, namely that retrograde or **backpropagating** impulses could summate with the depolarization of spine synapses and so enable them to detect input-output coincidences.

2.6.1.3. Dendritic spines as logic gates

Thirty years after Gray's confirmation of Ramón y Cajal's proposal about the function of spines, Shepherd and Brayton (1987) demonstrated how synapses at the end of the spines of simulated dendritic compartments could perform AND, OR, and AND-NOT gating operations according to the settings of a handful of parameters. The overall layout of Shepherd and Brayton's compartments is depicted in Fig. 2.31. The 'trunk' of the dendrite is the vertical column of circles, each of which is a compartment obeying the dynamics of the Hodgkin-Huxley equations. Spines project off to the left and right of the dendritic trunk – three out of four to the right. The parameters of these compartments and their interconnections are set to mimic as closely as possible the available electrophysiological measurements.

For the simulation of an AND gate, compartments 1 and 2 are subject to a simultaneous pulse of increased membrane conductance that depolarizes the membrane for 2.1 ms. Compartments 1 and 2 respond almost immediately with a spike up to about –10 mV. Compartment 3, then 4, respond within a few tenths of a millisecond with their own spikes up to about 0 mV, showing the spread of the original postsynaptic response through the immediate vicinity of the dendrite. Compartments 5 and 6 respond concurrently with compartment 4,

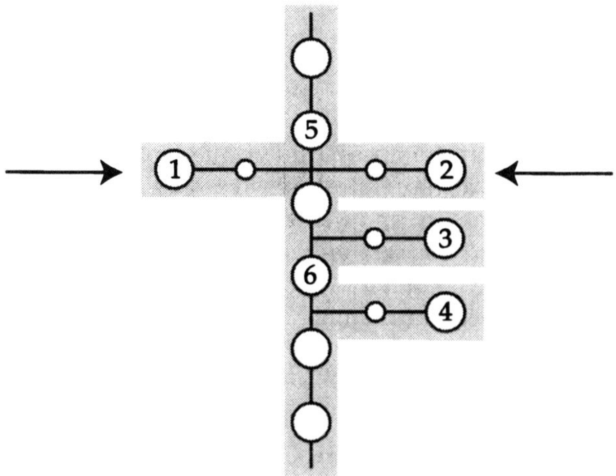

Figure 2.31. Layout of simulated dendrite showing sites of excitatory input for AND. Comparable to Shepherd and Brayton, 1987, Fig. 2.

with 'humps' of depolarization between 45 and 30 mV; see Shepherd and Brayton's Fig. 2 for a plot of the action potentials.

An OR gate is simulated in this paradigm by a single excitatory input to either compartment 1 or compartment 2, with the difference that the input must be doubled in order to reach threshold. The resulting profile of a response spreading to nearby compartments is almost identical to that of the AND gate.

Finally, an AND-NOT gate is achieved by means inhibition. An excitatory input is applied at compartment 1, along with a larger inhibitory input at the small circle between compartment 1 and the trunk of the dendrite, a location known as the "neck' of the spine. This configuration effectively squelches the expected OR-response from compartment 1 and its neighbors.

2.6.2. Synaptic stability

Following Segal (2001), one of the conclusions that can be drawn from the summary of research into calcium compartmentalization in dendritic spines is that the century-old belief that spines are stable storage sites of long-term memory has been overturned by the recent flurry of observations using novel high-resolution imaging methods of living cells in culture in favor of a dynamic structure, one which undergoes fast morphological changes over periods of hours and even minutes.

This conclusion has been taken to heart by Bartlett Mel and collaborators and become one of the major ingredients in their critique of the classical model. For instance, Poirazi and Mel, 2001, pp. 779-80, review recent research that calls into question two assumptions of the classical theory of the stability of the synapse. They find evidence that (i) synaptic response varies widely on short time scales, (ii) synaptic response has a very low resolution on longer time-scales (maybe just on or off), (iii) active membrane mechanisms can lead synaptic responses to depend on the ongoing activity of other synapses, and (iv) learning-induced changes remodel the physical structure that interfaces between axons and dendrites, namely dendritic spines. As Poirazi and Mel put it, these findings "... suggest that the setting of finely graded connection strengths between whole neurons may not provide the exclusive, or even the primary form of parameter flexibility used by the brain to store learned information."

2.6.3. The alternative of synaptic (or spinal) clustering

The alternative that Mel and his collaborators explore is that synapses, or the spines that bear them, form clusters based on their correlated activity. Such clusters act as computational subunits, so that a neuron's emission of an action potential may actually be the response to the output of several compartmentalized subcomputations scattered across the neuron's dendritic arbor. Mel's lab has published several reports of simulations that illustrate varies facets of how this type of computation could work, see Archie and Mel (2000), Mel (1992, 1994, 1999), and Poirazi and Mel (2000, 2001).

The main objection that could be raised to this approach is that it has not been observed in nature. However, there are several sources of intriguing indirect evidence for it. One is the theory of dendritic electrotonic compartmentalization reviewed above, which permits a partially isolated branch to perform its own computation. A case in point is the Shepherd-Brayton demonstration of how nearby spines could cooperate to perform logical operations. Another obvious piece of evidence is the mere existence of dendritic arbors themselves. Why would the brain devote expensive metabolic resources to the elaboration of such extravagant forms if they served no useful function? A case in point can built on any of the various stellate dendritic arbors. As originally proposed in Koch et al. (1983), the star-shaped branching of retinal ganglion dendrites creates the ideal structure for the isolation of individual branches for the performance of individual calculations.

We would like to add one more bit of indirect evidence, not mentioned as far as we can tell in Mel's work. It is the series of studies carried out by Marin-Padilla and his collaborations on the distribution of dendritic spines. The most interesting from our perspective is Marin-Padilla et al. (1969), in which the number of spines along the apical dendrite of human layer 5 pyramidal cells was counted and their distance from the soma measured. Plotting the number of spines at a given distance from the soma revealed a bell-shaped distribution, with the peak falling roughly at the center of the apical dendrite. Marin-Padilla et al. proposed that this distribution was actually the superposition of smaller overlapping Gaussian distributions, "... as if some cortical factors were 'aiming'

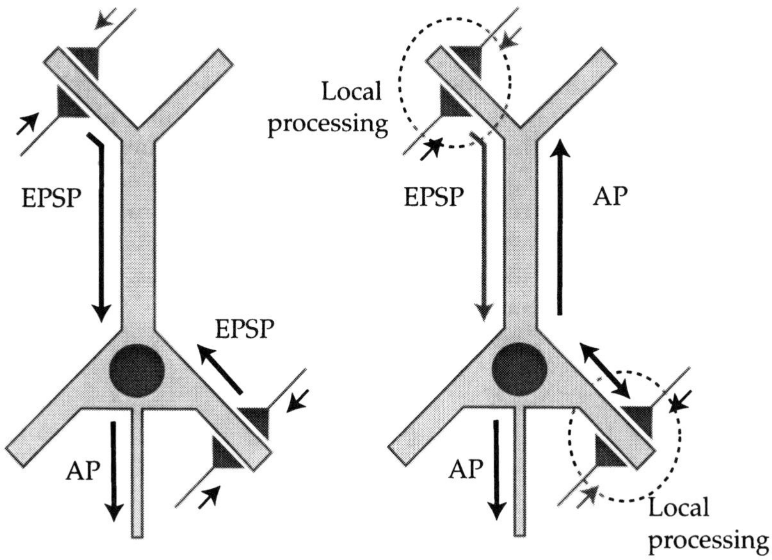

Figure 2.32. Classical vs. expanded neural processing. Comparable to Shepherd, 1999, Fig. 13.1-4.

to produce spines at the mean cortical depth, but the spines were deflected from the mean by random small causes." (p. 493)

Marin-Padilla et al. devised a computer program that tried to match a series of overlapping Gaussians to the observed distribution. The best fit had ten overlapping clusters, but it has no physiological interpretation. However, the fit of five overlapping clusters was almost as good, and it has an obvious physiological interpretation as clusters of inputs from each of the layers of cerebral cortex, with layer 4 and 5 absorbed into the same population. Thus Marin-Padilla et al.'s results supply indirect confirmation of the correlation-sorted clustering of inputs postulated by Mel: the inputs from a given cortical layer are presumably correlated among themselves, while being independent from the inputs from other layers. It follows that if a dendritic arbor is sensitive to input correlation, it should segregate correlated clusters to different regions – read different compartments – on it.

2.7. SUMMARY AND TRANSITION

This chapter is dedicated to explicating the generation and propagation of the action potential, the main signal exchanged among neurons. The bulk of the chapter develops what we have referred to as the classical model, and in particular the firing-rate version thereof. The final section expands this model to cover more recent experimental and computational modeling results. The two paradigms are contrasted pictorially in Fig. 2.32. The crucial contrast between them is the existence of local or clustered processing in the latter but not the former. The models developed in Chapter 5 for logical coordination follow the classical model and do not avail themselves of this possibility; however, the novel model developed in Chapter 7 to capture the statistics of natural language semantics and in particular the correlations on which logical coordination and quantification are grounded, does avail itself of dendritic subcomputations, and in particular on the topography of dendritic spines. But before we can take up these models, we first introduce some mathematical tools in Chapter 3 that will be fundamental for analyzing in Chapter 4 the patterns created by the logical coordinators.

Chapter 3

Logical measures

This chapter introduces the various branches of mathematics that can be called upon to transduce patterns into a format that is amenable to neurological processing, namely statistics, probability, information theory, and vector algebra. It also lays out the initial definitions of the logical operators as idealized patterns in these mathematical domains.

3.1. MEASURE THEORY

The proposal of this chapter is that the logical operators are measures, in the mathematical sense. In fact, they are signed measures, but let us first consider what a mathematical measure is.

3.1.1. Unsigned measures

In real analysis, a measure assigns sizes, volumes, or probabilities to subsets of some set. Krifka, 1990, p. 494, explains this assignment in the most perspicuous manner that we have seen:

> A **measure function** is a function from concrete entities to abstract entities such that certain structures of the concrete entities, the **empirical relations**, are preserved in certain structures of the abstract entities, normally **arithmetical relations**. That is, measure functions are homomorphisms which preserve an empirical relation in an arithmetical relation. For example, a measure function like °C 'degrees Celsius' is such that the empirical relation 'x is colder than y' is reflected in the linear order of numbers, as it holds that $°C(x) < °C(y)$.

Measures are defined over a sigma algebra in real analysis, see for instance Halmos, 1950, p. 30-31, but for our purposes it is more appropriate to define them over a simpler structure. The next paragraphs outline two.

An algebra consists of a set together with one or more operations on it which satisfy certain axioms. The simplest algebra on which an unsigned measure function can be based is assembled from one binary operator, union, one unary operator, complementation, and a zero element Ø, plus the afore-mentioned non-empty set S. Taken together, they produce the ordered quadruple $\langle S, \cup, ', \emptyset \rangle$, which we refer to as A. This is approximately the analysis of Applebaum, 1996, pp. 28ff. Formally, A is defined by the properties of (3.1).

Table 3.1. Axioms and theorems for operators. For any *a*, *b*, *c* in a set *X*...

Status and name	$\cup$	$\cap$
A1. associativity	$a \cup (b \cup c) = (a \cup b) \cup c$	$a \cap (b \cap c) = (a \cap b) \cap c$
A2. commutativity	$a \cup b = b \cup a$	$a \cap b = b \cap a$
A3. distributivity	$a \cup (b \cap c) =$ $(a \cup b) \cap (b \cup c)$	$a \cap (b \cup c) =$ $(a \cap b) \cup (a \cap c)$
A4. complementation	$a \cup a' = 1$	$a \cap a' = 0$
A5. bounding	$a \cup 0 = a$	$a \cap 1 = a$
A6. idempotency	$a \cup a = a$	$a \cap a = a$
A7. absorption	$a \cup (a \cap b) = a$	$a \cap (a \cup b) = a$
T1. bounding	$a \cup 1 = 1$	$a \cap 0 = 0$

3.1 a) The empty set Ø is in *A*.
b) If *a* is in *A* then so is the complement of *a*, *a'*.
c) If *a* is in *A*, and *b* is in *A*, then $a \cup b$ is also in *A*.

An unsigned measure μ can be defined as a function which assigns to every element *a* of *A*, a value $\mu(a)$, which is a non-negative real number. More succinctly, $\mu : A \rightarrow \Re^+$. The following properties have to be satisfied:

3.2 a) The empty set has measure zero: $\mu(\emptyset) = 0$.
b) If *a* and *b* are disjoint sets in *A*, $\mu(a \cup b) = \mu(a) + \mu(b)$.

(3.2b) is known as **additivity**. It says that the measure of a collection of mutually exclusive elements – represented as their union – is the sum of the measure of each individual element. It permits a group measure to be derived from the sum of its parts. If μ is a measure on *A*, then the members of *A* are called **μ-measurable sets**, or **measurable sets** for short. The ordered triple $\langle S, A, \mu \rangle$ is called a **measure space**.

Many other properties of μ can be derived from (3.2), but the one that interests us the most is how to calculate the measure of the complement of an element, *a'*. The standard way is to find the largest element in *A* and subtract $\mu(a)$ from its measure. The largest element of a set *A* is known as the **supremum** or least upper bound of *A*. It is denoted by *sup(A)* and is defined to be the smallest set that is larger than or equal to every set in *A*. The supremum helps us to define the measure of a complement as so:

3.3. $\mu(a') = \mu(\sup(A)) - \mu(a)$.

That is, by additivity, the complement of a measure of *a* is what is left after subtracting the measure of *a* from the measure of the largest set in *A*.

Invoking these notions without writing them into the axioms of *A* is a bit of a cheat, but to write them in adequately presupposes elaborating *A* to the full

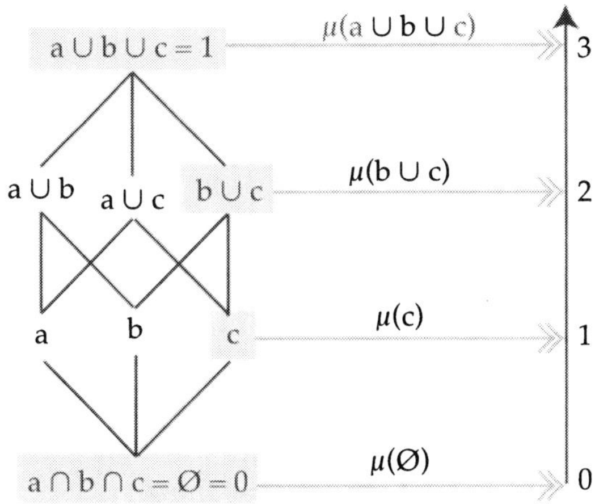

Figure 3.1. An unsigned measure mapped from an 'unsigned' lattice.

power of a **Boolean algebra** B, which consists of the ordered sextuple $\langle S, \cup, \cap, ', \mathbf{0}, \mathbf{1}\rangle$ for which axioms 1 to 5 in Table 3.1 hold; see for instance Stoll, 1979, p. 248ff, among many others. From these axioms, one can prove various other properties, for instance, that the zero or smallest element **0** and the unit or largest element **1** are unique, that every element has only one complement, that $\mathbf{1}' = \mathbf{0}$ and $\mathbf{0}' = \mathbf{1}$, and T1 in Table 3.1. With a particular simplification to be mentioned below, this is the algebra adopted in Krifka (1990) for the analysis of linguistic measures like "sixty tons".

A Boolean algebra lends itself to a convenient visualization as a type of lattice. Algebraically, a **lattice** L is a triplet $\langle S, \cup, \cap\rangle$, where S is a non-empty set, and $\cup$ and $\cap$ are binary relations which satisfy the axioms of associativity, commutativity, idempotency, and absorption in Table 3.1. A Boolean lattice is further constrained to be distributive and complemented. Note that it is somewhat more popular to use $\vee$, **meet**, and $\wedge$, **join**, for union and intersection in lattices, but for the sake of notational economy we retain the latter for both.

The visualization of a lattice is accomplished by the drawing of its **Hasse diagram**, which is a graph whose vertices are the elements of L and whose edges represent the binary relations of L. By way of illustration, Fig. 3.1 draws the lattice based on $S = \{a, b, c\}$ on its left side. The edges joining the vertex a and the vertex b to the vertex $a \cup b$ can be seen as the two halves of the union operation and could even be labeled as such, if $a\cup$ and $\cup b$ were considered well-formed expressions.

What interests us more than the lattice-theoretic visualization of a Boolean algebra is the measure that can be taken of this object. Fig. 3.1 adds the mapping from the lattice to the measure on its left side. Each set highlighted in the lattice on the left maps to its cardinality on the scale on the right. Note how this mapping manifests Krifka's explanation of the homomorphic nature of a measure function in the introduction to this topic: the empirical fact that a group of two entities $a \cup b$ is composed of single entities a and b is preserved in the arithmetical relation of 2 being greater than 1.

3.1.2. Unsigned measures in language and the problem of complementation

We now have enough background to consider an example. Say we toss three coins at the same time and want to answer the (measure) question, "How many coins turned up heads?" An examination of Fig. 3.1 should uncover all of the information that is necessary to formulate a correct answer. If coin *a* turns up heads, the answer given by the measure function is "one"; if coins *a* and *b* turn up heads, the answer given by the measure function is "two"; and if all three coins turn up heads, the answer given by the measure function is "three".

More interesting is to consider how to account for the coins that do not turn up heads. Once again, imagine that coins *a* and *b* turn up heads. By working out the complement of $\{a, b\}$ – its union with $\{a, b\}$ produces **1**, or its intersection with $\{a, b\}$ produces **0** – we find $\{c\}$, so we could say "Two came up heads, and one did not. (It came up tails.)" Moreover, if all three coins turn up tails, the corresponding set is Ø, and its complement is **1**, so we must say ??"Zero coins came up heads" or "(All) three coins did not come up heads. (They came up tails.)" However, English and many other languages prefer a negative quantifier to these expressions: "No coins came up heads". If negation is the linguistic expression of complementation, how is it licensed for the putatively 'positive' description of this state of affairs symbolized by Ø?

It gets worse. Imagine that the coins were tossed into the air and then fell through the gaps in the grate of the floor furnace in my old house and could not be retrieved. "How many coins turned up heads?" Well, I don't know, but the lattice would appear to impose the null set on this outcome – after all, none of them turned up heads. Yet if you assert this as a measure by saying "So, none of them turned up heads", I can truthfully deny your assertion by saying "No, they didn't." One is forced to conclude that the null set describes two truth-conditionally different outcomes.

And herein lies to the crux of the matter. The complement is a derived object, the result of a computation performed on a lattice. Does this assumption capture one's intuitions accurately? If one sees three coins laying on the floor, two of which are heads and one of which is tails, does one need to mentally traverse a lattice to fund out how many of the coins are tails? We think not. It is more plausible to encode the complementary items directly, in their own area of the set-theoretic object. Moreover, once the complementary items are cordoned off in their own subset, Ø can be called upon to represent the outcome of the coin toss which is neither positive nor its complement, namely the one in which we do not know at all. This more complex object is more easily approached through

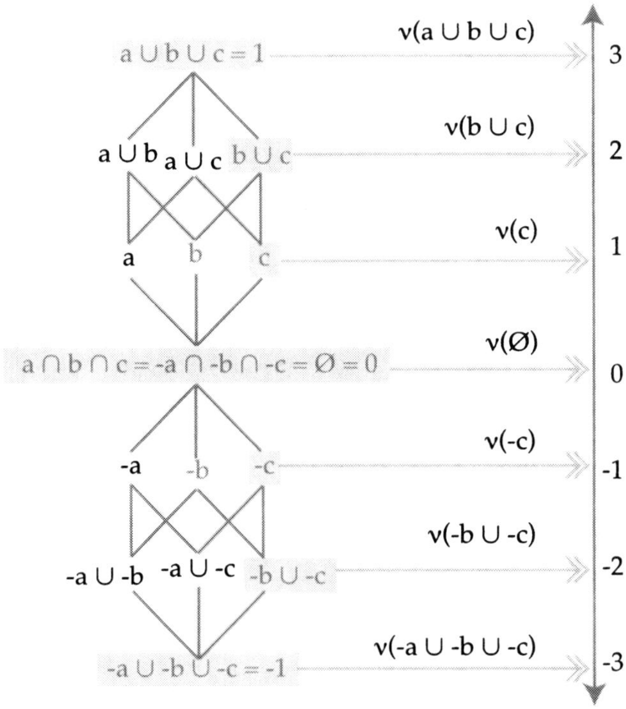

Figure 3.2. A signed measure mapped from a 'signed' lattice.

measure theory, as a signed measure. We will then work backwards to construct the algebra and lattice from which it maps.

3.1.3. Signed measures, signed algebras, and signed lattices

A signed measure adds the possibility of a measure being less than zero. As a first approximation, suppose that β is an algebra on a non-empty set ∑. A function $\nu : \beta \rightarrow \Re$ is called a **signed measure** on β if it has the properties of (3.4):

3.4 a) $\nu(\emptyset) = 0$.
b) Either one of the following is true:
(i) $\nu(\beta) < \infty$, for all $a \in \beta$;
(ii) $\nu(\beta) > -\infty$, for all $a \in \beta$.
c) If a and b are disjoint sets in β, $\nu(a \cup b) = \nu(a) + \nu(b)$.

The innovation with respect to (3.2) is the introduction of the negative half of the real number line -∞, along with the prohibition against mixing the two signs in (3.4b). A signed measure can be visualized by extending the scale of the right side of Fig. 3.1 down from 0 to negative numbers. The challenge is to draw the lattice from which it maps.

Fig. 3.2 offers the diagram that is most consistent with the desiderata sketched at the end of the previous subsection. The claim embodied in it is that the sets that map to the negative half of the measure replicate the sets that map to the positive half, with the only difference being that the former are prefixed with the negative sign. This claim is endowed with graphic form by duplicating the lattice and pivoting the duplicate's top vertex down to the bottom of the figure around the fixed point of **0**, so that the duplicate's **1** rotates from the top node in the lattice to the bottom one, **-1**. In this way, a **signed lattice** Λ is created.

The Boolean axioms apply to a signed lattice in much the same way that they apply to an unsigned one, so that β can be called a **signed Boolean algebra**, ⟨∑, ∪, ∩, **0**, **1**, **-1**⟩, where the set ∑ can consist of positive, +∑, and negative, -∑, elements. The members of ∑ are gathered together with respect to some property *P* by (3.5a). It is not clear that the complementation operator has any work left to do, since ∑ consists of signed elements which introduce complementation explicitly, as a primitive. Let us refer to this as **negation** and introduce an unary operation '–' on those members of ∑ that do not share *P*, as stated in (3.5b):

3.5 a) For all $a \in \sum$, $P(a)$ or $\neg P(a)$.
b) For all $a \in \sum$, $-a \in \{z \in \sum : z \notin P\}$.

The only other differences between β and an unsigned Boolean algebra *B* have to do with keeping track of the contrast in sign.

The most important difference is to enforce the separation of sign that (3.4b) accomplishes for a signed measure. While there are undoubtedly several ways to achieve this goal, perhaps the most general is to prevent the binary operations from applying to mixed elements. Adopting the notation that $\pm a \bullet \pm b$ is read as $a \bullet b$ or $-a \bullet -b$, the statement of (3.6) makes segregation of sign explicit:

3.6. For all $\pm a, \pm b \in \sum$, $\pm a \cup \pm b$, and $\pm a \cap \pm b$.

With the aid of (3.6), the axioms of Table 3.1 should generalize to a signed algebra and lattice in the intended manner.

A second question is how the introduction of sign extends to the 'top' and 'bottom' elements, especially given that their intuitive geometric standing no longer appears quite so intuitive. In view of the fact that the negative sublattice sports its own supremum **-1**, β should clearly include it, which is why it already does. With respect to the Boolean axioms, the '±' notation should be extended to **1** so that bounding works in the expected fashion. What is not so clear is the status of **0**. Fig. 3.2 unhesitatingly marks it as both positive and negative by sandwiching it between the two sublattices at the point where $\bigcap +\Sigma = \bigcap -\Sigma$.

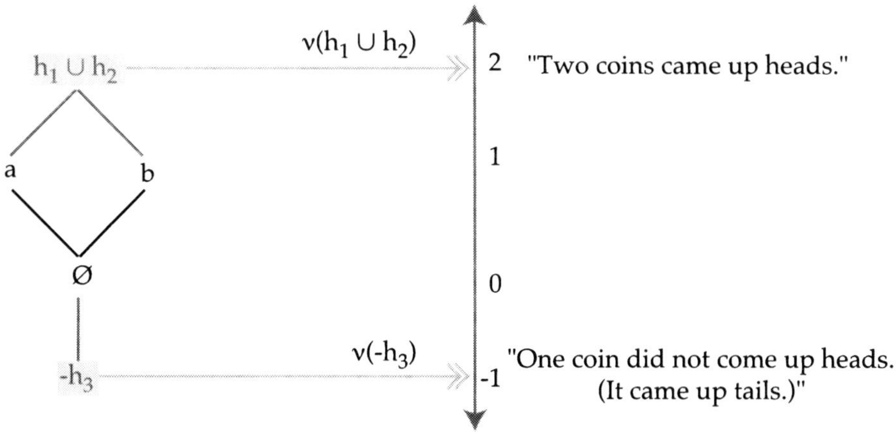

Figure 3.3. Signed lattice-measure mapping for the coin-toss example.

With respect to the measure of Λ, that 0 is both positive and negative is indeed the assumption, see Halmos, 1950, p. 121. Yet the Boolean axioms have been augmented specifically to prevent a measurable set from being both positive and negative, so we cannot adopt it quite so lightly. In fact, let us make a virtue out of this particular necessity and define the signed **0** to be the pathological case of equivalent countersigned sets:

3.7. $$0 = \grave{E} = \left(\bigcap +\Sigma\right) = \left(\bigcap -\Sigma\right)$$

Given that $+\sum$ and $-\sum$ are disjoint through their dependency on the property P, they can only become equivalent by being shorn of their members. This makes **0** useful for representing circumstances under which P is undefined.

By way of illustration of this welter of definitions, the coin toss example of two heads and one tails can be represented as $\sum = \{h_1, h_2, -h_3\}$. That is to say, $\sum$ contains two positive heads and one negative one. A signed lattice and measure function for this example are depicted in Fig. 3.3. The linguistic expressions on the right side demonstrate the fact that, despite the unitary nature of the lattice, the measure function obtains two results from it. These two results support two distinct linguistic objects, with and without negation. Fortunately, we can usually find evidence in the context of the linguistic utterance to prefer one measure to the exclusion of the other.

3.1.4. Response to those who do not believe in signs

Before moving on, there is one empirical objection and one theoretical objection to the notion of a signed algebra that should be dealt with, or the reader will resist taking it seriously.

On the empirical side, imagine that I want to know how many coins I have in my pocket. A simple way to find out is to stick my hand in my pocket and examine everything that I take out. If I do not find any coins, what measure do I use to communicate this outcome to you? It should not be $\nu(\emptyset)$, for there is no reason to think that this situation is undefined. Nor should it be a positive measure, because no coins were produced. By elimination, it should be a negative measure, but what could the measurable set be?

This is a weighty philosophical question, the only answer for which that we have space to entertain here is to assume that there is some contextually relevant set of existing or potential coins that could be in my pocket, but are not. The missing measure is the measure of the supremum of this negative set, whose only linguistic realization makes use of a logical quantifier: "There are no coins in my pocket", or "There are not any coins in my pocket".

Besides the symmetry of the overall system, the one argument in favor of this analysis that can be offered here is to engage in another thought experiment. Imagine that my wife has sewn my pockets shut, perhaps so that I could not slip any money into them to spend. Under these circumstances, there are no coins in my pocket because it is no longer a well-defined place to hold them. There is consequently no measurable set, with the result that the measure is correctly zero. Since this situation appears to be truth-functionally different from the case of the 'normal' pocket devoid of coins, it deserves a different analysis, one more along the lines of the absence of potential coins that was sketched above.

The other disturbing aspect of the notion of a signed algebra and lattice is that, since the introduction of lattice theory into linguistic semantics by Link (1983), no one else has proposed such a structure for any semantic phenomenon. What is more, most of the (unsigned) lattices that have been proposed lack the bottom element **0**. Landman, 1991, p. 302, in explaining the lattice-theoretic approach to plurality, formulates the most concise rationale for this choice that we know of:

> Our operation $\vee$ will be used for conjunctions at type *e* [*entity*]: *John and Bill* will be interpreted as $j \vee b$, the sum of *j* and *b*. There is no problem with this if we structure *e* as a join semilattice. However, if *e* is a full Boolean algebra, then $\vee$ is only one of the operations on *e*. We have the other Boolean connectives available as well. That leads us to expect that we could interpret *John or Mary* as $j \wedge m$ and *not Mary* as m'.
>
> But that leads to interpreting *John or Mary* as the zero element, and making it equivalent with *Mary or Sue*. Similarly, *not Mary*

> will be interpreted as 'everybody but Mary' (i.e. the sum of all atoms, except for Mary).

We agree with Landman when he goes on to say that *or* and *not* do not have these interpretations.

Beginning at the end, we have already proposed a solution for such pernicious complements by removing complementation from the Boolean algebra and reintroducing it as the negative elements of $\sum$. Under this interpretation, -*a* does indeed answer to a well-formed formula. As for the former objection, the analysis of coordination sketched in this chapter and developed more fully in the next two does not rely on the meets of individuals but rather the measures of certain measurable sets, so it will not lead to denotations for *or* in 0.

3.1.5. Bivalent vs. trivalent logic

The reader may have surmised that if we intend to use a signed measure as a foundation for logical-operator semantics, any inferences drawn using these objects must go beyond the confines of two-valued logic. The standard logical tradition which stems from Aristotle is called "two-valued" or "bivalent" because it only countenances two evaluations of a proposition, true or false. In a signed measure, these would correspond to the positive and negative halves of the scale, leaving 0 with no clear interpretation.

The possibility that a logical system could have evaluations beyond true and false is usually credited to Jan Lukasiewicz. In work done around 1920, Lukasiewicz agreed that a proposition should be true if it asserts a fact which is already determined and false if it asserts a fact which is determined negatively, that is, a fact whose negation is determined. Lukasiewicz's innovation was to argue that there is a third option, namely "undecided", for undetermined future states of affairs. This third option was subsequently developed in several directions.

One extremely useful tack for linguistics and computer programming resolves a problem with reasoning about terms which do not exist, such as "The king of France is bald". Since there is no king of France, we cannot say that the sentence is true, but it seem unfair to consider it false in the same way that, say, "The preceding sentence is in bold face" is false. It is much more intuitively satisfying to sweep it under the rug of 'undecided' and worry about its truth value if and when someone claims the throne of France.

Does this notion of undecidability characterize the zero degree of a signed measure? We believe that it does, but not without further elaboration. The elaboration comes from the work of Pieter Seuren and collaborators. By way of introduction, allow us a long quote from Seuren et al., 2001, p. 554-5:

> Suppose a quiz master asks the question: Which of these four was the youngest president ever of the United States: Reagan, Jefferson, Kennedy, or De Gaulle? The correct answer is, of

> course, 'Kennedy'. But of the three incorrect answers, one is somehow more incorrect than the other two. The answer *De Gaulle was the youngest president ever of the US* is somehow 'worse' than the answers that mention Reagan or Jefferson, because De Gaulle does not even fulfill the preliminary condition of having been president of the US. It is possible, or thinkable, to exploit this difference theoretically by distinguishing two kinds of satisfaction conditions, the PRECONDITIONS and the UPDATE CONDITIONS. ... Failure to satisfy the preconditions results in RADICAL FALSITY (F_2). Failure to satisfy the update conditions results in MINIMAL FALSITY (F_1). Satisfaction of all conditions results in TRUTH (T). The preconditions, moreover, determine the PRESUPPOSITIONS of the sentence in question. In this perspective, the sentence *De Gaulle was the youngest president ever of the US* presupposes that De Gaulle was president of the US. Since this presupposition is false, the sentence is radically false.

In the perspective of a signed measure, the positive scale represents truth, the negative scale represents minimal falsity, and 0 represents radical falsity.

If measure theory is indeed the right ontology for reasoning about linguistic expressions and in particular about sentences with logical operators, it should provide some motivation for the pathology of 0 which is described so eloquently by the notion of radical falsity. One candidate would be that at 0, a measure is both positive and negative – the interpretation argued for in the preceding subsections. Since no proposition can be both true and false, it follows that no proposition can take on the zero value, or perhaps it is more accurate to claim that some semantic pathology of the proposition forces it out of both the positive and negative spaces and into the quarantined space of 0.

The exact diagnosis of this semantic pathology thus takes on foundational importance. The proposal of this monograph relies on correlation, which, as will be explained below, is a signed measure. The exact statement of the conjecture is (3.8):

3.8. A logical operator is a measure of correlation between two semantic entities. A positive measure (or correlation) denotes TRUTH (T), a negative measure (or anticorrelation) denotes MINIMAL FALSITY (F_1), and a zero measure (or uncorrelation) denotes RADICAL FALSITY (F_2).

For the specific case of Seuren et al.'s youngest-president example, (3.8) works out as so: logical coordination holds between the four names and the predicate "be youngest U.S. president"; "Kennedy" is correlated with it, "Reagan" and "Jefferson" are anticorrelated with it, and "De Gaulle" is uncorrelated it.

The reason why we rely on correlation is broached in the next subsection.

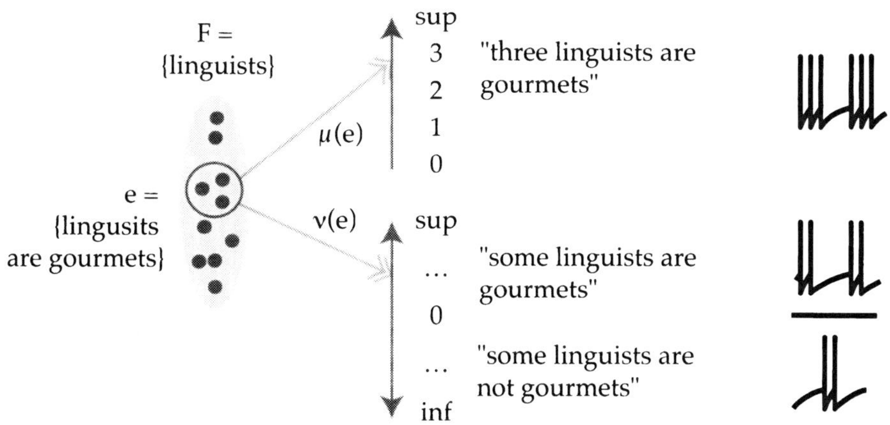

Figure 3.4. Unsigned vs. signed measures of linguists who are gourmets.

3.1.6. An interim summary to introduce the notion of spiking measures

Let Ω be a set of linguists and F be an algebra on it, with members e. Let $\mu(F)$ and $\nu(F)$ be measures of linguists, in a way that we introduce by means of Fig. 3.4. The measure $\mu(F)$ at the top center of Fig. 3.4 produces a scale from 0 to some supremum. $\mu(e)$ picks out the measure "three (linguists are gourmets)" on this scale, which is the expected linguistic instantiation of the unsigned measure $\mu(e) = 3$.

With this uncontroversial rendition as background, consider the corresponding signed measure $\nu(F)$ in the bottom center of Fig. 3.4 – namely a scale centered at 0 with a supremum and an infimum. $\nu(e)$ picks out the point labeled "some (linguists are gourmets)", which is our proposed quantificational instantiation of the signed measure $\nu(e) = 3$ in this context. As further illustration, the quantificational instantiation of the negation of this measure is appended in its place on the negative half of the scale.

The reader may have noticed that this exposition does not exhaust the graphical content of Fig. 3.4 – there still remains a series of squiggles across the right side to be accounted for. These squiggles look not unlike the spike trains that were introduced in the previous chapter, and this is indeed what they are meant to depict. The unsigned measure is associated with a single spike train, which is taken to be the product of the population of neurons that encodes the brain's representation of numerical quantity. Presumably each unsigned measure of quantity is slightly different from the others.

More pertinent to our own interests are the spike trains associated with the signed measure. Both signs are marked by different trains, in fact, by trains that are anticorrelated with each other. That is to say, where there are spikes in the positive train, there are none in the negative, and vice versa. Anticorrelation appears to be the neurological implementation – or the motivation – for the axioms that prohibit the two signs of signed algebras and measures from intermixing. We assume that such temporal anticorrelation arises from lateral inhibition, whereby the neuronal population that encodes the positive measure inhibits the population that encodes the negative measure, thus forcing any active neurons in the latter to be pushed into the gaps in the spike train of the former.

Fig. 3.4 also makes allowance for a 'flat' neuronal response associated with zero on the signed scale. Such a 'spikeless train' could result from the absence of input to the neuronal ensemble, or by symmetrical responses from the two signs canceling each other out.

A final observation attendant on Fig. 3.4 is that the measurable set e appears to be drawn from two sources. One is the set F of linguists, which for our purposes is more convenient to call X; the other is a set of gourmets, Y. As will become apparent in the upcoming analysis, there is an asymmetry between these two sets. The negation of X is never invoked, while the negation of Y can always be invoked. Moreover, a legitimate Y is constrained to having some functional relation based on X. For instance, for the logical quantifiers, Y must be a property of X, such as being the predicate of which F is the subject. In the previous examples, X has been the property of coming up heads or tails and the property of being or not the youngest President of the US.

Since there are a variety of functional relationships that can hold between linguistic entities – the next chapter reviews many of them – let us reserve a noncommittal notational device to signal the peculiar constitution of Y: $\boldsymbol{Y|X}$. This is read as "the conditional relation of Y given X". Whatever relationship between Y and X that is required by a particular operator has to satisfy the measure of (3.9):

3.9. $$\nu(X \cap Y|X) \in [-1\ 1].$$

This measure will be some variation on correlation.

The reader familiar with probability theory will recognize the analogy to conditional probability in $Y|X$, which is treated in more detail in Sec. 3.2.3.2 below. The reader familiar with generalized quantifier theory may be reminded by the expression $X \cap Y|X$ of Conservativity, a notion that is treated in more detail in Chapter 6. Thus (3.9) implicitly establishes an association between conditional probability and Conservativity, a proposal that we will have a chance to elaborate on at several points in the upcoming discussion.

We now have seen enough of measure theory to delve into the examples that interest us most in this monograph.

Table 3.2. A broad sample of logical operators.

POS	*MAX*	*MIN*	*NEG*
some/a	all	none/no	not all
something	everything	nothing	–
somebody	everybody	nobody	–
one (of them)	both (of them)	neither (of them)	–
(either) α or ß	(both) α and ß	(neither) α nor ß	–
NP too	only NP	no N′	–
somewhere	everywhere	nowhere	–
sometimes	always	never	–
already	still	not yet	no longer
continue	start	stop	–
possibility	necessity	impossibility	–
possible	certain	impossible	–
possibly	certainly	in no case/way	–
may	must	must not/may not	–
can/may	must/need	cannot	–
permission	command	prohibition	–
permit/allow	require	forbid/bar	–
let NP VP	make NP VP	keep NP from VP	–
think possible	believe	rule out	doubt
accept	claim	refuse	renounce
compatible	implies	contrary	–
satisfiable	tautological	contradictory	disputable
right	duty	–	–

3.1.7. The logical operators as measures

A diverse series of lexemes has been claimed to be logical operators. Table 3.2 lists those mentioned in Löbner, 1987, pp. 58-60, and Horn, 1989, Section 4.2, which draws heavily on Jesperson, 1917, Chapter 8, and 1924, pp. 324-5. It is organized so that each sort of operation occupies a row. In this monograph, we are principally concerned with the rows headed by *some/a* and *(either) α or ß*. However, let us dedicate a few words to the overall layout of the table.

As the columns demonstrate, there is a repeating pattern of meaning for each sort of operator. Löbner (1986), extending the work of Horn (1976), reduces this pattern to the scale of resistance to lexicalizability set forth in the terms of Table 3.2 as in (3.10):

3.10. POS < MAX < MIN < NEG

Löbner's scale is based on two pieces of evidence. One is that no language known to him lacks an existential quantifier, but some languages lack the other three. Japanese and Chinese, for example, "...use complex expressions in most cases of universal quantification." (p.62) The other observation is that morphemes for MAX and MIN are often derived from a basic existential root.

The Indo-European languages are held to be exceptional in having simple words for MIN, such as English *no, never, none, neither, nothing*, etc.

There is consequently a great deal of asymmetry in the lexicalization of logical operators, a conclusion which Löbner, 1987, pp. 65 summarizes in the following conjecture:

> Natural language quantifiers can be classified into four types. Type 1 [**POS** in Table 3.2] contains all existential quantifiers (maybe among others), type 2 [**MAX**] contains all universal quantifiers, type 3 [**MIN**] all negated existential quantifiers, and type 4 [**NEG**] all negated universal quantifiers. The type assignment is unique. Natural language exhibits significant differences with respect to the extent of the four subclasses and to the average complexity of the expressions used in the four subclasses. The number of lexical items decreases, and the complexity of the expressions increases from type 1 through type 4 with each step.

One sub-goal of this monograph is to reduce this observation to independent properties of the logical operators. As a first step in doing so, let us take up the various candidates for the representational ontology of these elements.

3.2. LOGICAL-OPERATOR MEASURES

There are five main sources of measures for the logical operators, set-theoretic cardinality, statistics, probability, algebraic topology, and information theory. They are not unrelated, though we shall initially take them up as if they were, and then bring them together in a final synthesis.

3.2.1. Conditional cardinality

The simplest method of measuring something is to count its units, which produces the counting or cardinality measure. What could the units of the logical operators be? We maintain that they are the members of the intersection of the two sets X and $Y|X$, as anticipated with respect to Fig. 3.4. The cardinality of this intersection is simply $\#(X \cap Y|X)$, which can be called its **conditional cardinality**. Conditional cardinality provides us with a measure for the logical operators. In accord with the preceding section, there are two potential versions, an unsigned measure $\#_2$, from which inferences are drawn by means of bivalent logic, and a signed measure $\#_3$, from which inferences are drawn by means of trivalent logic.

They are contrasted through the medium of Venn diagrams in Fig. 3.5. $Y|X$ is split into its component polarities in accord with the desideratum that the two signs never intermingle. The unsigned measures across the top half of Fig. 3.5. are arranged with Y above and its complement directly below in order to indicate that the two have the same measure regardless of which one

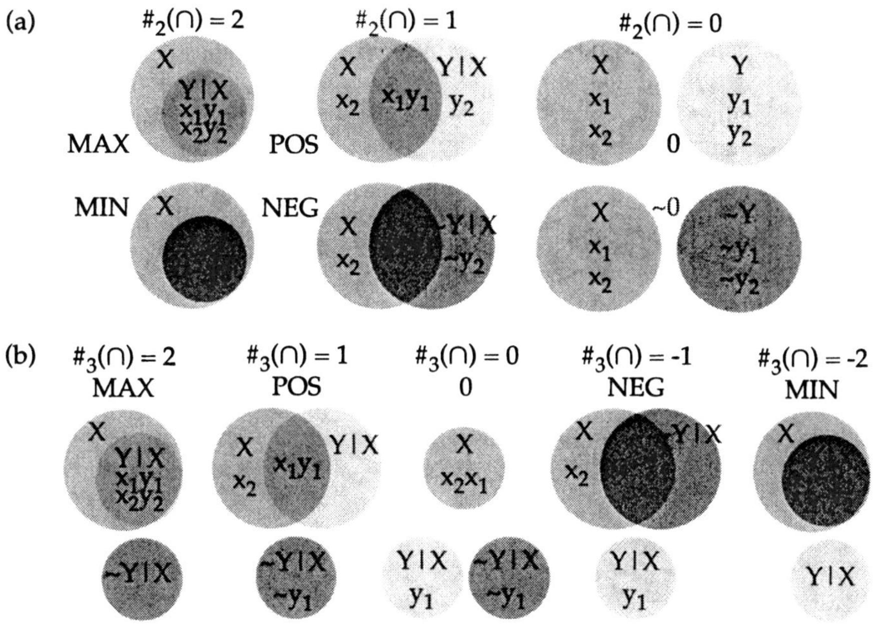

Figure 3.5. Venn diagrams of $\#(X \cap Y \mid X)$ for $\#(X) = 2$.

conditional cardinality embraces. This prevents the unsigned measure from distinguishing positive from negative operators of the same cardinality. As for the signed measures across the bottom half of the figure, they do not incur this ambiguity but rather distinguish complementary *Y* sets quite accurately. Thus the signed measures supply an unambiguous signal for distinguishing positive from negative operators.

This descriptive accuracy points the way to visualizing the space of values accepted by a logical operator as a signed mapping from the intersection of *X* and $Y \mid X$ to their conditional cardinality. Fig. 3.6 introduces two ways of drawing this mapping. The cube on the left displays the full mapping. The *xy* plane represents each $x \in X$ and each $y \in Y \mid X$ by a number $\langle x, y \rangle$, whose location on the scales is marked by a circle. It thus represents all possible intersections of *X* and $Y \mid X$ in the given range. The *x* number measures the cardinality of *X*, for $\#(X) \in [2\ 7]$. The *y* number orders each *y* from 1 to $sup_3(Y \mid X)$. The remaining axis, that of *z*, lays out $\#_3(X \cap Y \mid X)$ for each intersection.

From the visual perspective of Fig. 3.6, the cube is a challenge to understand. It is not at all clear that each plus sign marking a conditional cardinality is aligned correctly above or below the corresponding intersection of *X* and $Y \mid X$.

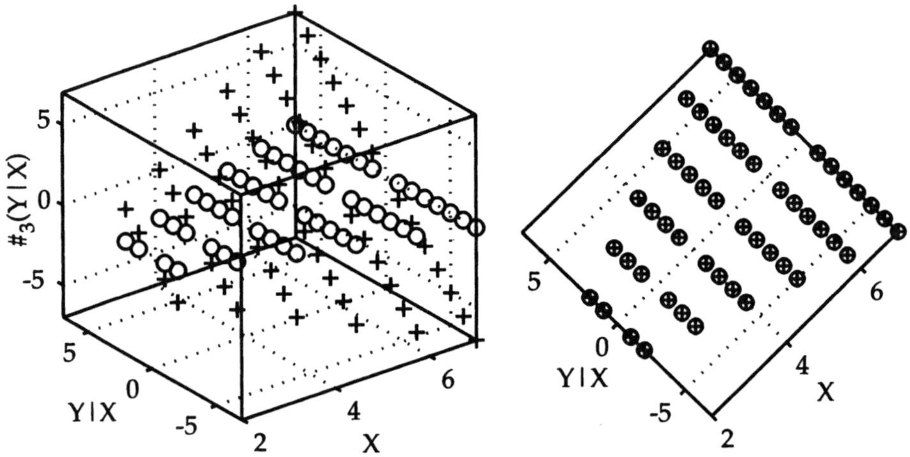

Figure 3.6. v(X ∩ Y | X). o = X ∩ Y | X; + = v. **(p3.01_condcard_cube.m)**

The rhombus on the right alleviates this problem by viewing the cube from directly above, so that the measures are seen to stand on top of the circles representing their measurable set.

The fact that the rhombus is still perspicuous despite the visual neutralization of the z axis suggests that there is considerable redundancy in the cube. Upon closer inspection, it becomes apparent that $\#_3(X \cap Y \mid X)$ is redundant with the numerical format of $Y \mid X$. For instance, if $X \cap Y \mid X$ is <6, -3>, its conditional cardinality is -3. It stands to reason that great gains in notational economy can be made by letting $\#_3(Y \mid X)$ stand in for $\#_3(X \cap Y \mid X)$.

Fig. 3.7 demonstrates the superiority of this proposal by marking the sequences of the rhombus accepted by a given logical operator with plus signs and the sequences rejected with circles. In this format, MAX, for instance, accepts the top diagonal of cardinalities. The graph is read in the following way: start at some cardinality of X, say four, and scan up the column to find a plus sign. Then look across to $\#_3(Y \mid X)$, which tells you that it must be four, too. The other logical operators are laid out in the other sub-diagrams of Fig. 3.7. For POS, starting at $\#_3(X) = 4$ leads to a choice among any of the four cardinalities {1, 2, 3, 4}. NEG covers any intersection of ~*Y* with *X*, so it extends down from zero to cover the complement {-1, -2, -3, -4}. Finally, MIN requires that every x_i find a conditional cardinality with ~*Y*, so it extends out the bottom diagonal.

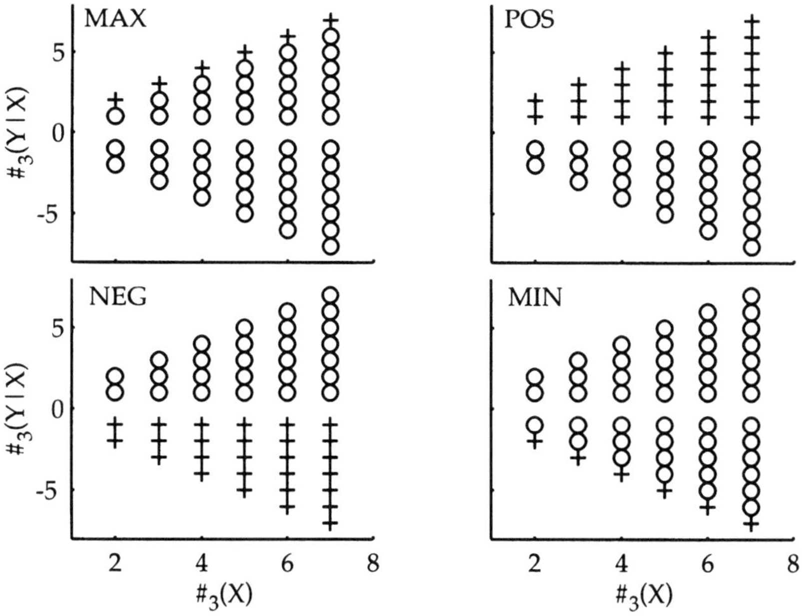

Figure 3.7. Signed conditional cardinalities of the four logical operators. (**p3.02_condcard_LOGOP.m**)

3.2.1.1. Cardinality invariance

The preceding discussion argues that signed conditional cardinality produces the proper ontology of logical-operator meanings. Nevertheless, the infinite extension of Fig. 3.7 off the right edge of the page suggests that we have not yet found the most compact representations of these meanings. In the terms that were found useful for describing early vision in Chapter 1, we may speculate that the conditional cardinality representation of Fig. 3.7 contains redundant information about logical patterns that is stripped out by linguistic processing. Exactly what the redundant information is, is not immediately clear, but Fig. 3.7 provides several clues. The most obvious is that the logical operators appear to be invariant to certain numerical relations. MAX accepts those cardinalities that are invariant for positive equivalence, $\#_3(X) = \#_3(Y|X)$, while MIN accepts those cardinalities that are invariant for the complementary equivalence, $\#_3(X) = \#_3(\sim Y|X)$. Likewise, POS accepts those cardinalities that are invariant for positive sign, $\#_3(Y|X) > 0$, while NEG accepts those cardinalities that are invariant for negative sign, $\#_3(Y|X) < 0$. These four

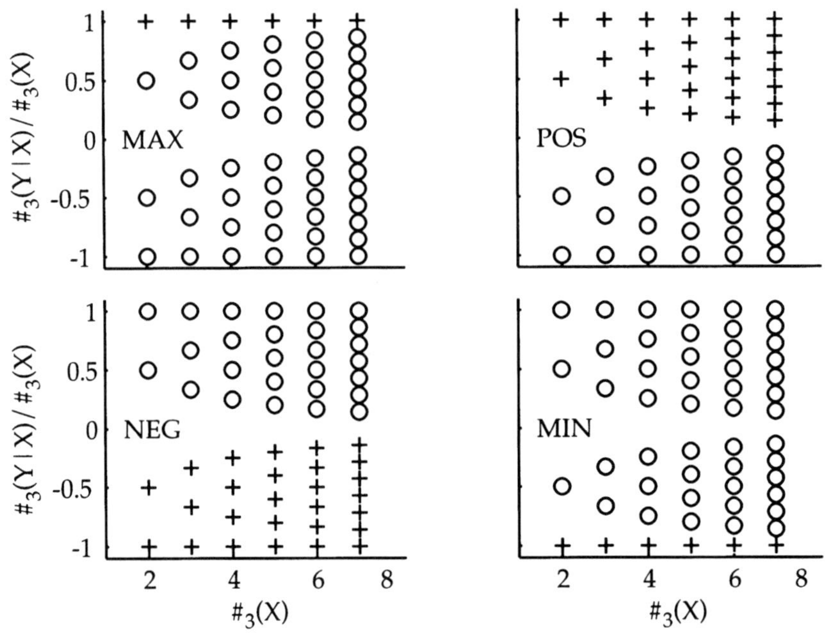

Figure 3.8. Normalized signed conditional cardinalities of the four logical operators. (**p3.03_norm_condcardLOGOP.m**)

relations have the effect of making the logical operators invariant to a specific cardinality.

Of course, if these statements are offered as the definitions of the corresponding operators, we still have to count up to $\#_3(Y \mid X)$ before deciding whether the operation is an instance of MAX or MIN – and thereby run afoul of the hundred-step limit in many cases. This conundrum is unavoidable for a cardinality measure, which is the reason why the upcoming sections explore alternatives. In the meantime, let us try to reduce Fig. 3.7 to a more economical format.

The simplest way to make $\#_3(Y \mid X)$ less dependent on the specific cardinality of a measure is to divide it by $\#_3(X)$. Performing this division on the sample values of Fig. 3.7 produces the graph of Fig. 3.8. The seven values for MAX and MIN have each now been distilled down to a single value, 1 and –1, respectively. This pares down the number of outputs that conditional cardinality must compute, and effectively squashes its range to the interval [-1 1]. The values for POS and NEG have also been reduced to less variant measures, though there are many more of them. The result is the extraction of a more parsimonious representation for the logical operators. Since dividing by some function of *X* will be met in many guises in the upcoming paragraphs, it will be helpful to have a name for it. **Normalization** is one that attains the requisite level of generality.

The fact that conditional cardinality normalization still permits the logical operators to be distinguished accurately suggests that natural language ignores the 'raw' cardinality of a logical operation in favor of this more invariant measure. The richer mathematical frameworks examined in upcoming sections will enable distillation of additional invariant properties.

3.2.2. Statistics

Within statistics, the patterns just uncovered for the logical operators have a precisely defined meaning as linear correlation between two quantitative variables. In other words, the two dimensions of Fig. 3.7 and 3.8 correspond to the two variables required by bivariate statistical correlation, and the linear nature of the patterns found in this space is required by the linearity of statistical correlation. Thus it behooves us to examine statistical correlation as a source of methods for understanding the patterns traced by the logical operators.

The is also a compelling neurological reason for considering correlation as a method for understanding logical operatorhood, namely the recently discovered ability of the brain to detect temporal correlations. Recalling the brief discussion of Sec. 2.3.2.1, a temporal correlation is understood as the coincidence of inputs in time. This ability opens the door to a radically new way of looking at the function performed by the logical operators, already anticipated in Fig. 3.4, which is to see them as detectors of coincidence between their input and presumed output, in the following way.

Imagine that each set *X*, *Y*, and ~*Y* is represented by a single neuron that emits one spike per singleton element of the set within some temporal window. For instance, if $\#_3(X) = 2$, the *X* neuron emits two spikes within the temporal window. To represent MAX, the *Y* neuron must also emit two spikes within the same window. Moreover, to represent intersection, the *Y* spike train must overlap with the *X* train in some fashion. Our assumption is that they overlap in time, which is to say that their spike trains have the same temporal phase – spiking in concert, or else not spiking at all. This assumption is given pictorial form under the MAX heading of Fig. 3.9. POS is under the same constraint, except that not every *X* spike must coincide with a *Y* spike. NEG evidently assimilates to POS in not requiring that every spike correspond to *X* spike, with the difference that the correspondence is to the ~*Y* neuron. And here a fascinating fact asserts itself. Given that ~*Y* is the set-theoretic complement of *Y*, the equivalent in dynamical systems theory should be that *Y* and ~*Y* cannot overlap temporally. That is to say, ~*Y* must be out of phase with *Y*. Since we have adopted the assumption that *Y* itself must be in phase with *X*, it follows that ~*Y* must be out of phase with *X*. Fig. 3.9 represents the out-of-phase spot in

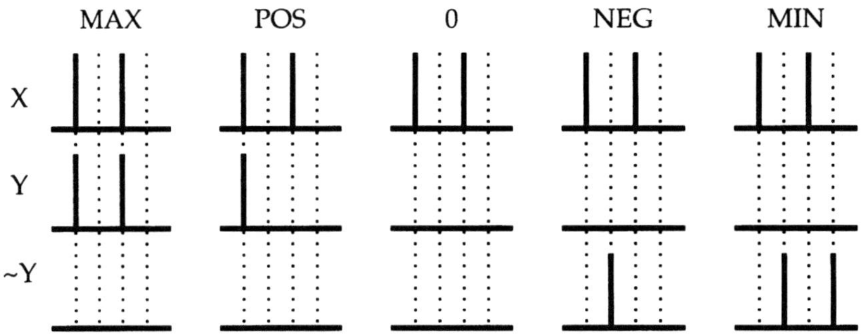

Figure 3.9. The logical operators as phasal coincidence of spiking neurons. Each horizontal bar represents a spike, and the dotted lines represent the phases of the *X* spike train.

a cycle by the dotted line that immediately follows an *X* spike. NEG is characterized by at least one out-of-phase ~*Y* spike, while MIN is characterized by an out-of-phase ~*Y* spike for every *X* spike. Finally, there is the logical possibility of no *Y* spike for any *X* spike, which produces the 0 operation in the center of Fig. 3.9.

In order to substantiate this hypothesis, let us first review the concept of correlation as defined in statistics, in order to gain a solid base for understanding the logical operators as correlation detectors.

3.2.2.1. Initial concepts: mean, deviation, variance

Consider a collection of observations of some feature x. The **mean** or **centroid** $\bar{x}$ of x is the average value of the occurrences of x in the sample. In summation notation, the format for its calculation is given in Eq. 3.11:

3.11. $$\bar{x} = \frac{\sum_{i=1}^{n} x_i}{n} \text{ or } \frac{1}{n}\sum_{i=1}^{n} x_i$$

The version on the left, with the summation operation in the numerator of the fraction, is difficult to read, if not confusing, so the alternative on the right is the one most often encountered.

Given that the samples of feature x collected rarely have the same value, it is convenient to have some way of measuring the spread or dispersion of their distribution. Having computed the mean for the samples of x, we can use it as a point of reference to measure the **deviation** of a sample x_i from $\bar{x}$. Such deviation can be measured simply by subtracting the mean from x_i:

3.12. $$\mathrm{dev}(x_i) = (x_i - \bar{x})$$

Knowing the deviation of each x_i, the **variance** of x should be its average deviation, that is, the sum of the deviations divided by their number. However, just summing up the deviations produces zero, because the positive and negative ones cancel each other out. Thus statisticians hit upon the procedure of summing the *square* of the deviations and then dividing by their number to find the variation in x:

3.13. $$\mathrm{var}(x) = \frac{1}{n}\sum_{i=1}^{n}(x_i - \bar{x})^2$$

The squaring of the deviations expands the variation greatly, and it also introduces a squaring of the units of measurement not found in the original data. The (estimated) standard deviation undoes the squaring by taking the square root of the variance. There are two versions:

3.14 a) $$\sigma(x) = \sqrt{\frac{1}{n}\sum_{i=1}^{n}(x_i - \bar{x})^2}$$

b) $$s(x) = \sqrt{\frac{1}{n-1}\sum_{i=1}^{n}(x_i - \bar{x})^2}$$

The difference between the two lies in a single parameter, namely whether the squared sum is divided by n or $n - 1$. The former is the true standard deviation, but is only appropriate when there are many observations. Otherwise, the estimated standard deviation of Eq. 3.14b is used.

Finally, a standard or z format for raw data is found by dividing the deviation by the standard deviation:

3.15. $$z(x_i) = \frac{x_i - \bar{x}}{s(x)}$$

These measurements are combined to formulate the statistics for correlation that interest us.

However, let us first summarize what has been said by applying it to the logical operators. Table 3.3 draws four sample logical values and classifies them as the logical operators standing at the head of each column. The mean and standard deviation for each operator are given at the foot of each column.

Table 3.3. Statistical analysis of sample logical operators (**p3.04_statlogop.m**)

	MAX	*POS*	*POS2*	*POS3*	*NEG*	*MIN*
$x_1\ y_1$	2, 2	3, 1	4, 1	2, 1	3, -1	2, -2
$x_2\ y_2$	3, 3	4, 2	4, 2	3, 1	4, -2	3, -3
$x_3\ y_3$	4, 4	5, 3	4, 3	4, 1	5, -3	4, -4
$x_4\ y_4$	5, 5	6, 4	4, 4	5, 1	6, -4	5, -5
mean(x, y)	3.5, 3.5	4.5, 2.5	4, 2.5	3.5, 1	4.5, -2.5	3.5, -3.5
s(x, y)	1.3, 1.3	1.3, 1.3	0, 1.3	1.3, 0	1.3, 1.3	1.3, 1.3
c(x, y)	1.67	1.67	0	0	-1.67	-1.67
r(x, y)	1	1	–	–	-1	-1
ρ(x, y)	1	1	0.5	0.5	-1	-1

In anticipation of the next section, the mean can be conceptualized geometrically as the center of the sample subspace, while each deviation can be conceptualized as the distance of the sample from its mean. This permits us to visualize the data sets and their statistics by means of Fig. 3.10. The samples for each operator are indicated by filled-in circles, and the mean is marked by an asterisk. The samples classified by an operator are grouping into the darkened ovals, whose boundaries are defined by deviation, though the deviations themselves are not indicated in any overt way.

This definition of the logical operators as deviations from the mean produces an organization of the samples that anticipates "clustering and associative pattern classification" discussed at length in Chapter 5, with the association holding between a sample and its operator label. However, there is also an evident linear patterning to the samples, so let us go on to consider statistical measures of linear bivariate relationships.

3.2.2.2. Covariance and correlation

The **covariance** or **dispersion** of two features measures their tendency to vary together, i.e., to co-vary. Mathematically, it is the average of the products of the deviations of feature values from their means:

3.16. $$c(x, j) = \frac{1}{n-1} \sum_{i,j=1}^{n} (x_i - \bar{x}_i)(x_j - \bar{x}_j)$$

Correlation imposes a standard range between 1 and –1 on covariance, by dividing each deviation by its standard deviation, which converts each multiplicand to its standard or *z* score:

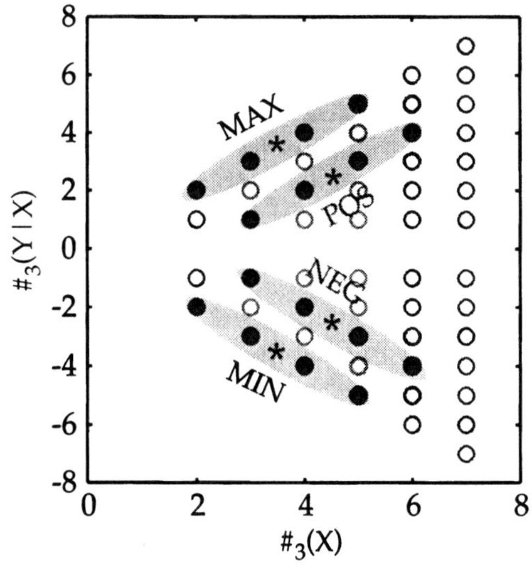

Figure 3.10. Sample logical operations from Table 3.3; * = mean and shading ≈ deviation.

3.17. $$r(x, y) = \frac{1}{n-1} \sum_{i,j=1}^{n} \left(\frac{x_i - \bar{x}_i}{s(x_i)}\right)\left(\frac{x_j - \bar{x}_j}{s(x_j)}\right)$$

This formulation is known as the **(Pearson product-moment) correlation coefficient**. Its values are constrained to lie between –1 and +1. A zero value indicates that the two variables are independent. A nonzero value indicates some dependence between them. A value of +1 indicates a perfect linear relationship with positive slope, and a value of –1 indicates a perfect linear relationship with negative slope. In other words, the sign (+ or -) of the correlation affects its interpretation. When the correlation is positive ($r > 0$), as the value of x increases, so does the value of y. Conversely, when the correlation is negative ($r < 0$), as the value of x decreases, the value of y increases. In accord with standard practice, the former is referred to as **correlation**, and the latter, **anticorrelation**.

Table 3.3 lists the covariance and correlation coefficients for the data samples. The two tests show four of the six samples to be strongly associated, with the first two samples being perfectly correlated and the last two being perfectly anticorrelated. The middle two, however, test out as being uncorrelated, even though they are instances of POS and therefore should show some degree of correlation.

This result brings up a flaw in using Pearson's r on these data sets, namely the fact that it requires that the data be normally distributed and not have any ties. A tie in this context means that no values of either variable should be duplicated. Sample POS2 contains duplicates the first variable, while sample POS3 contains duplicates of the second. In fact, these are the most pernicious cases, since they duplicate *all* instances of the variables in question.

Well aware of this problem, statisticians have devised various means of calculating correlation in the face of non-normally distributed and tied data. Two general approaches go by the names of Spearman's rho and Kendall's tau. The former works best for our data and is briefly explained here.

Following the exposition of Gibbons, 1993, p. 4ff, **Spearman's ρ (rho)** or **rank correlation coefficient** measures the strength of association between two variables by assigning a rank to each observation in each variable separately. That is to say, the first step is to rank the *x* elements in the paired sample data from 1 to *n* and independently rank the *y* elements from 1 to *n*, giving rank 1 to the smallest and rank *n* to the largest in each case, while keeping the original pairs intact. Then a difference *d* is calculated for each pair as the difference between the ranks of the corresponding *x* and *y* variables. The test statistic is defined as a function of the sum of squares of these differences *d*. The easiest expression for calculation is:

3.18. $$\rho(x, y) = 1 - 6\sum_{i=1}^{n} \frac{d_i^2}{n^3 - n}$$

The rationale for this descriptive measure of association is as follows. Suppose that the pairs are arranged so that the *x* elements are in an ordered array from smallest to largest and therefore the corresponding *x* ranks are in the natural order as 1, 2, ..., *n*. If the ranks of the *y* elements are in the same natural order, each $d_i^2 = 0$ and we have $\sum d_i^2 = 0$. Substitution into Eq. 3.18 shows that the value of ρ is +1. Therefore $\rho = 1$ describes perfect agreement between the *x* and *y* ranks, or a perfect direct or positive relationship between ranks. This is the case of *MAX* and *POS*. On the other hand, suppose that the ranks of the *y* elements are the complete reverse of the ranks of the *x* elements so that the rank pairs are (1, *n*), (2, *n*–1), (3, *n*–2), ..., (*n*, 1). Then it can be shown that $\sum d_i^2 = (n^3 - n)/3$. Substitution of this value in Eq. 3.18 ultimately gives a value of ρ as –1, which describes a perfect indirect or negative relationship between ranks. This can be called perfect disagreement. This is the case of NEG and MIN. Both agreement and disagreement are special kinds of associations between two variables.

However, the line in Table 3.3 for Spearman's rho does not contain exactly this calculation, for there still remains the problem of the ties of POS2 and POS3. These are resolved by assigning midranks to the ties, though the reader is referred to Gibbons' text for the mathematical details. Table 3.3 displays the results of augmenting Eq. 3.18 with this remedial mechanism. POS2 and POS3 now correctly show a partial correlation.

3.2.2.3. Summary

This section reviews the most popular tools for measuring statistical correlation. The results are somewhat disappointing for the analysis of the logical operators. While statistics supplies quite precise methods for calculating bivariate correlation, and the logical operators do indeed test out as expressing some degree of correlation, it is not the degree that we would expect. The main problem is that POS in Table 3.3 receives a measure equal to that of MAX, whereas we would expect it to have a measure less than MAX, such as that of POS2 and POS3. The same holds true of NEG and MIN, but in the negative direction of anticorrelation. Thus the Pearson and Spearman correlation coefficients do not even attain descriptive adequacy.

On the one hand, this negative result from the more complex statistics may not be entirely unexpected, since it is not at all clear how the Spearman correlation coefficient would be calculated neurologically, or even whether this is the kind of calculation that we would expect people to be biologically predisposed to performing in order to learn language. On the other hand, some of the less complex statistics do provide a first approximation to a classification of the logical operators, as was illustrated in Fig. 3.10 for the mean and variation. We turn to these less complex calculations, and especially their geometric interpretation as embodied by Fig. 3.13, after first examining probabilistic measures for the logical operators.

3.2.3. Probability

Perhaps the most well-known measure after counting is probability. As luck would have it, there already is an analysis of the logical quantifiers as a probabilistic measure in the work of Mike Oaksford and Nick Chater. This section provides just enough of an introduction to probability theory in order to extend Oaksford and Chater's proposals to our more general framework.

3.2.3.1. Unconditional probability

Probability is the branch of mathematics that deals with the calculation of the likelihood of the occurrence of an event. The probability of an event e, $P(e)$, is expressed as a number between 0 and 1. An event with a probability of 0 is considered an impossibility, while an event with a probability of 1 is considered a certainty. An event with a probability of 0.5 can be considered to have equal odds of occurring or not occurring. The canonical example is the toss of a fair coin resulting in "heads", which has a probability 0.5 because the toss is just as likely to result in "tails".

Formally, P is known as a **probability measure function**, which is a function satisfying the axioms of (3.19), which have the effect of assigning a real number to each event of a random experiment, see among many others Pfeiffer, 1995, pp. 2-3. This assignment takes two assumptions for granted. The first is that is there is a **sample space** S which constitutes the collection of all possible outcomes of the experiment. For example, if the experiment is tossing a coin, S = {heads, tails}. The second is that the **event space** E is a subset of S. For example, if the coin is tossed once and comes up tails, E = {tails}. Thus a **probability system** is a triple $\langle S, E, P\rangle$.

With this background, the axioms that govern P are stated as in (3.19):

3.19 a) $P(e) \geq 0$, for any event e in E.

b) $P(S) = 1$.

c) $P(\bigcup_{i=1}^{n} E_i) = \sum_{i=1}^{n} P(E_i)$, where E is a countable set of n disjoint events, $e_1, e_2, \ldots, e_n$

If the reader recalls the axiomatic definition of an unsigned measure in (3.2), then it should be clear that (3.19) is simply an unsigned measure instantiated by the particular conceptual requirements of probability. Axioms (3.19a) and (3.19b) are a matter of convention: it is convenient to measure the probability of an event with a number between 0 and 1, as opposed to, say, a number between 0 and 100. The notation of axiom (3.19b) is not quite that transparent, however. What it says is that the probability of the sample space is 1, which makes sense intuitively because one of the outcomes of S must occur. This axiom is often understood by taking the expression "one of the outcomes of S must occur" to be analogous to a proposition that is logically valid. Since such a proposition is true no matter what, we give it the highest measure of probability, namely 1 or certainty. In contrast to the other two, axiom (3.19c) or countable additivity, is fundamental, as was noted at the beginning of the chapter. It says that the probability of a collection of mutually exclusive events – represented as their union – is the sum of the probability of each individual event. The overall effect of the three axioms is to make the probability measure function P map events in E onto the interval [0 1].

Intuitively, the probability of an event should measure the long-term relative frequency of the event. Specifically, suppose that e is an event in an experiment that is run repeatedly. Let $\#_n(e)$ denote the number of times e occurred in the first n runs so that $\#_n(e)/n$ is the relative frequency of e in the first n runs. If the experiment has been modeled correctly, we expect that the relative frequency of e should converge to the probability of e as n increases. The formalization of this thought experiment is known as the Law of Large Numbers. While it is too peripheral to our concerns to be reproduced and explained here, it is pertinent to point out that the precise statement of the law uses the statistical measures of mean and variance (of the event space) that were introduced in Sec. 3.2. In this way, a connection is established between statistics and probability.

We now have enough preparation to express probability in its simplest mathematical form, as the number of occurrences of an event *e* divided by the total number of events in the experiment, *E*. *E* is conventionally given as the number of occurrences of *e*, plus the number of times that *e* fails to occur, *~e*:

3.20. $$P(e) = \#(e)/(\#(e) + \#(\sim e))$$

(3.20) expresses the idea that the likelihood of occurrence of an event does not depend on whether some other event occurs (or has occurred). It effectively normalizes all probabilities so that they will fit into the interval [0 1] stipulated by the axiomatic definition of (3.19).

3.2.3.2. Conditional probability and the logical quantifiers

As the reader may guess, there is also a counterpoised conditional probability, which expresses the idea that the likelihood of occurrence of an event *does* depend on whether some other event occurs (or has occurred). Conditional probability is the sort of probability in which the meaning of the logical operators can be stated. It is often understood to mean that the probability of an event is the probability of the event revised when there is additional information about the outcome of a random experiment. For instance, whether it rains (event *b*) might be conditional, i.e. dependent, on whether the dewpoint exceeds a certain threshold (event *a*), which is to say that the probability of rain may be revised when the dewpoint is ascertained.

The conditional probability of event *b* given event *a* is labeled **P(b | a)**. It is found by Eq. 3.21, provided $P(a) > 0$:

3.21. $$P(b \mid a) = P(b \cap a)/P(a)$$

This equation is derived from (3.20). If we know that *a* has occurred, then the event space *E* in the denominator of (3.20) is reduced from the entire space to the probability of *a*. Moreover, the probability of *a* must also be included in the numerator, since it has occurred, which is achieved via intersection with *b*. The result is equation (3.21).

Chater and Oaksford (1999) and Oaksford, Roberts, and Chater (2002) press this notion of conditional probability into service to analyze the logical quantifiers by postulating that the meaning of a quantified statement having subject term *X* and predicate term *Y* is given by the conditional probability of *Y* given *X*, or $P(Y \mid X)$. The entire gamut of conditional probabilities for the logical operators fall out as follows: MAX means that $P(Y \mid X) = 1$, POS means that $P(Y \mid X) > 0$, NEG means that $P(Y \mid X) < 1$, and MIN means that $P(Y \mid X) = 0$. As we ourselves have assumed, the probability interval for *some* includes that for *all*, and the probability interval for *some...not* includes that for *none*.

3.2.3.3. Signed probability and the negative quantifiers

These probabilistic measures are presumably more psychologically realistic than some of the others that have been considered above, but we do not wish to discard the guiding hypothesis of this chapter that logical operations express correlation. We therefore propose that the unsigned measures reproduced in the preceding paragraph be augmented to signed measures. The first step is to lay out the axioms of a signed probabilistic system P_3:

3.22 a) $P_3(\{\}) = 0.$

b) Either one of the following is true:
(i) $P_3(e) < 1, \forall e \in E;$
(ii) $P_3(e) > -1, \forall e \in E.$

c) $$P_3\left(\bigcup_{i=1}^{\pm n} e_i\right) = \sum_{i=1}^{\pm n} P_3(e_i).$$

The difference between (3.22) and the standard set of (3.19) lies in the expansion of signed probabilities to the interval [-1 1].

The signed probabilistic definitions of the logical operators can now be stated. In our terms, the signed conditional probability of a logical operator $<x, y>$, $P_3(y \mid x)$, is found by instantiating Eq. 3.21 as Eq. 3.23, where the bold face **y | x** refers to the conditional set for *Y* defined at the beginning of the chapter:

3.23. $P_3(y \mid x) = P_3(\mathbf{y \mid x} \cap x)/P_3(x).$

The numerator is just the relative frequency of each **y | x** of a given x. For example, if the operator is POS, and $x = 4$, then there are four possible events that make the operator true in an experiment. Since we have no way of knowing in advance which event will be the outcome, the assumption typically made is that they all have an equal chance – namely, 1/4. Turning to the denominator, $P_3(x)$ should be 1, given that an operator is evaluated with respect to a single value of x. Continuing with our example, an **y | x** of 4 is evaluated differently if x is 4 than if x is 5 – the former is true of MAX and POS, while the latter is only true of POS.

With this explanation behind us, MAX means that $P_3(Y \mid X) = 1$ and POS means that $P_3(Y \mid X) > 0$, just as in the Chater-Oaksford definitions. The innovation lies in taking NEG to mean that $P_3(Y \mid X) < 0$, and MIN to mean that $P_3(Y \mid X) = -1$. The upshot is that the probabilistic logical operators are isomorphic to the normalized conditional cardinalities illustrated in Fig. 3.8.

The conceptual content of the negative probabilistic measures is as follows. If MAX means that, given a member of *X*, a member of *Y* is certain to occur, then the signed understanding of MIN is that, given a member of *X*, a member of *Y* is certain to *not* occur. In other words, if the positive range expresses certainty, then the negative range expresses **anti-certainty**, the certainty that *Y* will not occur. True uncertainty, the lack of any knowledge at all about the occurrence of an event *b* given the occurrence of an event *a*, is expressed by zero, in accord with the other trivalent measures.

3.2.4. Information

Given the nature of natural language as a medium for communication, one would expect that information theory should also supply a pattern method for the logical operators. This is indeed the case, but the mathematical notion of information is rather particular. By way of explanation, undertake the thought experiment from Applebaum, 1996, p. 93, in which you are asked which 'statement' in (3.24) conveys the most information:

3.24 a) Xqwq yk vzxpu vvbgxwq.
b) I will eat some food tomorrow.
c) The prime minister and leader of the opposition will dance naked in the street tomorrow.

Applebaum hopes you will choose (3.24c), for the following reasons. (3.24a) is nonsensical and so appears to impart no information at all, but it does contain many rare English letter sequences which make it surprising. (3.24b) in contrast is meaningful, but it states what is generally true and so does not surprise us. (3.24c) is also meaningful, and its extreme improbability does surprise us. The result is a three-way classification: (i) surprise (low probability) but no meaning, e.g. (3.24a); (ii) meaning but no surprise (high probability), e.g. (3.24b); and (iii) meaning and surprise (low probability), e.g. (3.24c). Information theory usually takes the surprise element as its subject of study.

3.2.4.1. Syntactic information

The syntactic theory of information springs from the break-through work of Claude Shannon on characterizing the capacity of a communications channel such as a telegraph or telephone line, Shannon (1948) and Shannon and Weaver (1949), which was foreshadowed in Sec. 1.2.2.4.[18] Shannon information relies on

[18] The most-cited textbook is Cover and Thomas (1991), though van der Lubbe (1997) is more accessible and has solved problems. Within the more specific realm of computational neuroscience, Ballard (1997), Dayan and Abbott (2001), and Trappenberg

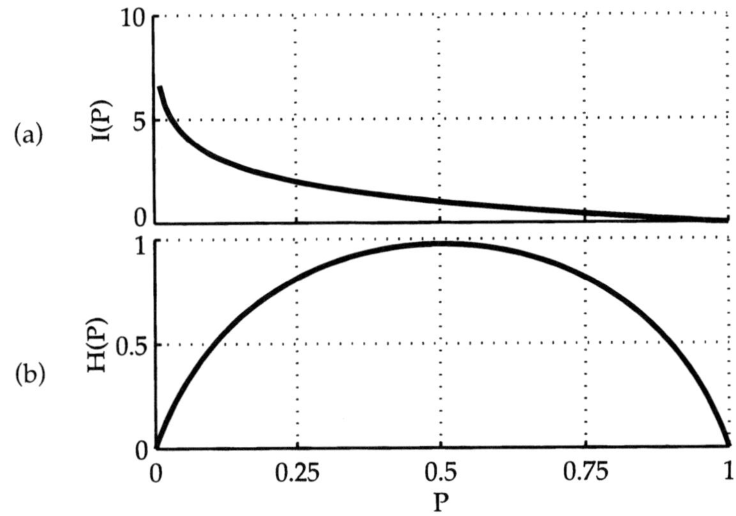

Figure 3.11. (a) Information; (b) entropy of a Bernoulli random variable. (**p3.05_info_theory.m**)

three common-sense assumptions: (i) there is no such thing as negative information, (ii) the information measured from the occurrence of two events together is the sum of the information of either event occurring separately, and (iii) the information measured from the occurrence of two events together is greater than the information of either event separately. The first two qualify information as a unsigned measure. The third makes it a decreasing function of probability. Shannon concluded that the only function that satisfies these three criteria is the logarithmic function. The most popular logarithm is that of base two, which gives the equation for Shannon information of Eq. 3.25:

3.25. $$I(e) = \log(1/P(e)) = -\log(P(e))$$

If the logarithm is taken in base 2, the units of measurement of this function are the well-known **bits**.

3.2.4.2. Entropy and conditional entropy

This measure still omits one crucial consideration. In a given experiment, we do not know which value *e* of the random variable will occur next, which prevents us from knowing how much information *I(P(e))* there is. Shannon

(2002) have good introductions, while Deco and Obradovic (1996) goes into considerably more detail.

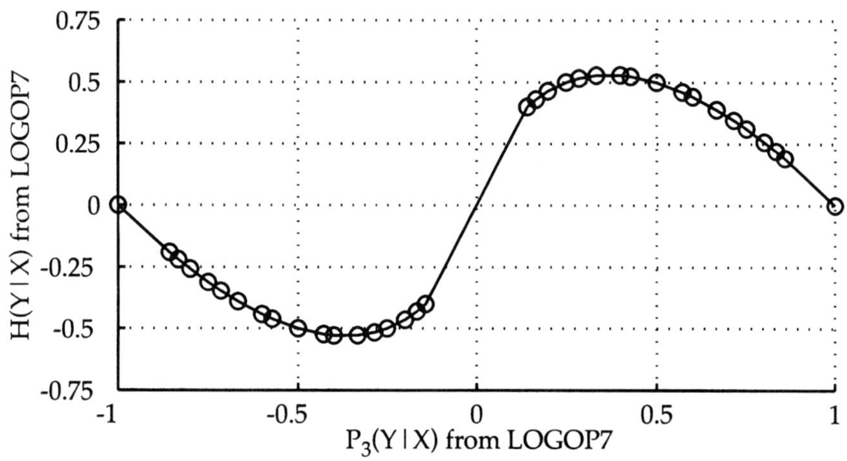

Figure 3.12. Conditional entropy of probabilistic LOGOP7 space. (**p3.06_cond_entropy_LOGOP.m**)

decided that the only recourse is to treat the information content of the entire variable *E* as a random variable, *I(E)* and find its mean information, which he named its **entropy**, *H(E)*:

3.26. $$H(E) = f(I(E)) = -\sum_{i=1}^{n} P(e_i) * \log(P(e_i)), \text{ where } n = \#(E)$$

Entropy is conceptualized as a measure of the uncertainty of a random variable. It is minimal precisely where we have the least uncertainty about *E*, which is the point at which one value is taken with certainty (i.e., *P(E)* = *0* or *1*). Conversely, it is maximal where we have the most uncertainty about *E*, which is the point at which all options are open (i.e., *P(E)* = *0.5*).

Entropy is of interest to us because it can be 'conditionalized' like probability to produce a measure that may be appropriate for the description of the logical operators. Called **conditional entropy**, this measure is the average degree of uncertainty of *Y* over all outcomes of *X*, *H(Y | X)*. One version of its definition is reproduced as Eq. 3.27:

3.27. $$H(Y \mid X) = -\sum_{i=1}^{n} P(x_i) * \sum_{j=1}^{m} P(y_j \mid x_i) * \log(P(y_j \mid x_i))$$

The conditional entropy of probabilistic LOGOP space is graphed in Fig. 3.12. One immediately perceives an insurmountable obstacle to promoting it as a measure: it produces the same output of 0 for the three input points at -1, 0, and 1. Consequently, any system that relied on conditional entropy to differentiate logical operations would confuse the three rather crucial operators at -1, 0, and 1.

This drawback is an inherent handicap of entropy, as can be appreciated from Fig. 3.11b: the input probabilities of 0 and 1 are collapsed into the same output of 0. We therefore have no recourse but to conclude that entropy is not even descriptively accurate for the representation of the logical operators.

3.2.4.3. Semantic information

There is an alternative to this 'syntactic' understanding of information in an approach founded by Bar-Hillel and Carnap, which is generally known as **semantic information.**[19] Bar-Hillel and Carnap, 1952, pp. 227ff, devise a theory of information which rests on the premise that a proposition is informative to the extent that it rules out possible states of the world. The content of a proposition is thereby identified with the set of possible states of the world which is excluded by it, where a state is excluded if it is incompatible with the truth of the proposition. This claim can ultimately be traced to Spinoza's dictum *omnis determinatio est negatio*, "every determination is a negation", and is often considered an operationalization of Popper's (1959) notion that a more informative scientific theory has more ways of turning out to be false.

Bar-Hillel and Carnap go on to derive two measures of the amount of information in a proposition *i* from a measure *P* of the probability of *i*:

3.28 a) $\text{inf}(i) = -\log(P(i))$
b) $\text{cont}(i) = P(\neg i) = 1 - P(i)$

Abstracting away from implementational details, *inf(i)* is the negative of the logarithmic value of the probability of *i*; it expresses how surprised we are to find this probability: we are very surprised to find a proposition with low probability, *inf(i)* $\approx$ *1*, while we are not surprised to find a proposition with high probability, *inf(i* $\approx$ *0*. It is identical to Shannon's measure of information in Eq. 3.25. *Cont(i)* is the complement of the probability of *i*; it expresses the content of *i* as the number of alternatives that it excludes.

[19] Floridi (forth.) further refines semantic information into the "weak semantic information" of Bar-Hillel and Carnap, in contradistinction to his own "strong semantic information". Unfortunately, Floridi's approach does not help us to understand the logical operators, so it is not reviewed here.

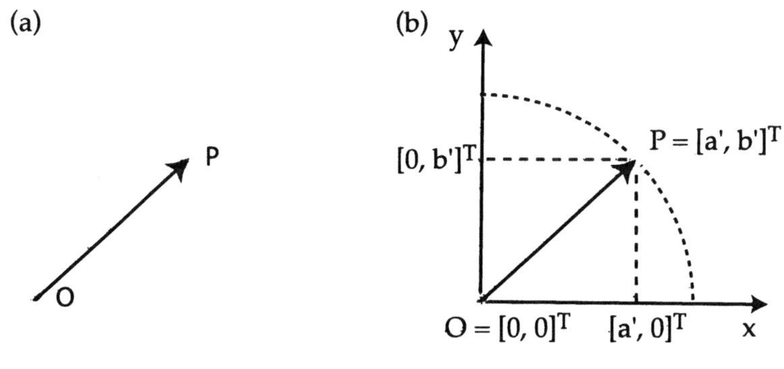

Figure 3.13. (a) The vector OP; (b) projection of OP onto a coordinate system.

It should be clear that *cont* can transform a measure of probability for a logical operator into a measure of its informativeness. If *cont* is made signed, it will generalize to the approach defended here:

3.29 Either one of the following is true:

a) $cont_3(i) = 1 - P_3(i)$, if $P(i) > 0$

b) $cont_3(i) = -1 - P_3(i)$, if $P(i) < 0$

Nevertheless, $cont_3$ still follows Shannon information in mapping MIN and MAX to the same values, namely 0. Thus Bar-Hillel-Carnap information also fails to qualify as a plausible classification of the logical operators. However, as will be demonstrated in the last section of the chapter, both sorts of information do help to explain how logical operators are used.

3.2.5. Vector algebra

Given the centrality of Fig. 3.8 to our representation, it behooves us to investigate its properties as thoroughly as possible. In this subsection, we examine Fig. 3.8 as a spatial object and explain how to effect measurements in the it. For most neuromimetic methods, measurements in a space are performed in terms of vectors, a concept which was introduced briefly in Chapter 2 and is developed more fully here. The study of vectors is undertaken in vector or linear algebra, which supplies the methods for this form of pattern classification.

3.2.5.1. Vectors

Geometrically, a vector is a line segment directed from some point O to another point P. It can be drawn as an arrow from O to P, as depicted in Fig. 3.13a. The vector OP has a **length**, symbolized as $|OP|$, and a **direction**, given

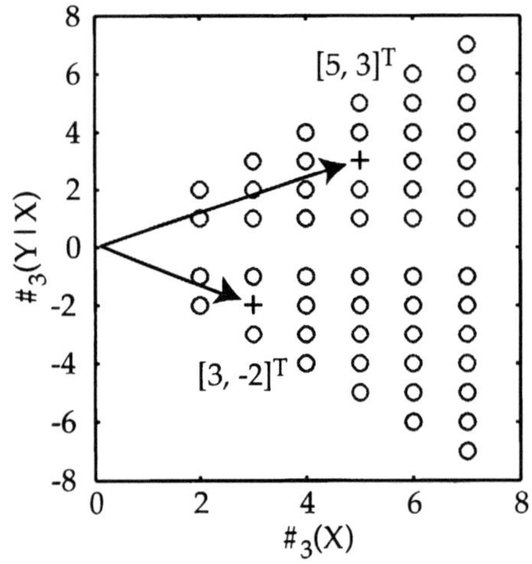

Figure 3.14. Two sample vectors in logical-operator space.

by the arrowhead. Any other directed line segment with the same length and direction is said to be equivalent.

Algebraically, a vector is an n-tuple of real numbers relative to a coordinate system. Such n-tuples are conventionally ordered as a column enclosed in square brackets, see the left side of (3.30), though for typographical convenience the **transpose** of the column to a row as in the right side of (3.30) is often encountered. Note that the name of a vector can be marked as such by means of an arrow appended above it:

3.30.
$$\vec{x} = \begin{bmatrix} a_1 \\ a_2 \\ \dots \\ a_n \end{bmatrix} = \left[a_1, a_2, \dots, a_n\right]^T$$

The real numbers $a_1, a_2, \dots, a_n$ are called the **components** of a vector. They arise naturally from the geometric description once it is projected onto a coordinate system. For instance, O is generally taken to be the origin at $[0, 0]^T$ and P the point in space given by the vector's length and direction from $[0, 0]^T$, say $[a, b]^T$, as illustrated in Fig. 3.13b.

The length or **magnitude** of vector v is the hypotenuse of a right triangle formed by the legs a and b and so is found by the Pythagorean theorem, namely by taking the square root of the sum of each leg squared:

3.31. $|[a, b]^T| = (a^2 + b^2)^{1/2}$

The angle θ can be found by the inverses of the cosine and sine functions:

3.32 a) if $\cos\theta = a / |[a, b]^T|$, then $\theta = \cos^{-1}(a / |[a, b]^T|)$

b) if $\sin\theta = b / |[a, b]^T|$, then $\theta = \sin^{-1}(b / |[a, b]^T|)$

These are the principle measurements needed to find points in the space of the logical operators.

By way of illustration, consider the difference between the vectors $[5, 3]^T$ and $[3, -2]^T$ in Fig. 3.14: $[5, 3]^T$ is longer than $[3, -2]^T$, but it lies at a much smaller angle than $[3, -2]^T$. More precisely, these two vectors have lengths of 5.8 and 3.6, respectively, and angles of 31° and -33.6° or 326°, respectively. The calculations themselves are:

3.33 a) $|[5, 3]^T| = \sqrt{5^2 + 3^2} = \sqrt{25 + 9} = \sqrt{34} = 5.8.$

b) $|[3, -2]^T| = \sqrt{3^2 + 0^2} = \sqrt{9 + 0} = \sqrt{9} = 3.6.$

c) $\theta = \text{cosine}^{-1}(5/5.8) = \text{cosine}^{-1}(0.862) = 30.5°.$

d) $\theta = -\text{cosine}^{-1}(3/3.6) = -\text{cosine}^{-1}(0.83) = -33.6°; 360° - 33.6° = 326.4°.$

Table 3.4 below presents a larger sample of magnitudes and angles for the logical operators.

Given the importance of angular measurements in the upcoming discussions, let us say a word about how they are calculated. The cosine of the angle between two vectors is calculated as follows from the dot product of the vectors divided by the product of their lengths:

3.34. $$\cos \angle(\vec{x}, \vec{y}) = \frac{\vec{x} \bullet \vec{y}}{|\vec{x}||\vec{y}|}$$

The **dot** or **inner product** of two vectors is found by summing their component-by-component products:

3.35. $$\vec{x} \bullet \vec{y} = x_1 y_1 + x_2 y_2 + \ldots + x_n y_n = \sum_{i=1}^{n} x_i y_i$$

Thus the full calculation of the cosine of two vectors is Eq. 3.36:

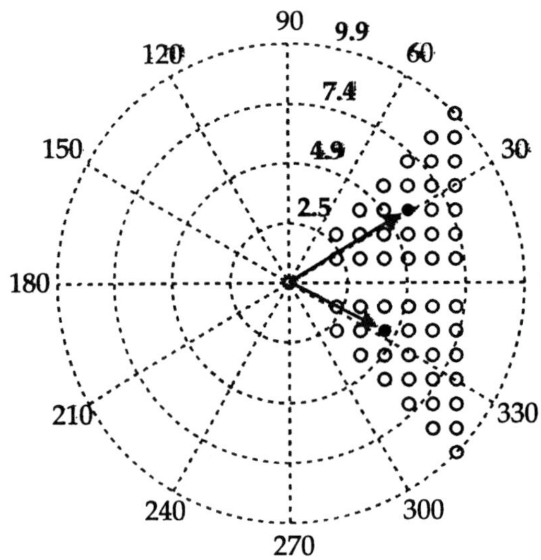

Figure 3.15. Polar projection of logical-operator space. (**p3.07_polarlogop.m**)

3.36. $$\cos \angle(\vec{x}, \vec{y}) = \frac{\vec{x} \bullet \vec{y}}{|\vec{x}||\vec{y}|} = \frac{\sum_{i=1}^{n} x_i y_i}{\sqrt{\left(\sum_{i=1}^{n} x_i^2\right)\left(\sum_{i=1}^{n} y_i^2\right)}}$$

This is the crucial calculation for measuring the similarity of two vectors.

3.2.5.2. Length and angle in polar space

If the length and angle of a vector have semantic relevancy, then we have no recourse but to use a representation that retains this information. The Cartesian plane used in Fig. 3.8 and the figures based on it supplies one possibility, but angle and magnitude are derivative properties in such a space. It would seem more accurate to choose a format that *only* encodes these two properties. The obvious choice is that of a **polar coordinate system**; see among many others Grossman, 1989, p. 492.

In a polar coordinate system, vectors are graphed in terms of their angle and magnitude. That is, a polar system is a plot of vectors [angle, magnitude]^T. Fig. 3.15 gives an example. The center of the graph is defined as the origin, at angle 0 and length 0. The concentric circles mark ever increasing lengths from the

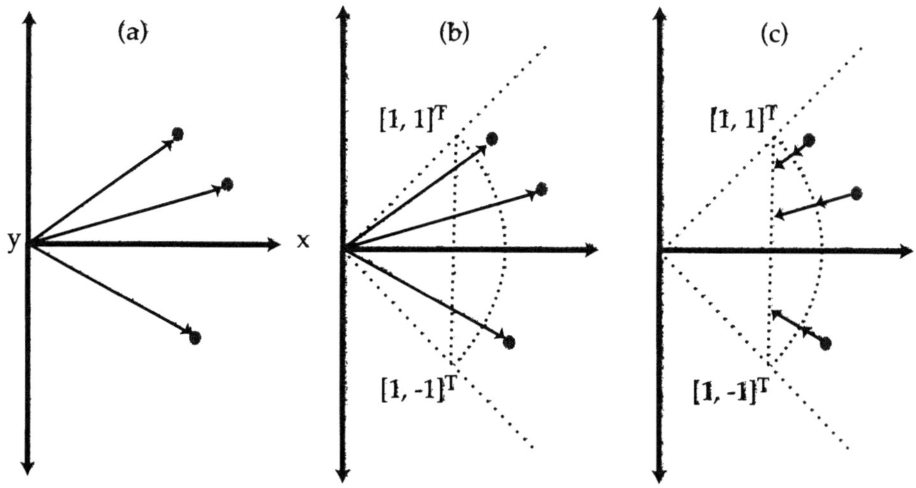

Figure 3.16 (a) Logical-operator locations as rays; (b) logical-operator rays cross lines across the two quadrants; (c) collapse of points onto a single line.

origin. In this case, they measure 2.5, 4.9, 7.4, and 9.9 units of magnitude, as indicated by the numbers to the right the 90° ray.

The two sample vectors are superimposed onto the graph as pointed out by the arrows. Note that, although the relative placement of the data points appears to be identical to that of the Cartesian version, the absolute measures are quite different. For instance, the sample vectors of $[5, 3]^T$ and $[3, -2]^T$ become $[31, 5.8]^T$ and $[-34, 3.6]^T$, respectively.

Even though the polar coordinate system represents the information that interests us here in a perspicuous fashion, we prefer to use the Cartesian coordinate system. The Cartesian coordinate system represents the 'raw' data more accurately, so that it is more difficult to lose track of the transformations that will be applied to it.

3.2.5.3. Normalization of logical operator space

A second reason for not using a polar coordinate system is that one of the most revealing transformations that can be applied to the raw data produces a sort of polar effect. The transformation that we have in mind is that of normalization, to which we now turn.

3.2.5.3.1. *Logical operators as rays*

The previous discussion takes for granted the following observation about the representation of a logical operator:

Table 3.4. Measures of LOGOP3. (**p3.08_samplelogopmeas.m**)

v#	*op*	*mag(op)*	$\theta^{rad}(op)$	$\theta°(op)$	*norm(op)*	$cos(\theta)$	$sin(\theta)$
1	[2, 2]T	2.83	0.79	45.00°	$[0.71, 0.71]^T$	0.71	0.71
2	[2, 1]T	2.24	0.46	26.57°	$[0.89, 0.45]^T$	0.89	0.45
3	[2, -1]T	2.24	-0.46	-26.57°	$[0.89, -0.45]^T$	0.89	-0.45
4	[2, -2]T	2.83	-0.79	-45.00°	$[0.71, -0.71]^T$	0.71	-0.71
5	[3, 3]T	4.24	0.79	45.00°	$[0.71, 0.71]^T$	0.71	0.71
6	[3, 2]T	3.61	0.59	33.69°	$[0.83, 0.55]^T$	0.83	0.55
7	[3, 1]T	3.16	0.32	18.43°	$[0.95, 0.32]^T$	0.95	0.32
8	[3, -1]T	3.16	-0.32	-18.43°	$[0.95, -0.32]^T$	0.95	-0.32
9	[3, -2]T	3.61	-0.59	-33.69°	$[0.83, -0.55]^T$	0.83	-0.55
10	[3, -3]T	4.24	-0.79	-45.00°	$[0.71, -0.71]^T$	0.71	-0.71

3.37. A logical operation is defined by a ray in the northeastern or southeastern quadrant of the Cartesian plane emanating from (or ending at) the origin.

Fig. 3.16a illustrates a set of such rays that could potentially define a coordination. Assuming the correctness of this representation implies that a single line drawn across the quadrant could serve to classify any quantifier at the points where the rays intersect the line. Fig. 3.16b traces two possible lines across the half-plane of Fig. 3.16a. The straight line maps the points above it onto the hypotheneus of the right triangle with legs $[1, 1]^T$ and $[1, -1]^T$. The curved line maps the points above it onto an arc of the unit circle – a circle with a radius of one. Either way, there is a representation on a line of all of the rays. If they all could be shrunk down to either distance from the origin, as in Fig. 3.16c, we would obtain an important result. The result is a tractable, one-dimensional representation of logical-operator space and of the patterns found within it.

The next few subsections define this shrinking calculation, known as normalization, and discuss its ramifications for the representation of logical operators. The first step is to explain the very simple mathematics of scalar multiplication of which normalization is one usage.

3.2.5.3.2. *Scalar multiplication*

One further property of vectors that now becomes important is that they can be multiplied by real numbers, known as scalars in this context. A two-dimensional vector v standing for $[a, b]^T$ is multiplied by a scalar s in the following way:

3.38. $$s \cdot v = [s \cdot a, s \cdot b]^T$$

Geometrically, where v represents the directed line segment *OP*, $s \cdot v$ will have a length $|s|$ times the length of *OP*.

Consider some simple examples drawn from logical-operator space, $[2, 2]^T$, $[2, 1]^T$, and $[2, -1]^T$. Multiplied by, say, 2, they take on the values given in (3.39):

3.39 a) $2 \cdot [2, 2]^T = [4, 4]^T; [2, 2]^T, [4, 4]^T \in \text{MAX}$
b) $2 \cdot [2, 1]^T = [4, 2]^T; [2, 1]^T, [4, 2]^T \in \text{POS}$
c) $2 \cdot [2, -1]^T = [4, -2]^T; [2, -1]^T, [4, -2]^T \in \text{NEG}$

Yet as the classifications appended to each line indicate, both input and the result are members of the same logical operation – a fact amply demonstrated in Chapters 4 and 6. This suggests that many of the values of a logical operator are scalar multiples of one another. A tremendous amount of redundancy could be stripped away by defining a standard value from which all the others could be calculated by the appropriate scalar multiplication. (3.40) indicates how this should work:

3.40 a) $1/4 \cdot [4, 4]^T = [1, 1]^T$
b) $1/4 \cdot [4, 2]^T = [1, 0.5]^T$
c) $1/4 \cdot [4, -2]^T = [1, -0.5]^T$

(3.40) shows that input values of four can be reduced to a unit value by multiplying them by a quarter. And as the fours go, so goes the rest of logical-operator space, if the correct fraction is used. This is what normalization does.

3.2.5.3.3. *Normalization of a vector, sines and cosines*

Normalization of a vector v assigns it a standard length of 1 by dividing it by its magnitude:

3.41. $\text{norm}(v) = v / |v|$

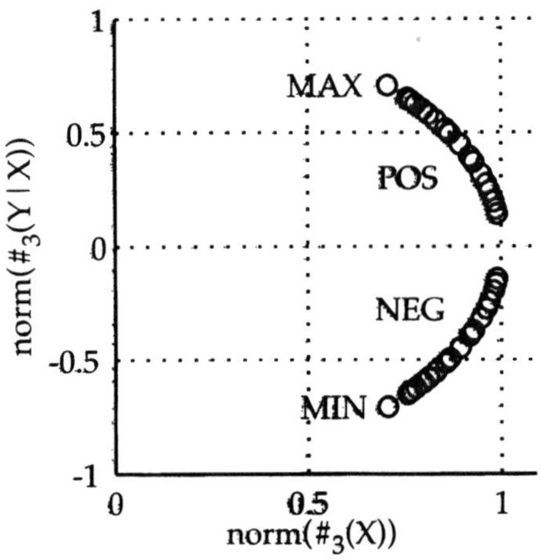

Figure 3.17. Plot of normalized LOGOP7. (**p3.09_normlogop.m**)

As an example, consider the ten vectors defined by $x \leq 3$ in logical-operator space, hereafter simply **LOGOP3**. They are listed in Table 3.4 under the column heading "op", with their normalized versions in the column labeled "norm(op)". This column constitutes the first ten points in the plot of Fig. 3.17, which extends the input up to seven, for 54 total vectors. The fact that all of these vectors now have a length of 1 maps them onto the unit arc in the two quadrants.

One of the happy by-products of normalization is that the x and y components of the normalized vectors are equivalent to the cosine and sine of the angle, respectively. The last two columns of Table 3.4 calculate the trigonometric values for the sample data. Comparison to the normalized column shows the corresponding sets of values to be identical. The utility of this reduction is that it supplies the simplest means possible for locating a vector, namely a one-dimensional scale based on either the cosine or sine values.

3.2.5.4. Vector space and vector semantics

Based on the semantics of spatial prepositions, Zwarts (1997) and Zwarts and Winter (2000) have argued that vectors are the primitive spatial entity in models of natural language. One by-product of this monograph is to reinforce Zwarts and Winter's conclusions indirectly by developing a neurologically grounded, vector-theoretic framework for the logical operators. With the aim of underscoring the compatibility of Zwarts and Winter's account with our own, let us briefly review their vector-space ontology. Such a review affords us the chance to tie together the various vector operations introduced above into an algebraic structure.

Zwarts and Winter's vector-space ontology consists of a vector space V of n Euclidean dimensions over the real numbers $\Re$, or $\Re^n$. The element $0 \in V$ is the zero vector, and the functions $+ : (V \times V) \to V$ and $\cdot : (\Re \times V) \to V$ are vector addition and scalar multiplication, respectively. Thus the basic vector space is simply the quadruple $\langle V, 0, +, \cdot \rangle$. Zwarts and Winter go on to augment this basic space with notions that are necessary for the expression of spatial relations, but they are not relevant to our concerns and can be omitted.

3.2.5.5. Summary

This subsection introduces the geometric interpretation of points in a space as segments directed from the origin to the point in question, with the particular aim of showing how this interpretation provides tools for paring logical-operator space down to just the information that natural language actually uses for the classification of the logical operators. In this endeavor, we have followed the lead of the first chapter in viewing the computational task of these semantic operations as analogous to that of early vision, namely the reduction of redundancy. The operations on the 'raw' vectors of logical-operator space that accomplish this reduction are either the calculation of vector angle or normalization. The next subsection returns us to statistical correlation in order to restate the vector-theoretic results in the terms of this framework.

On a final note, it should be pointed out that vector representations have a long history in neuroscience. Eliasmith and Anderson, 2002, p. 49, list the production of saccades, the orientation of visual input, orientation in space, the detection of wind direction, echo delay, and arm movement as systems for which vector representations have proved to be indispensable. One goal of this monograph is to convince the reader that logical operations should be added to this list.

3.2.6. Bringing statistics and vector algebra together

Wickens, 1994, p. 18ff., and in a more concise fashion, Kuruvilla et al. (2002), explain how to map between statistical variables and their vector geometry, and in particular how the notion of similarity expressed by the Pearson correlation coefficient is realized in a vector space.

The central insight is that the standard deviation of a variable and the magnitude of a vector are proportional to one another, since the calculation of both standard deviation and magnitude takes the square root of the sum of the elements squared. The calculations are arrayed side by side in Eq. 3.42a and 3.42b so as to highlight this commonality:

3.42. $$\text{(a) } s(x) = \frac{1}{\sqrt{n-1}} \sqrt{\sum_{i=1}^{n} (x_i - \bar{x})^2} \quad \text{(b) } |\vec{x}| = \sqrt{\sum_{i=1}^{n} (x_i)^2}$$

Wickens argues that the constant of proportionality $1/\sqrt{(n - 1)}$ is unimportant to most analyses, since every vector is based on the same number of observations, and so can be dropped. This result has the effect of equating the standard deviation of a variable to the length of its vector. They differ in that standard deviation centers the variable around its mean, which in geometric terms sets the origin to the mean. Due to this difference, the statistical and vector measures will not produce identical results if the mean is too large. For vectors with zero mean, the two measures yield identical results.

This correspondence can be made more explicit by pointing out the formal symmetry between the Pearson correlation coefficient and vector angle. Recall that this coefficient is based on the covariance of a pair of observations divided by their standard deviation, see Eq. 3.17, and the standard deviation itself is based on the square root of the number-adjusted deviation squared, see Eq. 3.14. Dropping the constant of proportionality from both operations as was suggested in the previous paragraph brings out the formal symmetry between Pearson correlation and the cosine of an angle in Eq. 3.43a and 3.43b. As in the case of the symmetry between standard deviation and vector magnitude, Pearson's r and the cosine differ in that covariance centers the variable around its mean, so that the two measures diverge if the mean is too large.

3.43 a) $$r(x_i, y_i) = \frac{\sum_{i,j=1}^{n} (x_i - \bar{x}_i)(x_j - \bar{x}_j)}{\sqrt{\left(\sum_{i=1}^{n} (x_i - \bar{x}_i)^2\right)\left(\sum_{i=1}^{n} (x_j - \bar{x}_j)^2\right)}}$$

b) $$\cos \angle(\vec{x}, \vec{y}) = \frac{\sum_{i=1}^{n} x_i y_i}{\sqrt{\left(\sum_{i=1}^{n} x_i^2\right)\left(\sum_{i=1}^{n} y_i^2\right)}}$$

The upshot is that we can let a measure of vector angle stand in for Pearson's (and Spearman's) measures of correlation when an operator sample does not meet their distributional criteria, as long as the mean of the vectors in question is not too large. Employing vector angle as a substitute for correlation avoids the problematic artifacts uncovered in Sec. 3.2.2.3, while still allowing us to talk of correlation among operator meanings. Moreover, the condition of small means is implicitly implemented in the learning rules introduced in Chapter 5, which work best on small subspaces of locally correlated vectors.

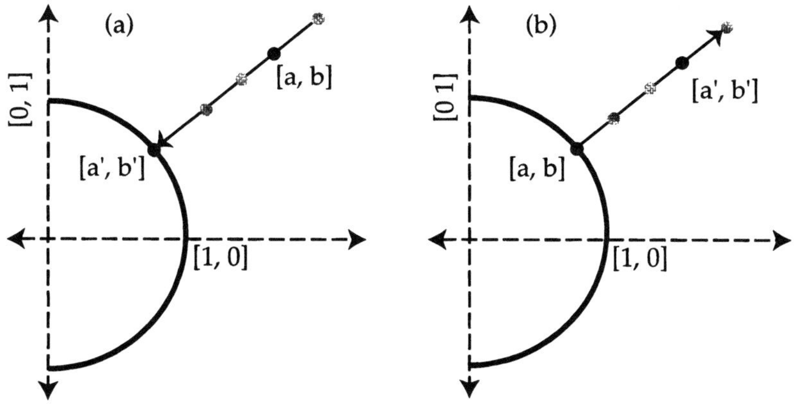

Figure 3.18. (a) Normalization onto the unit semicircle; (b) reconstitution from the unit semicircle.

3.3. THE ORDER TOPOLOGY OF OPERATOR MEASURES

The assumption that magnitude does not matter has one important exception: the highest or top point $[0.71, 0.71]^T$ must be distinguishable from the lowest or bottom point $[0.71, -0.71]^T$. What appears to be called for is some mechanism to strip away the magnitude of a point while leaving its place in the partial ordering of surrounding points. This can be done by establishing the order topology that underlies the Cartesian plane.

This question takes on additional transcendence once we inquire into the properties of normalization. The crucial one is that it is a function, that is, a correspondence between a point in the plane and a point on the unit semicircle. As is our wont, let us illustrate this with a picture. In Fig. 3.18a, the truth value of a point $[a, b]^T$ does not change when it is projected onto the unit semicircle at point $[a', b']^T$. However, many *other* points in the plane correspond to this single point on the unit semicircle, so normalization is a many-to-one, 'onto' or surjective function, from big semicircles in the plane onto the unit semicircle. This implies that its inverse, the mapping from the semicircle back out into the plane in Fig. 3.18b, which can be called **reconstitution**, is not a function. In other words, once the magnitude of an operator is lost through normalization, it is lost forever.

3.3.1. A one-dimensional order topology

In a set X arranged by the order relation $<$, two elements a and b that are ordered so that a is less than b define four subsets:

3.44 a) $(a\ b) = \{x \mid a < x < b\}$
b) $[a\ b) = \{x \mid a \leq x < b\}$
c) $(a\ b] = \{x \mid a < x \leq b\}$
d) $[a\ b] = \{x \mid a \leq x \leq b\}$

These four sets are referred to as the **intervals** determined by *a* and *b*. They can be called upon to define a collection of subsets of *X* that is the **order topology** on *X*:

3.45. Let X be a set with a simple order relation; assume that X has more than one element. Let α be the collection of all sets of the following types:

a) All **open** intervals *(a b)* in X.
b) All **half-open** intervals of the form $[a_0\ b)$, where a_0 is the smallest element (if any) of X.
c) All **half-open** intervals of the form $(a\ b_1]$, where b_1 is the largest element (if any) of X.

If *X* has no smallest element, there are no sets of type (3.45b); if *X* has no largest element, there are no sets of type (3.45c).

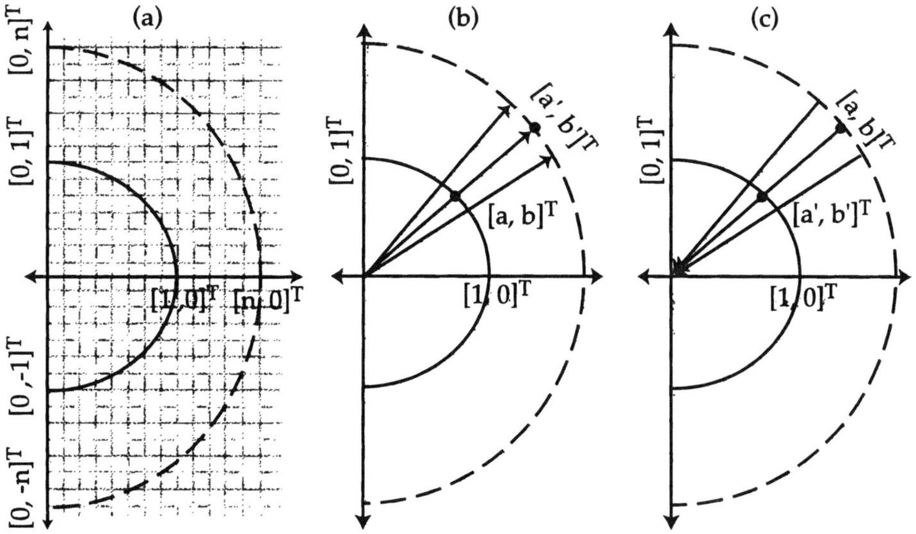

Figure 3.19 (a) Projection from the unit semicircle; (b) expansion; (c) contraction.

3.3.2. A two-dimensional order topology

Now consider the two-dimensional version of the order topology, $X \times Y$. The order relation on $X \times Y$ is defined as so:

3.46. $$[x\ y] \leq [z\ w] \leftrightarrow x \leq z \ \&\ y \leq w$$

In a plane $X \times Y$ arranged by the order relation '<', two pairs *(a b)* and *(x y)* that are ordered so that *(a b)* is less than *(x y)* define four subsets:

3.47 a) $(a\ b) = \{(x\ y) \mid (a_0\ b_0) < (x\ y)\}$
b) $[a\ b) = \{(x\ y) \mid (a_1\ b_0) \leq (x\ y)\}$
c) $(a\ b] = \{(x\ y) \mid (a_0\ b_1) \leq (x\ y)\}$
d) $[a\ b] = \{(x\ y) \mid (a_1\ b_1) \leq (x\ y)\}$

These four sets can be referred to as the **bands** determined by *(a b)*. They can be called upon to define a collection of square-shaped subsets of $X \times Y$ that is the **product order topology** on $X \times Y$:

3.48. Let $X \times Y$ be a set of ordered pairs with a simple order relation; assume that $X \times Y$ has more than one element. Let β be the collection of all sets of the following types:

a) All open bands $(a\ b)$ in $X \times Y$.

b) All half-open bands of the form $[a_1\ b_0)$, where a_1 is the largest element (if any) of X.

c) All half-open bands of the form $(a_0\ b_1]$, where b_1 is the largest element (if any) of Y.

As before, if X has no largest element, there are no sets of type (3.48b), and if Y has no largest element, there are no sets of type (3.48c).

Returning to the space that interests us specifically, the unit semicircle that sits in the square defined by $[0, -1]^T$, $[1, 0]^T$, and $[0, 1]^T$, as illustrated in Fig. 3.19a. Any larger semicircle can be cross-hatched in by setting the scalar multiple to n at the points $[0, -n]^T$, $[n, 0]^T$, and $[0, n]^T$, such as the dashed semicircle in Fig. 3.19a.

Thus any of the larger semicircles has the same order topology as the unit semicircle, which allows order-theoretic properties true of the unit semicircle to be preserved at higher scalar multiples, see Fig. 3.19b. Conversely, scalar contraction of a larger semicircle down to the unit semicircle allows any order-theoretic property true of the larger semicircle to be preserved on the smaller unit semicircle, see Fig. 3.19c. In particular, if we take any point p on either semicircle that lies on the interval $\{p \mid a < p < b\}$, we know that it maps to a corresponding, order-preserving interval $\{p' \mid a' < p' < b'\}$. The consequence is that the truth value of a point near $[a, b]^T$ in Fig. 3.18a does not change under projection (contraction or normalization) onto $[a', b']^T$.

3.3.3. The order-theoretic definition of a lattice

Order theory provides an alternative definition of a lattice that is isomorphic to the set-theoretic definition introduced at the beginning of the chapter. The strategy is to define a partial order ≤ on a set S, where a **partial order** is a binary relation obeying reflexivity, transitivity, and antisymmetry:

3.49. For all a, b and c in S,

a) $a \leq a$; (reflexivity)

b) if $a \leq b$ and $b \leq c$ then $a \leq c$; (transitivity)

c) if $a \leq b$ and $b \leq a$ then $a = b$. (antisymmetry)

≤ is often known as the **part-of** relation. $\langle S, \leq \rangle$ is called a **partially ordered set**, an **ordered set**, or a **poset**.

$\langle S, \leq \rangle$ is a lattice if the supremum and infimum of S are members of S. These two notions are understood order-theoretically as so:

3.50. For a set $X \subseteq S$ and an element a in S,

a′) a is an upper bound for X if $\forall x \in X, x \le a$.

a′) a is the supremum of X iff a is an upper bound for X, and $x \le b$ for all upper bounds b of X.

b) a is a lower bound for X if $\forall x \in X, a \le x$.

b′) a is the infimum of X iff a is a lower bound for X, and $b \le a$ for all lower bounds b of X.

That is to say, a set may have several upper or lower bounds, so the supremum is its least upper bound, and infimum is its greatest lower bound.

The algebraic lattice $\langle S, \cup, \cap \rangle$ and the order-theoretic lattice $\langle S, \le \rangle$ coincide, see for instance Landman, 1991, p. 237. This can be understood more intuitively by noting that the union of a and b is the smallest subset of S that a is part-of and b is part-of, see 3.51a. Likewise, the intersection of a and b is the largest subset of S that is part-of a and part-of b, see 3.51b:

3.51 a) $a \le a \cup b$, and $b \le a \cup b$.

b) $a \cap b \le a$, and $a \cap b \le b$.

In the Hasse diagram of a finite poset, the vertices are the elements of S and the ordering relation is indicated by the edges and the vertical positioning of the vertices. Element a is smaller than element b if and only if there is a path from a to b that always goes upwards.

Our final distillation of logical-operator space in Fig. 3.17 is described by a stronger requirement, namely that the logical operators form a chain. $\langle S, \le \rangle$ is a **chain** if …

3.52. $\forall a, b \in S$, either $a \le b$ or $b \le a$.

That is to say, a chain is a totally or linearly ordered set. The most useful linear ordering of LOGOP space is along its y axis, so that MAX is the largest element and MIN is the smallest.

3.4. DISCRETENESS AND CONVEXITY

The order-theoretic reduction of the unit semicircle developed in the preceding section leads to an even more interesting result, which depends on the notion of a linear continuum, see Munkres, 1975, p. 152:

3.53. A simply ordered set L having more than one element is called a **linear continuum** if (a) and (b) hold:

a) L has the least upper-bound property.

b) L has the intermediate-value property: If $x < y$, then there is a z between x and y, i.e. $x < z < y$.

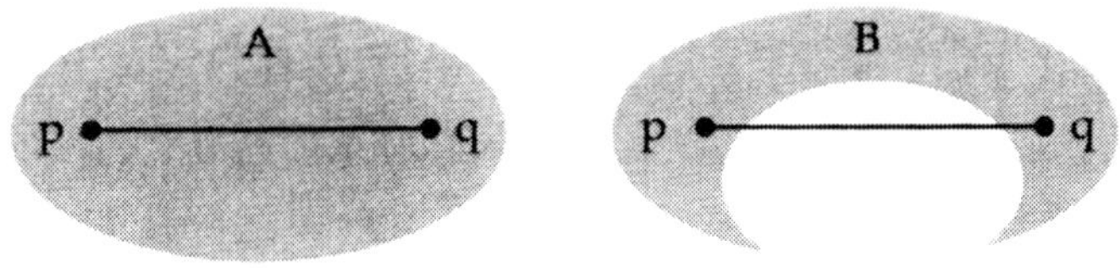

Figure 3.20. Shading shows (A) convex and (B) non-convex regions.

The unit arc of logical-operator space finds its least upper bound at 45°, to satisfy (3.53a). It satisfies (3.53b) by induction on its number-theoretic counterpart. Consider the two points for which $|x + y| = 1$, namely $[1, 0]^T$ and $[0, 1]^T$. The immediate successor shared by both points, namely $[1, 1]^T$, maps onto the unit arc at $[0.71, 0.71]^T$ or 45°. Thus if $x = [1, 0]^T$ and $y = [0, 1]^T$ in (3.53b), then $z = [0.71, 0.71]^T$. More generally, for any two half-open bands that instantiate $x < y$, the corresponding closed band is normalized as a point z between x and y.

Extending the intermediate value property to all of the points between a and b, defines the concept of convexity, see Munkres, 1975, pp. 152-3:

3.54. Let Y be a subset of L that equals either L or an interval or ray in L. The set Y is **convex** if a and b are any two points of Y and $a < b$, then the entire interval [a, b] of points of L is contained in the set Y.

For instance, a pattern that embraces the unit arc from ±1° to 45° is convex, which defines the range of the operator POS.

The reader may be acquainted with the notion of convexity from the more perspicuous definition of a convex region as one in which every point can be connected to every other point without leaving the region – a geometric instantiation of the intermediate-value property. In Fig. 3.20, the shaded region A is convex while the shaded region B is not, since the line joining points p and q stretches outside of it.

3.4.1. Voronoi tesselation

It will be convenient to have some method for dividing a space up into convex regions. Fig. 3.21 exemplifies a popular means of doing so. The lines divide the space into cells in which any vector in a cell is closer to the large black dot in the center of the cell than to a vector in another cell. Each black dot is termed the **centroid**, reference vector, or **prototype vector** of the cell. The centroid is often said to be the **nearest neighbor** of its companion vectors within a cell. The global partition of the plane is known as a **Voronoi tessellation**, from Voronoi (1908).

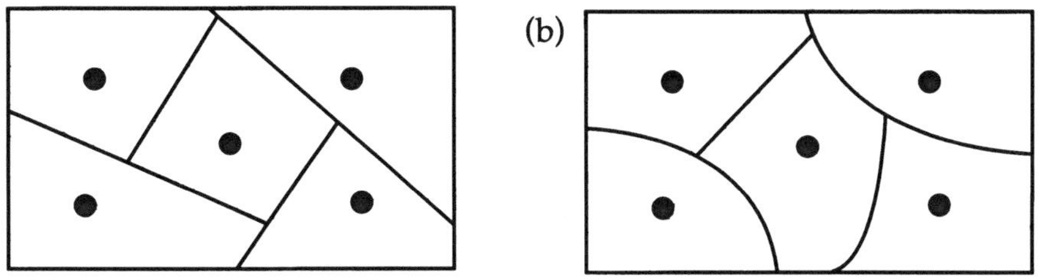

Figure 3.21. Voronoi tessellation (or vector quantization) of a plane into five regions around five centroids or prototype vectors. (a) Convex vs. (b) non-convex tessellation/quantization.

In accord with Klein, 1989, pp. 9-11, we can define these notions in the following manner. Let $S = \{p_1, ..., p_n\}$ be a set of n different points in the plane. For $p, q \in S$, $p \neq q$, let us define the perpendicular **bisector** of the line segment that joins p with q as $B(p, q)$ and the halfplane that contains p as $D(p, q)$:

3.55 a) $B(p, q) = \{z \in \Re^2 : |p - z| = |q - z|\}$

b) $D(p, q) = \{z \in \Re^2 : |p - z| < |q - z|\}$

The expression $|a - b|$ denotes the Euclidean distance between two points a and b. The demarcation of a single bisector defined in Eq. 3.55a is illustrated in Fig. 3.22a. The set of all points z that are closer to p than to any other element of S is called the **(open) Voronoi region of *p* with respect to *S*, *D(p, S)*,** defined in Eq. 3.55b and illustrated in Fig. 3.22b by drawing all of the bisectors for p. The union of all region boundaries is called the **Voronoi diagram of *S*,** already seen in Fig. 3.21. Crucially, as long as Euclidean distance is used, each region is a convex subset of the plane, see Okabe, Boots, and Sugihara (1992) for a lemma to this effect.

3.4.2. Vector quantization

Within the signal processing literature, Voronoi tessellation is one aspect of a more general concept known as vector quantization. Let us digress for a moment to consider this signal-processing perspective, condensing Gersho and Gray, 1992, p. 309ff.

Vector quantization can be viewed as a form of pattern recognition in which an input pattern is 'approximated' by one of a predetermined set of standard patterns, or, equivalently, it is matched to one of a set of stored templates or codewords. Depending on how the codewords are chosen, vector quantization can greatly reduce the complexity of the input, making it a powerful technique for data compression.

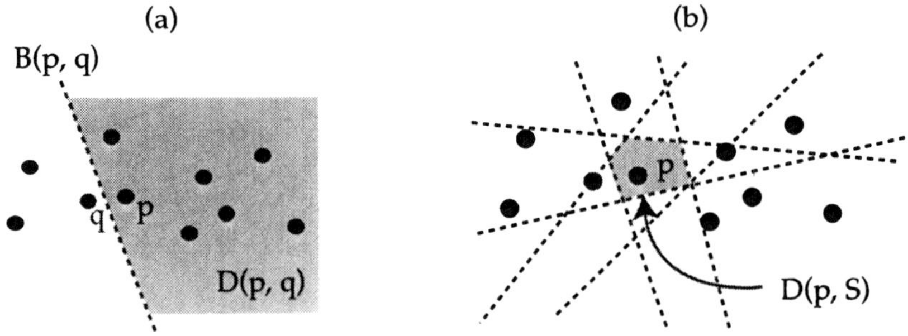

Figure 3.22. A Voronoi tessellation of the plane into convex sets by means of bisectors.

A vector quantizer Q of dimension k and size N is a mapping from a vector or point in k-dimensional Euclidean space, $\Re^k$, into a finite set C containing N output or reproduction points, called **code vectors** or **code words**:

3.56. $Q : \Re^k \rightarrow C$, where $C = (y_i, y_2, \ldots, y_N)$ and $y_i \in \Re^k$ for each $i \in I \equiv \{1, 2, \ldots, N\}$.

The set C is called the **codebook** or the **code** and has N distinct elements of $\Re^k$, which determines its **size**. Associated with every N-point vector quantizer is a **partition** of R^k into N regions or **cells**, R_i for $i \in I$. The ith cell is defined by:

3.57. $\Re_i = \{x \in \Re^k : Q(x) = y_i\}$

The union of all $\Re_i$ is the space $\Re^k$, and the individual $\Re_i$ are all disjoint, which is to say that their intersection is null.

By way of illustration, consider the two partitions of the two-dimensional Euclidean plane $\Re^2$ in Fig. 3.21a and 3.21b. In both cases, the codebook is of size 5, i.e. it contains five codewords, depicted here by the five dots, and they partition the plane into five cells. There is an important qualitative difference between the two, however. In Fig. 3.21a, any two points in a cell can be connected by a straight line that is also in the cell. As was mentioned above, such cells are said to be convex. Convexity generalizes readily to higher dimensions, so that a vector quantizer is called **regular** if it satisfies (3.58):

3.58 a) Each cell $\Re_i$ is a convex set, and

b) for each i, $y_i \in \Re_i$.

Fig. 3.21a illustrates a regular vector quantization. In Fig. 3.21b, in contrast, the curvature of the **faces** separating the cells introduces enough concavity to prevent them from qualifying as convex, so that Fig. 3.21b does not instantiate a regular vector quantization.

The encoding task of a quantizer is to examine each input vector x and identify in which partition cell i of $\mathfrak{R}^k$ it lies. The vector encoder simply identifies the index i of this cell, by means of the **selector function**, $S_i(x)$:

3.59. $S_i(x) = \{1 \text{ if } x \in \mathfrak{R}_i, 0 \text{ otherwise}\}$

The cells $\mathfrak{R}_i$ can be calculated in several ways, but the most popular is in terms of the **sum-squared distortion measure**, defined as the squared Euclidean distance between two vectors:

3.60. $d(x, y) = ||x - y||^2$

The idea is that each cell $\mathfrak{R}_i$ consists of all points x which have less distortion when reproduced with code vector y_i than with any other code vector y_j. A close approximation to the definition of such a **Voronoi** or **nearest-neighbor vector quantizer** is (3.61):

3.61. $\mathfrak{R}_i = \{x : d(x, y_i) \leq d(x, y_j) \text{ for all } j \in I\}$

With this background in artificial signal processing, we can return to natural signal processing, after one brief detour.

3.4.3. Voronoi regions as attractor basins

An alternative way to partition a vector space is to build a dynamical system which tiles the space with stable point attractors whose basins of attraction act as Voronoi cells. This is much better seen than said; the *xy* plane of Fig. 3.23 graphs the phase plane of a stable point attractor,[20] where the three swirling lines trace three trajectories that start at the edge of the graph and are 'pulled' into the point. This two-dimensional flow is often described by a three-dimensional analogy to how a ball released on the top edge of the basin would roll down its interior surface until it reached the bottom. Such a 3D basin of attraction is naturally convex and acts as a Voronoi region, several of which would effectively quantize the *xy* plane into areas of competing influence. In our simulations, the Voronoi quantization of the previous subsection is called upon

[20] The dynamical system is the well-known Lotka-Volterra predator-prey model, see Kaplan and Glass 2002, pp. 230ff.

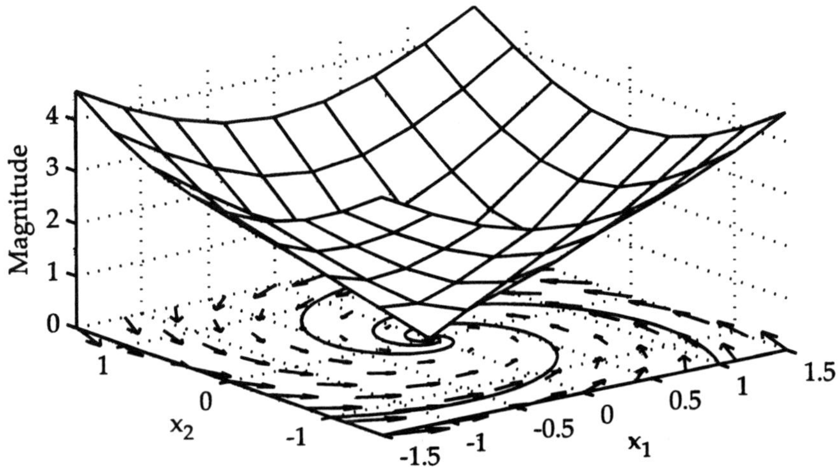

Figure 3.23. Attractor basin on top of phase plane with three sample trajectories. (**p3.10_lotvol_script.m**)

to the exclusion of the dynamical alternative, but the attractor-basin metaphor is perhaps more intuitive and may creep into our discourse from time to time.

3.4.4. Tesselation and quantization: from continuous to discrete

The partitioning of a vector space into Voronoi regions or attractor basins transforms the continuous input space into a set of discrete cells. O'Reilly et al. (1999) argue that this transformation is particularly advantageous for working memory, in that it allows items to be retained over delays and in the face of interference. With respect to this latter claim, imagine that some noise has crept into the representation of an item in working memory, perturbing it from its original state. If the item is coded with respect to a certain cell or prototype, then as long as the perturbation does not nudge it across a Voronoi boundary, the item will still be retained accurately. O'Reilly et al. speculate that discrete coding biases working-memory representations to be:

> more categorical, more easily verbalizable and generally accessible to other parts of the cognitive system, better for perceiving or performing a sequence of steps, and more "symbolic" in some respects.

Indeed, several of these properties have been found in experiments that address working-memory capabilities, as O'Reilly et al. show.

However, discreteness comes at the price of reducing the level of fine detail or graded information that can be encoded to that of the granularity of the quantization distortion measure. Those cognitive systems that do not labor under the presumed noise intolerance of working memory are therefore free to utilize more graded and less discrete encodings.

Having established discreteness as a desirable property of some crucial aspects of higher cognition, we can now wonder what particular form discreteness should take, which returns us to the notion of convexity.

3.4.5. Convexity and categorization

Convexity has an illustrious history in cognitive science in the realm of categorization, where many models assume that categories are convex regions of a similarity space, see Rosch (1978) and Fried and Holyoak (1984), as well as Chapter 5. More recently, it has become the locus of investigation into the structure of natural properties, chiefly through research of Peter Gardenförs starting in the early 1990's and culminating in a monograph published in 2000, see among others Balkenius (1999), Gardenförs (1990, 2000), and Mormann (1993). The principal claim of this work can be distilled in Criterion P, reproduced here as (3.62):

3.62. Criterion P: A natural property is a convex region of a domain in a conceptual space.

Some of the motivation for Criterion P is reviewed in Chapter 11; the analysis of the logical operators undertaken in this work can be taken as further evidence in its favor.

Gardenförs (2000) claims several "treasurable" computational properties for convexity. One that is repeatedly appealed to in his monograph is the fact that convexity, and especially Voronoi tessellation, provides a means for judging similarity and performing categorization, see for instance Gardenförs, 2000, pp. 84ff. For instance, all the vectors within a Voronoi region can be considered similar, and locating a vector within a Voronoi region can be considered a categorization of it. This notion is expanded upon considerably in the upcoming chapters on the neuromimetic learning of the logical coordinators and quantifiers.

One of the more interesting computational properties of convexity is developed in Balkenius (1999) under the rubric of "soft convexity". Herein we alter the details of Balkenius' explanation to suit our own exposition. Consider Fig. 3.24 below, in which three observation are made one after the other. After the observation of the first object, labeled *a*, is made, there is not any other information available to know whether it is part of any larger region, so it is classified within a closely circumscribing space. After an observation of the second object *b* is made, *a* and *b* are similar enough to be classified together, so they are enveloped within a common region. By convexity, all of the points between them are also attributed to the new region, which is why the shading connects the two.

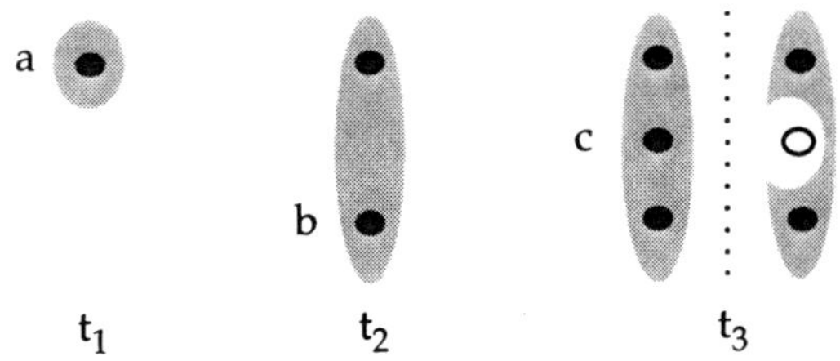

Figure 3.24. The postulation of convex regions for three observations.

The crucial case is the observation of *c*, which lies between *a* and *b* within the domain. Convexity leads us to guess that *c* should be similar enough to *a* and *b* to be classified with them. This happens automatically, as it were, and nothing more need be said. This is the image depicted on the left side of t_3 in Fig. 3.24. However, if it turns out that *c* is actually *not* a member of the category containing *a* and *b*, then the shaded region was postulated erroneously and must be redrawn, as on the right side of t_3. It is easy to see that this redrawn category is no longer convex – simply compare it to Fig. 3.20. Gardenförs would claim that *a* and *b* on the left side of t_3 in Fig. 3.24 form a natural category, while *a* and *b* on the right side do not.

To recapitulate briefly, the fundamental point is how convexity simplifies the effort to be expended to categorize object *c*. The expectation enabled by convexity is that *c* is a member of the emerging category. If it in fact is, then no effort was wasted in the default attribution; if it is not, then it counts as a highly informative counterexample to the emerging category. Convexity provides a place for it either way.[21]

3.5. SEMANTIC DEFINITIONS OF THE LOGICAL OPERATORS

We have exhausted the mathematical approaches to the classification of logical operators, so it is time to step back and take stock of what has been proposed. The first step is to define the logical operators in the most general way, then they need to be distinguished from one another.

---

[21] A more elaborate, but considerably more entertaining explication of this notion is provided at site available at the book's website, where it is argued more or less explicitly that that the simplest classifier, that one that best satisfies Occam's Razor, is convex.

3.5.1. Logical operators as convex regions

The most general definition of logical operatorhood is as a 'Gardenförsian' natural property. The natural extension of criterion P to the logical operators is stated in (3.63):

3.63. Monomorphemic logical operators denote natural properties, that is, they are convex subsets of the topological space defined by the order topology on the unit circle of the Cartesian plane.

To pin down the exact location of a logical operator on the unit semicircle, one more topological operation needs to be defined. It consists of a way to separate the space of the logical operators from the rest of the unit circle. This can be done by severing it at topological 'cut points' into two or more 'globs' or disjoint open sets, as described in Munkres, 1975, p. 147:

3.64. A **cut point** in a topological space X is a pair U, V of disjoint open subsets of X whose union is X. The space X is said to be **connected** if there is no cut point in X.

This provides the foundation for the definition of the space of logical operator denotations:

3.65. **LOGOP space** is a topological space defined by the order topology on the unit circle and separated from it by cut points at 45° and -45° (i.e. 215°).

3.5.2. Logical operators as edge and polarity detectors

3.5.2.1. Logical operators as edge detectors

Though all four logical operators lie between the cut points at 45° and -45°, two of them, MAX and MIN, are 'pure' cut points: they consist of no other points but cut points. Let us define these as "edges":

3.66. An **edge** is a topological subspace of LOGOP space that contains a cut point and no other point.

Given the anomaly of 0° within LOGOP space, we would like to distinguish it as another cut point. Let us therefore define two sorts of edges:

3.67. An **internal edge** is the edge created by the discontinuity at 0°; an **external edge** is an edge between LOGOP space and the rest of the unit arc.

A logical operator embraces a set on the unit circle somewhere between the internal edge and the closest external edge.

The notion of an edge could be pressed into service to define a binary feature for classifying logical operators, if there were some evidence for whether an edge or its complement is lexically marked. Fortunately, there is such evidence. Philosophers and psycholinguists have observed that higher values on a scale are unmarked with respect to lower ones. As one example, Givón (1970) observes that the unmarked form of *how* + adjective questions takes the 'large' value of polar adjective pairs:

3.68 a) How {many / #few} cupcakes do you have?
b) How {much / #little} logic do you know?
c) How {tall / #short} is Mary?
d) How {old / #young} is Mary?

The crosshatched adjectives are only felicitous if their proposition has already been asserted, e.g. (3.69):

3.69 a) A: I have very few cupcakes to give away today.
b) B: So, how few cupcakes do you have?

A simple statement of this observation is the conjecture in (3.70):

3.70. Lessor values (i.e. values further from an external edge) are more marked than greater values (i.e. values nearer to an external edge).

This conjecture can be attributed to information-theoretic notions that are taken up in the next section. It can be stipulated in the grammar by the definition of a binary feature [±blade], where "blade" is understood to mean the expanse of LOGOP space between two edges, by analogy to the blade of a knife being the part between two edges:

3.71. {MAX, MIN} ∈ [-blade]; {POS, NEG} ∈ [+blade].

This is only the first half of the story, however, since some means must be found to distinguish between the members of {MAX, MIN} and {POS, NEG}.

3.5.2.2. Logical operators as polarity detectors

At first glance, the choice of feature jumps out at us from the structure of LOGOP space, namely, the polarity of the *y* axis. However, just invoking positive or negative polarity obscures an important fact, which is that values are ordered in opposite directions according to their polarity. Values increase from 0 up or down, so -1 expresses a greater degree of anticorrelation than -0.5, not a lesser degree as would be expected from the normal ordering of negative numbers. This difference in direction in logical-operator space can be recovered from the order topology as in (3.72):

3.72. The direction from point x to point y on the unit arc is:
a) **antitone** (downward) if $y < x$, or
b) **monotone** (not downward) if $x < y$.
c) If $x = y$, then take y to be $^*x$ and apply (a) or (b).

The correlated and anticorrelated halves of logical-operator space can now be distinguished in these terms:

3.73. In logical-operator space $[x, y]^T$,
a) $\{y \mid y < 0\}$ is antitone; otherwise
b) $\{y \mid y > 0\}$ is monotone.

Logical operators can be classified according to the direction of their y value by (3.74):

3.74 a) An operator is [+down], read as **downwardly monotonic**, if its y value is antitone; otherwise,
b) an operator is [–down], read as **upwardly monotonic**, if its y value is monotone.

An unsought-after outcome of casting the two half-spaces as contrasting in direction is that the zero ray becomes undefined, since it has either no value for direction, or both. As we will have the chance to observe several times in the course of this chapter, logical-operator space embodies a trivalent logic drawn from {1, 0, -1}. Zero is not the half-way point between the other two; it is the statistical singularity of no correlation, and now, the order-theoretic singularity of no direction.

The binary feature [±down] classifies the logical operators as in (3.75):

3.75. $\{MAX, POS\} \in$ [-down]; $\{NEG, MIN\} \in$ [+down].

The resulting cross classification of the logical operators by [±blade] and [±down] is given in the last column of Table 3.5. It produces the ranking from least to most marked in (3.76):

3.76. MAX < POS, MIN < NEG

This order-theoretic implicational hierarchy does not predict the preferred sequence of lexicalization of logical operators stated in (3.10) and in Table 3.5, since lexicalization appears to depend more on pragmatic than semantic factors. The pragmatic factors are discussed in the next section.

Table 3.5. Summary of measures of the logical operators.

LOGOP	*Correlation*	*Angle*	*Norm*	*Order features*
MAX	$\rho(OP) = 1$	$\angle(OP) = 45°$	$n(OP) = [0.7, 0.7]^T$	[-blade, -down]
POS	$\rho(OP) > 0$	$\angle(OP) > 0°$	$n(OP) > [1, 0]^T$	[+blade, -down]
NEG	$\rho(OP) < 0$	$\angle(OP) < 0°$	$n(OP) < [1, 0]^T$	[+blade, +down]
MIN	$\rho(OP) = -1$	$\angle(OP) = -45°$	$n(OP) = [0.7, -0.7]^T$	[-blade, +down]

3.5.3. Summary and comparison to Horn's scale

The ultimate distillation of the various representational methodologies that have been reviewed in the preceding sections lies in the definitions of the logical operators summarized in Table 3.5 and in the labeling of Fig. 3.17. These formats track Horn's (1989) representation of logical operators rather closely, as we now show.

After 236 pages of historical review and analysis of the inferences that can be drawn from the logical operators, Horn (1989) develops a scalar representation of many sorts of operators which he describes with these words:

> The values on the positive scale range from 0 to +1; and those on the corresponding negative scale from 0 to -1. Each operator is ranked in accordance with its lower bound (i.e. *some* says 'at least some' and implicates 'at most some'), such that a simple proposition containing a scalar operator **P** will be true at all positions at or above the position assigned to **P**.

Levinson, 2000, pp. 79ff, agrees with this organization of scalar items – so much so that it becomes a cornerstone of his theory of entailment via what he dubs **Horn scales**. Horn scales are regimented within the confines of Fig. 3.25 – a diagram that will turn up again in our neuromimetic theory of inferencing presented Chapter 8 in the guise of the Square of Opposition. As for the crucial insight that negation reverses a scale, Levinson attributes it to himself, having first broached it in Atlas and Levinson (1981).

We too have divided the logical operators into a positive and a negative scale, though based on more explicit calculations. The substantive difference between our curved and Horn's square representation has to do with the location of the maximal value of the negative scale, namely -1. The orientation of Horn's square places it quite accurately at the top of the negative scale, on a par with the maximal value of the positive scale. The arc of our operator space places it counterintuitively at the bottom, making it the least value of the entire space. However, our suggestion at the end of the previous subsection to classify the logical operators in terms of their direction of monotonicity was made partly with the goal of rectifying this counterintuitive property of logical-operator space. The [+down] feature of the negative half of logical-operator space ensures

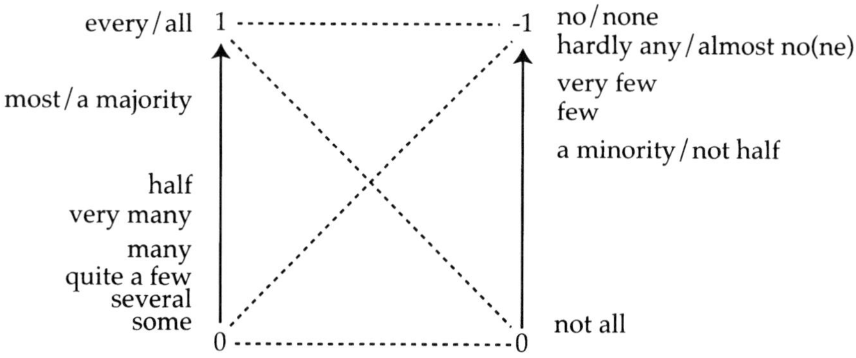

Figure 3.25. Paired positive and negative scales for quantifiers. Comparable to Horn, 1989, Fig. 53.

that -1 is indeed the maximal value of the negative operators, rather than the minimal value. In this way, Horn's square scales are can be cast as a notational variant of logical-operator space.

Let us add somewhat parenthetically at this juncture that Moxey and Sanford, 1993b, p. 48 come to a similar conclusion about the orientation of quantifier scales. Allow us a longish quote, which has the advantage of anticipating the ideas of Moxey and Sanford's that are examined in the next subsection:

> Functionally, a positive scale can be thought of as ordered from the weakest statement about a whole set (e.g. *a few people like ice-cream*) to the strongest statement about the whole set (*all people like ice-cream*). It would presumably be used when the communicator wishes to focus on *how many* is the case, relative to *all*. In contrast, the negative scale can be thought of as ordered from the weakest statement about the null set (*not quite everyone does this*) to the strongest (*no one does this*). It would be used when the communicator wishes to focus on *how few* is the case, relative to *none*. Thus, scales based on pragmatic considerations can be thought of as being about the strength of claim rather than being about number or proportions denoted.

The two scales are ordered in opposite directions, from the center at 0 outwards, in consonance with Horn's and our own conclusions. The difference is that Moxey and Sanford do not see the scales as quantifying numbers but rather the pragmatic notion of "strength of claim".

From the perspective of the order topology that undergirds both our own and Moxey and Sanford's claims, it makes no difference what units are being regimented. Our own units have the advantage of a transparent derivation from the entities that enter into the logical operation, while Moxey and Sanford do not offer any explanation of how to calculate "strength of claim".

3.5.4. Flaws in the word-to-scale mapping hypothesis?

In the preceding discussion, we have taken it for granted that the lexicalizations for the various logical operators map directly to some numerical value located on a scale. Linda Moxey and Anthony Sanford marshal two arguments that purport to show that this "word-to-scale" mapping hypothesis is flawed. The first is that there is great overlap in the numerical values that subjects assign to quantifiers. Moxey and Sanford (1993b) reports on a study of ten quantifiers[22] in which 450 participants were required to assign just one number to one quantifier on one occasion only, thereby depriving the subjects of an opportunity to develop a strategy of comparison between either quantifiers or situations. The results revealed that half of the quantifiers (*a few, only a few, not many, few,* and *very few*) were not distinguishable from one another.

The second drawback of the word-to-scale mapping hypothesis is that the values assigned by subjects to a quantifier vary according to the context of the quantification. Moxey and Sanford, 2000, p. 241, propound the following example:

> ... if an event has a high base-rate expectation, such as people enjoying parties, then the values assigned to (say) *many* in *many people enjoyed the party* is higher than it is for a low base-rate expectation (as in *many of the doctors in the hospital were female*).

The reader is referred to this same article for further evidence for the context-dependency of quantifiers. Exactly what Moxey and Sanford promote as an alternative to word-to-scale mapping is touched on in Sec. 3.5.3, but it is of little interest to us. Moxey and Sanford's objections cannot be validated, as is briefly explained in the next sub-subsection.

3.5.4.1. Vague quantifiers

Bartsch and Venneman (1972) began a popular tradition of analysis by claiming that that the interpretation of *few* and *many* can depend on a contextually given **comparison class**. In the following sentences, they ask, how many is *many*?

---

22 The ten quantifiers studied were: *very few, few, only a few, not many, a few, quite a few, quite a lot, many, a lot,* and *very many*. See Moxey and Sanford, 1993a, Chapter 2, for further background on this task.

3.77 a) Berlin has many inhabitants.
b) Mary has many children.

Certainly, what is 'many' for the inhabitants of a city is much more than what is 'many' for the number of children a woman can have. Bartsch and Venneman go on to note a second reading that is brought by the difference in paraphrase between *a great number* and *greater than average*:

3.78. Many students attend Peter's class.
a) A great number of students attend Peter's class.
b) A greater-than-average number of students attend Peter's class.

That is to say, the comparison class for *many* in (3.78a) may be a standard large number, while in (3.78b) it is an average calculated for the context at hand. Since Partee (1988), the former is known as the **cardinal** reading, and the latter, the **proportional** reading.

The fact that Moxey and Sanford found the least differentiation and the most context-dependence with various collocations of *few* and *many* is consequently to be expected from their ambiguous lexical meaning – especially since Moxey and Sanford's experiments do not control for this extremely confounding factor. We therefore do not take their results to undermine the word-to-scale mapping hypothesis, and will continue to assume it. We do take them to show that the vague quantifiers introduce an extra iota of meaning that makes pinning their denotation down rather challenging. For this reason, we forgo an analysis of them in this monograph.

3.6. THE USAGE OF LOGICAL OPERATORS

The previous section notes that the semantic definitions of the logical operators do not predict very well their ease of lexicalization. This section introduces the notions that predict the lexicalization cline more accurately, though they will not be understood fully until the analysis of neuromimetic inferencing in Chapter 8. Most of the upcoming discussion is devoted to the markedness of negation, since the literature on it is more extensive; only at the very end do we take up the asymmetry between edge and blade operators.

3.6.1. Negative uninformativeness

Givón, 1978, pp. 103-4, invites his reader to engage in the following thought experiment. Suppose that there is a world in which there are two individuals and one property that distinguishes them, such as having horizontal or vertical lines. The two squares in Fig. 3.26a depict one means of visualizing this world. In it, there is no way to judge whether one individual is more salient than the other, say for the purposes of figure-ground segmentation or the markedness of one version of the property over another.

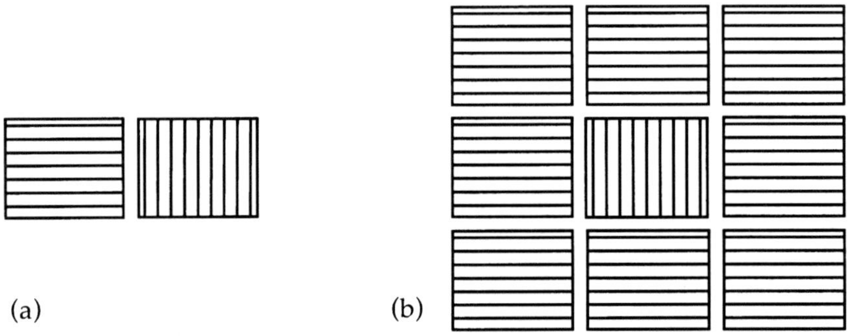

Figure 3.26. Surprise value. Comparable to Givón, 1978, Fig. 92 and 93.

Now, imagine another universe containing nine individuals which are also distinguished by the direction of horizontal or vertical lines – with the crucial difference that eight of them share the same orientation and only one is oriented in the other direction. Fig. 3.26b depicts an image of this world. As Givón says,

> In this universe, in terms of perceptual saliency or figure/ground relations, the single individual stands out on the background of the other [eight]. It is a BREAK IN THE PATTERN, it has SURPRISE VALUE, it can be SINGLED OUT. If one were now to either construe or report about this second universe, reporting the [eight] individuals which constitute the perceptual ground as having the property and single individual which constitutes the perceptual figure as not having it, is preposterously uneconomical enterprise, as compared to the converse procedure. Further, in terms of retrieval strategies, if the [eight] individuals were designated by the presence of the binary property, while the single one by its absence, then identifying that single individual who is different in this universe will be an extremely costly strategy, since one would have to proceed by eliminating all the "present" [eight] others first. On the other hand, if the single individual is coded with the presence of the property, the search procedure will be obviously much more efficient.

Givón draws several conclusions from these and other similar considerations, but the one that interests us the most is an embryonic information-theoretic explanation: information can be defined as 'surprise' or 'breaking the norm', so that perceptually our attention is drawn to a change/figure over the norm/ground. Negation can be incorporated into this dichotomy on the

Table 3.6. Measures of information for Fig. 3.26.

Description	*(a) Two individuals*			*(b) Nine individuals*		
	P_3	I_3	$cont_3$	P_3	I_3	$cont_3$
a) Some square has horizontal stripes.	1/2	0.7	0.5	8/9	0.12	0.11
b) Some square does not have horiz'l stripes.	-1/2	-0.7	-0.5	-1/9	-2.20	-0.89
c) Some square has vertical stripes.	1/2	0.7	0.5	1/9	2.20	0.89
d) Some square does not have vert'l stripes.	-1/2	-0.7	-0.5	-8/9	-0.12	-0.11

perceptually less salient half, so the ultimate claim takes the form of change/figure/positive over norm/ground/negative.

A next step in the elucidation of the markedness of negation comes from the **Principle of Negative Uninformativeness** proposed in Leech, 1981, p. 431:

> Negative propositions are generally far less informative than positive ones, simply because the population of negative facts in the world is far greater than that of positive facts. Consider the sentences:
>
> [3.79]a. Bogota isn't the capital of Peru.
> b. Bogota is the capital of Colombia.
>
> Both statements are true, but assuming a current United Nations membership of 132, (a) is 131 times less informative than (b). Hence to reconcile such a negative proposition with the first Maxim of Quantity, we must assume a context in which the negation of X is precisely as informative as required.

The first Maxim of Quantity mentioned by Leech is one of the sub-principles which regulates Grice's, 1975, p. 45, Cooperative Principle: "Make your conversational contribution such as is required, at the stage at which it occurs." The idea behind the Principle of Negative Uninformativeness is that a given negative proposition is consistent with so many states of affairs that it does not tell the hearer much unless the hearer assumes that it is the denial of a certain one of these states of affairs – the most readily available one of which is the one described by the corresponding unnegated proposition.

3.6.2. Quantifying negative uninformativeness

The reader may have been struck by how the preceding discussion throws around the word "information" without any attempt to explain what it means. We can call on the information-theoretic measures to calculate exactly how negative uninformativeness plays out in the verbal descriptions of the worlds of Fig. 3.26. Let us say that there are four potential statements, consisting of a positive sentence and its negation for each property. All four are listed down the left side of Table 3.6. The first three measures quantify the informativeness of

Table 3.7. Horn's **Q** and **R** principles, from Horn, 1989, p. 194.

The Q Principle (*Hearer oriented*)	*The R Principle* (*Speaker oriented*)
Make your contribution SUFFICIENT	Make your contribution NECESSARY
Say as much as you can, given R.	Say no more than you must, given Q.
Lower-bounding, inducing upper-bounding implicata	Upper-bounding, inducing lower-bounding implicata
Grice's QUANITY1, MANNER1,2	Grice's RELATION, QUANITY2, MANNER3,4

each statement for the world of Fig. 3.26a; the second three measures do the same for the world of Fig. 3.26b.

For the two-individual world, each description is just as probable as its alternative of the same sign and therefore just as informative. Information theory consequently provides no basis for choosing among them, though signed information theory does make it clear which should be positive and which should be negative. The situation is quite different for the nine-individual world.

Givón takes the information borne by a negated proposition to be its surprisal value I_3, while Leech takes it to be its content value $cont_3$. For Givón, a negative proposition is the most probable choice and therefore the least surprising. This desideratum picks out the negative value of I_3 closest to 0 as the better negation, which is –0.12 in Table 3.6d for the nine-individual world. For Leech, a negative proposition is drawn from the largest population of facts and so excludes the fewest alternatives. This desideratum picks out the lowest value of *cont* as the better negation, which is 0.11 in Table 3.6d for the nine-individual world. Either way, both measures classify *Some square does not have vertical stripes* as a better negation than *Some square does not have horizontal stripes* for the world in question.

Yet which of Table 3.6a/c if the better affirmation? Presumably, affirmation works as the inverse of negation, so that it is the most informative option that is the most highly valued. This conjecture chooses I_3 = 2.2 and *cont* = 0.89 in Table 3.6c as the better affirmation. The information-theoretic valuation therefore implicitly divides the universe of discourse into a figure (*Some square has vertical stripes*) and a ground (*Some square does not have vertical stripes*).

3.6.3. Horn on implicatures

Horn, 1989, pp. 198ff, situates Leech's principle in a broader pragmatic and philosophical discourse, and in particular within Horn's own interpretation of Grice's maxims. Horn distills Grice's maxims into two principles, **Q** and **R**, as set forth in Table 3.7. In prose, what these two principles mean is the following:

3.80 a) A **Q**-based implicature proceeds from a speaker's nonuse of a stronger or more informative form to the inference that the speaker did not know enough to have employed the stronger form. In saying P, the speaker implicates that for all s/he knows, 'at most P'.

b) An **R**-based implicature proceeds from a speaker's use of a weaker or less informative form to the inference that the speaker may have intended the stronger form. In saying P, the speaker implicates that 'at least P' is the case.

It is the **Q** implicature that is singled out to account for Leech's example of negative uninformativeness. Horn intuits two cases. Let us start with the more interesting case first, though it is the one that Horn takes up second.

The background is that "I assume it would be relevant for you to know not just that Bogota isn't the capital of Peru, but what it is the capital of". In such a context, I may still utter (3.79a), but "what I [**Q**-]implicate is that there is no stronger, more informative proposition that I could have uttered ... for all I know, [Bogota] may not be the capital of Colombia, that is, [...] I don't know for a fact that it is." If I do know, then my utterance of (3.79a) is unhelpful, misleading, or implausible. As for the other case Horn mentions, "if I suppose that you are only concerned with whether or not Bogota is the capital of Peru, my utterance of the negative proposition [3.79a] licenses no upper-bounding implicature." That is to say, there is no **Q** implicature.

Horn draws the global conclusion that the asymmetry between positive and negative statements is a pragmatic phenomenon, driven by the fact that a negative is prototypically less informative than a positive. Constrained by **Q**-based implicature, a negative statement is only felicitous when the speaker does not know the more informative positive statement to be true.

3.6.4. Quantifying the Q implicature

Let see whether weak semantic information can make Horn's dissection of Leech's Bogota example any more precise. For the case in which Horn did not find any **Q** implicature, the question is just to find out whether Bogota is the capital of Peru or not. This case is comparable to the world of Fig. 3.26a, with the difference that, instead of two individuals, there are two properties applied to the same individual. From this, it should be clear that Horn did not find any **Q** implicature because the positive and the negative statements are equi-probable and so do not give rise to a principled choice between them that would form the basis for the implicature.

The case for which Horn intuited a **Q** implicature is comparable to the nine-individual world of Fig. 3.26b: *Bogota is the capital of Colombia* has a probability of 1/132, while *Bogota isn't the capital of Peru* has a probability of 131/132, under the assumption that there are 131 countries, including Peru, that Bogota is *not* the capital of. Obviously, the information-theoretic evaluation picks the affirmative as the preferred verbalization of the situation. If the speaker instead utters the negative, the hearer can conclude that she does not actually know the affirmative to be true. Thus the quantification reverts to the equi-probable case, which provides no principled choice between the two. In this way, the utterance of the negative (**Q**) implicates the speaker's ignorance of the more informative positive case.

The overall result is that the Bar-Hillel-Carnap measures of semantic information formalize the Givón-Leech insight into the asymmetry between positive and negative statements that makes the calculation of which form to use much more perspicuous.

3.6.5. Rarity and trivalent logic

A more general approach to the problem of negative uninformativeness is the **rarity assumption** of Oaksford and Chater (1994, 1996):

3.81. **Rarity**: most terms apply only to a small number of objects and hence rarely cross-classify them.

For example, the probability that a table is a toupee is zero, so the only true statement that can be made about them is that "no toupees are tables". We interpret this assumption to mean that most facts about terms are negative and by their abundance have a high probability and so a low informative value. Therefore the rare cases of licit cross-classification have a low probability and a correspondingly high 'surprisal' value.

As defined by Oaksford and Chater, rarity is a bivalent concept that assumes that cross-classifying terms are improbable. To our way of thinking, this statement is accurate but imprecise. It can be made more precise by saying that most terms are sortally incompatible: their preconditions clash so that one cannot be predicated of another. And indeed, Oaksford and Chater's example "no toupees are tables" is a perfect case in point: it is not a case of minimal falsity but rather a case of *radical* falsity. A minimally false version of this example would be something like "no toupees are women's hairpieces", since a toupee *is* a sort of hairpiece, just not one worn by women. The inescapable conclusion is that rarity names the trivalent observation that most cross-classifying terms are uncorrelated, not anticorrelated. This paves the way for a few sortally compatible terms to cross-classify by correlation or anticorrelation. Either method results in a measure of probability that is higher than the zero measure derived from predication between a sortally incompatible subject and predicate.

3.6.6. Quantifying the usage of logical quantifiers

Chater and Oaksford (1999) and Oaksford, Roberts, and Chater (2002) take a novel approach to quantifier meanings by calculating an informativeness value for selected quantifiers from probability-density functions. On the basis of rarity, Chater and Oaksford (1999) argue that the highest density of true statements between any two terms taken at random will correspond to $P(Y|X)$ values of zero. That is to say, rarity implies that negative quantifications should be more

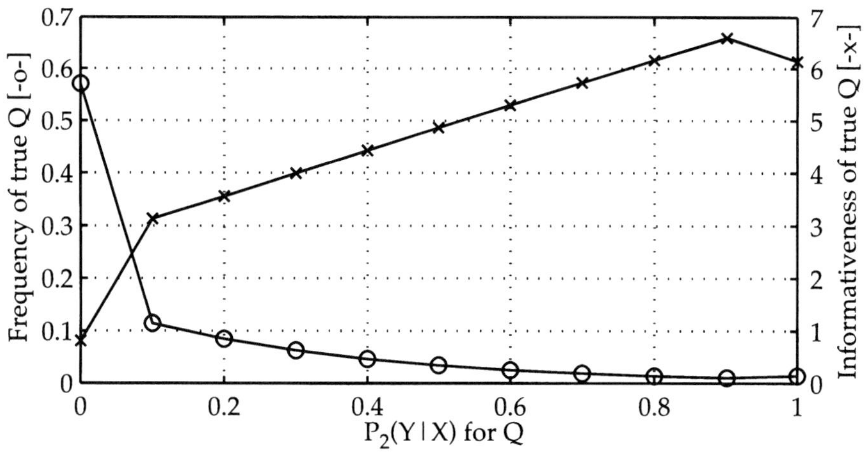

Figure 3.27. Frequency and informativeness of true Q. (**p3.11_CO_freq_info.m**)

frequent than positive quantifications. Chater and Oaksford (1999) use this assumption to calculate an idealized P_2 probability density function which we have replicated on the left scale of Fig. 3.27. It graphs the frequency of true quantified statements by integrating over $P_2(Y \mid X)$ at 0.1 intervals:

3.82. $$P(T) = x + 1.547 \int_{min}^{max} e^{-3P(Y|X)} dP(Y \mid X)$$

The *x* variable represents a quantity that is added to the integral to manipulate the two ends of the graph. For randomly selected terms, the large value at $P_2(Y \mid X) = 0$ indicates that true *no* statements should be very frequent, whereas the small value at $P_2(Y \mid X) = 1$ indicates that true *all* statements should be very infrequent.

To calculate the informativeness of a true quantified statement, its frequency is converted to bits of information using the surprisal or Shannon information formula of Eq. 3.25. The calculations from this formula are graphed on the right scale of Fig. 3.27. The tendency is in conformity with information theory: the lower the probability of making a true quantification, the more informative it is. Chater and Oaksford use these informativeness values to elaborate a theory of how people reason with syllogisms.

The problem with this approach is that *some* cannot be distinguished from *not all*, because they both lie somewhere in the middle of the graph. It would

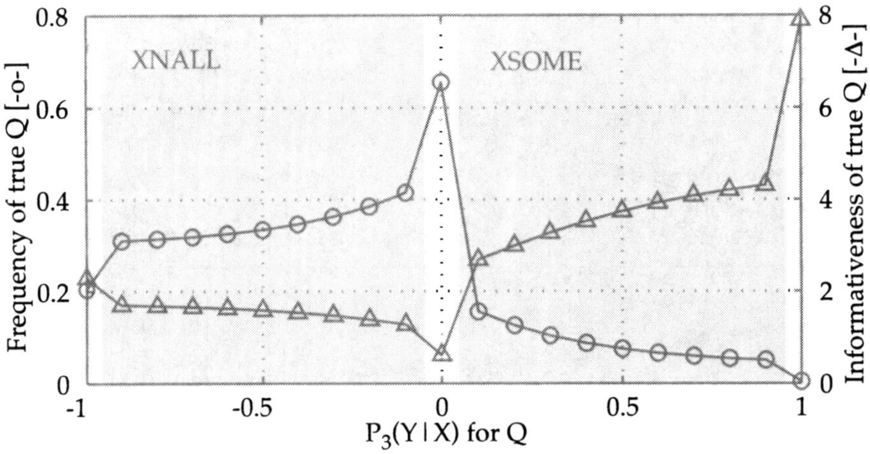

Figure 3.28. Frequency and informativeness of true Q. (**p3.12_P3_freq_info.m**)

help to resolve this problem to convert Fig. 3.27 to the P_3 format, but this cannot be accomplished without a certain amount of tinkering with the *x* variable of Eq. 3.82. On the one hand, we want to build in the rarity assumption, so that $P_3(Y|X) = 0$ has the highest frequency, with $P_3(Y|X) < 0$ being somewhat lower. On the other hand, we want to ensure that maximal values of $P_3(Y|X)$ are less frequent than non-maximal values of the same polarity. Both desiderata are met in Fig. 3.28 by arbitrary manipulations of *x* in Eq. 3.82.

The *y* values are now organized into five groups, which are highlighted by the alternating light and dark bands of the graph. Given that NO and ALL are stipulated to lie significantly outside of the range of NALL and SOME, respectively, NALL and SOME are prefixed with an 'X' to indicate that they exclude their expected endpoints. The ordering of the five groups by either measure is:

3.83 a) Frequency: ALL < XSOME < NO < XNALL < 0
b) Informativeness: 0 < XNALL < NO < XSOME < ALL

The frequency ordering is the closest to the preferred sequence of lexicalization of logical operators reproduced in (3.10) and in Table 3.5, differing only in the interchange of ALL and XSOME. Try as we may, we have not found any account of the order of lexicalization that is more accurate than this, though we return to the topic in Chapters 4, 6, and 8.

3.7. SUMMARY: WHAT IS LOGICALITY?

This chapter introduces various methods for finding patterns in logical-operator space. We have journeyed from measures of cardinality to measures of correlation to measures of angle and norm, and then tied them together via that observation that Pearson's r and the cosine are symmetrical, so the latter can stand in for the former as long as the mean of the observations is not too large. The happy result is that we can bring all the paraphernalia of vector algebra to bear on the description of the logical operators, which includes the vector-space ontology developed for spatial prepositions by Zwarts and Winters and the theory of natural categories developed by Gardenförs and collaborators. Along the way, we found it more accurate to reject the standard bivalent logic in favor of a trivalent one, and to briefly touch on the information-theoretic constraints that seem to govern how logical operators are used.

It turns out that we are about two-thirds of the way through a Marrian three-level analysis of truth-value assignment: correlated spiking produced by some appropriate dynamical system constitutes the implementational level, we do not yet know exactly what algorithm the dynamical system is following, and measure theory supplies the input-output function that constitutes a large part of the computational level.

If the measure-theoretic approach that underlies this chapter is on the right track, it should be able to answer the foundational question, what makes an operator logical? The only previous answer that we know of is elucidated in Westerståhl, 1989, p. 98. Westerståhl proposes that logical constants do not distinguish cardinal numbers, since "...such distinctions belong to mathematics, not logic." While we are sympathetic to Westerståhl's proposal – indeed, our own logical operators do not distinguish cardinal numbers even though they are defined over cardinalities – the issues raised in Chapter 1 sway our sympathy away from his explanation. To echo Ludlow, "...our concern should be with investigating the knowledge that underlies semantic competence, rather than drawing pre-theoretic boundaries around the various disciplines." To state it in pseudo-neurological terms, we find it implausible to believe that children come pre-wired with a distinction between logic and mathematics, and that they can immediately apply it to learning what is a logical operator in their ambient language and what is not.

No, it is the contention of this monograph that at most children come pre-wired with the ability to learn patterns, and the logical operators form a peculiar class because their patterns in LOGOP space are peculiarly salient. What is salient about them is their homogeneity or equi-probability. Consider first the edge detectors, MAX and MIN. For these operators, each y is (anti-)correlated with an x. No distinction is drawn among subsets; each xy (anti-)correlation is just as probable as every other. For the 'blade' detectors POS and NEG, each value of their respective subspace of LOGOP is just as probable as every other; no distinction is drawn between specific values. The vague quantifiers, to cite but one contrast, differ considerably in that their comparison class serves to carve LOGOP space up into non-equi-probable subsets: subsets of smaller values are more probable for *few*; subsets of larger values are more probable for *many*.

We therefore agree with Westerståhl's intuition that the logical operators are immune to cardinality; what we disagree with is his attribution of this observation to a pre-theoretic division between logic and mathematics. In our approach, such boundaries are drawn, if they are drawn at all, by the learner, not the theoretician.

With this background under our belts, we can finally turn to some real linguistic data in the next chapter.

Chapter 4

The representation of coordinator meanings

In this chapter, we substantiate the proposal of the previous chapter that logical operator meanings can be represented via correlation, for the specific case of the logical coordinators. Several advantages of the correlational approach are pointed out along the way, not the least of which is that is provides a common semantics for the coordination of both phrases and clauses, as well as for the discourse coherence of juxtaposed sentences.

4.1. THE COORDINATION OF MAJOR CATEGORIES

Every major category can be coordinated, which produces at least the five possibilities listed in (4.1):

4.1 a) nominal phrases: common noun phrases (NPs) and determined noun phrases (DPs)[23], the latter of which include names
b) verb phrases (VPs) and their inflected projections (IPs)
c) adjective phrases (APs), adjectival PPs, or adjectival clauses
d) adverb phrases (AdvPs), adverbial PPs, or adverbial clauses
e) clauses, also known as complementizer phrases (CPs)

One clarification is in order.

It is that the mixture of prepositional and other types of phrases are among the known counterexamples to the Coordinate Constituent Constraint, from Schachter, 1977, p. 90:

4.2. The constituents of a coordinate construction must belong to the same syntactic category and have the same semantic function.

This constraint springs from the observation made in Chomsky, 1957, p. 36, that (4.3a) is ill-formed, despite the well-formedness of the two conjuncts by themselves in (4.3a'). Further examples from Schachter follow:

[23] We adopt the Determiner Phrase hypothesis of Abney (1987) and much posterior work on the syntax and semantics of minor or functional categories. Within this framework, functional elements head their own category, so that a phrase like *the girl* has an NP embedded within a DP: [$_{DP}$ the [$_{NP}$ girl]].

4.3 a) *The scene of the movie and that I wrote was in Chicago.
a') cf. The scene of the movie was in Chicago, and the scene that I wrote was in Chicago.
b) *John ate quickly and a grilled cheese sandwich.
b') cf. John ate quickly, and John ate a grilled cheese sandwich.
c) *John ate with his mother and with good appetite.
c') cf. John ate with his mother, and John ate with good appetite.

Grosu (1985, 1987) elaborates Schachter's characterization of this phenomenon considerably. For our present purposes, all that is necessary is to point out that the Coordinate Constituent Constraint or some more detailed version thereof imposes an upper bound on the kinds of categories that can be coordinated.

4.2. PHRASAL COORDINATION

Our hypothesis of correlation requires two categories to be correlated. For the relevant coordinated phrases in (4.1), this would be (4.4):

4.4 a) nominal phrase <> a slot or argument position in a predicate
b) slot or argument position in a predicate <> nominal phrase
c) adverbial phrase <> verb
d) adjectival phrase <> nominal

Nominal and verbal phrases can be treated as two sides of the same coin, while adjectival and adverbial phrases can be united under the heading of modification.

4.2.1. The application of nominals to verbals, and vice versa

Under the assumption that nominals are arguments that saturate a slot in a verbal construction, and that verbal constructions are ill-formed if all their available slots are not saturated, nominals and verbal predicates must be explicated at the same time.

4.2.1.1. Verbal predicates as patterns in a space of observations

The first step is to explain how verbal meanings can be represented by vectors. Fig. 4.1 sketches how observations of several simplex or complex verbals can be classified in two dimensions. Without getting too bogged down in details that are beyond the scope of this monograph, this plane plots the direction of motion of the hands with respect to the body against whether the hands are released from contact with the object.

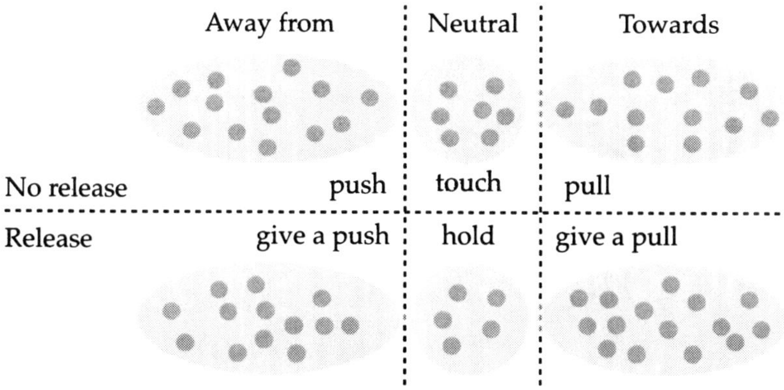

Figure 4.1. Release from contact by orientation of motion.

For the case of release, this lays out the series of punctual eventualities *give a push - touch - give a pull*, while the absence of release lays out the series of durative eventualities *push - hold - pull*. It follows that the meaning of these verbs can be encoded as association between a phonological form and a cluster of observations. No additional insight would be obtained from writing out specific definitions for the verbals, so we refrain from doing so for the sake of brevity.

4.2.1.2. Coordinated names and other DPs

The simplest case to characterize is that of coordinated proper nouns, since we can assume that they are single vectors. To depict them diagrammatically, it is straight-forward to project them onto a one-dimensional grid, as seen on the right side of Fig. 4.2a. To depict coordination, the plane of the coordinated names can be projected onto the plane of the verbal space by double-headed arrows between the coordinatees and their targets. The full diagram of Fig. 4.2a indicates what is needed for the clause *Marta and Chao (each) pushed hard*. The idea is that for AND to correctly recognize the pattern, each point on the input axis coincides with a point in the predicate cluster *pushed hard*. Fig. 4.2b depicts this idea from the perspective of the calculation of correlation.

There is an interesting variety of possibilities for coordinated determiner phrases:

4.5 a) a baboon and an orangutan
b) the baboon and the orangutan
c) the dominant baboon and most orangutans
d) Koko and every orangutan

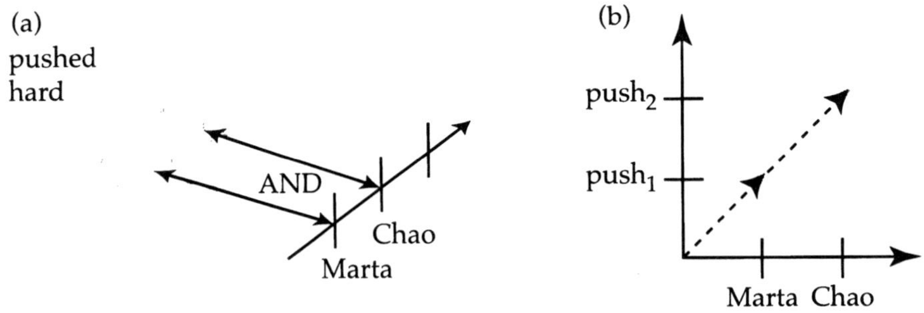

Figure 4.2. Correlation between coordinated names and a predicate.

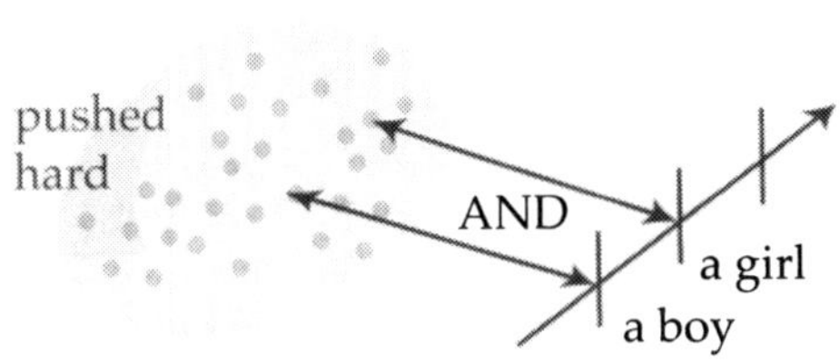

Figure 4.3. Correlation between coordinated DPs and a predicate.

e) some baboons and almost all orangutans
f) [some baboons] and [orangutans]
g) my baboons and your orangutans
h) my baboons and orangutans

Though the representation of quantifiers is not taken up until Chapter 6 – and articles and possessives are not discussed at all – the coordinative format that is most consistent with what has been said so far is to represent each DP on a separate scale, which then correlates to the corresponding predicate observations as in Fig. 4.3. The two inputs correlate with two events of pushing hard in the target space, just as with names.

The following formula defines the calculation:

4.6. $$\mathrm{COOR(DP)} := \cos(\overrightarrow{\mathrm{dp}_i}, \overrightarrow{\mathrm{pred}(x_i)}) \in [-0.71\ 0.71].$$

That is, a coordinated Determiner Phrase finds a correlation between the vector denoting it and the vector denoting the corresponding argument position

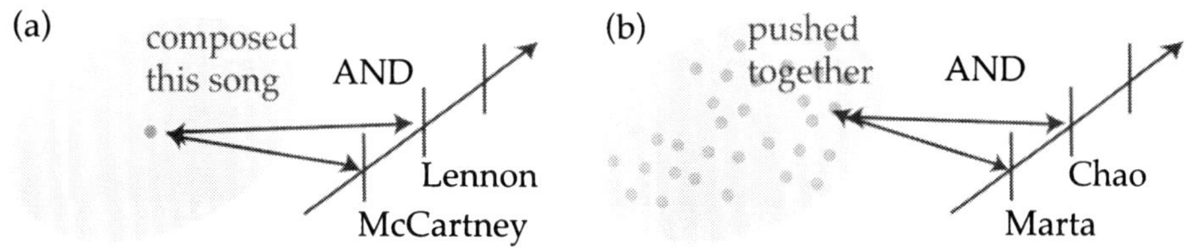

Figure 4.4. Correlation between (a) collective coordinated names and a predicate; (b) coordinated names and a collective predicate.

in a predicate, where the correspondence is given by coindexation with *i*. This correlation is measured by the cosine of the angle formed by the two vectors, to form an *xy* vector space equivalent to the ones considered in Chapter 3.

4.2.1.3. A first mention of coordination and collectivity

This distributive interpretation of the predicate is specifically called for by the usage of *each* in *Marta and Chao (each) pushed hard*, yet there are instances in which a collective interpretation of a coordination is called for. Consider AND coordinations in which the inference from left to right in (4.7) is blocked:

4.7. McCartney and Lennon composed this song. ↛ McCartney composed this song, and Lennon composed this song, too.

Here *McCartney and Lennon* are conceived of as a collective, so collaboration constitutes but a single event of song writing. Fig. 4.4a illustrates the difference between this and the preceding case. The diagram is no more than a illustration of the prose description: the two coordinatees coincide with a single element in the predicate.

Not only can arguments be collective, so too can predicates. A simple example of a collective predicate is *Marta and Chao pushed together*, diagrammed in Fig. 4.4b. The arrows converge on the same observation of *push* on the left side, because this is what adverbial *together* requires. The result is that collectivity can be enforced either by the subject or by the predicate, independently.

However, the calculation of correlation will not uphold the analyses depicted in Fig. 4.4, since the relationship between the two components of the vector denoted by either predication is not additive. The only way to resolve the disparity is to reduce the two entities to a single one, such as by the stipulation in (4.8) that mandates the length of their vector to be unity:

4.8. $\mathrm{COOR}^{\mathrm{COLL}}(\mathrm{DP}) := |\overrightarrow{\mathrm{dp}}| = 1 \;\&\; \cos(\overrightarrow{\mathrm{dp}_i}, \overrightarrow{\mathrm{pred}(x_i)}) = 0.71.$

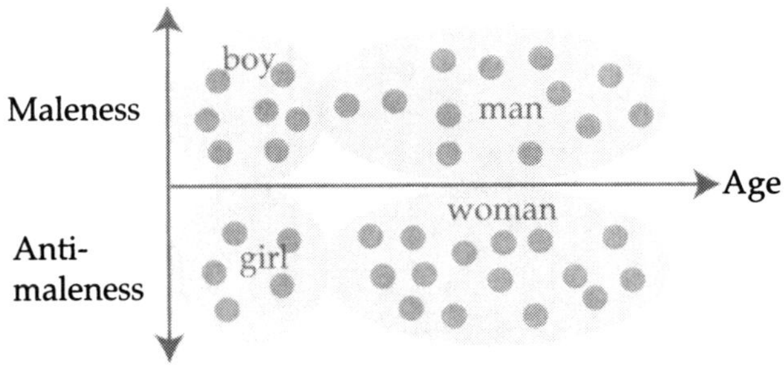

Figure 4.5. Age × maleness in English.

Chapter 9 delves deeply into the peculiarities of these constructions.

4.2.1.4. Common nouns as patterns in a space

As a first step, let us approach the meaning of a singular or plural noun as a problem in pattern classification. Take as an illustrative example the count noun *boy*. Its denotation is composed of values for at least two parameters, age and male gender. Let us say that age is measured on a scale from 0 to 1, where 0 indicates the least age and 1 the greatest (for humans), while male gender is given by a scale of maleness from 1 to -1, where 1 is most male and -1 is most anti-male, i.e. female (in a patriarchal culture, see Howard (2001) for further refinement). Then any observation of a boy in our 'model' can be plotted on a Cartesian coordinate system that takes the measure of age as the x axis and the measure of maleness as the y axis. A fictitious example of such a plot is given in Fig. 4.5, with the other gender-denoting terms of *girl, man* and *woman* included for the sake of comparison. This type of representation presupposes that common nouns have an internal semantic structure like that of vectors, exemplified in (4.9) for the presumed prototypical members of each category, where the components of the vectors are found in (age, maleness):

4. 9	a)	prototypical boy	↔	$[0.2, 0.4]^T$
	b)	prototypical girl	↔	$[0.2, -0.4]^T$
	c)	prototypical man	↔	$[0.65, 0.75]^T$
	d)	prototypical woman	↔	$[0.65, -0.75]^T$

This notion of an internal vectorial structure is put to good use in the upcoming discussion.

4.2.1.5. Coordinated common nouns

Common nouns can be coordinated with or without various modifiers:

4.10 a) baboons and orangutans
b) large baboons and small orangutans
b′) large [baboons and orangutans]
c) thirty baboons and twenty orangutans
c′) thirty [baboons and orangutans]
d) some baboons and all orangutans
d′) some [baboons and orangutans]

Let us assume that the propensity for coordination of a DP is actually inherited from its embedded NP in some systematic fashion that does not concern us here, but which should be based on the following definition:

4.11. $\text{COOR(NP)} := \cos(\overrightarrow{np_i}, \overrightarrow{pred(x_i)}) \in [-0.71\ 0.71]$.

For instance, the unprimed examples in (4.10) are coordinated DPs, which are derived from their embedded NPs.

The primed examples cannot be dispatched so hastily, however, for some of them have a collective reading:

4.12 a) thirty [baboons and orangutans] sleep in the treetops
b) ≠ thirty baboons sleep in the treetops, and thirty orangutans sleep in the treetops

Numerals should impose a length on the vector of the noun that they modify, defined in (4.13a), which serves to restrain the predicate vector to that same length, or less, through the calculation of correlation:

4.13 a) $\text{NUM(NP)} := |\overrightarrow{np}| = n$
b) $\text{COOR(NUM(NP))} := |\overrightarrow{np}| = n \ \& \cos(\overrightarrow{np_i}, \overrightarrow{pred(x_i)}) \in [-0.71\ 0.71]$.

This may be an idiosyncrasy of the numerals. Our intuitions are that an adjective like *large* is distributive.

4.2.1.6. Coordinated verbs

The next category is that of verb phrases, VP. We assume that (4.14) is an example of a coordination of VPs.

4.14. Anabel woke up, leapt out of bed and ran off to work.

These are the most difficult to fit into the framework constructed so far.

Given that singular proper names have not been conceptualized as anything more than an elementary point or vector, there are not three entities for the three coordinated verbs to connect to. That is, we cannot look at the

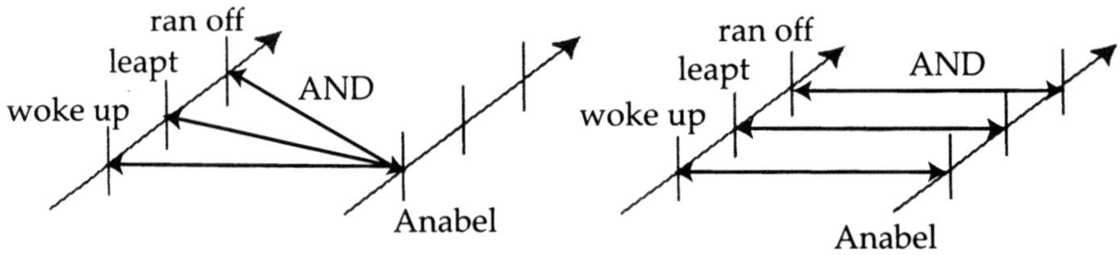

Figure 4.6. (a) Incorrect rendering of verbal coordination; (b) correct rendering of verbal coordination.

representation of *Anabel* in Fig. 4.6a and find three elements in it to fill out the vector $[3, 3]^T$, because the point representing *Anabel* has no internal structure.

It would be *ad hoc* to modify the analysis of coordination just to fit this datum, so let us modify the analysis of proper names instead. The simplest assumption is to assume that they have an internal temporal structure, as depicted in Fig. 4.6b. That is, the name is a proxy for a more complex object, in pattern-recognition terms, a high dimensional vector. This is compatible with the Montague Grammar definition of the extension of a proper noun as the set of sets of which the noun is a member, see for instance Chierchia and McConnell-Ginet, 1990, p. 417. In particular, Anabel is a member of the sets named by {... {*wake_up*, *leap*, *run_off*} ...} at the time referred to. The vectorial definition is the following:

4.15. $\text{COOR}(\text{VP}(x)) := \cos(\overrightarrow{dp_i}, \overrightarrow{vp(x_i)}) \in [-0.71\ 0.71]$.

In this way, one of the most superficially recalcitrant instances of coordination can be assimilated into the developing framework.

4.2.1.7. Coordination beyond the monovalent predicate

Up to now, we have made the simplifying assumption of treating predicates as if they were all intransitive. It would be derelict of us to leave the analysis in this stunted state, so let us take a moment to imagine what would be needed to open it up to the predicates with more than one argument. Consider the example of (4.16):

4.16. Anabel pushed Rukayyah and Chris (one after the other).

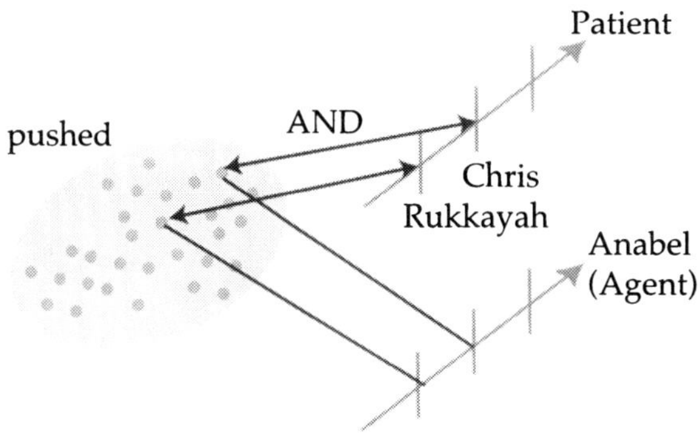

Figure 4.7. Coordination of non-subjects.

Its representation in Fig. 4.7 displays correlation between two events of pushing and two patients, over sequential sub-properties of the agent.

4.2.1.8. Multiple coordination and *respectively*

More complex alignments come from multiple coordinations in the same clause, of which (4.17) is a simple example. It has three readings which are listed beneath:

4.17. Anabel and Yetunde pushed Rukayyah and Chris.

a) both arguments are distributive, so that each pusher pushes each pushee, to give four events of pushing,

b) a collectively-coordinated agent pushes a distributively-coordinated patient, to give two events of pushing, and

c) a distributively-coordinated agent pushes a collectively-coordinated patient, to again give two events of pushing.

More generally, the number of events observed across a polyvalent predicate with coordinate arguments appears to have a greatest lower bound as the smallest coordination and a least upper bound at the product of all of the coordinations.[24]

[24] This suggests that the input from separate thematic roles is integrated by weighted summation, though we have no specific proposals to offer in what follows. The idea is that $P \times P$ is the sum of the number of connections in the fully connected circuit, which is reduced to smaller numbers by zero or negative weightings of individual connections. For

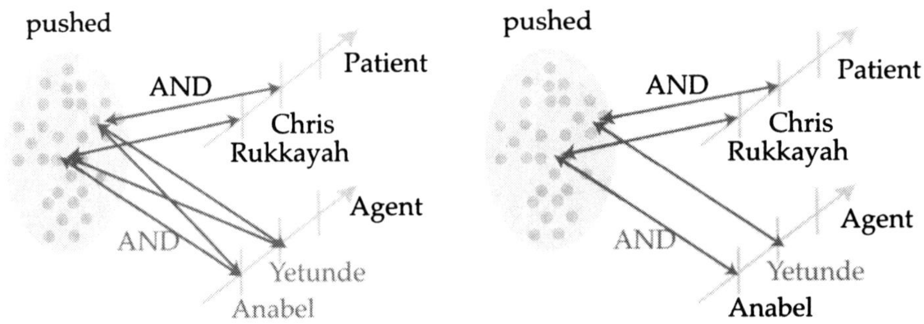

Figure 4.8. (a) Multiple coordination (collective agent on distributive patient); (b) *respectively* coordination.

Returning to specific examples, Fig. 4.8a illustrates (4.17b). Note how correlation permits the multiplication of events: the distributed patient must correlate to two events, while the collective agent need only correlate to one. To satisfy both constraints simultaneously, the collective agents each wind up correlating to separate events. Since correlation does not count the number of correlated pairs, it is not sensitive to whether other aspects of the clause duplicate the number of pairs, as long as the correlations are made correctly.

These considerations lay the groundwork for a quick glance at *respectively* coordination. Consider how a clause like *Anabel and Yetunde pushed Rukayyah and Chris, respectively* differs from the coordination seen in (4.17). *Respectively* reduces its three readings to just one, *Anabel pushed Rukayyah, and Yetunde pushed Chris,* represented in Fig. 4.8b. This appears to be the minimal way of satisfying the requirements of two coordinators: both the agent coordination and the patient coordination correlate to two events of pushing – they just regiment themselves so that the first event relates the first member of each coordination, and the second event relates the second member.

4.2.2. Modification

Having laid out the basics of a theory of correlational coordination of arguments and predicates, it remains to sketch a corresponding theory of coordination of modifiers.

4.2.2.1. Coordinated adjectivals

(4.18) adduces some examples of coordinated adjectivals:

---

instance, the *respectively* reading can be derived from the fully connected reading by zeroing out or lateral inhibition of the crossing connections.

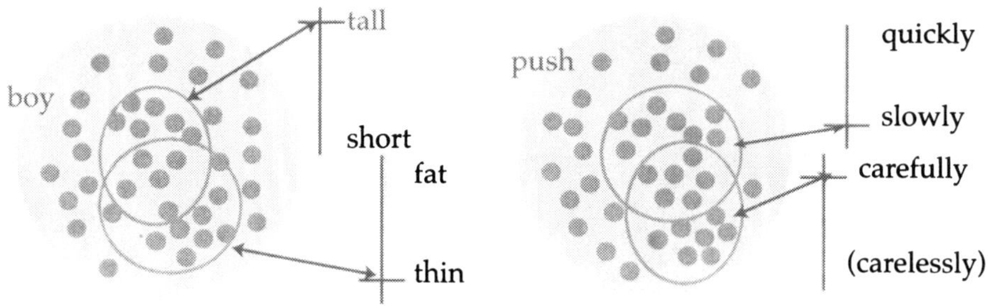

Figure 4.9. (a) tall and thin boys; (b) push slowly and carefully.

4.18 a) My goldfish is fat and happy.
b) My goldfish is happy and without a care in the world.
c) a fat and happy goldfish
d) a goldfish (that is) free from parasites and happy to be alive

At least for restrictive adjectivals, their coordination is played out over the representations of common noun meanings given above. Fig. 4.9 enlarges the boy cluster of Fig. 4.5 in order to show how 'tall and thin boy(s)' would be calculated. 'Tall' and 'thin' are both poles of two different scales of measurement, as depicted. The projection of each adjective onto the 'boy' cluster picks out a sub-cluster of boys. This is a one-to-one and positive mapping, which licenses AND, and more generally, any other logical coordinator, where "nomp" stands in for NP or DP:

4.19. $$\text{COOR(AdjP)} := \cos(\overrightarrow{\text{adjp}_i}, \overrightarrow{\text{nomp}_i}) \in [-0.71\ 0.71].$$

Thus the proper-name hypothesis of coordination is instantiated by a coordination of adjectivals as well.

The question that is begged by all of this is where on the map of age by maleness from which 'boy' has been plucked is the information for height and girth? The only answer is to assume that the map actually abbreviates a much higher dimensional space for the classification of the observations, which was pruned down to just those components that make the most contribution to the definitions of 'boy', 'girl', 'man', and 'woman'. The extra information for height and girth is thus hidden 'behind' the representation in Fig. 4.9a and can be retrieved by following the vectors in Fig. 4.9a out to their full extent.

4.2.2.2. Coordinated adverbials

The story for coordinated adverbials does not differ substantially from that of coordinated adjectivals. First, some examples:

4.20 a) Belinda lectured slowly and carefully.
b) Belinda lectured slowly and with great care.
c) John ate quickly and without much appetite.

Of course, adverbs also modify adjectives and other adverbs, but to the best of our knowledge, such constructions are not behave significantly different from the verbal case with respect to coordination

Just as restrictive adjectivals can be projected from a scale onto the cluster of a noun, so too can restrictive adverbials be projected from a scale onto the cluster of a verb, as depicted in Fig. 4.9b and defined below:

4.21. $$\text{COOR(AdvP)} := \cos(\overrightarrow{advp_i}, \overrightarrow{vp_i}) \in [-0.71\ 0.71].$$

As before, it must be assumed that only the most relevant components of the observations contribute to the delineation of the cluster, leaving the other components to play the secondary roll of supplying the information needed by a modifier.

4.2.3. Summary of phrasal coordination and vector semantics

The preceding eleven subsections offer a vector-based theory of the coordination of phrases which itself piggy-backs on the way in which various phrasal categories are applied to one another. The tools of model-theoretic semantics would allow us to elaborate our proposals more precisely, but since we have few ideas about the neurological grounding of these tools, we will not invoke them here. The principal claim is that, however a correlation is effected between two phrasal categories that enter into a grammatical relation, this correlation is taken advantage of to (anti-)correlate additional elements by way of coordination.

4.3. CLAUSAL COORDINATION

The reader may recall from the beginning of the chapter that there are five types of categories relevant to coordination, listed in (4.1). (4.1e), the clause, is the topic of this section. Having devoted so many pages to a vector-theoretic analysis of phrasal coordination, it is to be hoped that the results generalize to clausal coordination without too much stipulation. Actually, the results generalize with only the most natural of stipulations.

4.3.1. Conjunction reduction as vector addition

We know of no evidence that the meaning of the coordinators in their clausal usage differs from their meaning in their phrasal usage, *modulo* the different meaning potentials of clauses and phrases. This observation motivated early generative grammar to propose a rule of **conjunction reduction** by means of which to derive a phrasal coordination from the corresponding clausal coordination. For instance, Chomsky's usage of coordination as a diagnostic for phrase structure, Chomsky, 1957, p. 35, relied on the assumption that coordination operated at the sentential level. His idea is that...

> If we have two sentences *Z+X+W* and *Z+Y+W*, and *X* and *Y* are actually constituents of these sentences, we can generally form a new sentence *Z-X+and+Y-W*. For example, from the sentences [4.22a, b] we can form the new sentence [4.22c]:

4.22 a) the scene - of the movie - was in Chicago
b) the scene - of the play - was in Chicago
c) the scene - of the movie and of the play - was in Chicago

It was soon observed[25] that some phrasal coordinations cannot be derived from coordinated sentences, because the putative sources are not grammatical:

4.23 a) Jimmie and Timmie are a pair of fools.
b) *Jimmie is a pair of fools and Timmie is a pair of fools.

Chapters 8 and especially 9 will deal with what is wrong with this derivation; here we would like to concentrate on what is right with it, or at least with (4.22).

From our perspective, the natural hypothesis is to propose some way to map between the phrasal and clausal vectors. We have already explained what the phrasal vectors look like, so let us turn to the clausal representation. The initial assumption should be that each main clause is encoded as a separate vector, in order to formalize the fact that such clauses are independent or free-standing constituents. Recycling the example of *Marta and Chao pushed hard*, in a coordinate system in which x encodes the pushers and y the hard pushings, the clausal coordination takes on the form of Fig. 4.10a. The mapping of this representation into the phrasal format of Fig. 4.10b should be obvious: it is just the addition of the two clausal vectors. More generally, we may say that

---

25 In Peters (1966) and Smith (1969), for instance. Lakoff and Peters (1969) credit Curme, 1931, p. 162, with having noticed this phenomenon long before generative grammar. Note that most of this historical sketch is based on van Oirsouw, 1987, §1.1.

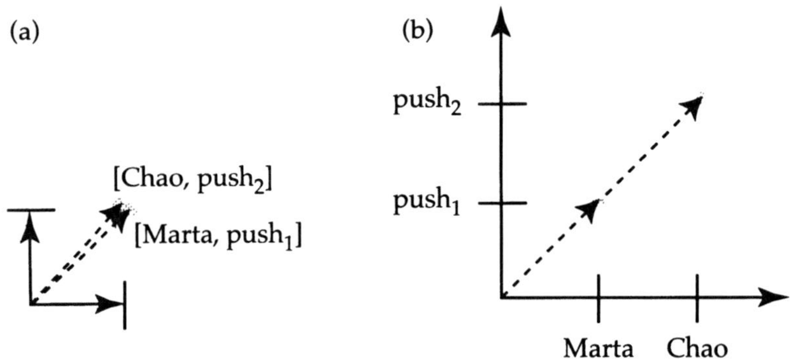

Figure 4.10. Conjunction reduction. (a) coordinated clauses; (b) coordinated phrases.

conjunction reduction is a linear transformation *CR* between two vector spaces of the same dimensionality *n* in which the sum of the *m* clausal vectors $\vec{s}_i$ gives the conjunctively reduced clause $\vec{S}_{cr}$, stated in (4.24).

4.24. $$CR : \Re^n \rightarrow \Re^n \text{ given by } \vec{S}_{cr} = \sum_{1}^{m} \vec{s}_i .$$

The rest of this section will be spent following up some of the ramifications of this hypothesis.

4.3.2. Coordination vs. juxtaposition and correlation

There is glaring omission in the discussion of coordinator meanings which should now be taken up before we move on to neuromimetic models of coordination, to wit, that we have not offered any direct evidence for the correlational hypothesis of coordination. Such evidence is forthcoming in this subsection, which shows how correlation supplies the correct mechanism for the differences between asymmetric coordination and discourse coherence, and ultimately, for the Common Topic Constraint.

4.3.2.1. Asymmetric coordination

It is a staple of propositional calculus that the conjunction operator '&' is commutative, which is to say that the order of its conjuncts does not matter. And indeed, for many cases of natural-language coordination, such as (4.25b) and (4.25b'), propositional calculus makes the right prediction:

4.25 a) Commutativity: p & q ≡ q & p.
b) Paris is the capital of France, and Rome is the capital of Italy.
b') = Rome is the capital of Italy, and Paris is the capital of France.

However, as Schmerling, 1975, p. 211, mentions, linguists and philosophers have long noted the existence of clausal coordinations with *and* that violate the prediction of propositional calculus. There are two general sorts, a temporal-precedence reading and a cause-and-effect reading.

The temporal-precedence relation among clausal coordinatees follows their linear order:

4.26 a) Harry robbed the bank and drove off in a car. (Lakoff and Peters, 1969, 123)
a') ≠ Harry drove off in a car and robbed the bank.
b) A republic has been declared and the old king has died of a heart attack. (Cohen, 1971, 3)
b) ≠ The old king has died of a heart attack and a republic has been declared. (" 4)
c) The Lone Ranger broke the window with the barrel of his gun, took aim, and pulled the trigger. (McCawley, 1971, 68)
c') ≠ The Lone Ranger pulled the trigger, took aim, and broke the window with the barrel of his gun. (" 69)
d) Harold opened his briefcase, and he ceremoniously pulled out his completed term paper. (Bar-Lev and Palacas, 1980, 1)
d') ≠ Harold ceremoniously pulled out his completed term paper, and he opened his briefcase.

This reading depends on the clauses being understood punctually. The cause-and-effect reading relaxes this constraint, so that the non-punctual – progressive or generic – clauses in (4.27) can also produce a violation of commutativity:

4.27 a) Tom has a typewriter, and he types all his own letters. (Cohen, 1971, 6)
a') ≠ Tom types all his own letters and he has a typewriter.
b) The police came into the room and everyone swallowed their cigarettes. (R. Lakoff, 1971, 40)
b') ≠ Everyone swallowed their cigarettes and the police came into the room. (" 42)
c) John raised the blinds, and the sun poured into the room. (Bar-Lev and Palacas, 1980, 2)
c') ≠ The sun poured into the room, and John raised the blinds.
d) The lights were off, and I couldn't see. (" 3c)
d') ≠ I couldn't see, and the lights were off.

Due to some historical accident that we have not been able to pin down, these unpredicted usages have come to be known as **asymmetric**, rather than the more accurate term 'non-commutative'.

Grice (1975) initiated the modern analysis of asymmetric coordination by proposing that *and* means what it means in propositional calculus, but that it can be used to implicate additional, asymmetric meanings because people obey certain maxims of cooperative conversation when the speak. For this particular case, a speaker obeys the maxim "Be orderly", which motivates the speaker to put the coordinatees in the order in which they actually take place.

Grice's account provoked considerable response, both in its general intent to prove the usefulness of conversational implicature and in its specific treatment of asymmetric coordination. We have already mentioned the reduction of Grice's maxims to Horn's **Q** and **R** principles in the previous chapter. We will not review the literature on asymmetric coordination – the reader is welcome to piece it together from the references that we cite – but rather cut to the chase as quickly as possible.

To our way of thinking, the crucial flaw in the previous approaches to asymmetric coordination is that they do not explain why it licenses the chronological and causal readings to the exclusion of any others. To be charitable, this may be due to the belief that there is an imponderable number of alternative meanings that are not licensed, so there is no practical algorithm for examining them to show way they are excluded. However, this 'straw man belief' is erroneous. In fact, a list of potential meanings does exist, and it is not very long. It is founded on what Hume (1955[1748]) calls "connection among ideas", refined in Hobbs (1979, 1982, 1990) in order to understand the relationships that hold between clauses that enable them cohere together as a text, and reaches its most recent distillation in Kehler (2002). Given that our theory of coordination as correlation constrains rather drastically the coherence relationships that coordinatees can enter into, all we need to do is check Kehler's list to see which relationships require correlation and which ones should not. The coherence relationships that require correlation should be the ones licensed by asymmetric coordination, and the coherence relationships that are incompatible with correlation should be the ones not licensed by asymmetric coordination.

4.3.2.2. Kehler's coherence relations

Kehler draws an initial distinction among Resemblance, Cause-Effect, and Contiguity. Let us run through them with the goal of discovering their core constituents and then make a second pass to analyze them in terms of correlation. It will save time to point out at the beginning that all of Kehler's coherence relations are stated between two sentences, S_1 and S_2, and that many of them involve applying a predicate p_1 to a set of entities $a_1, \ldots, a_n$ in S_1 and a predicate p_2 to a set of entities $b_1, \ldots, b_n$ in S_2.

Our vector-theoretic reformulation of Kehler's ideas starts from the assumption that predicates and arguments of S_1 and S_2 are represented together by a single vector $\vec{s}_1$ and $\vec{s}_2$ for each sentence. A passage consisting of S_1 and S_2 is coherent if $\vec{s}_1$ and $\vec{s}_2$ are (anti)correlated:

4.28. $\text{Coherence}(S_1, S_2) := \cos(\vec{s}_1, \vec{s}_2) \in \{-1, 1\}.$

Incoherence then falls out as all the other values, though 0 would seem to be an especially salient candidate.

Having defined coherence in a general way, a pressing question is whether any specific coherence relation licenses coordination. It turns out that there are three that do, and they are described by variations on the following theme:

4.29. $S_1 \text{ and } S_2 := \cos(\vec{s}_1, \vec{s}_2) = 1 \;\&\; \vec{s}_1 \leq \vec{s}_2$

That is, the vectors denoted by the two sentences must correlate, and the first vector must precede the second in some ordering. After laying out the evidence for this claim, we will attempt to explain it in a separate section.

4.3.2.2.1. The data structure

Finally, a few words should be said about the vector space that encodes the knowledge of American politics that informs many of Kehler's examples, both because the reader may not be familiar with it and because our formal renderings of Kehler's examples are stated within its confines. The basis of the space are three axes:

4.30 a) x: oppose(y, z) = $[-1, 0, 0]^T$, support(y, z) = $[1, 0, 0]^T$.
b) y: hi-ranking Republican = $[0, -1, 0]^T$, hi-ranking Democrat = $[0, 1, 0]^T$.
c) z: George W. Bush = $[0, 0, -1^T]$, Al Gore = $[0, 0, 1]^T$.

This space itself is displayed in Fig. 4.11. The prototypical vector – the largest arrowhead – encodes *support(high-ranking-Democratic-politician, Gore)* as $[1, 1, 1]^T$. For the sake of illustration, let us say that the there are two assertions which are close, but not identical, to the prototype. For instance, *organized-rallies-for(Gephart, Gore)* can take on the value $[0.98, 0.99, 1]^T$, and *distributed-pamphlets-for(Daschle, him)* can take on the value $[0.99, 0.98, 1]^T$. These are two vectors that flank the prototype Fig. 4.11.

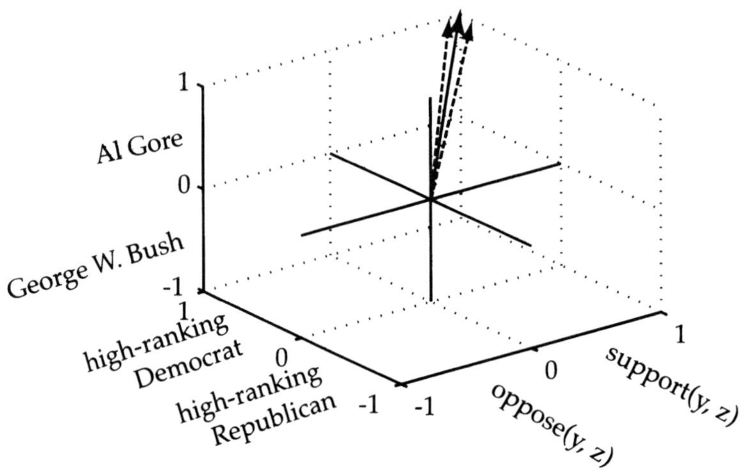

Figure 4.11. The illustrative data space for coherence relations, showing two highly correlated vectors clustered around a prototype vector.

With this background, we can turn to Kehler's coherence relations, starting with the family dubbed Resemblance.

4.3.2.2.2. Coherence relations of Resemblance

The archetypal instance of a Resemblance relation is Parallel, defined by Kehler as in (4.31a) and exemplified as in (4.31b). (4.31c) uses Kehler's explication in prose on p. 16 to instantiate the variables of the definition with the linguistic constituents of the example:

4.31 a) Infer $p(a_1, a_2, ...)$ from the assertion of S_1 and $p(b_1, b_2, ...)$ from the assertion of S_2, where for some property vector $\vec{q}$, $\vec{q}_i(a_i)$ and $\vec{q}_i(b_i)$ for all i.

b) Dick Gephart organized rallies for Gore. Tom Daschle distributed pamphlets for him.

c) Infer *support(Gephart, Gore)* from S_1 and *support(Daschle, him)* from S_2, where *high-ranking-Democratic-politician(Gephart)* and *high-ranking-Democratic-politician(Daschle)*.

The challenge of Parallel is to ascertain the common relation p and the property vector $\vec{q}$. We know that p is as we have stated it in (4.31c) because Kehler says so ("...the common relation p that subsumes [*organized rallies for* and *distributed pamphlets for*] might thus be roughly the relation denoted by *do something to support"*). We have chosen *high-ranking-Democratic-politician* for $\vec{q}$ because it is suggested by Kehler's analysis of upcoming relations.

Saying all of this with vectors is considerably simpler, since p and $\vec{q}$ are not distinguished by different types of formal objects. (4.32a) states our definition, where i refers to one or more vector components. The two lines of (4.32b) illustrate the definition with the relevant vectors drawn from the space of Fig. 4.11. (4.32b′) supplies the putative prototype or common relation. (4.32c) demonstrates that Parallel lends itself readily to clausal coordination:

4.32 a) Parallel(S_1, S_2) := $i \in \vec{s}_1, \vec{s}_2 \; \cos(\vec{s}_1(i), \; \vec{s}_2(i)) = 1$.

b) $\vec{s}_1$ = [organized_rallies_for$_1$, Gephart$_2$, Gore$_3$]^T = $[\approx 1, \approx 1, 1]^T$.

$\vec{s}_2$ = [distributed_pamphlets_for$_1$, Daschle$_2$, him$_3$]^T = $[\approx 1, \approx 1, 1]^T$.

b′) [support$_1$, high-ranking_Democratic_politician$_2$, Gore$_3$]^T = $[1, 1, 1]^T$.

c) Dick Gephart organized rallies for Gore, and Tom Daschle distributed pamphlets for him.

It is not clear to us that Parallel need stipulate a common relation such as that of (4.32b′). The correlated vector components might automatically activate a more general relation, or they might not; it would seem to depend on the knowledge base. In any event, this is an empirical question which is too peripheral to our concerns to pursue here. What is crucial is that the two assertions be similar enough to be highly correlated, which is certainly the case of the representations in (4.32). Such correlation licenses the clausal *and* coordination of (4.32c), in conformity with the claim of (4.29) and with the proviso that the order is '='.

A second sort of Resemblance relation attends to points of contrast, of which Kehler recognizes two subsorts. Contrast1 stipulates an opposition among the predicates. (4.33) puts it into the format of (4.31):

4.33 a) Infer $p(a_1, a_2, \ldots)$ from the assertion of S_1 and $\neg p(b_1, b_2, \ldots)$ from the assertion of S_2, where for some property vector $\vec{q}$, $\vec{q}_i(a_i)$ and $\vec{q}_i(b_i)$ for all i.

b) Gephart supported Gore. Armey opposed him.

c) Infer *supported(Gephart, Gore)* from S_1 and ¬*supported(Armey, Gore)* from S_2, where *high-ranking-politician(Gephart)* and *high-ranking -politician(Armey)*.

Contrast2 stipulates an opposition among the entities that participate in the predication:

4.34 a) Infer $p(a_1, a_2, \ldots)$ from the assertion of S_1 and $p(b_1, b_2, \ldots)$ from the assertion of S_2, where for some property vector $\vec{q}$, $\vec{q}_i(a_i)$ and $\neg\vec{q}_i(b_i)$ for all i.

b) Gephart supported Gore. Armey supported Bush.

c) Infer *supported(Gephart, Gore)* from S_1 and *supported(Armey, Gore)* from S_2, where *[Democratic-politician, Democratic-candidate](Gephart, Gore)* and *¬[Democratic-politician, Democratic-candidate] (Armey, Bush).*

We prefer to lump Contrast 1 and 2 together, because the vectorial representation does not need a special apparatus to keep the predicate and its arguments separate. One apparatus fits all:

4.35 a) $\text{Contrast}(S_1, S_2) := i \in \vec{s}_1, \vec{s}_2 \cos(\vec{s}_1(i), \vec{s}_2(i)) = -1.$

b) $[\text{supported}_1, \text{Gephart}_2, \text{Gore}_3]^T = [1, 1, 1]^T.$

$[\text{opposed}_1, \text{Armey}_2, \text{him}_3]^T = [-1, -1, 1]^T.$

c) Gephart supported Gore, {??and/but} Armey opposed him.

d) $[\text{support}_1, \text{Gephart}_2, \text{Gore}_3]^T = [1, 1, 1]^T.$

$[\text{support}_1, \text{Armey}_2, \text{Bush}_3] = [1, -1, -1]^T.$

e) Gephart supported Gore, {and/but} Armey supported Bush.

The two lines of (4.35b) represent the predicate contrast of (4.33b); the two lines of (4.35d) represent the entity contrast of (4.34b). Clausal conjunction in (4.35c) and (4.35e) is preferentially marked with *but*. To the extent that *and* is felicitous, it is licensed by an alternative Parallel reading of the passage. For instance, (4.35e) can be read as a generalization over the entity components to recover a prototype vector illustrated here:

4.36. $[\text{support}_1, \text{high-ranking\_politician}_2, \text{pres\_cand}_3]^T.$

That is to say, the reading neutralizes party affiliations, so Gephart and Armey share the property of being high-ranking politicians, and Gore and Bush share the property of being presidential candidates. This correlation is what licenses *and*, just as in the previous case of Parallel.

The third instance of Resemblance that Kehler posits is Exemplification:

4.37 a) Infer $p(a_1, a_2, \ldots)$ from the assertion of S_1 and $p(b_1, b_2, \ldots)$ from the assertion of S_2, where b_i is a member or subset of a_i for some i.

b) Young aspiring politicians often support their party's presidential candidate. For instance, Bayh campaigned hard for Gore in 2000.

c) Infer *support(young_pol, pres_cand)* from S_1 and *support(Bayh, Gore)* from S_2, where Bayh is a young aspiring politician, and Gore is Bayh's party's presidential candidate.

The idea is that some element of the second sentence is a member of the first.

In vector-theoretic terms, the vector representing the second sentence points to the matrix or set of vectors, $\vec{S}_1$, representing the first sentence:

4.38 a) $\text{Exemplification}(S_1, S_2) := \vec{i} \in \vec{S}_1 \cos(\vec{i}, \vec{s}_2) = 1.$

b) $\vec{S}_1 = [\text{support}_1, \text{young\_pol}_2, \text{pres\_cand}_3]^T =$
$[0.8{:}1.2, 0.9{:}1.1, 0.9{:}1.1]^T.$
$\vec{s}_2 = [\text{campaign\_for}_1, \text{Bayh}_2, \text{Gore}_3]^T = [1.1, 0.95, 1]^T.$

c) #Young aspiring politicians often support their party's presidential candidate, and Bayh campaigned hard for Gore in 2000.

Under the assumption that a sentence must be exhaustively parsed into a single vector, coordination should be infelicitous. This prediction is validated, as (4.38c) adduces.

The coherence relation of Generalization reverses the order of sentences stipulated by Exemplification:

4.39 a) Infer $p(a_1, a_2, \ldots)$ from the assertion of S_1 and $p(b_1, b_2, \ldots)$ from the assertion of S_2, where a_i is a member or subset of b_i for some i.

b) Bayh campaigned hard for Gore in 2000. Young aspiring politicians often support their party's presidential candidate.

c) Infer *support(Bayh, Gore)* from S_1 and *support(young_pol, pres_cand)* from S_2, where Bayh is a young aspiring politician, and Gore is Bayh's party's presidential candidate.

The semantic relation remains the same, however, so Generalization is covered by what was said about Exemplification.

4.40 a) $\text{Generalization}(S_1, S_2) := \vec{i} \in \vec{S}_2 \cos(\vec{s}_1, \vec{i}) = 1.$

b) $[\text{campaign\_for}_1, \text{Bayh}_2, \text{Gore}_3]^T = [1.1, 0.95, 1]^T.$
$[\text{support}_1, \text{young\_pol}_2, \text{pres\_cand}_3] = [0.8{:}1.2, 0.9{:}\ 1.1, 0.9{:}\ 1.1]^T.$

c) #Bayh campaigned hard for Gore in 2000, and young aspiring politicians often support their party's presidential candidate.

The coherence relations of Exception 1 and 2 introduce negation into the constraints of Exemplification and Generalization, respectively:

4.41 a) Infer $p(a_1, a_2, \ldots)$ from the assertion of S_1 and $\neg p(b_1, b_2, \ldots)$ from the assertion of S_2, where b_i is a member or subset of a_i for some i.

b) Young aspiring politicians often support their party's candidate. However, Rudy Giuliani supported Mario Cuomo in 1994.

c) Infer *support(young_pol, party's_cand)* from S_1 and *support(Giuliani, Cuomo)* from S_2, where Giuliani is a young aspiring politician, and Cuomo is not Giuliani's party's candidate.

This first sort of Exception tracks Exemplification, with the difference that the relation inferred from the second sentence is negated. The example may not be as transparent as some of the others, since it relies on the reader knowing that Giuliani is a Republican, and Cuomo, a Democrat. It licenses Exception1 because Giuliani does not support the candidate of his party.

With respect to the vector approach, the vector denoted by the second sentence points away from the space denoted by the first sentence.

4.42 a) $\text{Exception1}(S_1, S_2) := \vec{i} \in \vec{S}_1 \cos(\vec{i}, \vec{s}_2) = -1$.

b) $[\text{support}_1, \text{young\_pol}_2, \text{party's\_cand}_3]^T = [1, 0.9{:}\ 1.1, 0.9{:}\ 1.1]^T$.
$[\text{support}_1, \text{Giuliani}_2, \text{Cuomo}_3]^T = [1, -0.95, 1]^T$.

c) #Young aspiring politicians often support their party's candidate, and Rudy Giuliani supported Mario Cuomo in 1994.

Coordination under the exceptional reading is predicted to be thwarted, which (4.42c) confirms. Note that (4.42c) is not infelicitous at face value, since it can instantiate Parallel under the construal of Cuomo as the candidate of Giuliani's party. Yet this construal is counterfactual, or at least not well-informed.

Exception2 tracks Generalization, with the second relation negated:

4.43 a) Infer $p(a_1, a_2, \ldots)$ from the assertion of S_1 and $\neg p(b_1, b_2, \ldots)$ from the assertion of S_2, where a_i is a member or subset of b_i for some i.

b) Rudy Giuliani supported Mario Cuomo in 1994. Nonetheless, young aspiring politicians often support their party's candidate.

c) Infer *support(Giuliani, Cuomo)* from S_1 and *support(young_pol, party's_cand)* from S_2, where Giuliani is a young aspiring politician, and Cuomo is not Giuliani's party's candidate.

Despite the change in linear order from Exception1, the semantics remains the same, and we have nothing further to add.

4.44 a) Exception2(S_1, S_2) := $\vec{i} \in \vec{S}_2 \cos(\vec{s}_1, \vec{i}) = -1$.

b) [$support_1$, $Giuliani_2$, $Cuomo_3$]^T = [1, -0.95, 1]^T.

[$support_1$, $young\_pol_2$, $party's\_cand_3$]^T = [1, 0.9:1.1, 0.9:1.1]^T.

c) #Rudy Giuliani supported Mario Cuomo in 1994, and young aspiring politicians often support their party's candidate.

The final sort of Resemblance, Kehler terms Elaboration. As its name implies, an Elaboration restates an event in greater detail. The inference that binds its two halves together is that both describe the same state of affairs:

4.45 a) Infer $p(a_1, a_2, ...)$ from the assertions of S_1 and S_2.

b) A young aspiring politician was arrested in Texas today. John Smith, 34, was nabbed in a Houston law firm while attempting to embezzle funds for his campaign.

c) Infer *was_arrested*(*young_pol*) from S_1 and S_2.

Kehler suggests "that is" as a conjunction for marking Elaboration, but it is not particularly felicitous in this passage.

The vectorial treatment of Elaboration is to stipulate that the two vectors are identical, with the second adding additional dimensions of description to the first.

4.46 a) Elaboration(S_1, S_2) := $\vec{s}_1 \subset \vec{s}_2$.

b) [$was\_arrested_1$, $young\_pol_2$, in_Texas_3]^T.

[was_nabbed_1, $John\_Smith_2$, $in\_Houston\_law\_firm_3$]^T.

c) #A young aspiring politician was arrested in Texas today, and John Smith, 34, was nabbed in a Houston law firm while attempting to embezzle funds for his campaign.

That is to say, the contribution of S_2 is to locate S_1 in a larger space. If the two sentences denote the same vector at different levels of resolution, coordination is again undefined (or at least, the trivial correlation of a vector to itself is too uninformative to warrant linguistic realization). Indeed, coordination is infelicitous in (4.46c) under the Elaborative reading. It is only felicitous to the extent that a Parallel reading can be imposed on it, which forces the two sentences to be understood as two separate events involving two different people, though in parallel circumstances.

4.3.2.2.3. Coherence relations of Cause-Effect

Kehler characterizes a Cause-Effect relation of coherence as drawing a path of implication between the propositions recoverable from S_1 and S_2, where "implicate" is used to mean "could plausibly follow from", rather than the stronger sense of classical logic. Kehler details four such relations.

Result is perhaps the purest such relation:

4.47 a) Infer P from the assertion of S_1 and Q from the assertion of S_2, where normally $P \rightarrow Q$.

b) George is a politician. Therefore he is dishonest.

c) Infer *George is a politician* from S_1 and *he is dishonest* from S_2, where normally *being a politician* → *being dishonest*.

The idea is simply that some aspect of S_1 implicates S_2.

This does not give the vectorial approach much to grab hold of. The subject components of $\vec{s}_1$ and $\vec{s}_2$ are trivially identical, while the predicate components appear to be uncorrelated – *a priori, politician* and *dishonest* do not appear to share any sememe that would motivate a relation between them, correlated or not. Thus we are forced to delve more deeply into the semantics of implication than Kehler would have us believe is necessary.

In classical propositional logic, '→' or material implication is a truth-functional connective, so its definition is no more than its truth table. The truth table of material implication is simple enough to be reduced to a slogan: $P \rightarrow Q$ is false only when P is true and Q is false. However, it became clear that this degree of simplicity came at the price of several paradoxes. Stalnaker (1968) and Lewis (1973) independently proposed a possible worlds semantics for conditionals that avoids these paradoxes. In a nutshell, *if A then B* is true under this approach just in case B is true in the possible world most like the real world among those in which A is true. By "possible world", we mean what Lewis, 1973, p. 84, means in the following quote:

> It is uncontroversially true that things might have been otherwise than they are. I believe, and so do you, that things could have been different in countless ways. But what does this mean? Ordinary language permits the paraphrase: there are many ways things could have been besides the way they actually are. On the fact of it, this sentence is an existential quantification. It says that there exist many entities of a certain description, to wit, "ways things could have been." I believe permissible paraphrases of what I believe; taking the paraphrase at its face value, I therefore believe in the existence of entities which might be called "ways things could have been." I prefer to call them possible worlds."

While the Stalnaker-Lewis theory of conditionals has become the standard analysis of conditional semantics, it also leads to an account of causation.

The theory of causation that stems from it understands a causal statement of the form "event e_1 caused event e_2" in terms of counterfactual conditionals of the form of (4.48), see Lewis (1973), Noordhof (1999), and Menzies (2001):

4.48 a) If e_1 had not occurred, then e_2 would not have occurred.
b) If e_1 were to occur, then e_2 would occur.

The joint holding of these conditionals imply that e_1 is both necessary and sufficient for e_2 to occur.

Though there are flaws in Lewis' original formulation of counterfactual causation, some of which are touched on in the essays by Noordhof and Menzies cited previously, there does appear to be a core set of notions that most researchers agree characterize causation. Hausman, 1998, p. 1, collates a handy list of eleven characteristics, couched in terms of asymmetries between causes and their effects. Perhaps the most well known is *time order*: a cause precedes its effect, but an effect does not precede its cause. Two others give us insight into how to proceed:

4.49 a) Connection dependence: if one were to break the connection between cause and effect, only the effect would be affected.
b) Robustness: the relationship between cause and effect is invariant with respect to the frequency of the cause or with respect to how the cause comes about but not with respect to the frequency of the effect or with respect to how the effect comes about.

From Connection Dependence we may conclude that some components of the effect vector depend on the cause vector, since breaking the dependency alters the composition of the effect vector. From Robustness we may conclude that this dependency of the effect's components on the cause's components is invariant to the cause's frequency. For instance, infrequent causes will have the same effect as frequent causes.

For our purposes, from these two local conclusions it seems safe to draw the global conclusion that an effect is correlated to its cause, though in a special way that is not at all easy to pin down. In our analytic framework, this means at a minimum that the cosine of the angle between an effect and its cause in the asserted world @ is unity. The counterfactual theory inspires us to add that in some nearby possible world w, pointing $\vec{s}_1$ and $\vec{s}_2$ in the opposite direction also produces correlation. Yet this is not enough, for both of these calculations are symmetric. The asymmetries alluded to by Hausman are left unaccounted for. The only simple solution that we see is to stipulate the order, with the effect that the coherence relation of Result is licensed when the ordering and both correlations hold:

4.50 a) Result(S_1, S_2) := $\cos_@(\vec{s}_1, \vec{s}_2) = 1$ & $\cos_w(-\vec{s}_1, -\vec{s}_2) = 1$ & $\vec{s}_1 < \vec{s}_2$.

b) [politician$_1$, George$_2$]^T; [dishonest$_1$, he$_2$]^T.

c) George is a politician, and (therefore/so/etc.) he is dishonest.

This formula is a rough initial approximation, since it probably needs to be relativized to certain components and not the entire vectors, but it is sufficient for our rather modest requirements here. It licenses *and* coordination transparently, since the correlation and order stipulations of (4.29) are satisfied.

The order of sentences is crucial for Result, for reversing it produces a passage that coheres through Explanation:

4.51 a) Infer *P* from the assertion of S_1 and *Q* from the assertion of S_2, where normally $Q \rightarrow P$.

b) George is dishonest. He is a politician.

c) Infer *George is a politician* from S_1 and *he is dishonest* from S_2, where normally *being a politician* → *being dishonest*.

In the light of our proposal for Result, Explanation should be defined simply by reversing the order of sentences:

4.52 a) Explanation(S_1, S_2) := $\cos_@(\vec{s}_1, \vec{s}_2) = 1$ & $\cos_w(-\vec{s}_1, -\vec{s}_2) = 1$ & $\vec{s}_2 < \vec{s}_1$.

b) [dishonest$_1$, he$_2$]^T; [politician$_1$, George$_2$]^T.

c) George is dishonest, {#and/because} he is a politician.

It is to be presumed that the misalignment between the asserted order and the canonical order introduces a processing burden that is resolved by marking the unexpected S_2 with a reserved lexical item, *because*. It is to be further presumed that this misalignment blocks coordination, though it strikes us as odd that a minor alteration of linear precedence would change the syntactic relation from coordination to subordination.

The theory of signed measures leads us to expect that a positive formula should be mirrored with its negative complement, and this is exactly what Kehler finds in the relations of Violated Expectation and Denial of Preventer. Violated Expectation asserts the negation of the second sentences in canonical serial order:

4.53 a) Infer *P* from the assertion of S_1 and *Q* from the assertion of S_2, where normally $P \rightarrow \neg Q$.
b) George is a politician. He is honest.
c) Infer *George is a politician* from S_1 and *he is honest* from S_2, where normally *being a politician* → ¬*being honest.*

Our version follows Kehler's lead by changing correlation to anticorrelation:

4.54 a) Violated_Expectation(S_1, S_2) := $\cos_@(\vec{s}_1, \vec{s}_2) = -1$ & $\cos_w(-\vec{s}_1, -\vec{s}_2) = -1$ & $\vec{s}_1 < \vec{s}_2$.
b) [politician$_1$, George$_2$]^T; [honest$_1$, he$_2$]^T.
c) George is a politician, {#and/but} he is honest.

Denial of Preventer asserts the negation of the second sentences in reverse serial order:

4.55 a) Infer *P* from the assertion of S_1 and *Q* from the assertion of S_2, where normally $Q \rightarrow \neg P$.
b) #[George is honest. He is a politician.]
c) Infer *George is honest* from S_1 and *he is a politician* from S_2, where normally *being honest* → ¬*being a politician.*

Again, our version involves the expected substitution of anticorrelation for correlation:

4.56 a) Denial_of_Preventer(S_1, S_2) := $\cos_@(\vec{s}_1, \vec{s}_2) = -1$ & $\cos_w(-\vec{s}_1, -\vec{s}_2) = -1$ & $\vec{s}_2 < \vec{s}_1$.
b) [honest$_1$, he$_2$]^T; [politician$_1$, George$_2$]^T.
b) George is honest, {#and/even though} he is a politician.

The result is an elegantly symmetrical system of four contrasts created by the cross-classification of two choices for serial order and two choices for sign.

4.3.2.2.4. Coherence relations of Contiguity

The final family of coherence relations expresses a sequence of eventualities centered around some system of entities, which Kehler dubs Contiguity. Contiguity has only one member, Occasion, for which Kehler follows Hobbs (1990) in delineating the two versions of (4.57a, a'):

4.57 a) Infer a change of state for a system of entities from S_1, inferring the final state for this system from S_2.

a′) Infer a change of state for a system of entities from S_2, inferring the initial state for this system from S_1.

b) George picked up the speech. He began to read.

Unfortunately, no means are offered to illustrate the distinction between the two versions, so it is not even clear to us which one passage (4.57b) is meant to exemplify – perhaps both.

Kehler does not go into any further formal detail, for he deems Occasion to be too heavily dependent on human experience to afford any greater precision. By way of support, he cites an example from Hobbs:

4.58. A flashy-looking campaign bus arrived in Iowa. Soon afterward, George W. Bush gave his first speech of the primary season.

As Kehler, p. 23, explains...

> ... understanding passage [4.58] as a coherent Occasion requires inferences beyond the asserted information that the events occur in temporal progression, such as that Bush was on the bus and the speech was delivered in Iowa. In general, assumptions will be required that allow the final state of the first sentence to be identified as the initial state of the second, and hence temporal progression in the absence of a common scenario connecting the events is insufficient in and of itself.

To our way of thinking, this principled identification of the final state of S_1 with the initial state of S_2 means that the two are correlated at the point of overlap. We therefore offer (4.59) as a more accurate definition, in which the *i* variable indexes the relevant script or prototypical situation, and the predicates *end* and *begin* find the final and initial states of their respective sentences:

4.59 a) Occasion(S_1, S_2, i) := $\vec{s}_1$, $\vec{s}_2 \in$ i, cos(end($\vec{s}_1$), begin($\vec{s}_2$)) = 1.

b) George picked up the speech, and (then) he began to read.

In the 'bare' case of (4.57b), *i* is null, so no overarching situation-type is called, and the correlation is made strictly in terms of temporal progression. In the more experientially loaded case of (4.58), *i* calls a particular sequence, parts of which the two clauses have to instantiate in the order stipulated by the correlation. The invocation of correlation should license *and* coordination,

Table 4.1. Vector-theoretic definitions of Kehler's coherence relations.

Relation(S_1, S_2)	*Definition*	*and*
Parallel	$i \in \vec{s}_1, \vec{s}_2 \cos(\vec{s}_1(i), \vec{s}_2(i)) = 1$	yes
Contrast1, 2	$i \in \vec{s}_1, \vec{s}_2 \cos(\vec{s}_1(i), \vec{s}_2(i)) = -1$	no
Exemplification	$\vec{i} \in \vec{S}_1 \cos(\vec{i}, \vec{s}_2) = 1$	no
Generalization	$\vec{i} \in \vec{S}_2 \cos(\vec{s}_1, \vec{i}) = 1$	no
Exception1	$\vec{i} \in \vec{S}_1 \cos(\vec{i}, \vec{s}_2) = -1$	no
Exception2	$\vec{i} \in \vec{S}_2 \cos(\vec{s}_1, \vec{i}) = -1$	no
Elaboration	$\vec{s}_1 \subset \vec{s}_2$	no
Result	$\cos_{@}(\vec{s}_1, \vec{s}_2) = 1$ & $\cos_{w}(-\vec{s}_1, -\vec{s}_2) = 1$ & $\vec{s}_1 < \vec{s}_2$	yes
Explanation	$\cos_{@}(\vec{s}_1, \vec{s}_2) = 1$ & $\cos_{w}(-\vec{s}_1, -\vec{s}_2) = 1$ & $\vec{s}_2 < \vec{s}_1$	no
Violated_Expectation	$\cos_{@}(\vec{s}_1, \vec{s}_2) = -1$ & $\cos_{w}(-\vec{s}_1, -\vec{s}_2) = -1$ & $\vec{s}_1 < \vec{s}_2$	no
Denial_of_Preventer	$\cos_{@}(\vec{s}_1, \vec{s}_2) = -1$ & $\cos_{w}(-\vec{s}_1, -\vec{s}_2) = -1$ & $\vec{s}_2 < \vec{s}_1$	no
Occasion(S_1, S_2, i)	$\vec{s}_1, \vec{s}_2 \in i, \cos(end(\vec{s}_1), begin(\vec{s}_2)) = 1$	yes

where the requisite partial order is implicit in the ordering of *end* and *begin*. (4.59) shows this prediction to be correct.

4.3.2.2.5. *Summary*

The preceding subsections have argued for the vectorial representations of Kehler's thirteen coherence relations summarized in Table 4.1. It is quite striking that only three of them lend themselves to a paraphrase with clausal coordination, and these are exactly the three characterized by correlation of exhaustive single vectors and canonical clause order. Moreover, the second two delimit the two sorts of asymmetric coordination.

This survey of coherence relations reveals asymmetric coordination not to be exceptional, but rather to fill a predictable slot in the classification as a particular constellation of independently motivated coherence features. The asymmetric relations of Result and Occasion combine the constraint of correlation with the particular ordering constraints of Cause-Effect and Contiguity. The outcome is that a grammar augmented with the vectorial coherence relations does not need to stipulate asymmetrical coordination. All it needs to do is define coordination and to stipulate Cause-Effect and Contiguity, and the asymmetric readings of Result and Occasion follow directly. One can best appreciate the power of this theory by comparing it to an alternative.

4.3.2.3. Asymmetric coordination in Relevance Theory

In more recent work within the framework of Relevance Theory, see Sperber and Wilson (1986) among many others, Blakemore and Carston (1999) elucidate several restrictions on the relations which hold between coordinated clauses – relations which stand out most clearly when coordinated clauses are contrasted to their juxtaposed counterparts. The most vivid restriction prevents the second conjunct from helping to explain the first – a relation often understood to hold between juxtaposed clauses. The key minimal pair is reproduced in (4.60), originally cited by Gazdar (1979) from Herb Clark:

4.60 a) John broke his leg. He tripped and fell.
b) John broke his leg and tripped and fell.

As Blakemore and Carston put it ...

> While there is a possible, though not very likely, interpretation which is shared by the conjoined utterance and the non-conjoined sequence – namely, the one in which John broke his leg (say, by falling out of a tree) and then tripped and fell (say, when he tried to get up) – there is a more accessible interpretation for [4.60a] that **cannot** be recovered from the conjoined utterance [4.60b], that is, the one in which the information communicated by the second sentence is understood as an explanation for the event described in the first.

In other words, in the juxtaposed version it is possible to present first the leg-breaking event and then the tripping-and-falling event, even though they will be understood as having happened in the reverse order.

Blakemore and Carston's relevance-theoretic account of this contrast relies on two notions. The first is that the interpretations which are exclusive to the juxtaposed alternatives are only possible where a juxtaposition expresses two propositions each of which is processed individually for relevance. As a case in point,

> ... in [4.60a] the second segment is relevant as an explanation for the state of affairs represented in the first; in other words, it is an answer to the question, 'Why?' or 'How?', which will be understood to have been raised by the first segment. Questions and answers are by their very nature planned as separate utterances each carrying the presumption of relevance individually.

The second explanatory notion is the proposal of Blakemore (1987) that in coordinated clauses, the presumption of relevance is carried by the global proposition rather than by each constituent proposition. For a narrative such as (4.60b), the global proposition is often a script:

> A conjoined utterance in which events are narrated may achieve relevance because its conjuncts represent components of a scenario which itself is an instance of a more general stereotypical scenario; that is, its conjuncts are instances of propositions which are stored together in memory as a single cognitive unit or schema.

Thus the burden of explanation for the contrast in (4.60) is shouldered by the distributive (juxtaposed) vs. collective (coordinated) processing of the two clauses.

We find many points to praise in the Relevance-theoretic approach to asymmetric coordination, but we cannot overlook the handicap of laboring in the absence of a precise formalization of the various explanatory principles. A much healthier degree of precision can be achieved just by replacing "relevance" with "correlation" in the passages that have just been quoted.

In particular, the cohesion relation of the juxtaposed clauses in (4.60a) is Explanation, which Blakemore and Carston realize, though not in so many words. Thus where they say that the second sentence's relevance lies in answering the question "Why?" or "How?" about the first, we say that the passage comes closest to satisfying the constraints on correlation which define what it means to explain. With respect to the coordinational alternative of (4.60b), where they say that that the individual clauses' relevance lies in instantiating some tripping-and-falling-down schema, we say that it comes closest to satisfying the constraints on correlation which define an Occasion (with or without a supporting schema). Where they seem to advocate an infinitely variable field of coherence relations, we advocate a calculus based on a principled number of variations in the correlation of two clauses. We believe our approach to be superior in falsifiability, empirical coverage, and explanatory adequacy, but a convincing demonstration would take us too far afield, so let us return to one other related empirical oddity of clausal coordination.

4.3.2.4. The Common-Topic Constraint

AND can in principle conjoin any two propositions, but in practice it requires some tie of relevance between its conjuncts. R. Lakoff, 1971, p. 116, presents the following cline of acceptability as evidence thereof:

4.61 a) John eats apples, and John eats pears.
b) John eats apples, and his brother drives a Ford.
c) ?John eats apples, and many New Yorkers drive Fords.
d) ?John eats apples, and I know many people who never see a doctor.
e) ??Boys eat apples, and Mary threw a stone at the frog.

f) *John is a strict vegetarian, and he eats a lot of meat.

Lakoff argues that natural language conjunction is only licensed if all the conjuncts have a constituent that can be reduced to partial or complete identity, and that this constituent must be what the sentence is about. She refers to it as the **common topic.**[26]

Lowe (1984) attributes the common-topic requirement of AND to the fact that conjunction creates lists, and lists are defined by some common property. For instance, at first glance one may believe that (4.62) is just a jumble of disparate characteristics:

4.62. 1971, good condition, only 9000 miles on a reconditioned engine and gearbox, one year's MOT

Nevertheless, if it is pointed out that this is an ad for selling a second-hand car in Britain, then the jumble resolves itself into a well-formed list of conjuncts answering the question, "What are the selling points of this car?"

The reader should have enough background to bring the Common-Topic Constraint up to date without our intervention, but we will give a nudge in the right direction anyway. Lakoff's and Lowe's examples all instantiate the coherence relation of Parallel, and Lakoff's best examples are the ones that have the fewest conflicting components. Thus the Common-Topic Constraint can be viewed as evidence in favor of augmenting Parallel with a specific indication of a schema or prototype such as was done for Occasion, but we will leave this idea for another day.

4.3.3. Summary of clausal coordination

The claim of the correlation approach to clausal coordination is that the coordinatees correlate with one another in a manner specified by the coordinator. This section has argued that this approach is best understood by a comparison to the discourse coherence relations that clausal coordination competes with. We have shown that the relations of Parallel, Result, and Occasion are the ones that lend themselves to coordinative expression, due to the compatibility of the constraints that they impose on correlation with 'regular' coordinative correlation. The rest of the chapter is devoted to cleaning up some of the ends left dangling in our discussion.

---

[26] Lakoff makes similar claims for OR, but for the sake of brevity, this sections focuses on AND.

Table 4.2. The sixteen patterns of truth values over two variables, with their putative linguistic instantiations labeled. Shading indicates those patterns which have the same values for rows 2 and 3.

p	q	*A* OR	*B*	*C* IF	*D* NAND	*E* IFF	*F*	*G*	*H*	*I*	*J* XOR	*K* AND	*L*	*M*	*O* FLS	*V* TR	*X* NOR
1	1	1	1	1	-1	1	-1	-1	1	1	-1	1	-1	-1	-1	1	-1
1	**-1**	1	1	-1	1	-1	-1	1	-1	1	1	-1	1	-1	-1	1	-1
-1	**1**	1	-1	1	1	-1	1	-1	1	-1	1	-1	-1	1	-1	1	-1
-1	**-1**	-1	1	1	1	1	1	1	-1	-1	-1	-1	-1	-1	-1	1	1

4.4. LEXICALIZATION OF THE LOGICAL OPERATORS

The elaboration of various measures to express the patterns formed by the logical operators calls into question their traditional logical definitions. As was mentioned in Chapter 1, the standard format for representing the meaning of a logical operator is drawn from propositional logic. It is most easily illustrated with two propositions, such as the two in (4.63), to which we have suffixed their possibilities for evaluation:

4.63 a) "Anabel is a gourmet" evaluates to True or False
b) "Yetunde is a gourmet" evaluates to True or False

In accord with Shastri's deduction that scalar numbers are the easiest message to communicate, it would be convenient to have some numerical substitute for *true* and *false*. And indeed, there is a long tradition of representing *true* numerically as one, and *false* as zero. It is rather enlightening – or vexing – that the neuromimetic considerations of this monograph immediately call this simple assumption into question. One of the claims advanced here is that the positive operators are grounded neurophysiologically in correlation, which naturally presupposes that the negative operators should be grounded in anticorrelation. To ground truth values in these terms requires that the value of *true* as '1' be opposed to the value of *false* as '-1'. However, as was set forth in Chapter 3, between correlation and anticorrelation there lies a third degree, namely no correlation, or '0'. Working backwards to truth values, the addition of '0' between '1' and '-1' makes for a three-valued system, which is the claim of Seuren et al. (2001) – and one that we cannot do justice to in this monograph.

Fortunately, the 'extra' truth value has almost no role to play in the logical-operator system, so that for all practical purposes we will be dealing with only two truth values, '1' and '-1'. Recall form the synopsis of correlation in Chapter 3 that a lack of correlation simply means that there is no relationship between the variables in question. Consequently, there is no basis for a logical operation to apply to them, and such cases are simply irrelevant. It follows that there is no reason to include uncorrelated data in any usage of the logical operators, so there is no reason to include the requisite '0' values in the upcoming truth tables. Let us therefore leave the topic for the nonce and begin to grapple with the data. We will come back to it several times over the course of the monograph.

4.4.1. The sixteen logical connectives

As is well-known, if two propositions p and q are assigned the truth value *true* or *false* as in (4.63), the resulting four combinations of possible valuations can be evaluated in sixteen ways, listed in Table 4.2 from Bochenski (1959). An immediate problem with these evaluations is that only six or seven of them are attested with any consistency in natural languages. They are indicated by introducing a sub-heading in Table 4.2 with the most common English abbreviation. How are these sixteen pared down to the three or four observed coordinators that have been the subject of this chapter? The next few paragraphs offer some guidelines.

4.4.2. Conversational implicature: from sixteen to three

Gazdar and Pullum (1976) and Gazdar, 1979, Chap. 4, offer a conversational implicature account for the small number of attested truth-functional connectives that is worth pursuing for a moment for the contrast that it provides with the neuromimetic analysis. Their treatment is couched in a two-valued logic, which is converted to three values in our summary to maintain consistency with our own representational format. Nothing hinges on this alteration.

Gazdar/Pullum make the initial syntactic assumption that the base component is unordered, so that the connectives which linearize their conjuncts by differing values for (1, -1) and (-1, 1) cannot be stated therein. Gazdar/Pullum execute this assumption by defining the truth-functional connectives as those functions C which take as their sole argument the set of nonempty subsets of the truth-value set {1, -1}, i.e. {{1}, {1, -1}, {-1}}. Since the 'intermediate' set {1, -1} is unordered, it excludes those connectives in Table 4.2 which disagree on (1, -1) and (-1, 1). This reduces the sixteen connectives in Table 4.2 to the eight in Table 4.3. The net result of this hypothesis is to make the truth-functional connectives commutative.

Having made this initial cut, Gazdar/Pullum next exclude all of those connectives which are negative, on the grounds that negatives are difficult to process. This exclusion is accomplished by postulating a principle of **confessionality**, see Gazdar, 1979, p. 76:

4.64. A connective $c \in C$ is **confessional** iff $c(\{-1\}) = -1$.

Confessionality rules out NAND, IFF, FALSE, and NOR, leaving OR, AND, XOR, and TRUE. OR and AND are uncontroversially attested in many languages. The other two are suspect.

Table 4.3. Truth-functional connectives with same valuation for rows 2 and 3 of Table 4.2. Shading indicates those connectives which do *not* cover continuous regions of LOGOP space.

p q	*OR*	*NAND*	*IFF*	*XOR*	*AND*	*FLS*	*TR*	*NOR*
{1}	1	-1	1	-1	1	-1	1	-1
{-1, 1}	1	1	-1	1	-1	-1	1	-1
{-1}	-1	1	1	-1	-1	-1	1	1

TRUE is ruled out as a violation of Grice's maxim of relevance. Since TRUE is true no matter what its arguments are, they make no contribution to the truth valuation of the sentence as a whole and so are irrelevant. Finally, XOR is taken to be derived from OR, a position which is examined in the next section, where it is concluded that the inclusive option of OR is superior to the exclusive option of XOR.

4.4.3. Neuromimetics: from sixteen to four

The Gazdar/Pullum analysis works adequately for a theory that is not grounded in any more basic phenomena, but it is not adequate for a theory such as the one espoused in this monograph which attempts to ground linguistic phenomena in neurological processing. Moreover, there is empirical evidence which calls into question whether the crucial syntactic assumption of an unordered base makes the right semantic predictions. The evidence has already been presented, as the phenomenon of asymmetric coordination. Whereas the correlational analysis of coherence relations is compatible with base commutativity, specific coherence relations such as those that license asymmetric coordination require the canonical order of conjuncts. It consequently seems *ad hoc* to claim that there is a syntactic level of representation at which conjuncts are unordered and a semantic level at which they are ordered, and that the syntactic level takes precedence over the semantic for calculating the meaning of a coordinator.

A reformulation of the Gazdar/Pullum invocation of an unordered base is straightforward within the neuromimetic framework. As Chapter 1 emphasizes and Chapter 3 operationalizes, it is simply that evaluation takes place in parallel, via some calculation of correlation such as finding the cosine of an angle:

4.65. Parallelism: a monomorphemic operator evaluates its inputs in parallel, so it cannot draw distinctions in truth value from the linear order of its inputs.

Parallelism rules out all of those truth assignments which the Gazdar/Pullum analysis attributed to the unordered base, which is to say that it also reduces Table 4.2 to Table 4.3. The difference is that it is immune to the counterexample of asymmetric coordination, because the ordering relations which characterize it are calculated in addition to its truth-functional evaluation.

A second constraint arises from consideration of where the natural boundaries are in the topological space of the logical operators, namely at the cut points defined in Chapter 3:

4.66. Convexity: a monomorphemic operator has natural boundaries, which is to say that it is a single convex subset of the unit circle, separated from it at 0° plus one other cut point.

Convexity rules out those operators with discontinuous topologies, such as IFF, which includes both edges while excluding both middle regions, and its converse XOR, which includes both middle regions across the cut point at 0°, while excluding both edges.

It is worthy of note that convexity may also exclude the two "trivial" connectives, TRUE and FALSE. If it does, the reasoning is the following. TRUE includes all of LOGOP space, thus including the two discontinuous convex subsets constituting either polarity. Finally, FALSE accepts none of LOGOP space, but since LOGOP space consists of two halves, FALSE winds up accepting two convex subsets for either region, Ø for correlation and Ø for anticorrelation.

If the reader finds this reasoning specious, it is preferable to fall back on the basic tenets of signed measure theory, which forbid the mixing of the two signs. TRUE indisputably mixes both signs, while FALSE may also do so in the sense of evaluating both to the same outcome. In any event, we have the chance to revisit this topic in Chapter 6, in the guise of the Generalized-Quantifier constraint of Triviality.

The end result is that Parallelism and Convexity – and maybe signed measure theory – reduce the sixteen connectives definable in propositional logic down to the four coordinators that we have been discussing all along. The problematic status of NAND is taken up in Chapter 8.

4.5. OR VERSUS XOR

Our final task is to justify the inclusive meaning of *(either...)or* against the commonly attested exclusive sense, abbreviated by XOR. This task is broached here because the data mix both clausal and phrasal constructions.

There is some controversy surrounding the OR/XOR distinction because it is not always easy to demonstrate the range of meanings of *(either...)or*. By way of illustration, consider the clause *(Either) Anabel, Yetunde, Chris, or Rukayyah is a gourmet*. With singular agreement, only one of them is a gourmet, an exclusive reading. What we would like to know is how many of them are gourmets with plural agreement: *(Either) Anabel, Yetunde, Chris, or Rukayyah are gourmets*. Unfortunately, the construction is ungrammatical, so the question cannot be answered directly.

The closest that we can come is to observe that partitive constructions let us pick out any subset of a disjunction:

4.67 a) Any three of Anabel, Yetunde, Chris, or Rukayyah can lift this piano.
b) Any four of Anabel, Yetunde, Chris, Rukayyah, or Vaneeta would make up a formidable string quartet.

The crucial question is whether a partitive can include the entire group:

4.68 a) All four of Anabel, Yetunde, Chris, or Rukayyah can lift this piano.
b) All five of Anabel, Yetunde, Chris, Rukayyah, or Vaneeta would make up a formidable string quintet.

Certainly, using the maximum number in a partitive is not false in (4.68), as would be the case if the meaning of *(either...)or* were XOR; it is just awkward to the point of calling into question the speaker's ability to add.

Secondly, there is a use of *either* that means *both*:

4.69 a) There are benches on either side of the river.
b) ≈ There are benches on both sides of the river.

Since *both* highlights the inclusiveness of *and* for two conjuncts, the fact that *either* has a synonymous usage suggests that it is not inconsistent with inclusiveness either – nor should the coordinator *or* that it is often used in construction with.

Thirdly, Pelletier (1977) and McCawley, 1981, p. 33 and p. 230, point out that *(either...)or* is often understood as excluding the entire group for reasons having nothing to do with its semantics:

4.70 a) Today is either Monday or Tuesday.
b) Either there is a God or there isn't.
c) Shirley visited either Ayuddha or Lopburi last year.
d) On the $11.25 lunch you can have either a soup or a dessert.
e) You can use either the hall closet or the attic to store your books.

For instance, in (4.70a) 'today' can only be one day of the week. Consider how this plays out in McCawley's exposition of (4.70d):

> ...the offer need not entitle the recipient to make the proposition true in whatever way he pleases: the generosity is only broad enough to make the recipient entitled to more than what linguistically simpler alternatives entitle him to.

> [4.70d] entitles the hearer to take a soup and entitles him to take a dessert (since if he were not entitled to one of them, a linguistically simpler alternative such as *On the 11.50 lunch you get a soup* would express the full generosity of the offer) but does not entitle him to take both. (McCawley, 1981, p. 130)

Such observations suggest that the exclusive sense can be overridden by conditions that make the inclusive sense more plausible. (4.71) relates our attempts at ameliorating (4.70) with such conditions, plus the addition of a few new examples:

4.71 a) Either there is a God, or there isn't, or both: there was a God, but she died.
b) On the $11.25 lunch you can have either a soup, a dessert, or both, if it looks like all the soup will not be consumed today.
c) To store your books, you can use either the hall closet, the attic, or both, if you really have a lot of books.
d) Shirley visited either Buda or Pest last year, or both, if she crossed the bridge between them.
e) A prize will be awarded to the biggest or the juiciest tomato, or both, if we collect enough prize money from the entrance fees.
f) I would marry Yetunde, or Rukayyah, or both, if I were a Mormon.

Thus the exclusive reading of *(either…)or* is taken to fall out from pragmatic principles. In fact, Horn, 1989, p. 225 and sec. 6, after reviewing the literature on XOR, argues that *p or q* conversationally implicates that the stronger *p and q* does not hold, i.e. that as far as the speaker knows, ~(*p and q*).

This account avoids the algorithmic problem of treating XOR as a basic coordinator presented in McCawley, 1981, pp. 77-8 and pp. 153-4. According to McCawley's calculation of its truth table, XOR should be true if an odd number of disjuncts is true and false if an even number of them is true, even though specific instances of XOR are usually understood to be true if one of any number of disjuncts is true. The conclusion is that that *(either…)or* has the inclusive meaning of OR that we have been attributing it throughout this chapter.

4.6. SUMMARY

This chapter fleshes out the analysis of coordination introduced in terms of abstract logical operators in the preceding chapter. It is argued that all coordination can be understood as expressing a degree of correlation between (at least) two vectors. For the coordination of phrases, one vector is the denotation of the coordinated linguistic category and the other is the vector denoting the linguistic category which the coordinated category applies, which is given by the grammar. For the coordination of clauses, the vectors are those denoted by the two (or more) coordinated clauses. Substantiating this latter claim led us into a discussion of the difference between clausal coordination and coherence established through clausal juxtaposition, in which the power of the correlational theory was demonstrated by the ease with which it can formalize the various coherence relations and predict exactly which ones support clausal coordination.

The choice of correlation is mandated by the hypothesis that it is implemented neurologically by Hebbian learning and perhaps by synchronization of oscillating cells or cell assemblies. With this grounding, increasing the number of coordinatees to be correlated only makes the recognition process take a marginally longer period of time, e.g. four cell assemblies will correlate practically as quickly as two assemblies. A theory of coordination based on correlation thereby satisfies the parallelism desideratum of neuromimetic computation and finesses the problem of serialism discussed in the first chapter. The next chapter puts all of these results to good use by showing how neuromimetic systems can learn them.

Chapter 5

Neuromimetic networks for coordinator meanings

In the previous chapters, we used the phrasal verb *pick out* several times, in locutions such as "such-and-such a coordinator *picks out* this pattern". What does it mean to pick a group of vectors out of the surrounding space? In this chapter, several neuromimetic architectures are reviewed which explain how the patterns that organize observations of coordinator meanings can be learned. Seven architectures are introduced that perform some instantiation of logical vs. associative and hyperplane vs. clustering classification: the McCulloch and Pitts neuron, single-layer perceptrons, multilayer perceptrons or backpropagation networks, instar networks, competitive networks, learning vector quantization networks, and finally, a novel algorithm based on correlation among dendritic spines. Only perceptrons, LVQ networks, and the dendritic algorithm are observationally adequate for coordination, but the perceptrons are not explanatorily adequate, for they implement a biologically unrealistic non-local learning rule.

5.1. A FIRST STEP TOWARDS PATTERN-CLASSIFICATION SEMANTICS

There are two basic ways of picking out a pattern, by lines and by clusters. Both methods start with a finite set of examples A which is disjoint from one or more finite set of counterexamples B or C in the same n-dimensional space $\mathfrak{R}^n$. They go on to define a function f that apportions $\mathfrak{R}^n$ into a region R whose interior contains A and whose closure is disjoint from B (and C). f evaluates the patch of observations within R to some value and the patches of observations outside of R to some other value. The act of defining f is the process of **pattern classification**, and once f is defined, it is said to **recognize** A and B (and C).

The linear method calculates a line – or a set of lines connected into a polygon – that separates R from the rest of $\mathfrak{R}^n$. Fig. 5.1a presents a simple example. This is perhaps the most popular method in contemporary neuromimetic modeling. The cluster method calculates the distance from each point to a reference point, so that those points that make up the pattern cluster around one reference point, and those that do not belong to the pattern cluster around another. Fig. 5.1b presents a simple example, based on the patterns of Fig. 5.1a.

The two methods can be further subdivided by how they deal with the non-R space. In non-associative pattern classification, f evaluates the observations within R to 1 (true), i.e. $f(a) = 1$ for all a in A, and observations outside of R to 0 (false), i.e. $f(b) = 0$ for all b in B. The associative alternative differs in that there is

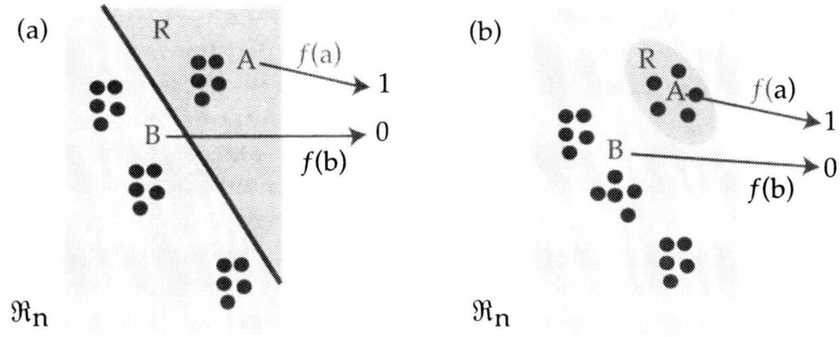

Figure 5.1. Nonassociative pattern classification. (a) Linear; (b) clustering.

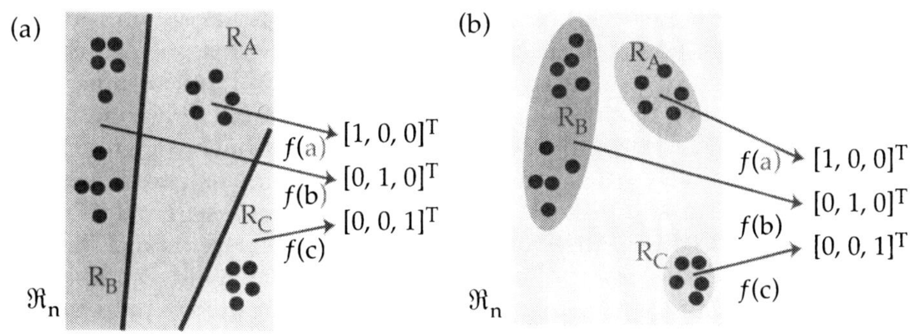

Figure 5.2. Associative pattern classification. (a) Linear; (b) clustering.

no 0 output – every set of examples is mapped to a different index, which is to say, a different kind of 1. This permits a larger number of patterns to be distinguished by the same function and so is the alternative for defining f in cases that have more than two outcomes. The closure of each region is still disjoint from every other region; it is just that now there are more regions recognized by f because f assigns each one a different label. The linear alternative is exemplified in Fig. 5.2a, and the clustering alternative in Fig. 5.2b.

It is an empirical question which format is more accurate for the learning and recognition of linguistic patterns. We endeavor to answer this question in the following chapters. Yet even from the simple distinction drawn in the preceding few paragraphs, it should be obvious that the associative format is more efficient at classifying the input space $\mathfrak{R}_n$. It already draws the three-way distinction depicted in Fig. 5.2, whereas the nonassociative format would need two more functions in addition to those of Fig. 5.1 to impose the same partition.

In either case, pattern classification allows us to traffic in numbers – the input to the classification function is a number, as well as its output. Thus pattern classification is consistent with Shastri's desideratum of intelligent computation (1.15d), that messages be communicated with as little internal structure as possible. For nonassociative pattern classification, the message is a scalar, i.e. a magnitude with no internal structure. For associative pattern classification, the message is a vector all of whose components are zero but for one. It follows that pattern classification should be highly ranked for explanatory adequacy, so it behooves us to investigate as thoroughly as possible the potential formats for pattern classification of the logical operators.

5.2. LEARNING RULES AND CEREBRAL SUBSYSTEMS

Now that we have a typology of pattern classification, let us look at the algorithms which learn these patterns. There is such a wide variety of learning rules discussed in the neuromimetic computing literature that one can easily become bogged down in a morass of competing algorithms and results. One taxonomy that we have found helpful in imposing a simplifying conceptual organization on this variety is offered by Doya (1999), which categorizes all learning rules into one of the three types: (i) unsupervised learning, (ii) reinforcement learning, and (iii) supervised learning or error-correction. They are illustrated in terms of information flow through components in Fig. 5.3. This classification is due to the nature of the teaching signals that guide learning: no such signal in unsupervised learning, a measure of reward in reinforcement learning, and a measure of error in supervised learning. The rest of this chapter is devoted to applying these paradigms to our representations of coordinator meanings, so it is premature to explain them here in detail. A few words of introduction suffice.

In the unsupervised paradigm, the goal is to find some interesting mapping of the input onto an output. Since no information other than the input, and the architecture of the learning component, intervenes in the construction of this mapping, the interesting part lies in the discovery of some interesting statistical structure in the input. In the reinforcement learning paradigm, a learning agent makes an action in response to the state of the environment, which results in a change of the state and the delivery of a reinforcement signal or reward. The goal is to find a policy which maximizes the cumulative sum of the rewards. In the supervised learning paradigm, the goal is to construct an input–output mapping that predicts the output for an input data point. This mapping minimizes an error measurement that compares the learned output to the desired output associated with the input data points.

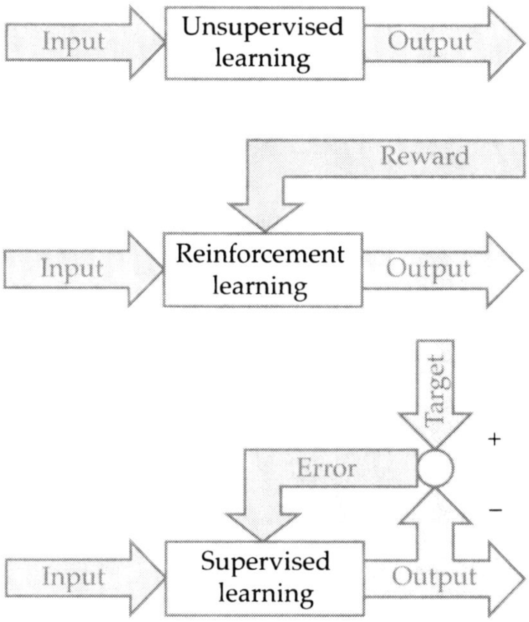

Figure 5.3. The three types of learning rules. Comparable to Doya, 1999, Fig. 2.

The usefulness of this taxonomy of learning rules is augmented dramatically by Doya's claim that three fundamental cerebral subsystems implement one of these learning rules: (i) the neocortex implements unsupervised learning, (ii) the basal ganglia implement reinforcement learning, and (iii) the cerebellum implements supervised learning.[27] The location of and pathways between these subsystems are laid out schematically in Fig. 5.4. What we see is sensory input entering from the brain stem and passing through the thalamus and on to the basal ganglia and neocortex. From these two sites, information flows to the other sites in a way that is not easy to describe in just a handful of words. At some point in this reciprocal circuit, a motor response will enter the thalamus and exit through the brain stem.

---

[27] Rolls and Treves (1998) concurs with Doya with respect to the anatomical specialization of the three learning rules, but it does not offer the overall theory that Doya does.

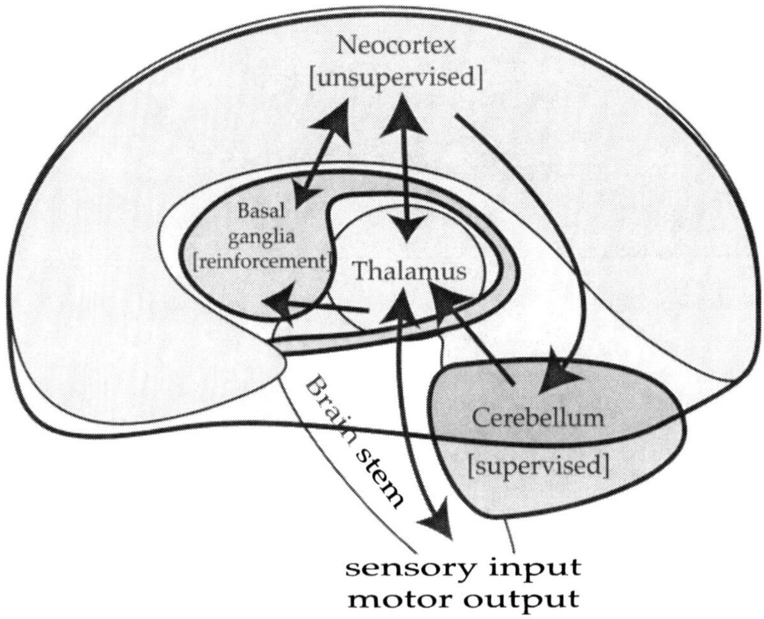

Figure 5.4. Schematic left-side view of brain showing the systems for which Doya posits a computational specialization. Comparable to Doya, 1999, Fig. 2.

Doya's computational theory marks a sharp departure from the traditional understanding of these subsystems, which characterizes them in terms of a functional specialization. Thus Delcomyn, 1998, p. 65, lists their functions as so: (i) the cerebral hemispheres (neocortex) produce "higher" mental functions, sensory processing, and motor control; (ii) the basal ganglia produce motor planning and control; (iii) the cerebellum aids in motor coordination and learning. This is at best vague, and at worst, contradictory, since motor control is shared between all three subsystems.

Doya takes this criticism a step further by following Bower (1997) and labeling any such functional specialization as ill-posed, since all of these functions depend on each other in a normal behavioral context. This mutual dependence stands out in Fig. 5.4 in the form of the various arrows that reciprocally connect each subsystem. Given that reciprocally connected areas tend to be active simultaneously, it becomes difficult to differentiate their roles by mere observation of their activity. Even more damagingly, recent research finds that the cerebellum and the basal ganglia contribute to non-motor tasks as

Figure 5.5. A McCulloch and Pitts neuron with two weighted inputs.

well, so the traditional functional specialization may be worse than ill-posed; it may just be wrong.

To be charitable, Doya, 1999, p. 961, attributes much of the persistence of the functional characterization to the "paucity of alternative theories that would enable us to comprehend the way the cerebellum and the basal ganglia participate in sensory or cognitive tasks". In support of his own theory of computational specialization, Doya points to the uniform anatomical organization of each subsystem, which suggests that each one is organized to execute a different learning rule. He sketches just how each anatomical wiring plan subserves a particular learning rule.

Interconnecting the three learning modules creates the outlines of a goal-oriented behaving system. Within the overall system, each subsystem becomes responsible for a different kind of representation. The supervised learning module in the cerebellum creates an internal model of the environment. The reinforcement learning module in the basal ganglia enables action selection by an evaluation of the cerebellum's environmental representations. The unsupervised learning module in the cerebral cortex provides statistically efficient representation of the states of the environment and the actions selected.

5.3. ERROR-CORRECTION AND HYPERPLANE LEARNING

Having gone to the trouble of introducing Doya's careful analysis of the cerebral distribution of learning rules, we are immediately confronted with the quandary that the oldest and most popular type of rule does not fit into Doya's system. It is a supervised or error-driven rule, which Doya localizes to the cerebellum, but it has been used to model cortical learning for decades. Since this *is* the oldest and best-known learning paradigm, we cannot help but mention it here, but we do so with the caveat that it is not biologically plausible, a fact which is returned to at the end of the section.

The rule introduced in this first section learns to divide accepted from rejected observations by a linear object: a line in two-dimensional space, a plane in three-dimensional space, or a hyperplane in higher dimensions. The three architectures examined below also happen to effect their classifications in terms of two-valued logic, with 0's and 1's, but this is not a necessary property.

Table 5.1. McCulloch and Pitts computation of OR and AND, $w_i = 1$.

Input at t		$\Sigma(w_i x_i)$	*Output at t+1*	
w_1x_1	w_2x_2	$w_1x_1 + w_2x_2$	$\Sigma \geq 1$ *(OR)*	$\Sigma \geq 2$ *(AND)*
0	0	0	0	0
1	0	1	1	0
0	1	1	1	0
1	1	2	1	1

Table 5.2. McCulloch and Pitts computation of NOR and NAND, $w_i = -1$.

Input at t		$\Sigma(w_i x_i)$	*Output at t+1*	
w_1x_1	w_2x_2	$w_1x_1 + w_2x_2$	$\Sigma \geq 0$ (NOR)	$\Sigma \geq -1$ (NAND)
0	0	0	1	1
–1	0	–1	0	1
0	–1	–1	0	1
–1	–1	–2	0	0

5.3.1. McCulloch and Pitts (1943) on the logical connectives

McCulloch and Pitts (1943) demonstrated that two-state neurons could perform all of the computations necessary to describe the Boolean operators, and in particular, the logical coordinators.[28] A two-state neuron has two inputs, say x_1 and x_2, whose value can be 0 (inactive) or 1 (active). Moreover, each input is weighted by some value w_i which is unique to that input. After a stipulated delay, the neuron sums the two weighted inputs together. If this sum meets or exceeds a certain threshold θ, the neuron becomes active and emits an output of 1. If not, the neuron remains inactive. Thus both the inputs and the outputs of a McCulloch-Pitts neuron are drawn from two-valued logic, and the resulting classification is nonassociative. Schematically, such a neuron looks like that of Fig. 5.5. This diagram is meant to manifest the following equation:

5.1. $$\text{if} \left(\sum_{i=1}^{n} w_i x_i \right) \geq \theta,\ f\left(\sum_{i=1}^{n} w_i x_i \right) = 1, \text{ otherwise } f\left(\sum_{i=1}^{n} w_i x_i \right) = 0.$$

[28] This discussion draws heavily on Demuth and Beale, 1994, pp. 3.5-6, though the discussions in Fausett, 1994, pp. 59ff., and Anderson, 1995, pp. 220ff., were also consulted.

Table 5.3. McCulloch and Pitts parallel computation of AND, $w_i = 1$.

Input at t			$\Sigma(w_i x_i)$	*Output at t+1*
w_1x_1	w_2x_2	w_3x_3	$w_1x_1 + w_2x_2 + w_3x_3$	$\Sigma \geq 3$ (AND)
0	0	0	0	0
0	0	1	1	0
0	1	0	1	0
1	0	0	1	0
0	1	1	2	0
1	0	1	2	0
1	1	0	2	0
1	1	1	3	1

Table 5.4. McCulloch and Pitts binary computation of NOR, $w_i = -1$.

Input at t		$\Sigma(w_i x_i)$	*Output at t+1*
w_1x_1	w_2x_2	$w_1x_1 + w_2x_2$	$\Sigma \geq 0$ (NOR)
0	0	0	1
Input at t+2		$\Sigma(w_i x_i)$	*Output at t+3*
NOR	w_3x_3	NOR + w_3x_3	$\Sigma \geq 0$ (NOR)
1	0	1	0

McCulloch and Pitts neurons can be connected into larger networks, but for the purposes of representation of the logical coordinators, a single one is sufficient.

Let us illustrate this device for OR and AND, whose threshold θ is set in Table 5.1 by hand to values of one and two, respectively. The inputs are those of a binary truth table, where 0 indicates false and 1 indicates true. Notice how manipulating the threshold permits the two neurons to distinguish between OR and AND. For NOR and NAND, the weights must be set to negative one, and the threshold to zero and negative one, respectively, as done in Table 5.2. Thus the system is quite accurate for binary inputs.

The problem is how to handle a larger number of coordinated elements. OR and NOR are invariant for additional inputs, but the threshold of AND and NAND must be incremented by 1 for each additional input, which gives the impression that different lexical items are needed for every increment of the coordinatees. Table 5.3 illustrates the problem for AND with three coordinatees. The alternative is to break down the input sequence into a series of iterations of the basic binary format. However, this leads to a insurmountable empirical problem.

Halbasch (1975) points out that an odd number of conjuncts can have different truth values depending on their bracketing. Consider the variation in

Table 5.5. McCulloch and Pitts parallel computation of NOR, $w_i = -1$.

Input at t			$\Sigma(w_i x_i)$	*Output at t+1*
w_1x_1	w_2x_2	w_3x_3	$w_1x_1 + w_2x_2 + w_3x_3$	$\Sigma \geq 0$ *(NOR)*
0	0	0	0	1
0	0	–1	–1	0
0	–1	0	–1	0
–1	0	0	–1	0
0	–1	–1	–2	0
–1	0	–1	–2	0
–1	–1	0	–2	0
–1	–1	–1	–3	0

bracketing between (5.2a) and (5.2b), in which NOR conjoins the three clauses *p*, *q*, and *r*:

5.2.

	p q r		
a)	((0 0) 0)	= ((1) 0)	= 0
b)	(0 (0 0))	= (0 (1))	= 0

In (5.2a), the bracketing proceeds from left to right, but the binary outcome is false. Commuting the order of application does not help. In (5.2b), the bracketing proceeds from right to left, and the outcome is again false. The binary McCulloch and Pitts network does not fare any better, as calculated in Table 5.4. No matter which way the inputs are processed, the outcome is false.

All is not lost, though. The parallel approach in Table 5.5 turns out to evaluate to true, as Halbasch predicts. This is a strong argument for the parallel processing of coordination and against serial processing.

5.3.2. Single-layer perceptron (SLP) networks

The next parallel unit is the Perceptron of Rosenblatt (1958, 1961). A perceptron sums up its weighted inputs, compares this sum to a threshold, and passes the result through a hard limit activation function. This equation for this process is given in Eq. 5.3:

5.3.
$$y = \text{hardlim}\left(\sum_{i=1}^{n} w_i x_i - \theta\right)$$

If the sum is high enough, the function outputs a 1; otherwise, it outputs a 0. Fig. 5.6 restates this calculation graphically.

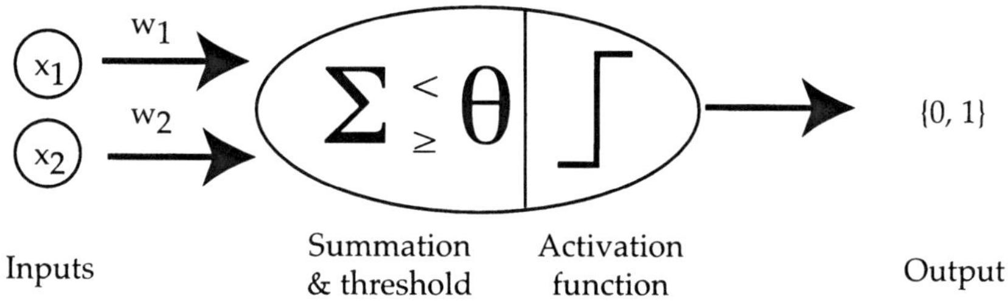

Figure 5.6. Early perceptron.

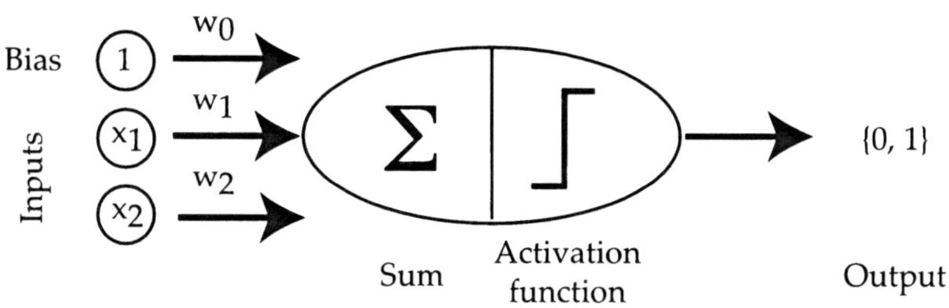

Figure 5.7. Current perceptron.

It is mathematically clumsy and confusing to include the extra step of computing the threshold, so it is incorporated as the weight of a special input node called the **bias**, whose activation is always 1. Eq. 5.4 lays out the modified version, which corresponds to the diagram in Fig. 5.7.

5.4. $$y = \text{hardlim}\left(\sum_{i=1}^{n} w_i x_i + w_0 b\right)$$

In this way, the algorithm for updating the weights of the inputs is extended automatically to change the threshold.

5.3.2.1. SLP classification of the logical coordinators

The patterns described by the logical coordinators are so simple that the smallest perceptron network imaginable can be trained to recognize them, that of two inputs and one neuron, plus its bias, which is depicted in Fig. 5.8. The next few paragraphs walk us through a sample classification.

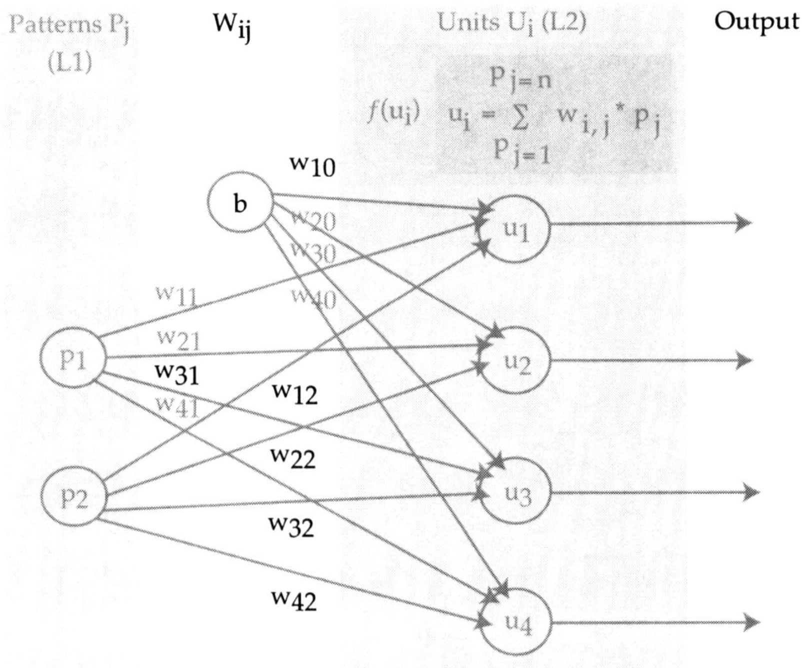

Figure 5.8. SLP network.

The first step is to sum up the inputs and bias to give a number called **net input** or simply **net**. Eq. 5.5a gives the general form of the equation for doing so, and Eq. 5.5b gives the specific formulation for two inputs, which is the one that interests us here.

5.5 a) $\text{net} = b + W * P$

b) $\text{net} = (1 * b) + (p_1 * w_{1,1}) + (p_2 * w_{1,2})$

The bias is handled separately since it is treated as an input that is always on, i.e. 1. As an initial choice of weights, *b* can be set to 0.15 and *W* to $[-0.2, 0.3]^T$, which will create an interesting error. Substituting these values into Eq. 5.5b gives Eq. 5.6:

5.6. $\text{net} = (1 * 0.15) + (p_1 * -0.2) + (p_2 * 0.3)$

For *P*, let us use the values of OR for two and three coordinatees. As for the specific representational format of a coordinator, we start with just plain

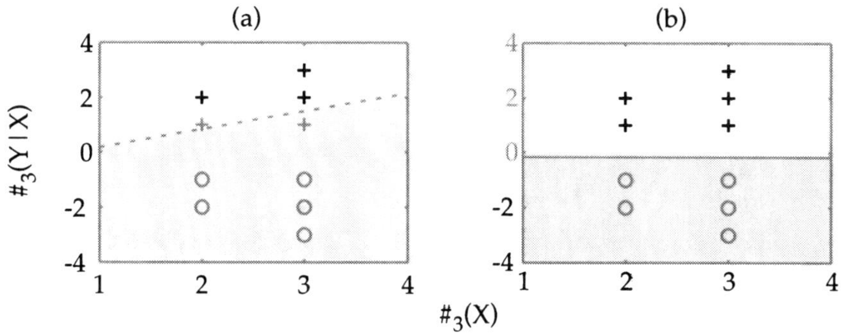

Figure 5.9. Plot of decision boundaries for the SLP test on the data space for OR. (a) Initial; (b) Final. Shading marks those values that are rejected by the SLP. (**p5.01_SLP_OR.m**)

cardinalities, in order to make the patterns to be learned a little more complex and so illustrate the learning algorithms more effectively. In particular, let us use $[3, 1]^T$. Substituting it into Eq. 5.6 gives:

5.7. $\quad$ net = (1 * 0.15) + (3 * –0.2) + (1 * 0.3) = 0.15 – 0.6 + 0.3 = –0.15

This sum is then passed through the hard limit activation function, introduced in Fig. 2.30. The reader may recall that this function evaluates any input of one or greater to 1, and all others to 0:

5.8. $\quad$ output classification = hardlim(net) = hardlim(–0.15) = 0

It outputs a classification of 0 for the negative value for *net*. Referring to Fig 5.9a for the classification imposed by OR on the data set, we see that 0 is not correct; the point $[3, 1]^T$ should have been accepted, since it is marked with a plus sign indicating an accepted value. This figure also invokes a graphic device to highlight the evaluation of the initial choice of weights for all of the data, the dotted line known as a **decision boundary**.

The decision boundary is defined much like a nullcline in a phase-plane plot: it is result of setting the equation in question to 0 and solving for one of the variables. The equations in question here are those of the net output, Eq. 5.5a, which are set to 0 in the first two lines of Eq. 5.9. Solving the second for p_2 produces Eq. 5.9c:

5.9 $\quad$ a) $\quad$ 0 = b + W * P

b) $0 = (1 * b) + (p_1 * w_{1,1}) + (p_2 * w_{1,2})$

c) $$p_2 = -\frac{w_{1,1}}{w_{1,2}} p_1 - \frac{b}{w_{1,2}}$$

This instantiates the equation for a line, $y = ax + b$. Passing the values for the x axis in Fig. 5.9a through Eq. 5.9c defines the dotted line, which divides the plane into two regions, those above the line, for which equation Eq. 5.9c produces positive results and so are evaluated by the hard limit function to 1, and those below the line, and shaded, for which Eq. 5.9c produces negative results and so are evaluated by the hard limit function to 0. The former are accepted, and the latter rejected. Under the choice of weights used here, only point $[3, 1]^T$ is classified incorrectly as rejected; the other six are classified correctly.

Such errors are to be expected from an arbitrary initial weighting. Fortunately, they are easy to recognize, for we know what the correct output should be. We can even quantify them, by defining an error as the target output minus the calculated output. Eq. 5.10a shows the general equation, and Eq. 5.10b conveys the results of applying it to the case that was just discussed:

5.10 a) error = target – calculated
b) error = 1 – 0 = 1

The challenge is to correct them, so that posterior iterations of the perceptron perform closer to the targets. The algorithm used to effect this correction is generally called the **learning rule**, though it is perhaps more accurate to refer to it as a **training algorithm**.

5.3.2.2. SLP error correction

The obvious thing to change in order to increase the accuracy of the perceptron is the value of the weights – in fact, this is the only thing that can be changed. The network architecture should not be changed, because it is this particular architecture that we want to test as a model of the acquisition of the logical coordinators.

Intuitively, the perceptron corrects misclassifications by moving the decision boundary to accept only the true points and reject all the false points. The essence of the perceptron learning rule is that this pivoting of the decision boundary in space involves using the error measure to classify the input by multiplying the two together. The result of this operation can be conceptualized as the amount that the weights should be changed by, **dW**, as defined by Eq. 5.11:

5.11 a) dW = (target – calculated) * input = error * input
b) $W^{new} = W^{old} + dW$

Let us try this procedure out on our example:

5.12 a) $dW = (1 - 0) * [3, 1]^T = 1 * [3, 1]^T = [3, 1]^T$

$$W^{new} = W^{old} \pm \text{input}$$

b) $$W^{new} = \begin{bmatrix} -0.2 \\ 0.3 \\ 0.15 \end{bmatrix} \pm \begin{bmatrix} 3 \\ 2 \\ 1 \end{bmatrix} = \begin{bmatrix} -3.2 \\ -1.7 \\ -0.85 \end{bmatrix}$$

Note that the bias is included at the top row of each matrix as the 'zeroth' weight. Using these new weights to recalculate the net output produces (5.13a), which is passed through the hard limit activation function in (5.13b) to produce a classification of 1:

5.13 a) net = (1 * 1.15) + (3 * 2.8) + (1 * 1.3) = 1.15 + 8.4 + 1.3 = 10.85
b) output classification = hardlim(net) = hardlim(10.75) = 1

The data point is now classified correctly.

As a additional elaboration, it is convenient to include a **learning rate,** *lr*, to modulate the change brought about by learning. Usually drawn from the interval [0 1], the learning rate enters the equation for weight change as an additional multiplicand:

5.14. dW = lr * error * input

It specifies how large a change will be made during weight updates. The larger the learning rate, the larger the change, so a large enables a network to correct an error quickly, but it also raises the chance that a weight update will overshoot the target and so lead to further errors. Thus the learning weight is usually set to a relatively small value.

The algorithm described in the preceding paragraphs can be applied iteratively to all of the inputs until the weights reach a steady state, that is, they do not change in subsequent iterations. Rosenblatt (1961) showed that this algorithm permits a perceptron to converge on a solution in a finite number of iterations, if a solution exists.

5.3.2.3. SLPs and unnormalized coordinators

We now have enough background to attempt a simulation of the four logical coordinators using a 2x1 perceptron network. Yet this brings us face to face with a crucial empirical question: how many coordinate elements should be included in the data set? I know of no discussion of this issue in the linguistics literature, outside of early generative comments on the recursivity of coordination, see Chomsky, 1965, p. 212 among others, that suggest that the number of conjuncts is infinite.

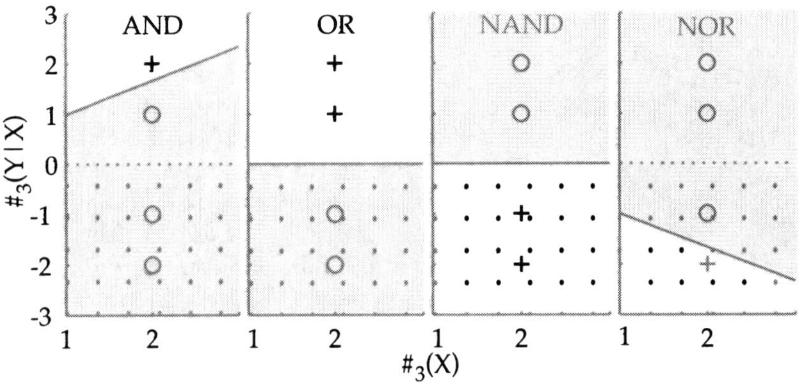

Figure 5.10. Plots of initial and final decision boundaries for COOR2. Initial state: bias = 0; W = [0, 0]^T. Dotted line marks initial decision boundary, stippling marks initially rejected area. Solid line marks final decision boundary; shading marks finally rejected area. (**p5.02_SLPCOOR.m**)

Table 5.6. Logical coordinators for | x | = 2, or COOR2.

v#	*COOR*	*norm*	*AND*	*OR*	*NAND*	*NOR*
1	2, 2	[0.71, 0.71]^T	1	1	-1	-1
2	2, 1	[0.89, 0.45]^T	-1	1	-1	-1
3	2, -1	[0.89, -0.45]^T	-1	-1	1	-1
4	2, -2	[0.71, -0.71]^T	-1	-1	1	1

It is unlikely that children learn the meaning of the logical coordinators from coordinations of an infinite number of elements. In fact, it is unlikely that children learn the meaning of the logical coordinators from coordinations of more than five elements, since such constructions are textually rare, if not difficult to understand. It is much more likely that children learn the meaning of the logical coordinators from coordinations of two or three elements. We therefore construct the data and training sets for the perceptron simulations as illustrated in Table 5.6, from | *x* | = 2, or COOR2.

The results of training a 2x1 (two inputs, one neuron) perceptron network on COOR2 are reproduced in Fig. 5.10 as graphs of the initial and final decision boundaries. They have all converged to a state in which the accepted and rejected values are correctly separated by the decision boundary. The weights that derive these boundaries are listed in Table 5.7. We therefore have our first neuromimetic model of the logical coordinators.

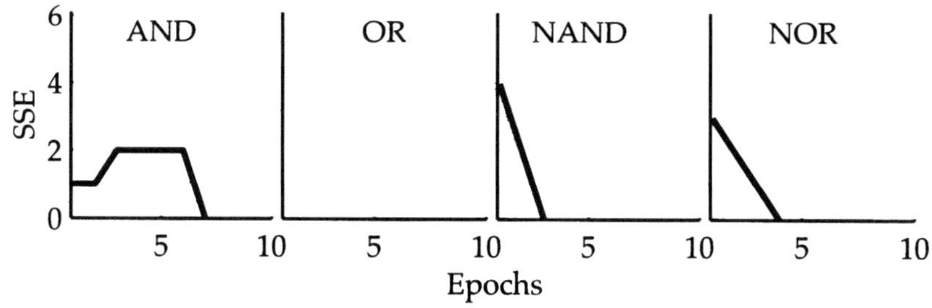

Figure 5.11. Sum-squared error per epoch of SLP, lr = 0.1. (**p5.02_SLPCOOR.m**)

Table 5.7. Initial and final weights of SLP for COOR2.

	initial	*AND*	*OR*	*NAND*	*NOR*
w1:	0	-0.20	0	0	-0.20
w2:	0	0.30	0	-0.30	-0.30
bias:	0	-0.10	0	0	0
epochs:	0	7	1	3	4

There is one point glossed over in the preceding exposition that warrants further discussion. It is how the network actually knows that a steady state has been reached, since it can always go on changing the weights by zero forever. The ideal solution is to make the halting of the execution of the program conditional on some measure of the error. A common measure is the **sum-squared error** (SSE), found by squaring the error e for each input, taking its square root $(e^2)^{1/2}$, and then adding together all $(e^2)^{1/2}$for a single presentation of the data set – a single **epoch**. Using the square of the error is necessary to avoid having the positive and negative errors cancel each other out. When the SSE reaches zero, the simulation halts.

Under this condition, the perceptrons for the four data sets evolve as summarized in Fig 5.11. They all halt within seven epochs, though NAND halts in half that time, and OR halts immediately. OR halts so quickly because the initial weighting of $[0, 0, 0]^T$ already defines it, even though this state is the most neutral that can be chosen.

This differential sensitivity of the learning algorithm to the initial state of the system sparks one's imagination with vistas in which the setting of the initial conditions create computational difficulties which mimic difficulties that humans have in learning linguistic phenomena. Unfortunately, before letting our imaginations run wild, we must still do the dull science thing of checking these initial results to see how well they generalize to larger data sets.

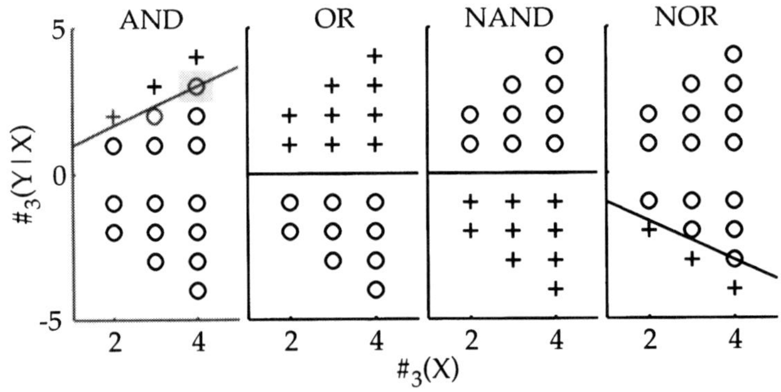

Figure 5.12. Plot of final decision boundaries for COOR2 projected onto COOR4; shaded regions are misclassified. (**p5.02_SLPCOOR.m**)

Most interesting would be to test these results against a data set which tests the limits of human short-term memory, which is usually taken to be 7 ± 2 items, see Miller (1956). A quick glance at the graphs of the decision boundaries calculated for COOR2 projected onto the larger data set in Fig. 5.12 shows why. The decision boundaries learned for COOR2 do not necessarily lie at the proper angles to generalize to higher numbers of coordinatees. Thus at first glance, these single-layer perceptrons simulate human performance in that their accuracy falls off as the number of coordinates increases.

5.3.2.4. SLPs for the normalized logical coordinators

The next question to be answered is how normalization affects the single-layer perceptron. Table 5.6 displays the normalized version of coordinator input for COOR2 to the right of the raw input. Using these new numbers as input, the SLP classifies them as in the top row of Fig 5.13. There is no surprise here; the network is able to discover the correct classifications.

Moreover, projecting the final decision boundary for COOR2 to COOR4 as done in the bottom row of Fig 5.13 reveals a tendency towards misclassification similar to the one seen in Fig. 5.12. The reason is clearer with normalization: there are gaps between the data points at COOR2 which are filled in at COOR4, so what was an imprecision due to lack of data for COOR2 is an error for COOR4.

The distribution of errors is not qualitatively different from that of the unnormalized version. The only differences are that AND is learned more quickly and with fewer errors, while NOR is learned less quickly and with more errors. The facilitation of AND is presumably due to the greater number because

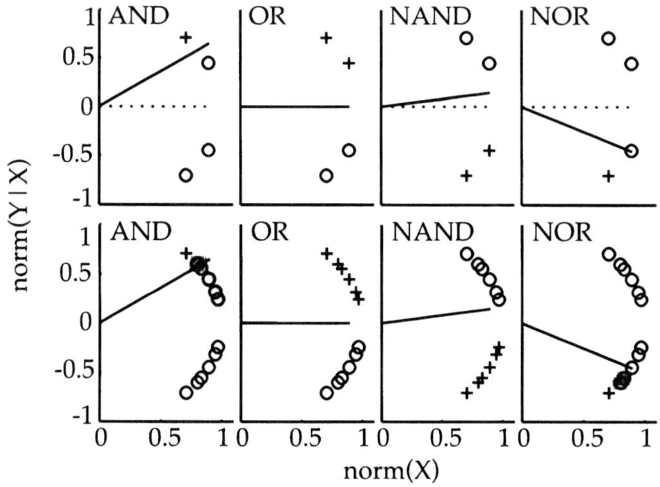

Figure 5.13. Top: initial (dotted) and final (solid) decision boundaries for the normalized COOR2. Bottom: extrapolation of final decision boundaries up to COOR4. Shading indicates erroneously classified points. (**p5.03_SLPCOORnorm.m**)

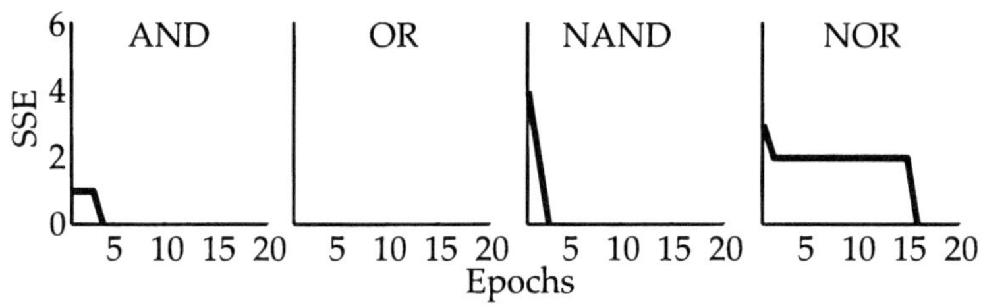

Figure 5.14. Sum-squared error per epoch for top row of Fig 5.13, lr = 0.1. (**p5.03_ SLPCOORnorm.m**)

the distance between the accepted and rejected points is much smaller with the latter than the former.

5.3.2.5. Linear separability and XOR

So far, the classification of coordinator meanings has involved nothing more complicated than the correct placement of a single straight line. And while this is good enough for the logical operators as presented in this book, there are

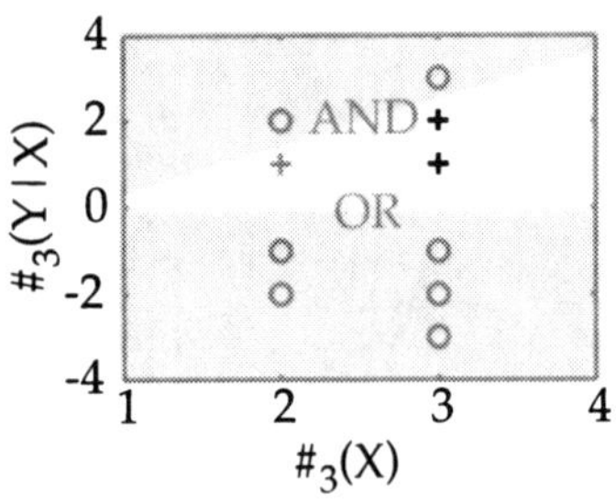

Figure 5.15. XOR = OR – AND. (**p5.04_XOR_plot.m**)

many other possible natural language meanings that the system under development here may want to generalize too.

The most well-known is that of exclusive *or*, or XOR, which excludes the AND meaning from OR. Its pattern is plotted in our data set in Fig. 5.15. The reader is welcome to take a straight-edge and try to draw a single straight line that separates the o's from the +'s in the figure, but you will soon find it to be impossible. In the jargon of linear algebra, the accepted region for XOR is not **linearly separable** from the rejected region. We know that a single-layer perceptron cannot learn such a pattern, because it can only place a single straight line. XOR, and XNAND, for that matter, needs two straight lines to separate out both sets of rejected values. Fig 5.15 sketches the two decision boundaries whose combination would give XOR, namely, OR and the removal of AND. How could a SLP be augmented to draw two lines?

5.3.3. Multilayer perceptron (MLP) and backpropagation (BP) networks

Well, if a single SLP draws a single line, maybe two SLPs combined can draw two straight lines to effect the classification. Fig. 5.16 schematizes the architecture of the simplest such network, where many labels have been suppressed for clarity. Unit u_1 can draw AND, unit u_2 can draw OR, and unit u_3 can combine the two as a kind of AND. Such multi-layer perceptrons do in fact lead to the correct solution, though it was several decades before an adequate implementation was attained.

5.3.3.1. Multilayer perceptrons

The challenge of the MLP is to update the weights W_{ij} – now **hidden** between the input layer and the output layer – from an error signal that is only directly observable at W_{jk}. The perceptron learning rule cannot do this, for two reasons. The obvious one is that updates are only calculated for the weights that lead into the output unit, u_3. There is no allowance made for changing the weights

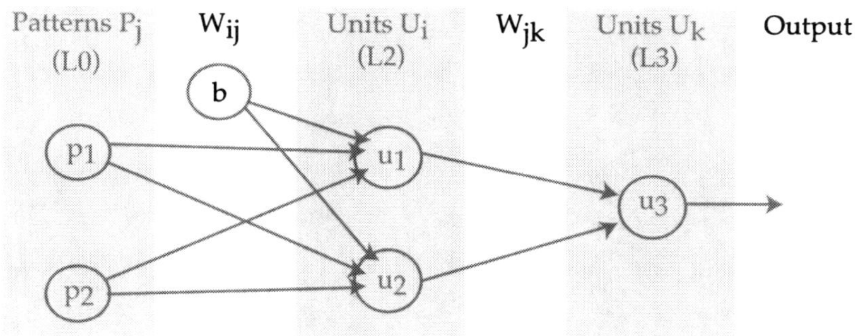

Figure 5.16. 2 x 1 MLP network.

upstream of u_3 at u_1 and u_2. The other difficulty lies in the all-or-nothing behavior of the hard limit transfer function.

The output of this function only differentiates two inputs, positive and not-positive. In other words, from an output classification of 1, all you know about the net input is that it was positive. It would be much more informative – and much easier to propagate the error backwards from classification to net input – if the output could be differentiated at all values between 0 and 1. It follows that the first step to be taken in building a MLP is to find a new, differentiable transfer function.

5.3.3.2. Sigmoidal transfer functions and the neurons that use them

One of the most effective is a sigmoid, or S-shaped function, illustrated in Fig. 2.30. At 0 input, the log-sigmoid outputs 0.5, while approaching 0 at smaller inputs and 1 at larger. Thus in the middle of the curve, the classificatory output can be differentiated into all values of net input. Neurons that use this activation function make up perhaps the most generic of all connectionist neurons. Here the output is in the range of 0 to 1, as opposed to just being a member of the set {0, 1}.

5.3.3.3. Learning by backpropagation of errors

Given a differentiable activation function, it is possible to calculate how much the error at the output layer depends on the weights of the neurons that lead into it. This is done by calculating the sensitivity of the error at the output node to the contribution of the weights that produce it. Upon obtaining these results, the weights leading into the output node are adjusted so as to minimize the ones that produce most of the error. The algorithm then backs up to the previous layer and calculates the difference between using the new weights and the previous weights to find the error for that layer. It then iterates the process of calculating sensitivities and optimizing the input weights to reduce the error, and then moves another layer backwards, until the input layer is reached. Then

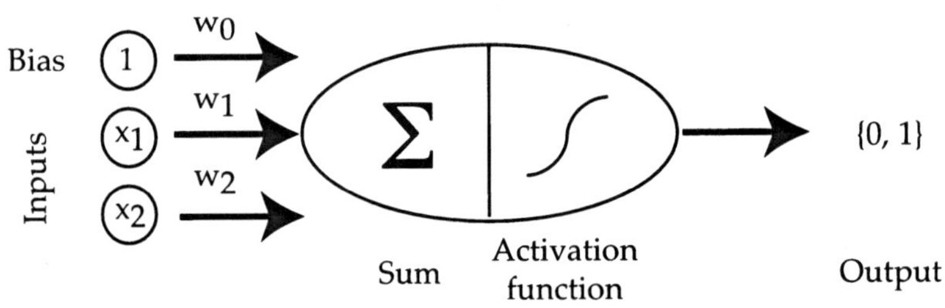

Figure 5.17. MLP/backpropagation neuron.

a new pattern is submitted to the network for analysis. As Kartalopoulos, 1996, p. 76, says, this **backpropagation** algorithm "... is an involved mathematical tool", and we cannot begin to do it justice in this brief review, especially since it lacks neurological plausibility.[29]

5.3.4. The implausibility of non-local learning rules

As luck would have it, error-correcting algorithms have proven to be so useful for engineering applications that have no need for neurological plausibility that they are by far the most widely used – so much so, in fact, that the phrases "neural network" and "neural-network learning" are practically synonymous with error-correcting algorithms: stochastic gradient descent (on an error surface), more commonly known as the back-propagation (of error) algorithm.

Despite this engineering advantage, for our purposes this family of learning rules suffers from a serious flaw. From a neurophysiological perspective, one notices that the perceptron and backpropagation learning rules are driven by information that is external to the weights being updated: weights are updated in proportion to the error measure, which is calculated at some remove from the layer containing the weights. Such learning rules are known as **non-local** rules, since the decision to change the weights is not made at the layer at which the weights reside. In contrast, the learning rules based on neurological investigation are all **local**, in the sense that weights are updated in proportion to factors present in the input itself. , see Grossberg (1987), Massaro (1988), Stork (1989), Hinton (1989), and Levine, 2000, p. 217ff.

[29] The original statement of the algorithm is in Werbos (1974). It was rediscovered independently by Parker (1982) and Rumelhart et al. (1986); see Duda, Hart and Stork, 2001, p. 334, for more historical background.

To be charitable, those that use the backpropagation algorithm to simulate human cognition are not unaware of its biological implausibility. In their defense, they claim that the brain optimizes a given learning task, so that an idealized learning rule such as backpropagation of error can yield valuable insights into how the brain itself represents the solution to a learning task, as explained in Golden, 1996, p. 20. However, it strikes us as odd that a procedure that could introduce so many erroneous factors would be resorted to when there are so many more realistic alternatives available.

5.3.5. Summary

This first section examines a class of hyperplane classifiers and finds them to not implement a neurologically-plausible learning rule. If there were no other alternatives, we would be obliged to use them anyway, but there are many other architectures that do not have these failings. The next section discusses a learning rule that marks the transition between hyperplane and cluster classification and is the first neurologically-plausible learning rule to be introduced.

However, that is not to say that we have not learned anything useful. In view of the arguments from the previous chapter that XOR is not a basic coordinator, this result suggests that basic (logical) coordinators are limited to linearly separable spaces:

5.15. HYPERPLANE CONJECTURE: A monomorphemic coordinator classifies its input space into two linearly separable sub-spaces.

This is due to the fact that linearly separable regions are convex subsets of the plane. One may conclude that single-layer perceptrons provide observationally and descriptively adequate models for the logical coordinators, but they do not attain the higher degree of explanatory adequacy due to the reliance on a non-local learning rule.

5.4. UNSUPERVISED LEARNING

Unsupervised learning algorithms can be divided into three kinds, see Becker (1995) and Becker and Plumbley (1996):

5.16. a) information-preserving algorithms
b) probability-density estimation algorithms
c) invariance learning

The first kind attempts to preserve as much of the input information as possible in the output, often by maximizing the information in the output that is also in the input, known as mutual information. The second only tries to preserve 'important' information, especially an estimate of the statistical structure of the input, known as its probability distribution. Finally, the third kind of unsupervised algorithm tries to find particular information in the input by building into the algorithm assumptions about what the output should look like.

For the logical coordinators, it is not necessary to build any specific assumptions about what the output should look like into the training algorithm in order to the coordinator patterns accurately. That is to say that the first two types of unsupervised learning suffice to capture the range of coordinator meanings. This may be due to the fact that coordinator meanings are so simple and easy to establish on the basis of a handful of samples that more complex kinds of unsupervised learning are unnecessary. Or it may be a general constraint on human grammar that learning rules can not come predisposed, i.e. genetically pre-programmed, to look for certain grammatical entities. In any event, the upcoming pages will exemplify one information-preserving algorithm, that of Hebbian learning, and one probability-density estimation algorithm, that of competitive learning.

5.4.1. The Hebbian learning rule

If an error cannot be directed backwards to modulate the input weights, then what else is left? The only parameters left are the input weights themselves, so they must be able to modify themselves directly from the input that they are processing. This is the essence of Hebb's postulate (Hebb, 1949, p. 62):

> When an axon of cell A is near enough to excite a cell B and repeatedly or persistently takes part in firing it, some growth process or metabolic change takes place in one or both cells such that A's efficiency, as one of the cells firing B, is increased.

Slightly more succinctly, Hebb's postulate says that if two neurons on either side of a synapse are activated simultaneously, the synapse will get stronger.

This conjecture can be put in mathematical form as in Eq. 5.17, where Δ is read as "the change in":

5.17. Δ synapse = post-synaptic activation * pre-synaptic activation

Thus if either neuron is inactive, there will be no change; if both are only slightly active there will only be a slight change, while if both are greatly active there will be a great change.

In terms of a neuromimetic implementation, the change in synaptic strength Δw_{ij} is represented as the product of the activations of the post-synaptic neuron *i* and the pre-synaptic neuron *j*, as in Eq. 5.18a, which is then added to the current weight to get the new one, as in Eq. 5.18b:

5.18 a) $\Delta w_{ij} = a_i * a_j$

b) $w_{ij}^{new} = w_{ij} + \Delta w_{ij}$

It is convenient to include a learning rate *lr* in order to modulate the pace of change:

5.19. $\Delta w_{ij} = lr * (a_i * a_j)$

Note that there is nothing in this formulation that imposes a limit on how large the weights can grow when both neurons are active, so they could increase without bounds.

Such incessant growth can be slowed in two ways: each weight might decay at a rate proportional to its current strength, or each weight might decay at a fixed rate, independent of its strength. The former is known as multiplicative decay, while the latter is known as subtractive decay. Multiplicative decay is more interesting, see Miller and MacKay (1994) for a mathematical review, and Turrigiano (1999) for a review of the more general issue of the biological substrate for the homeostasis of synaptic plasticity implied by weight decay. Multiplicative decay is incorporated into Eq. 5.18b by multiplying a positive constant γ, the **decay rate**, by the current weight to make the decay proportional to the current synaptic strength, and then reducing the current weight by this proportion, as shown in Eq. 5.20a:

5.20 a) $w_{ij}^{new} = (w_{ij} - \gamma * w_{ij}) + \Delta w_{ij}$

b) $w_{ij}^{new} = (1 - \gamma) * w_{ij} + \Delta w_{ij}$

Thus as the current weight is increased by Δw_{ij}, it is decreased by decay through γw_{ij}, in an attempt to maintain a balance between the two. Eq. 5.20b simplifies the terms for the final version.

This type of learning is local in the sense introduced above, in that the information that leads to a change in weights is located at the same site as the weights, namely the synapse. It is also characterized as **activity-dependent**, since it is driven directly by the level of activity across the synapse, and **correlation-based**, since it only takes place when both sides of the synapse are active at the same time.

One of the curious properties of a network that evolves under the influence of this purest form of Hebbian learning is that it is often useful for the output of the network to be the same as the input! This usage characterizes a network as **autoassociative**, whereas the type of network that was assayed with the perceptron, in which the output is different from the input is known as **heteroassociative**. The next subsections demonstrate more heteroassociative networks, in which the output is a classification of the input.

5.4.2. Instar networks

Grossberg (1982) invented two sorts of artificial neurons, the instar and the outstar, to explain certain aspects of visual learning in humans and other

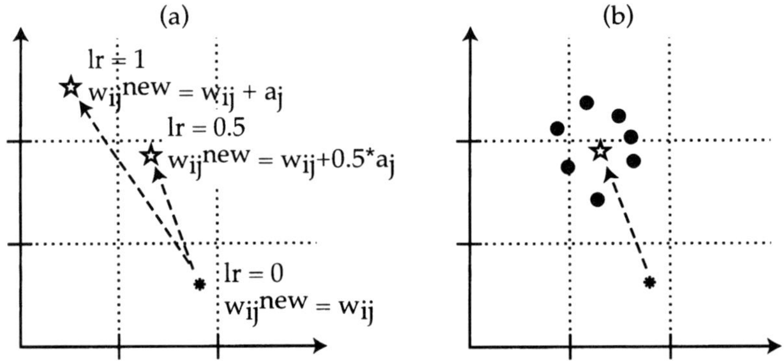

Figure 5.18. (a) Instar learning as a multiple of the learning rate; (b) An instar neuron moves to the center of a cluster.

animals. The instar neuron learns to recognize a vector, while the outstar learns to produce one. In this section, we examine the viability of using instars to recognize the patterns of logical coordinators. They are not viable, but they provide a necessary background for understanding the successful competitive networks taken up in the next section.

5.4.2.1. Introduction to the instar rule

One of the drawbacks of weight decay is that associations tend to be forgotten without repetition of the stimuli. Grossberg reasoned that this deficiency could be mitigated by allowing weights to decay only when a neuron is active, that is, when it is producing output, i.e. $activation_i > 0$. His idea can be implemented by subtracting the decay term γ from the product of the input and output, as in Eq. 5.21:

5.21. $$\Delta w_{ij} = lr(a_i * a_j^T) - \gamma(a_i * w_{ij})$$

This can be simplified by setting γ equal the learning rate *lr*, so that old values decay at the same rate that new values are learned. Eq. 5.22a replaces γ with *lr*, and Eq. 5.22b simplifies:

5.22 a) $$\Delta w_{ij} = lr(a_i * a_j^T) - lr(a_i * w_{ij})$$

b) $$= lr * a_i(a_j^T - w_{ij})$$

This new learning equation is most easily understood by examining how it performs when the neuron *i* is active, which is to say that $a_i = 1$. Eq. 5.23a substitutes 1 for a_i, the result of which is simplified to Eq. 5.23b:

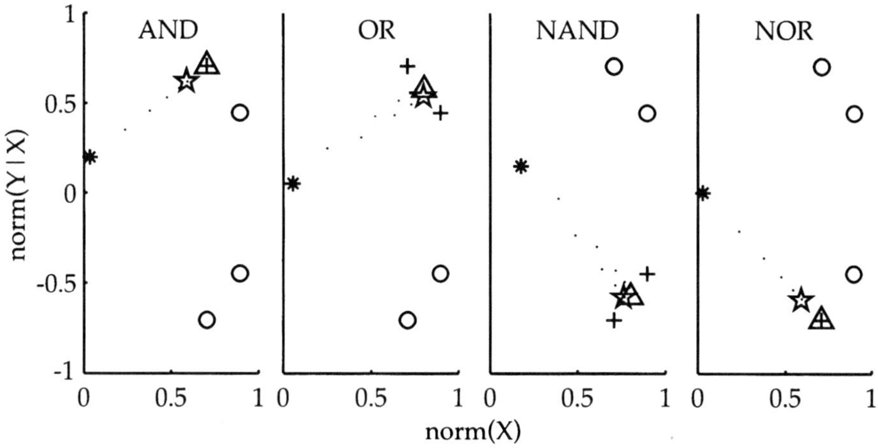

Figure 5.19. Instar learning of COOR2. 'o' = rejected point, '+' = accepted point, 'Δ' = mean, '*' = initial weight, dot = intermediate weight, star = final weight. (**p5.05_INCOOR.m**)

Table 5.8. Final weights of INCOOR.

COOR	*AND*	*OR*	*NAND*	*NOR*
x	0.69	0.78	0.83	0.69
y	0.69	0.57	-0.48	-0.68

5.23 a) $\Delta w_{ij} = lr * 1\,(a_j^T - w_{ij})$

b) $\quad = lr(a_j^T - w_{ij})$

To further refine our understanding of this new rule, let us incorporate it into the equation for weight update from Eq. 5.18b to give Eq. 5.24 and consider the effect of the learning rate.

5.24. $w_{ij}^{new} = w_{ij} + \Delta w_{ij} = w_{ij} + lr(a_j^T - w_{ij})$

Eq. 5.25 illustrates how Eq. 5.24 is solved if the learning rate is 0:

5.25 a) $w_{ij}^{new} = w_{ij} + 0 * (a_j^T - w_{ij})$

b) $w_{ij}^{new} = w_{ij}$

Table 5.9. Test of INCOOR; out = satlin(W_{final} * COOR2^T).

#	COOR2	*AND*		*OR*		*NAND*		*NOR*	
		Tar	*Out*	*Tar*	*Out*	*Tar*	*Out*	*Tar*	*Out*
1	$[0.71, 0.71]^T$	1	0.86	1	0.95	0	0.13	0	0.03
2	$[0.89, 0.45]^T$	0	0.82	1	0.96	0	0.43	0	0.30
3	$[0.89, -0.45]^T$	0	0.29	0	0.47	1	0.95	0	0.81
4	$[0.71, -0.71]^T$	0	0.02	0	0.18	1	0.95	1	0.84

The new weight equals the old, so there is no learning. Eq. 5.26 illustrates how the equation is solved if the learning rate is 1:

5.26 a) $w_{ij}^{new} = w_{ij} + 1 * (a_j^T - w_{ij})$

b) $= w_{ij} + a_j^T - w_{ij} = a_j^T$

The new weight now equals the input, so there is complete learning. Any learning rate between these two extremes will move w_{ij} part of the way between itself and the input, as diagrammed in Fig. 5.18a. The current weight is moved towards the input, according to the percentage given by the learning rate. If there were many inputs clustered together, as illustrated in Fig. 5.18b, the instar neuron would move to the center of the cluster. The cluster of inputs that the instar neuron becomes sensitive to can be called its **neighborhood**. The instar neuron itself, or more accurately, the point in space defined by its weights, can be called the **prototype** vector of the cluster.

It is here that the effort expended in Chapter 3 to relate statistical and vector measures of the logical operators bears its fruit. Even though the instar representation is that of a vector space, the learning rule itself explores this space in a statistical manner. That is to say, the neuronal weights tend to converge on the center of the statistical distribution, since this is the point of maximum correlation between input and output activity driven by Hebbian learning, see Intrator, 1995, p. 220. What this means is that the prototype vector of a cluster is the vector at which the cluster's members are most correlated spatially and so give rise to the highest level of neuronal input, which the Hebbian rule uses as a signal to increase the neuronal output. The end result is an increase in the correlation between the two.

5.4.2.2. An instar simulation of the logical coordinators

Given that an instar neuron can learn to approximate multiple inputs, it is appropriate to use an instar network to learn the patterns of the distributive coordinators. Simulation **p5.05_INCOOR** illustrates the instar learning process for a single neuron per coordinator, trained on COOR2. The trajectory of the training is depicted in Fig. 5.19. Despite the apparent success of the training process as diagrammed therein, it is actually more like a draw. Table 5.9

organizes the outcome of testing the network against the COOR2 data set. Every coordinator accepts every data point to a certain degree. The reason can be grasped by a consideration of what the final weights for each neuron laid out in Table 5.8 means. All the weights are non-zero, so any point drawn from COOR space and multiplied by a particular pair of weights will produce an non-zero output, thus incorrectly accepting the input.

There is one way to fix the instar network. It is to add more instar neurons in order to divide the space into more manageable portions. This route leads to greater transparency of analysis, but it also results in the network memorizing the input space, where by memorization, we mean that each data point is covered by at least one neuron. It is unlikely that such an approach will generalize correctly to new cases, so we are consequently obliged to search for an alternative kind of learning that preserves the instar advantage of moving weights through space, but provides more context for the evaluation of these weights.

5.4.3. Unsupervised competitive learning

The instar networks proposed in the preceding section give an accurate account of the learning of COOR space, but it seems overly generous of the neurological system to spend its resources on four independent networks. It would be much more economical for there to be but a single network that takes any coordinator pattern as input to produce the correct coordinator, plus some kind of signal about that part of the input space which must be ignored. This may appear to lead to a contradiction, because a Hebbian learning rule only responds positively to positive input and so cannot calculate negative weightings directly. Fortunately, this contradiction can be avoided by mapping the entire potential input space onto many neurons, so that only those that correspond to non-zero input will become active. The others will either 'die' or alter their weights to approximate other zones of the input space. When a vector is submitted to such a network, the neuron whose weight vector is closest to it receives the most activation. This neuron is said to 'win' on this cycle and its weights are updated to be more similar to the input. It therefore becomes more likely to win on the next input of the same or a similar vector, and less likely to win on inputs of different vectors. As this process is repeated for every neuron in the network, the network gradually learns to represent different regions of the space where input vectors occur. This procedure is often called the **winner-take-all (WTA) function**, see among others Didday (1976), Grossberg (1976), Amari and Arbib (1977), and Yuille and Geiger (1995), the **competitive learning rule**, Rumelhart and Zipser (1986) and Intrator (1995), or the **Kohonen learning rule**, Kohonen (1989).

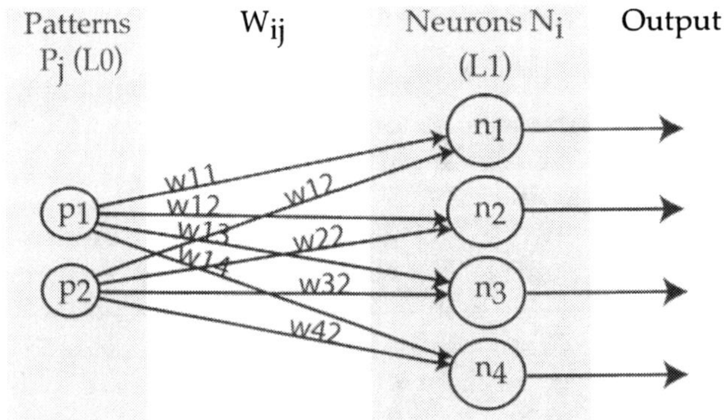

Figure 5.20. A competitive network for the logical coordinators.

5.4.3.1. A competitive simulation of the logical coordinators

To depict a competitive network that learns the logical coordinators, let us simply add enough neurons to effectively cover the input space. Trial and error has shown that four is enough to cover COOR space, which results in the network of Fig. 5.20. Fig. 5.21 displays a simulation using such a network to learn each coordinator in COOR2.

The initial weights of the neurons are assigned to fixed positions outside of COOR space so that their paths will not cross. If competitive neurons are initialized to positions from which their paths would cross, the algorithm is so efficient that one of the crossing neurons would always wind up losing to the other and so become otiose.

The results of testing each neuron's final weights against COOR2 are tabulated in Table 5.10. The numbered columns show the competitive output of the network to each input listed under the COOR2 heading. This test shows that each neuron responds maximally to the data point closest to it, but that does not necessarily lead to an accurate rendition of the coordinators. For instance, neuron 1 responds maximally to the point that defines NOR, #4, while neuron 2 responds maximally to the other negative point, #3. Given the winner-take-all algorithm, neuron 2 will lose out to neuron 1 over point #4, so there is no way to define NAND in terms of these receptive fields. The same holds for OR in the positive half of COOR space. The consequence is that the algorithm does not succeed at learning the logical coordinators. Note that the algorithm cannot be 'fixed' by relaxing it so that a neuron will respond to, say, its two nearest inputs, since then each pair of neurons will respond to both inputs of the appropriate polarity, effectively making them both existential.

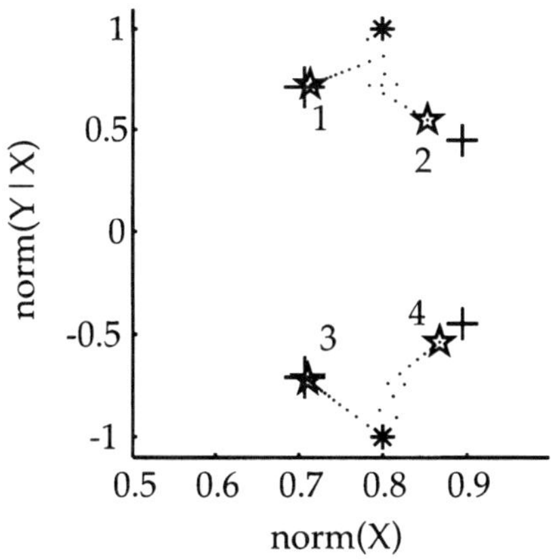

Figure 5.21. Unsupervised competitive learning of COOR2. '+' = accepted point, '*' = initial weight, dot = intermediate weight, star = final weight. (**p5.06_COMPCOOR.m**)

Table 5.10. Test of COMPCOOR: classification = W * COOR2^T.

#	COOR2	n_1	n_2	n_3	n_4
1	$[0.71, 0.71]^T$	1.01	0.99	-0.01	0.23
2	$[0.89, 0.45]^T$	0.96	1.01	0.31	0.54
3	$[0.89, -0.45]^T$	0.32	0.52	0.96	1.02
4	$[0.71, -0.71]^T$	0.00	0.22	1.01	0.99

However, the algorithm does come close to learning the logical coordinators. What is missing is some means of combining neurons into larger groups and sharing neurons between output classes. We will supply the missing link in the next section, but for now let us delve somewhat deeper into the workings of competitive learning.

5.4.3.2. Quantization, Voronoi tesselation, and convexity

Kong and Kosko (1991) and Kosko (1991) prove that competitive learning adaptively quantizes the space of patterns that serves as its input. Quantization is the procedure introduced in Chapter 3 which reduces a set of continuous values to a single discrete value. A convenient visualization of quantization is offered by Voronoi tessellation, also introduced in Chapter 3. With the picture afforded by Voronoi tessellation in mind, quantization can be viewed as the reduction of the continuous values of the vectors within a cluster to the single discrete centroid vector. This reduction is adaptive in the sense that it takes place incrementally, during each epoch of a simulation. In the unsupervised competitive learning of the logical coordinators, each competitive neuron comes to occupy the place of a centroid of some number of observations. The number of observations within a cell depends on the number of neurons competing to tessellate COOR space – the more the neurons, the fewer the observations per cell.

We can go a step further in the characterization of the competitive neuron by recalling that each cell of a Voronoi tessellation answers to the definition of a convex region. Thus a competitive network implements Gärdenfors' postulation of convex regions as the ontology of natural predicates. In this way, competitive learning provides a bridge from the theoretical postulates of cognitive science to the linguistic data of logical coordination.

5.5. SUPERVISED COMPETITIVE LEARNING: LVQ

The adaptive quantization of COOR space performed by competitive learning already uses up all of the statistical information available there, so it is not clear how the various coordinators can be kept separate if their patterns are to be mixed together in a single network. Fortunately, there still remains at least one source of information that has not made any contribution to the learning of coordinator patterns, namely, the fact that each coordinator meaning is associated with a phonological form. This extra information can be pressed into service to supply a "teacher" that will help to separate patterns even when mixed together. With this augmentation, the unsupervised competitive learning of the previous section becomes supervised competitive learning, which is more commonly known as **learning vector quantization** or **LVQ**, introduced in Kohonen (1986) and developed in considerable further work by Kohonen and his students.

5.5.1. A supervised competitive network and how it works

In the simulations described below, the competitive layer has four neurons, while the supervising layer also has four neurons, one for each coordinator. Thus the unsupervised competitive learning network of Fig. 5.20 is augmented to that of Fig. 5.22. The details of the new elements W2 and L2 are taken up as part of the upcoming prose explanation of how supervision works.

In a standard LVQ layout, initialization of the supervised half of the network sets the W2 connections so that each competitive neuron is connected with a weighting of 1 to one L2 neuron, and with a weighting of 0 to all the other L2 neurons. Once these connections are set, they cannot be altered. In the particular case of our simulations, such a standard layout would look like (5.27a). However, Fig. 5.22 does not match this description, because both n_1 and n_4 are connected to *two* L2 neurons, a visual format that answers more to the mathematical implementation of (5.27b):

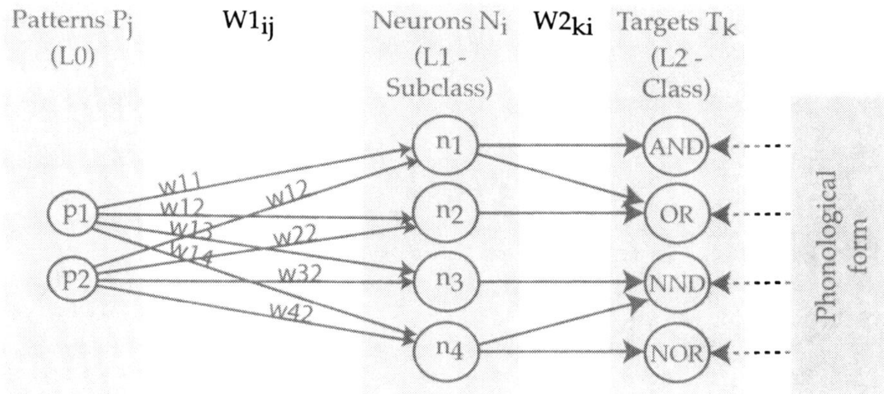

Figure 5.22. Learning vector quantization network; W2 = 0 are suppressed.

5.27. (a) $W2 = \begin{bmatrix} 1, 0, 0, 0 \\ 0, 1, 0, 0 \\ 0, 0, 1, 0 \\ 0, 0, 0, 1 \end{bmatrix}$ (b) $W2 = \begin{bmatrix} 1, 0, 0, 0 \\ 1, 1, 0, 0 \\ 0, 0, 1, 1 \\ 0, 0, 0, 1 \end{bmatrix}$

This alteration of the LVQ layout is motivated by the need to represent shared inputs to OR and NAND.

This representational alteration provokes an algorithmic alteration, namely, the rejection of the notion that the W2 weights do not change after initialization. Our alternative is to increase those W2 weights that connect a L1 neuron to the proper L2 neuron. In other words, initialization sets the W2 to zero and decides which L2 neuron will represent which coordinator. When an L1 neuron wins the competition for an input, the algorithm checks which coordinator it instantiates and increases the W2 between it and the winning L1 neuron. No attempt is made to constrain this weight in a realistic manner, so it keeps accumulating throughout the simulation. At the end of the simulation, each W2 is divided by the number of times it was incremented to produce a percentage – or probability – of winning.

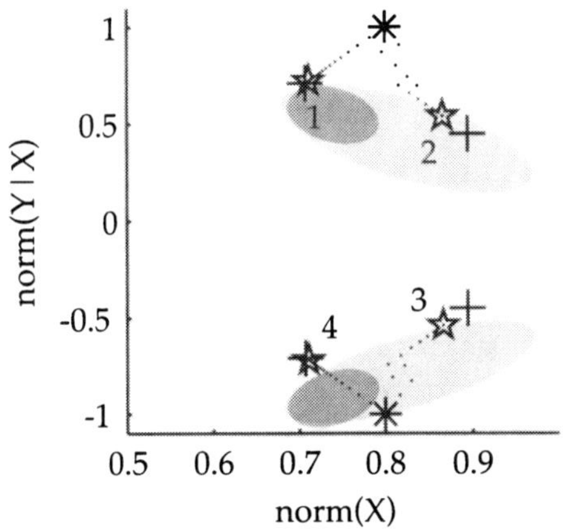

Figure 5.23. Supervised competitive learning of COOR2. '+' = accepted point, '*' = initial weight, dot = intermediate weight, star = final weight. Shading indicates class – see text for details. (**p5.07_LVQCOOR.m**)

Table 5.11. Final state of LVQCOOR: W1 * COOR2^T and W2.

#	COOR2	n_1	n_2	n_3	n_4	L1	AND	OR	NND	NOR
1	$[0.71, 0.71]^T$	1.01	0.99	0.23	-0.01	n_1	0.8	0.5	0	0
2	$[0.89, 0.45]^T$	0.96	1.02	0.54	0.31	n_2	0.2	0.5	0	0
3	$[0.89, -0.45]^T$	0.31	0.53	1.02	0.96	n_3	0	0	0.5	0.2
4	$[0.71, -0.71]^T$	-0.01	0.23	0.99	1.01	n_4	0	0	0.5	0.8

The end result is that, not only do the competitive neurons adaptively quantize their input space, as in unsupervised learning, but they also become subclasses of the classification imposed by L2. As Fig. 5.22 suggests, the L2 classes are supervised by connections from the phonological component, so that L2 acts as the interface between semantics and phonology. It thus learns an association between a semantic form, the output of the competitive network, and a phonological form, the output of the gray box on the right of Fig. 5.22, though we have nothing substantive to say about this latter process in this monograph.

5.5.2. An LVQ simulation of the logical coordinators

Since LVQ builds on a competitive network, the outward form of our LVQ simulation is identical to that of the competitive simulation of Fig. 5.21, as depicted in Fig. 5.23. We have attempted to manifest the contribution of the classificatory network by highlighting the four coordinator classes through shading, with the lighter shade belonging to the existential coordinator of either polarity, and the darker shade belonging to its universal counterpart.

The accuracy of the competitive subnetwork is identical to that of the previous simulation, conveyed in Table 5.11 in the columns labeled with an 'n' subscripted with the number of the L1 neuron. The accuracy of the classificatory network is more difficult to gauge, since it acts in a probabilistic fashion. For our illustrative simulation, the final W2 are given in Table 5.11 under the columns labeled with a coordinator. The rows represent input from the competitive (L1) neurons, in the order given. Thus competitive neuron 4 contributes to 50% of the input to OR and to 80% of the input to AND. This is a perhaps unreasonably high contribution to OR, which is due to the undoubtedly unreasonable symmetry in the data. However, for the purposes of illustration, such symmetry aids considerably in explaining how the simulation works. Thus the activation of neuron 4 activates OR and AND in the ratio of 5/8. AND becomes more active, though the classificatory network does not put the two into competition. This is an accurate method for overcoming the inability of a competitive network to combine or share its neurons in order to create more complex classifications.

5.5.3. Interim summary and comparison of LVQ to MLP

To summarize briefly, in these sections, seven artificial neural network architectures are investigated as potential models for the logical coordinators. Only the last, the learning vector quantization network, was found to be descriptively and explanatorily adequate. An LVQ network consists of a unsupervised competitive layer whose output goes to a linear layer that learns from a supervising target set. Each neuron of the competitive layer, L1, learns a prototype vector that permits it to subclassify a convex region of the input space P. These convex subclasses are then grouped into classes by the linear layer, L2, supervised by the training data. Since the linear layer groups regions from the competitive layer in any combination, it can produce decision regions that are not contiguous and thus are not convex.

Nevertheless, logical coordination does not need the full power of the linear network, as suggested by the reworking of the HYPERPLANE CONJECTURE into a form more compatible with our overall results:

5.28. COORDINATOR CONJECTURE: A monomorphemic coordinator classifies its input into two convex regions.

As was mentioned earlier, this conjecture dovetails with Gärdenfors' theory of the structure of natural categories. It suggests that LVQ is not constrained enough – more exactly, that the linear layer should be constrained to producing convex regions out of the convex regions formed by the competitive layer. This effect could be achieved by imposing a topological structure on the connections from the competitive to the linear layer in such a way that only nearby

Table 5.12. Comparison of perceptron and supervised competitive networks

	Property	*Perceptron network*	*Supervised competitive network*
a)	*input:*	COOR(x, y)	COOR(x, y)
b)	*output:*	0 or 1	$[1, 1, 1, 1]^T$ * probability
c)	*analyzes input:*	into hyperplanes	into hyperellipsoids
d)	*COOR accepted:*	in hyperplane	in hyperellipsoid
e)	*COOR rejected:*	not in hyperplane	in another hyperellipsoid
f)	*architecture:*	homogenous (perceptrons)	heterogeneous (competitive + linear)
g)	*function computed:*	same for all	nonlinear + linear
h)	*hidden layers:*	unlimited	1
i)	*hidden function:*	P * W (inner product)	distance(P, W) (Euclidean distance)
j)	*learning:*	homogenous (supervised)	hybrid (unsupervised + supervised)
k)	*learning:*	global (backpropagation)	local (competitive)
l)	*input-output:*	global approximation	local approximation
m)	*input-output:*	distributed coding	localist coding

competitive neurons project to the same linear neuron. Such a limitation is not explored here due to the large number of other topics that still remain to be considered.

Along the way to drawing this conclusion, we have had the opportunity to contrast two classes of ANN architectures, whose properties are summarized in Table 5.12, drawing on discussions in Haykin, 1994, pp. 262-3; Bishop, 1995, pp. 182-3; Intrator, 1995, p. 220; and Lowe, 1995, p. 780. Despite these contrasts, it is difficult to find empirical effects based on the meanings of individual coordinators that bear on a decision between the two architectures. More general neurological considerations, such as the implausibility of error backpropagation, are what tip the balance in favor of LVQ.

5.5.4. LVQ in a broader perspective

It is hoped that it does not require too strenuous an exercise in imagination on the reader's part to appreciate how the LVQ architecture begins to recapitulate the organization of the visual system outlined in Chapter 1, and in particular in the layout of Fig. 1.34. Like the visual system, LVQ possesses a feedforward direction of processing in which smaller features are combined into larger features under some constraint of similarity. This direction appears to be governed by the principal of redundancy reduction, though the patterns created by the logical coordinators are so simple that there is little redundancy to remove. LVQ also possesses a feedback direction of processing, in that the linear layer L2 is ultimately vertebrated by phonological forms. Thus a more complete analysis of L2 should incorporate phonological input to guide the formation of L2 classes from L1 subclasses in a Bayesian manner, though again we leave this additional elaboration for some other venue.

5.6. DENDRITIC PROCESSING

The LVQ analysis of logical coordination and quantification presupposes that the underlying situations come conveniently pre-digested into a representation of coordinative/quantificational invariants that can be feed into an LVQ network in a straightforward manner. As was mentioned in Chapter 1, this mastication of the raw data into a tractable form is known as preprocessing. An objection was raised there to any account that relies as heavily on preprocessing as the LVQ analysis does. Körding and König, 2001, p. 2825 point out that preprocessing requires that the constructor of a network have specific *a priori* knowledge about which variables the processing is supposed to be invariant of – and thus omits any explanation of why these invariants were chosen over some others or how these invariants could be extracted from the raw data in a more principled fashion. An answer to these questions may be as enlightening to the problem at hand as the discovery of the appropriate patterns in the preprocessed data.

In this section, we make the first stabs at the discovery of coordinative and quantificational patterns in an unpreprocessed version of the data. The results will be shown to have a certain similarity to the extraction of invariants preformed by early vision, so that this section stands as the first attempt to characterize the statistical structure of natural-language semantics, or at least the situations from which natural-language semantic processing extracts the semantic patterns of a given language.

A second contribution of this section is to demonstrate an algorithm for the extraction of semantic invariants that relies on dendritic processing. The reason for this shift from traditional algorithms which rely on the change of the synaptic weights of a spatially undifferentiated neuron has the advantage of increasing the neurologically plausibility of the model. Yet what is equally important, the type of dendritic processing that is developed here traffics exclusively in correlation, and in particular, in a measure of correlation as the distance between two synapses on a dendrite. As has been mentioned time and time again in the preceding pages, the most obvious neurophysiological grounding of natural-language coordination and quantification is in the extraction of correlations in the environment by the central nervous system. In the following pages, we explore the hypothesis that the central nervous system transduces correlations in the (semantic) environment into dendritic synapses that are close to one another.

5.6.1. From synaptic to dendritic processing

The previous sections illustrate a richly detailed theory of learning as modification of the strengths of connections between neurons. This theory is grounded physiologically on long-term synaptic plasticity, as set forth in Chapter 2, and most notably on long-term potentiation and depression. In practical terms, it has been applied successfully to difficult learning-related tasks, including problems in pattern recognition, associative memory, clustering, and map formation – applications which serve as the inspiration for the architectures that have been introduced in this chapter. In the words of Poirazi and Mel, 2001, p. 779, "Taken together, these physiological, theoretical, and practical considerations form a mutually reinforcing collection of ideas, founded on the core principle that in networks of neuron-like units, learned information is encoded in the patterning of synaptic weight values."

Nevertheless, as we learned at the end of Chapter 2, there is a growing body of evidence that suggests that synaptic efficacy may not be the only, or even the principle, means of knowledge retention. For more than a decade, Bartlett Mel and his coworkers have investigated an alternative in which knowledge is encoded in clusters of synapses or dendritic spines, rather than in the strength of their connections.

This approach to structural plasticity is implemented by correlation-based sorting of synaptic contacts on their postsynaptic targets, a mechanism which Poirazi and Mel (2001) trace to Shatz (1990). Poirazi and Mel, 2001, p. 780, divide this mechanism into the three steps in (5.29):

5.29 a) Synapses are initially formed between axons and dendrites in a random activity-independent fashion;

b) newly formed synapses begin their life cycle in a probationary, or "silent," phase (i.e., containing only NMDA channels) that leaves them unable to unilaterally activate their postsynaptic targets

c) silent synapses that are frequently coactivated with mature (non-silent) synapses within the same post-synaptic compartment are structurally stabilized and thus retained, perhaps via the insertion of AMPA receptors, while those that are poorly correlated with their neighbors may be eliminated.

Poirazi and Mel admit that such a scheme could be used in a way that is not inimical to the standard notion of synaptic efficacy, by dynamically regulating the overall connection strength between any two neurons through a balance of learning-induced synapse formation and elimination. However, given the nonlinear dendritic physiology that was discussed in Sec. 2.6, Poirazi and Mel go on to argue that changes in the addressing of synaptic contacts onto existing dendritic subunits, or formation of entirely new subunits, could constitute forms

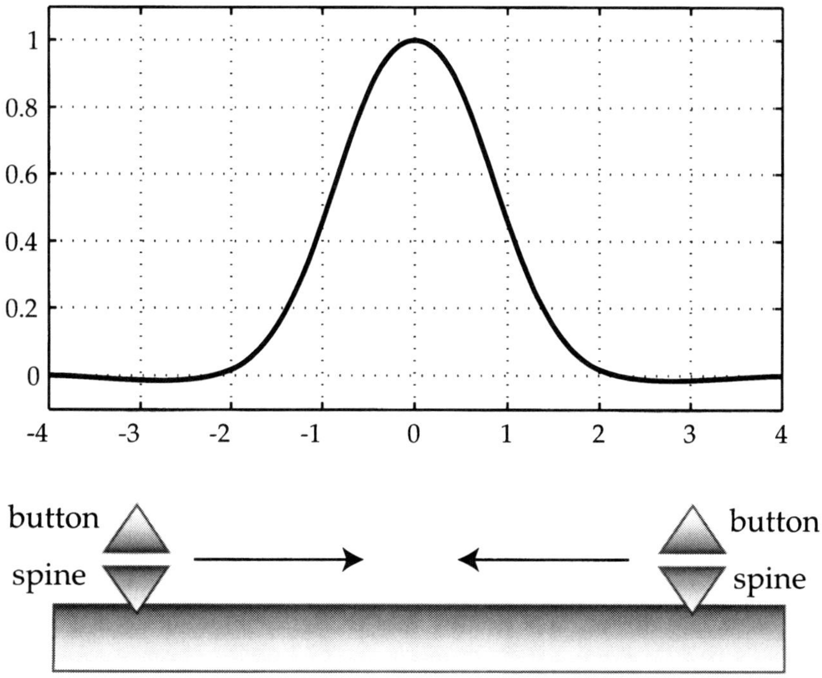

Figure 5. 24. Dendritic representation of correlation between two features as correlation in space of two spines. The closer the two spines come, the higher their output is on the Gaussian function graphed above the dendrite.

of plasticity that cannot be expressed in terms of simple weight changes from one neuron to the next.

5.6.2. Clustering of spines on a dendrite

The organizing principle which we wish to pursue in our novel network architecture arises out of notion of correlation-based sorting reviewed in Poirazi and Mel (2001). In particular, we develop a mechanism by which correlated inputs cluster together on a dendrite. We mean this quite literally: a correlation between two features is 'learned' by moving their dendritic connections closer together as a function of the degree of correlation between them, where the dendritic connections are conceptualized as dendritic spines. The output of the branch that hosts the two features increases as the spatial correlation between them is established, but it does not increase linearly. In accord with Mel's results, the output increases supralinearly, but not quite at the exponential rate that Mel advocates. Instead, it increases in a sigmoidal fashion that can be represented by the upstroke of a Gaussian function.

For two features, the simplest visualization that illustrates their possibilities for spatial interaction is to imagine them lying at some distance from one another on a dendritic branch like that of the bottom of Fig. 5.24, in such a way that their separation aligns with a value on the opposite ends of a two-dimensional Gaussian function. As the caption to Fig. 5.24 puts it, the closer together the two spines come, the higher their output will be to the dendritic branch via the Gaussian function. The result of this process is that two or more features that are correlated in some abstract feature space become spatially correlated (physically closer together) on a dendrite. The overall process is functionally similar to the AND gate simulated through compartmental modeling of two spines in Shepherd and Brayton (1987), which was briefly reviewed at the end of Chapter 2.

In our implementation of this model, the distance measure for a spine is the distance of the spine from the point midway between it and its neighbor. By convention, one of the two spines is assigned a negative distance, so that each of them will align with a leg of a two-dimensional Gaussian function, as depicted in Fig. 5.24. The input to a spine is either 0 or 1: 0 if the synapse is inactive, and 1 if it is active. Synapses do not take on any intermediate value, in accord with Poirazi and Mel's (2001) review of recent evidence against the classical synapse summarized at the end of Chapter 2. It should be pointed out, however, that it is our hunch that synapses in this model could take on a few intermediate values without perturbing the global results. The activation of a spine *s* is thus the product of its distance and its input passed through the Gaussian function. The output of a branch *b* is the sum of each spine, for which our simple simulation just uses two. Eq. 5.30a compacts this prolix verbal explanation down to a single mathematical line, with the help of the summation notation:

5.30 a) $$\text{output}_b = \sum_{s=1}^{2} \text{gauss}(\text{distance}_s * \text{input}_s)$$

b) $$\text{activation}_n = \sum_{b=1}^{2} \text{output}_b$$

The activation of a neuron *n* is the sum of the outputs of its branches, as stated in Eq. 5.30b. Again, the initial simulation only uses two.

Intuitively, some operation in addition to Eq. 5.30 appears necessary, since Eq. 5.30 by itself is not sufficient to distinguish, say, ALL from SOME. Given that ALL is a value for SOME, it is conceivable that a neuron built on Eq. 5.30 will wind up responding equally well to both quantifiers.

An obvious solution is to impose a threshold on the neuron's output. With the background of the formal methods introduced in Chapter 3, one attractive threshold would be the mean activation of the neuron. Letting W stand for the number of times that neuron n has won the competition for an input, the output of neuron n takes the form of Eq. 5.31, where k is a constant gain that modulates the contribution of the mean activation.

$$5.31. \qquad \text{output}_n = \text{activation}_n \pm k \frac{1}{W} \sum_{w=1}^{W} \text{activation}_w$$

Trial and error has shown that k should be around 0.9.

The learning algorithm begins by initializing each branch to a low activation (a value on the abscissa of the Gaussian function that produces an output of about 0.1 on the ordinate) and zero threshold. A data point is submitted to the set of neurons, and all outputs are calculated. The neuron with the highest activation is chosen as the winner. Any tie is broken by the random selection of a single neuron. The winner's parameters are updated in accord with its activation. In particular, the distance between the x and y spines of the coordination is decreased by a fraction k of the output of the branch, Eq. 5.32a. As mentioned above, the distance is measured from the center of the Gaussian out to the relevant edge, so that each spine is the same distance from the center. By convention, the left edge has been assigned negative, and the right, positive, so that (positive) output is added to the negative left side to decrease it, and subtracted from the positive right side to decrease it:

$$5.32 \quad a) \qquad \text{distance}_s = \text{distance}_s \pm (k * \text{output}_b)$$

$$b) \qquad \text{threshold}_n = \sum_{b=1}^{2} \left(\frac{1}{W} \sum_{w=1}^{W} \text{activation}_w \right)$$

The threshold is updated directly as the sum of the mean activation of each branch, Eq. 5.32b.

This is the algorithm for the active synapses; for every active y input there is also an anticorrelated y input. By the definition of anticorrelation, when one spine in such a relationship is active, the other must be inactive. Given that the spines representing correlated features tend to grow together, it is to be expected that the spines representing anticorrelated features should tend to grow apart. And this is exactly what is implemented in the model: when a neuron wins a competition, the distance of the inactive y spine from the center of the Gaussian is increased. No effort was made to effect this growth in a plausible manner (especially since the exact mechanism for spine growth is not known), so the distance of the anticorrelated spine on a winning neuron grows without bound. It would perhaps have been more realistic for the spine to 'die' by being reabsorbed into its dendrite, but having it move in the opposite direction

Table 5.13. Sample results of spatial correlation. (**p5.08_gauss_cluster.m**)

COOR	*x,x,y,y,-y,-y*	*N*	*Distance*	*Gauss*	*Threshold*
AND	1 1 1 1 0 0	2	-0.23 0.23 36	0.90 0.90 0	2.84
			-0.23 0.23 36	0.90 0.90 0	
"	1 1 1 1 0 0	2	"	"	"
"	1 1 1 1 0 0	2	"	"	"
NOR	1 1 1 0 0 0	1	-0.23 47 0.23	0.90 0 0.90	2.88
			-0.23 47 0.23	0.90 0 0.90	
"	1 1 0 1 0 0	1	"	"	"
"	1 1 1 0 0 0	1	"	"	"
OR	1 1 1 1 0 0	5	-0.23 0.08 8.46	0.90 0.90 0	2.03
			-0.23 0.08 8.46	0.99 0.99 0	
"	1 1 0 1 0 0	5	"	"	"
"	1 1 1 0 0 0	5	"	"	"
"	1 1 0 1 0 0	5	"	"	"
"	1 1 0 0 -1 0	5	"	"	"
"	1 1 0 0 0 -1	5	"	"	"
"	1 1 0 0 -1 0	5	"	"	"
NAND	1 1 0 0 0 -1	6	-0.25 8.17 0.25	0.89 0 0.89	1.96
			-0.26 8.17 0.26	0.87 0 0.87	
"	1 1 0 0 -1 -1	6	"	"	"
"	1 1 0 0 -1 0	6	"	"	"
"	1 1 0 0 0 -1	6	"	"	"
"	1 1 0 0 -1 -1	6	"	"	"
"	1 1 0 0 -1 -1	6	"	"	"
"	1 1 0 0 -1 -1	6	"	"	"
"	1 1 0 0 -1 -1	6	"	"	"

illustrates the purposed algorithm of spatial correlation rather perspicuously. Since the anticorrelated spine is inactive, it does not make any contribution to the calculation of the neuron's activity or threshold.

A typical result of this algorithm is reproduced in Table 5.13, for which seven neurons competed to represent each input through four presentations of the data set. The first column labels each input with the coordinator that it instantiates. The second column lists the inputs, paired from left to right by sort: first a pair of *x*'s, then two pairs of *y*'s. To distinguish the latter two pairs, the second is labeled negative to show that it is anticorrelated with the first. The third column relates the number of the neuron that won the competition for that particular data point. The fourth column lists the raw distance of each input. From left to right are listed the three different sorts of spine, and from top to bottom, the two branches. Note that the anticorrelated *y* spine – the one that is zero in the input – can increase to attain a very large value. The fifth column

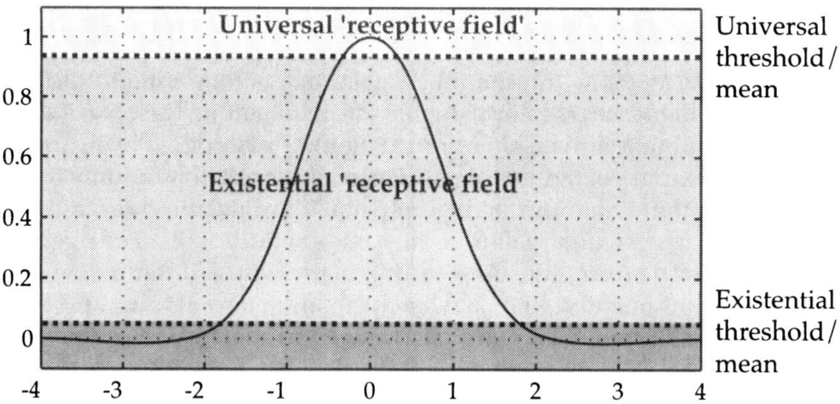

Figure 5.25. Thresholding divides the Gaussian spatial correlation function into separate receptive fields.

passes these distances through the Gaussian activation function. The sixth and final column reproduces the neuron's threshold. The fact that each neuron corresponds to one and only one coordinator indicates that the network learned the data set perfectly.

We hasten to add that, while Table 5.13 relates the most accurate final state of the network, it is unfortunately not the only one. A second, less common final state is for two neurons to divide the whole data set among themselves by polarity. That is, one neuron captures all of the positive coordinations, and another captures all of the negative ones. It also happens with much less frequency that there is overlap with a polarity, in which the neuron that responds to all of the universal coordinations also responds to some of the existentials (especially the ones that are ambiguously universal). Finally, the terminal state is not necessarily stable, in the sense that increasing the number of epochs permits a certain amount of turnover among the winning neurons, in which some stop responding to previously won input and others start responding to it.

Much of this instability can be lessened by making the Gaussian function steeper. The reason for this sensitivity to the slope of the function is intuitively obvious: the steeper the slope, the more the function distinguishes between nearby inputs. Nevertheless, it does not escape our attention that some more elaborate connection between the neuron's threshold and each branch's activation needs to be sought, in order to make the algorithm more stable. Such an endeavor must await further research, however.

Despite its primitive nature, the activation/threshold relation that we have proposed is remarkably accurate within the parameter space that we have investigated. A few words should be said about how this occurs before we leave the topic.

Since its threshold rises with a neuron's mean activation, a highly active neuron will attain a higher threshold than a less active neuron. This provides a parameter to discriminate universal from existential operators. Given that a universal operator expresses the maximal degree of correlation, a 'universal' neuron approaches the maximum activation that the data provide and the highest threshold. An existential value of the same polarity will also trigger a response from such a neuron, but the resulting activation will fall below the universal threshold / mean and so not activate the neuron. For instance, in Table 5.13, the universals AND and NOR reach an average threshold of 2.86, while the existentials OR and NAND only average 1.98. Fig. 5.25 attempts to communicate this collaboration between the Gaussian function and the threshold by superimposing idealized thresholds onto the function to show how they divide its domain into two receptive fields. The real thresholds would have to be divided by two and passed through the function to locate them in this space.

This Gaussian algorithm takes us deep into uncharted territory of neurological function, perhaps deeper than we need to go to establish the neurological foundations of the simple constructions that are dealt with in this monograph. Let us therefore turn to the second area of linguistic data that we wish to explicate, that of the logical quantifiers.

5.7. SUMMARY

This chapter introduces the reader to the principal algorithms for neurologically plausible pattern classification using the patterns formed by logical coordination of a handful of coordinatees. We finally settled on LVQ quantization as the best standard algorithm, both for its descriptive adequacy and its explanatory adequacy in bringing together various notions of vector quantization and convexity. We rounded out the discussion with the illustration of an algorithm of our devising, based on notions drawn from the theory of dendritic processing, that does nothing but find correlations.

We have argued to the extent possible that associative pattern classification is superior to nonassociative pattern classification. As its name suggests, associative pattern classification is a way of finding an association between a pattern and a label. With respect to this monograph, at the most general level our interests lie in the association between a semantic pattern, in particular a measure of correlation, and a phonological form. The next chapter extends what we have posited about the logical coordinators to the logical quantifiers.

Chapter 6

The representation of quantifier meanings

This chapter extends the pattern classification analysis of the logical coordinators to the four logical quantifiers, ALL, SOME, NALL, and NO. As has been suggested throughout the previous pages, a suitable generalization of coordinator space provides a natural space for the statement of quantifier meanings. In this way, we can account for ALL as a 'big AND' and SOME as a 'big OR', in the felicitous phrasing of McCawley, 1981, p. 191.

6.1. THE TRANSITION FROM COORDINATION TO QUANTIFICATION

A similarity between the logical quantifiers and the logical coordinators has long been noticed. The next few subsections sketch several ways in which coordinators and quantifiers pattern together.

6.1.1. Logical similarities

Horn, 1976, pp. 75ff, points out that logicians have observed that for any universal quantification of the form (6.1a), a semantically equivalent conjunction of the form in (6.1b) can be constructed:

6.1 a) $\forall x[P(x)]$, where $x \in \{x_1, x_2, ..., x_n\}$
b) $P(x_1)$ & $P(x_2)$ & ... & $P(x_n)$

Likewise, for any existential quantification of the form (6.2a), a semantically equivalent disjunction of the form in (6.2b) can be constructed:

6.2 a) $\exists x[P(x)]$, where $x \in \{x_1, x_2, ..., x_n\}$
b) $P(x_1) \vee P(x_2) \vee ... \vee P(x_n)$

Moreover, McCawley, 1981, p. 191, provides a list of theorems for coordination and quantification in which the format of a universal quantifier parallels that of conjunction and the format of an existential quantifier parallels that of disjunction, along with some further discussion.

6.1.2. Conjunctive vs. disjunctive contexts

Ross (1973) distinguishes contexts which license embedded questions into two sorts, conjunctive versus disjunctive, according to four criteria. We can use this notion to organize several disparate observations about contexts that favor AND/ALL over OR/SOME, or vice versa,

Conjunctive contexts reject OR/SOME as appositive phrases:

6.3 a) It's {clear/wild/surprising/odd/fascinating} what contains DDT – (namely) coffee {and/*or} tea.
b) It's {clear/wild/surprising/odd/fascinating} what contains DDT – (namely) {every /*some} imported beverage .

A similar point is observed in Rooth & Partee (1982:355), which we can expand into an entire paradigm:

6.4 a) Mary is looking for a maid and a cook, namely Jane and Harry.
b) Mary is looking for a maid and a cook; *she's not sure which.
c) Mary is looking for a maid and a cook, *but I don't know which.

6.5 a) Mary is looking for everyone, namely Tom, Dick and Harry.
b) Mary is looking for everyone; *she's not sure who.
c) Mary is looking for everyone, *but I don't know who.

Horn, 1976, pp. 75ff, based on McCawley (1972), points out that AND and ALL but not OR and SOME are found as the subject of, and often as the object of, performatives:

6.6 a) Ralph {and/*or} I hereby promise(s) to give you $6.
b) {All/*Some/*One} of us hereby promise(s) to give you $6.

6.7 a) The Pope hereby excommunicates Daniel {and/*or} Philip.
b) The Pope hereby excommunicates {all/*some} radical American Jesuits.

In addition, AND and ALL but not OR and SOME are found as the object of the pseudo-imperative quasi-verb discussed in Quang (1971):

6.8 a) {Goddamn/Fuck/Screw/Down with} Nixon, Brezhnev, {and/*or} Mao!
b) {Goddamn/Fuck/Screw/Down with} {all/*some} of those imperialist butchers!

Thus both of these contexts qualify as conjunctive contexts.

Disjunctive contexts reject AND/ALL as appositive phrases. (6.9) gives Ross' examples and (6.10, 6.11) give those that can be constructed from Rooth and Partee's observation:

6.9 a) It was {a mystery/unclear/not remembered/not known} to whom she sent it – to Jim {*and/or} to Bill.
b) It was {a mystery/unclear/not remembered/not known} to whom she sent it – to {*everyone/someone} on the list.

6.10 a) Mary is looking for a maid or a cook, *namely Jane and Harry.
b) Mary is looking for a maid or a cook; she's not sure which.
c) Mary is looking for a maid or a cook, but I don't know which.

6.11 a) Mary is looking for someone, namely Tom, Dick and Harry.
b) Mary is looking for someone; she's not sure who.
c) Mary is looking for someone, but I don't know who.

The fact that the two contexts treat AND on a par with ALL and OR on a par with SOME demonstrates that linguistic constructions find some similarity in their make-up. One of the goals of this chapter is to ascertain what this similarity is.

6.1.3. When coordination ≠ quantification

Having argued for the similarity of the logical coordinators and quantifiers, it would be remiss of us not to point out the differences that we know about. The most general one is that coordinators specify sets by enumeration, whereas quantifiers specify them by description. As McCawley, 1981, p. 192, notes, the difference lies in the fact that quantified sentences do not always have available an enumeration of the objects that are 'relevant'. Indeed, the context of the sentence may imply that it is not known exactly how many objects are relevant. McCawley offers (6.12) as an example of such a sentence:

6.12. Everyone who has ever set foot in Saint Peter's Basilica has been astonished at its magnificence.

Furthermore, even if one did have a complete list of those who have ever set foot in Saint Peter's, the content of (6.12) would not be represented accurately by the conjoined proposition of (6.13):

6.13. Pope Pius IX was astonished at the magnificence of Saint Peter's, and Jacqueline Onassis was astonished at the magnificence of Saint Peter's, and Msgr. Umberto Quattrostagioni was astonished at the magnificence of Saint Peter's, and ...

This is because (6.13) does not include the information that Pope Pius IX, Jacqueline Onassis, Msgr. Umberto Quattrostagioni, et al. are all the persons who have ever set foot in Saint Peter's. A person who has set foot in Saint Peter's and was not astonished at its magnificence would be counterexample to (6.12), but not to (6.13) unless he happened to be on the list. What it takes to falsify (6.13) is to show that one of the persons on the list was not astonished at the magnificence of Saint Peter's, irrespective of whether that person ever set foot in Saint Peter's; what it takes to falsify (6.12) is to show that someone who has set foot in Saint Peter's was not astonished at its magnificence, irrespective of whether that person is mentioned in (6.13).

Nevertheless, from the perspective of phrasal coordination developed in Chapter 4, it is apparent that part of the asymmetry adduced by McCawley is attributable to the fact that the clausal coordination of (6.13) is too 'coarse' a level to recapitulate all the information asserted in the phrasal quantification of (6.12). Two complementary repair strategies suggest themselves. One is to restate (6.12) with an appositive phrasal coordination to supply the relevant list:

6.14 Everyone who has ever set foot in Saint Peter's Basilica, namely Pope Pius IX, Jacqueline Onassis, Msgr. Umberto Quattrostagioni, and ..., has been astonished at its magnificence.

The other is to restate (6.13) with an additional clause that supplies the information missing from the quantified phrase:

6.15 Pope Pius IX was astonished at the magnificence of Saint Peter's, and Jacqueline Onassis was astonished at the magnificence of Saint Peter's, and Msgr. Umberto Quattrostagioni was astonished at the magnificence of Saint Peter's, and .... These are all the people that have ever set foot in Saint Peter's Basilica.

Either strategy vitiates McCawley's objection by demonstrating that the clausal coordinative periphrase was distilled from the phrasal quantificational source without attending to all of the criteria asserted by the quantification. Falsification of the two repaired examples is now comparable: it is achieved by finding someone on the list who was not astonished at the magnificence of Saint Peter's. This is the commonality of coordination and quantification that is investigated in this chapter.

Dougherty, 1970, p. 856, points out that the definite subject in (6.16a) is more like the universally quantified subject in (6.16b) than the list of conjoined names in (6.16c), because both (6.16a, b) indicate that there are no boys in my class without beards:

6.16 a) The boys in my class have beards.
b) All the boys in my class have beards.
c) The boy in my class has a beard; the boy in my class has a beard; the boy in my class has a beard; etc.

Where does the implication of exhaustiveness come from that unites (6.16a, b) against (6.16c)?

Our guess is that it comes from the pragmatic indeterminacy of magnitude of coordination. Even though both AND and ALL apply to 100% of their first argument, one more coordinatee can always be added to AND – say you accidentally left one out – whereas ALL is insensitive to magnitude and so 'always' applies 100%, which is to say that it implies an exhaustive listing. In this way, a potential counterexample to our claim of an equivalent representation of coordination and quantification can be shown on deeper analysis to actually support it.

6.1.4. Infinite quantification

The evaluation of a universally quantified noun as a list of conjuncts has a long history in logic. Martin, 1987, p. 112-9, 170ff, attributes its first systematic formulation to William of Ockham, a fourteenth century scholastic philosopher. But this **substitutional** hypothesis came under fire in the first decades of the twentieth century, as logicians concerned with mathematical reasoning began to recognize that some natural language predicates denote sets with an infinite number of members. The most obvious examples are based on classes of numbers:

6.17 a) All real numbers are useful.
b) All irrational numbers are fractions.
c) All numbers between zero and one are my favorites.

ALL, as a big AND, requires that every number named by the subject DP be correlated with the predicate, which is encoded in our pattern-recognition system by a calculation that ultimately has to know the length of the DP vector. Yet how can this evaluation proceed if the length cannot be measured, because it is infinite? A convincing answer to this question dovetails with conditions on finiteness assumed in the Generalized Quantifier framework, introduced in the upcoming section.

6.2. GENERALIZED QUANTIFIER THEORY

The upcoming analysis is based on the standard theory of quantifier meanings in current formal or model-theoretic semantics, known as generalized quantifier theory. This section introduces the theory, first by explaining the problems that it attempts to resolve and then by reviewing the initial set-theoretic definitions of the logical quantifiers.

6.2.1. Introduction to quantifier meanings

How many quantifiers are there in a natural language? Most natural languages are like English in having only about fifteen simple quantifier words, plus a scheme for the recursive construction of numerals from a small set of primitives:

6.18 a) the logical quantifiers: all / each / every, some / any, no
b) the duals: both, neither
c) the vague quantifiers: few; a few, several, many, most
d) the numerals: one, two, three, ..., twenty, twenty-one, ...
e) the comparatives: less ... than, as many ... as, more ... than

There are also a handful of adverbials that modify these forms:

6.19 a) at least, exactly, at most [five]
b) less / fewer than [five], as many as [five], more than [five]
c) finitely, infinitely, uncountably [many]
d) only [five]
e) exceptive but: [none / all] but Chris

This is too many for traditional logical analyses.

Traditional predicate logic only countenances two quantifiers, the universal ∀ and the existential ∃. They can be pressed into use for describing the content of natural language quantifiers like *all* and *some* only with certain distortions of surface structure:

6.20 a) All linguists are gourmets. $\forall x\, [Lx \rightarrow Gx]$
b) Some linguists are gourmets. $\exists x\, [Lx\ \&\ Gx]$

For instance, there is no surface correlate of the 'if...then' biclausal structure of the logical translation of (6.20a). Accompanying these syntactic distortions are certain semantic distortions. (6.20a) claims that all things are, if linguists, then gourmets. This does not seem to be exactly the same as saying that all linguists are gourmets. Let us quote Horn, 1989, p. 466, on what is at issue here:

> In effect, then, a proposition like *All ravens are black*, as analyzed into $\forall x$ *(raven (x) → black (x))*, is not about ravens at all. It is in fact about everything, it states that every individual *x* that if *x* is a raven, then *x* is black, that is, either everything is either black or a non-raven. More generally, as Sommers, 1970, p. 38 observes, "A statement of the form 'All S is P' in quantificational transcription is not about all S or any S; it is about all things and affirms of them 'is either un-S or P'". A general statement is thus entirely distinct from a singular statement ... in its logical form.

Likewise, (6.20b) is not about linguists or gourmets, but rather is about anything at all, and claims that there exists things which are linguists and gourmets.

These observations may seem too trivial to justify abandoning the usage of ∀ and ∃ altogether, but what does justify abandoning them is an attempt to extend the sentential operator form to quantifiers such as *most*:

6.21 a) Most linguists are gourmets.
b) $\mu x\,[Lx\ ?\ Gx]$

None of the sixteen possible truth-functional connectives listed in Table 4.2 can be inserted for the question mark in (6.21b), so traditional logic makes the implicit prediction that a quantifier like *most* should not exist in natural language. This prediction is obviously false.

6.2.2. A set-theoretic perspective on generalized quantifiers

The solution to these problems proposed by Barwise and Cooper (1981) is to represent all quantifiers as binary relations between two sets, the set which denotes the nominal which they modify and the set which denotes the predicate which the nominal is an argument of. If these two sets are abbreviated by the terms *N* and *P* respectively, then the interpretations of some of the quantifiers in (6.18) can be given as in (6.22), where unsigned cardinality is represented by the standard convention of $|\bullet|$:

6.22	a)	each/every/all NP	$N \cap P = N$ (or $N - P = \emptyset$ or $N \subseteq P$)
	a')	some NP	$N \cap P \neq \emptyset$
	a")	no NP	$N \cap P = \emptyset$
	b)	both NP	$N \cap P = N\ \&\ |N| = 2$
	b')	neither NP	$N \cap P = \emptyset\ \&\ |N| = 2$
	c)	most NP	$|N \cap P| > |N - P|$
	d)	three NP	$|N \cap P| \geq 3$
	d')	exactly three NP	$|N \cap P| = 3$
	e)	more N than N' P	$|N \cap P| > |N' \cap P|$

This Generalized Quantifier framework is so general that it easily expands to embrace definite articles and demonstratives, which can be grouped together with the quantifiers in (6.18) under the syntactic category of Determiners. However, to maintain consistency with the discussion of coordination we concentrate on the logical quantifiers in (6.18).

6.2.3. QUANT, EXT, CONS, and the Tree of Numbers

If a quantifier were to relate just any two sets in a universe, then it could refer to all of the sets or their elements indicated in the Venn diagram of Fig. 6.1. As Westerståhl (1989), points out, this leads us to expect to find $2^{\aleph_0}$ quantifiers in English. Even counting these adverbials as 'quantifiers', (6.18) and (6.19) are nowhere near the $2^{\aleph_0}$ predicted forms, so some means must be

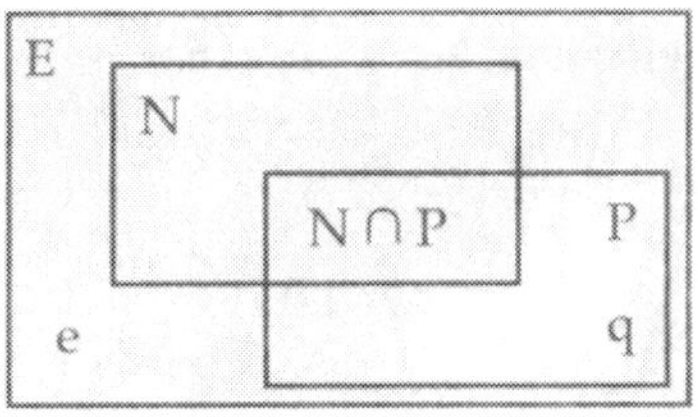

Figure 6.1. Any two sets in the universe.

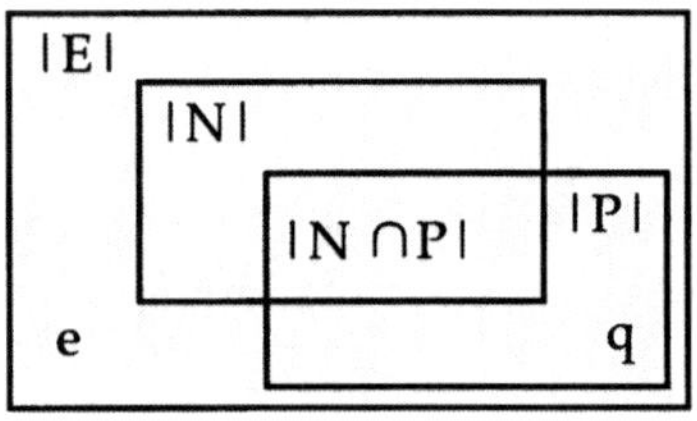

Figure 6.2. The effect of QUANTity.

found to pare the possible forms down to the handful of attested ones. Generalized quantifier theory proposes several conditions or constraints which perform the necessary reduction. The next few sections sketch the fundamental constraints of Quantity, Extension, and Conservativity.[30]

6.2.3.1. Quantity

The initial constraint is to exclude all non-numeric information from the sets in Fig. 6.1. This is achieved by Weak QUANTity or Permutation. In the diagram of Fig. 6.2, QUANT removes any dependence on anything but the number of entities in the sets mentioned. It is defined as so:

6.23. For a quantifier *Q*, all elements in universe *E*, and every permutation π of the individuals of *E*: Q_E NP iff Q_E $\pi[N]$ $\pi[P]$.

[30] There are several reviews of constraints on quantifier denotations which were consulted for these subsections and should be consulted for additional discussion: Westerståhl, 1989, pp. 63ff; Partee, ter Meulan and Wall, 1990, pp. 375ff; and van der Does and van Eijck, 1996, pp. 6ff.

Permutation is to be understood here as a reference to qualities of objects, as opposed to their quantity. Quantifiers which satisfy QUANT are "topic-neutral", or insensitive to individual traits of objects. Almost all quantifiers satisfy QUANT; the only exceptions are context-sensitive readings of the vague quantifiers.

Vague quantifiers such as *few* and *many* are understood with respect to a contextual parameter that chooses a standard of comparison for the quantification, as was mentioned briefly in Sec. 3.5.4.1. This parameter can be brought out by contrasting usages of *many* in the same context that nevertheless rely on different standards, such as (6.24):

6.24 a) Many students in the evening class play soccer.
b) Many students in the evening class got an A.

With twenty students in the class, the many "soccer-playing students" of (6.24a) could be more than half, whereas the many "A students" of (6.24b) may only reach four. In the former, this is "many" for students in general, that is, in the universe, while in the latter, this is "many" for the number of students in the class. Numerically, the two readings correspond to the first and second ways of choosing the comparison classes *c* from (6.25):

6.25 a) $\text{many}_E\ NP \leftrightarrow |N \cap P| \geq c|E|$
b) $\text{many}_E\ NP \leftrightarrow |N \cap P| \geq c|N|$

The challenge for QUANT comes in those cases in which the number of entities is the same, but the comparison classes are different. Returning to the example of (6.24), let us lower the number of soccer players to the number of A getters, i.e. four. Then $|N \cap P_a| = |N \cap P_b|$, but (6.24a) is no longer true. QUANT is violated because information additional to the cardinality of the sets determines the quantification: the nature of the contextual standard must also be taken into consideration.[31]

6.2.3.2. Extension

The second principal QG constraint excludes the universe *E* from consideration. This is achieved through EXTension, a constraint on 'context-neutrality'. It can be stated with the help of Fig. 6.3 which expands the universe *E* to *E'*. Intuitively, the idea is that the $|N|$ and $|P|$ boxes, and their overlap, do not vary as the universe expands. As a consequence, the potential relevance of entities in *E* which fall outside of the interpretation of the nominal and the predicate, such as *e*, can be left out of consideration. The result

---

[31] This is the essence of the example in Partee, ter Meulan, and Wall, 1990, p. 395.

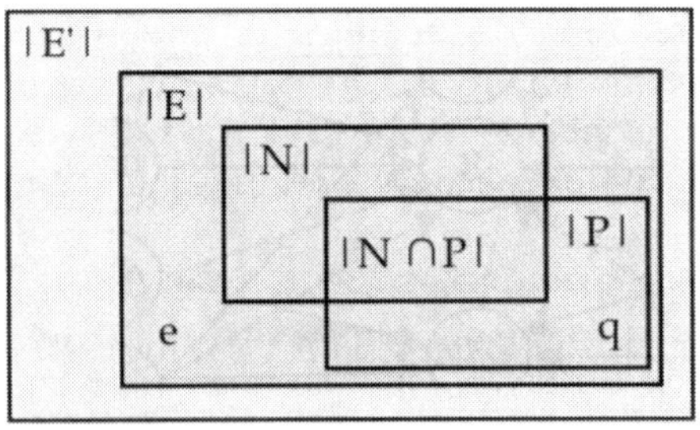

Figure 6.3. Expansion of the universe to E'.

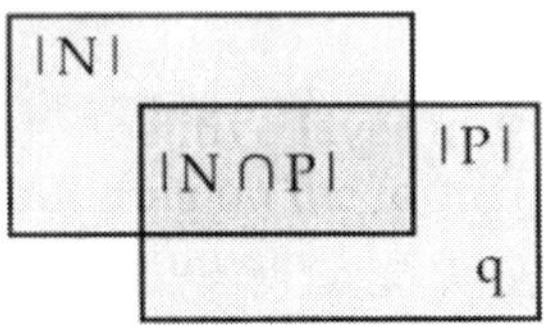

Figure 6.4. The effect of EXTension.

of applying EXT to Fig. 6.3 is to remove E (and E'), paring it down to Fig. 6.4. Formally, EXT is defined to mimic the form of permutation in (6.23), i.e. the interpretation of a quantifier stays the same as something else, here the size of the universe, changes:

6.26. For a quantifier Q, all elements in universe E, and any set N, P $\subseteq$ E $\subseteq$ E': Q_E NP iff $Q_{E'}$ NP.

Once again, almost every quantifier satisfies EXT. The only exception is the (6.25a) reading of *many* that depends on the size of the universe.

On a final note, care must be taken to distinguish the expansion of the universe mentioned in (6.26) from its contraction. If EXT were to include contraction, then all of the cardinality quantifiers such as *five* would fail it, since they are not defined for universes smaller than the number they name.

6.2.3.3. Conservativity

The third GQ constraint that reduces the number of binary quantifiers excludes the part of the predicate denotation that does not overlap with the nominal. In other words, we may safely ignore entities such as q in Fig. 6.4 in order to evaluate a quantifier, which reduces Fig. 6.4 to Fig. 6.5. This constraint, known as **CONServativity**, restricts the predicate denotation to the (union of) the nominal denotation. It can be brought out by inferences such as those of (6.27):

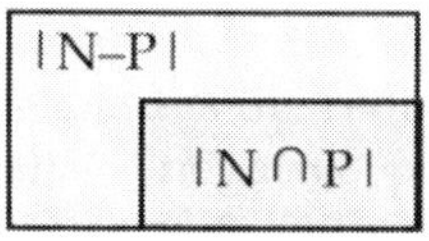

Figure 6.5. The effect of CONServativity.

6.27 a) Every athlete eats Wheaties ↔ Every athlete is an athlete and eats Wheaties
b) Most Dutch are morose ↔ Most Dutch are morose Dutch
c) Few women are bald ↔ Few women are women who are bald

In other words, any quantification which replaces the nominal individuals with those of an entirely different set is ruled out. An example of such an unintuitive quantification would be a universal quantifier which interpreted *all linguists*, say, as denoting *all anthropologists*, by replacing each linguist as it is related to the predicate with an anthropologist. (6.28) provides a sample of Conservativity-violating inferences for (6.27):

6.28 a) Every athlete eats Wheaties ↔ Every athlete is an intellectual and eats Wheaties
b) Most Dutch are morose ↔ Most Dutch are morose Australians
c) Few women are bald ↔ Few women are men who are bald

Though perhaps some of the inferences from right to left go through, none of the inferences from left to right are acceptable, since they add information that is not found on the left side.

CONS is defined as follows:

6.29. For a quantifier Q, all elements in universe E, and any set N, P of E: Q_E NP iff Q_E N(N ∩ P).

This accounts for the privileged role of the nominal in a quantifier statement: it "sets the stage" for the evaluation. The only quantifier that violates CONS is, once again, *many* under certain readings, such as that of (6.25a) in which it refers to the universe *E*.

6.2.3.4. The Tree of Numbers

Van Benthem, 1986, p. 26, points out the combined effect of EXT, QUANT, and CONS is to make a quantifier equivalent to the set of couples of cardinalities (x, y) which it accepts. If all the possibilities for x and y are plotted against one other, the resulting array should be able to depict any

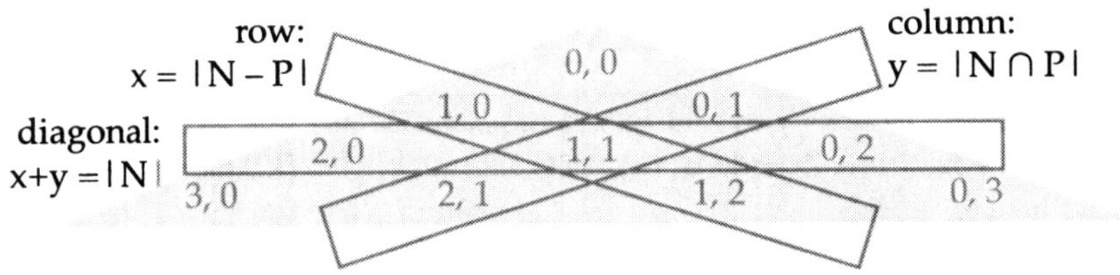

Figure 6.6. The Tree of Numbers.

quantifier meaning, see van Eijck (1984), van Benthem, 1986, §2.2, and Westerståhl, 1989, pp. 90ff. Conversely, representability in this array implies QUANT, EXT, and CONS.

In order to design such an array, the infinite extent given by *x* and *y* must be reduced to a range that is representable in a finite manner. Thus we need a constraint of finiteness:

6.30. FIN: Only finite universes are considered.

FIN tells us that we can use a finite number of the pairs *x* and *y* to depict quantifier meanings; see Westerståhl, 1989, pp. 82-3, for the source of FIN and more discussion. We take up the neuromimetic explanation for FIN at the end of this chapter.

Under FIN, the diagram in Fig. 6.6 begins a plot of *Q(x, y)*, which is traditionally known as the **Tree of Numbers**. The Tree is articulated by arithmetic progression in three directions, down the row, column, and diagonal. A row is any sequence of cells whose first number is the same. For instance, the first row of the Tree is the one whose first or *x* coordinate is zero, i.e. (0, 0), (0, 1), (0, 2), etc. Thus rows run from the top left to the bottom right. A column is the converse: a sequence of cells whose second or *y* coordinate is the same. For instance, the first column of the Tree is the one whose second number is zero, i.e. (0, 0), (1, 0), (2, 0), etc. Thus columns run from the top right to the bottom left. Technically, each point *(x, y)* in Fig. 6.6 has two **immediate successors** *(x+1, y)* and *(x, y+1)*, which in turn are the **immediate predecessors** of the point *(x+1, y+1)*. Finally, a diagonal runs straight across the page, parallel with the top and bottom edges.

Perhaps the most useful characteristic of the Number Tree is that it lets numbers stand in for sets. One can consequently talk about relationships among sets without having to worry too much about the actual set-theoretic formalization of these relationships. In particular, there are often more perspicuous representations in the number-theoretic format of the Tree of Numbers than in the set-theoretic format of Venn diagrams.

0, 0
1, 0 0, 1
2, 0 1, 1 0, 2
3, 0 2, 1 1, 2 0, 3
4, 0 3, 1 2, 2 1, 3 0, 4

Figure 6.7. ALL in the Tree of numbers.

0, 0
1, 0 0, 1
2, 0 1, 1 0, 2
3, 0 2, 1 1, 2 0, 3
4, 0 3, 1 2, 2 1, 3 0, 4

Figure 6.8. NO in the Tree of Numbers.

Quantifiers can be visualized in the Tree of Numbers by highlighting the nodes which indicate the value for the quantifier on each diagonal. For instance, the universal quantifiers *each, every, all* select the shadowed nodes in Figure 6.7. The pattern obviously continues indefinitely down the right periphery of the Tree. It can be expressed as the number-theoretic equation:

6.31. $\text{ALL}(x, y) \leftrightarrow x = 0$

Let us pause for a moment to imagine how this pattern is derived.

For the sake of argument, assume that the clause to be described is *All linguists are gourmets*, and I know four linguists. The *x* argument is calculated by subtracting the set of gourmets, G, from the set of linguists, L: L – G, which produces the set of linguists who are *not* gourmets. In my little world, this produces the null set, since all the linguists I know are simultaneously gourmets. The cardinality of the null set is zero, i. e. $|L - G| = 0$, which is the value of *x* in this case, and in every other case of universal quantification. The *y* argument is calculated by the intersection of the set of gourmets with the set of linguists: $L \cap G$, which produces the set of linguists who *are* gourmets. All of the linguists I know are in the set of gourmets, so the intersection of the linguists and the gourmets is just the set of linguists, $L \cap G = L$, whose

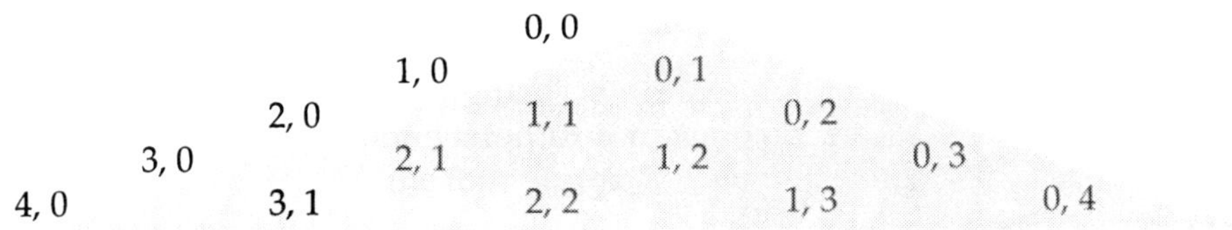

Figure 6.9. SOME in the Tree of Numbers.

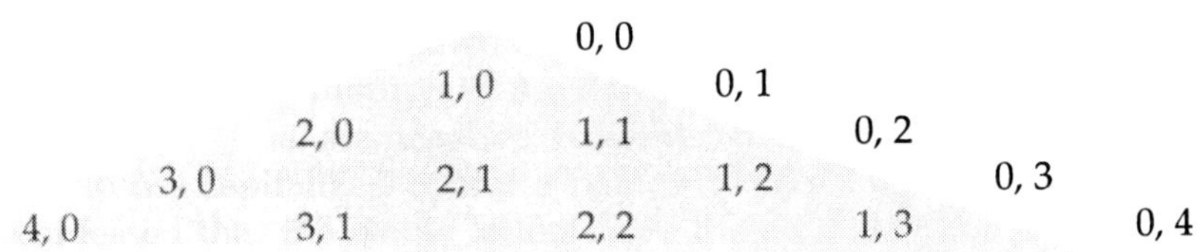

Figure 6.10. NALL in the Tree of Numbers.

cardinality is four, i.e. $|L \cap G| = |L| = 4$. The entire set is now pegged at (0, 4), and it will only increase or decrease at *y* according to the number of linguists.

The mirror-image of ALL is found with NO. NO produces a left-peripheral pattern in the Tree of Numbers, as seen in Fig. 6.8. The number-theoretic equation which describe this pattern is:

6.32. $\text{NO}(x, y) \leftrightarrow y = 0.$

This result can be worked out just as with ALL in the previous paragraph, and so is left to the reader.

The quantifier *some* is intuitively the one that includes all values but those for NO. It is diagrammed in Figure 6.10. The corresponding number-theoretic equation is (6.33):

6.33. $\text{SOME}(x, y) \leftrightarrow y > 0$

By parity of reasoning, one would expect there to be a quantifier that excludes all the values for which ALL are true. In the Tree of Numbers, this would look like Fig. 6.10, which is referred to here as NALL. English does not lexicalize a single morpheme for this quantification, just as it does no lexicalize a single morpheme for the coordinator NAND.

By way of summary, the four logical quantifiers partition the Tree among themselves in complementary patterns, depicted in Fig. 6.7-10 show. The

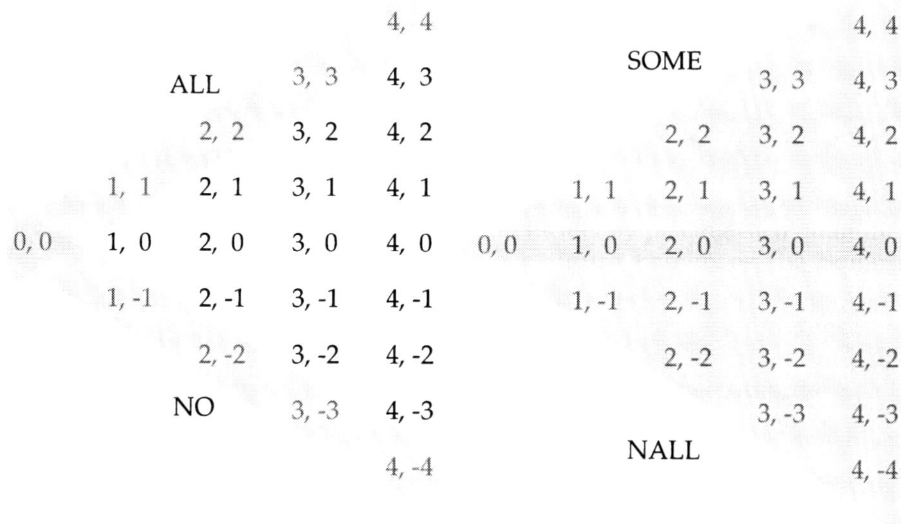

Figure 6.11. The space |N| x |P|, i.e. Q space, with the accepting zones of the logical quantifiers shaded in.

accepted and rejected values in each pattern can be separated by a straight line. This line is the decision boundary seen in the previous chapter, and its orientation is what distinguishes the various quantifiers. Much of the technical contribution of this monograph is to ascertain how this boundary is calculated, but there are still some issues concerning the Tree that need to be clarified.

6.2.4. The neuromimetic perspective

In generalized-quantifier theory, quantifier representations defined in the Tree of Numbers are thought of as a kind of curiosity, only useful for visual proofs of certain theorems. In contrast, from a neuromimetic perspective, quantifier representations can only be defined in a numerical format such as the Tree of Numbers, so it is their set-theoretic equivalencies that are curiosities without a biological foundation. The Tree of Numbers is consequently pivotal in allowing us to examine natural language quantification from a neuromimetic point of view, while retaining both the insights of generalized-quantifier theory and the insights of neural network design.

As a first step, the Tree of Numbers needs to be located in the representational format of the previous chapters, namely the quadrant of the Cartesian plane bounded by 45° and –45°. To do so, let us simply redefine the conversion from sets to numbers accomplished in the Tree so that x in $Q(x,y)$ is $|N|$ and y is $|P|$. This redefinition obeys QUANT, EXT, and CONS, as long as y itself obeys CONS, which was an assumption implicit in the representation of coordinator meanings. Under this conversion, the logical quantifiers trace the patterns shaded in Fig. 6.11.

6.2.4.1. $|N–P| \times |P \cap N|$ vs. $|N| \times |P|$

If the Tree of Numbers provides the right ontology for quantifier meanings, we would expect there to be a straightforward mapping from a quantified expression like *All linguists are gourmets* to its number-theoretic representation. This expectation is not fulfilled, however. Even if the intermediate translation is something amenable like *all(linguist, gourmet)*, this expression does not map into the Tree directly as *all(|linguist|, |gourmet|)*, but rather indirectly as *all(|linguist – gourmet|, |linguist ∩ gourmet|)*. Moreover, it is not obvious what number-theoretic principle would account for the deformation of *all(|linguist – gourmet|, |linguist ∩ gourmet|)* to achieve its surface form.

It would be easy to just assert that a representation like *all(|linguist – gourmet|, |linguist ∩ gourmet|)* is so far removed from the surface linguistic expression of a quantified clause as to be an implausible candidate for its semantic representation, but we would like to pause for a moment to review an argument in favor of this rather fundamental assertion. It comes from the rule of Quantifier Raising as set forth in May (1985).

6.2.4.2. The form of a quantified clause: Quantifier Raising

May (1985) adopts Barwise and Cooper's ideas for the interpretation of quantifier-variable structures at the level of analysis in Principles-and-Parameters Grammar known as Logical Form. May's model is based on a non-null domain *D*, from which the various sets which instantiate variables are drawn. There are two such variables, *X*, the set denoted by an *n*-level projection of a lexical category, and *Y*, the set denoted by the quantifier's scope. May requires a one-to-one correspondence between the syntactic constituents and the sets that they denote as in (6.34), which conflates May's separate syntactic and semantic schema:

6.34. $[\,Q\text{-}X^{n}_{i}\ [_{\beta} \ldots e_{i} \ldots]]: Q(X, Y)$

The scope of the quantifier *Q* is represented here by the open sentence $[_{\beta}\ldots e_{i}\ldots]$, i.e., the maximal domain in which e_i is not bound to a quantifier expression. β is established at Logical Form by a rule of **Quantifier Raising**, which adjoins a quantified NP to the IP (S) node of a phrase marker, leaving a coindexed trace in the S-structure position of the quantified NP. For instance, the sentence *John saw every picture* is represented at LF as Fig. 6.12. The raised NP can be treated as a variable, whose scope is given by the following definition:

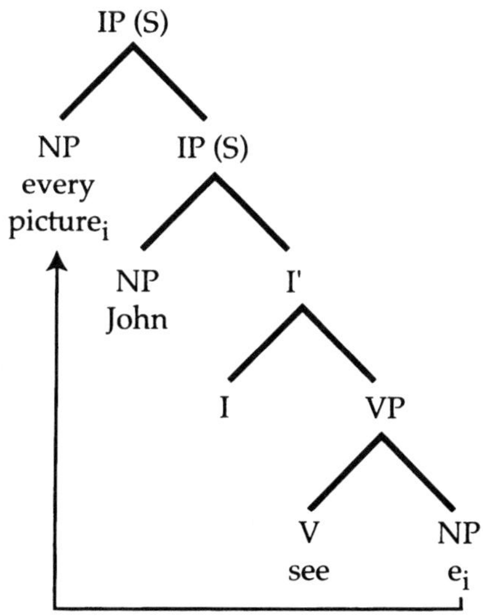

Figure 6.12. Quantifier Raising.

6.35. The scope of α is the set of nodes that α c-commands at LF, where α c-commands ß iff the first branching node dominating α dominates ß (and α does not dominate ß).

The relation between the quantified phrase and its trace is subject to Trace Theory, thus bringing quantification under the sway of general principles of syntactic well-formedness. For example, Trace Theory requires that antecedents c-command their traces, so downgrading quantification, in which an NP adjoins to a lower position in a tree, are proscribed.

The set-theoretic formula *Q(X, Y)* is evaluated for truth (1) or falsehood (0) according to:

6.36. $$Q(X, Y) = 1 \text{ iff } \Psi$$
$$= 0 \text{ otherwise.}$$

The symbol *Ψ* stands in for a function from *X* and *Y* onto subsets of *D*, given by the lexical requirements of the quantifiers.[32] May details six such functions:

[32] Notice that the requirements of particular lexical items are not relevant until after LF.

6.37	a)	every(X, Y)	$= 1$ iff $X \cap Y = X$, otherwise 0.
	b)	some(X, Y)	$= 1$ iff $X \cap Y \neq \emptyset$, otherwise 0.
	c)	not all(X, Y)	$= 1$ iff $X \cap Y \neq X$, otherwise 0.
	d)	no(X, Y)	$= 1$ iff $X \cap Y = \emptyset$, otherwise 0.
	e)	n(X, Y)	$= 1$ iff $\lvert X \cap Y \rvert = n$, otherwise 0.[33]
	f)	the(X, Y)	$= 1$ iff $X \cap Y = X = \{a\}$, for $a \in D$, otherwise 0.

The *X* variables translate into the set denoted by the X' constituent in the LF structure, and the *Y* variables translate into the set denoted by the scope of the quantifier.

An illustrative derivation should help to clarify the representation. The LF bracketing of the example in Fig. 6.12 corresponds to (6.38):

6.38. $[_{IP} [_{NP_i}$ every $[_{N'}$ picture$]]$ $[_{IP}$ John saw $e_i]]$

Substitution of these expressions for the variables in (6.37a) gives:

6.39. every({x | picture(x)}, {y | saw(John, y)})
= 1 iff {x | picture(x)} ∩ {y | saw(John, y)} = {x | picture(x)},
= 0 otherwise.

This derivation demonstrates how the accuracy of a semantic representation depends on the accuracy of its syntactic source. It also demonstrates, as was said, that Number Tree representations do not map directly into standard linguistic units. The two Number Tree variables $|N - P|$ and $|N \cap P|$ do not map onto the domain and range of the quantifier, respectively. We would rather use two variables that match the syntactic structure more accurately, in order to make the mapping between components as simple and transparent as possible.

Fortunately, there is an alternative that does preserve the surface form in a more transparent fashion. It is to map *Q(N, P)* directly to *Q(|N|, |P|)*. In our on-going example, this works out as the much more perspicuous *all(|linguist|, |gourmet|)*. This alternative is already depicted in Fig. 6.11. the obvious name for it should be **quantifier space** or simply **Q space**, on analogy to coordinator space. It can be normalized to give a space that is identical that of COOR space, except with many more points.

---

[33] This formulation only encodes the "exactly" sense of the numerals. There is also an "at least" sense, arrived at by using $\geq$ for =, and an "at most" sense, gotten by using $\leq$.

6.2.4.3. Another look at the constraints

None of the logical constraints introduced in Section 6.2.3 explain anything; they merely formalize particular observations about quantifiers within a larger system for drawing inferences. Our neuromimetic analysis, grounded in the study of the logical coordinators, attempts to supply the missing explanation by appealing to some aspect of brain function.

6.2.4.3.1 Quantity and two streams of semantic processing

For the case of QUANT, we would wish to posit two pathways of semantic processing, in analogy to the two pathways of visual processing mentioned in Chapter 1. The reader may recall our mention of the fact that visual information splits into two streams after V1, a ventral stream that parses an object's properties and a dorsal stream that locates the object in space. These two streams are mnemonically nick-named the *what* and *where* pathways, respectively.

Perhaps not fortuitously, QUANT also draws a distinction between two types of processing, where processing is understood as permutation. Those meanings which change under permutation define qualitative properties of entities and so are analogous to the *what* pathway in vision. Those meanings which do not change under permutation define quantificational properties of entities. The fascinating question is whether they are analogous to the *where* pathway in vision.

In a representational sense, we have already argued that they are. Our chosen representational format is a quadrant of the Cartesian plane, Q or COOR space, and a specific coordination or quantification corresponds to a location in one of these spaces. Unfortunately, nothing is known about the neurology of quantificational properties beyond what is posited in this book, so there is no way at present to test our semantic hypothesis against the broad range of neurological observations of its putative visual analog. However, in the interests of falsifiability, the next expression offers the most explicit statement of what we have in mind:

6.40. Semantic processing splits into two streams in order to distinguish permutable from non-permutable properties, the former of which provide the biological grounding of QUANT.

In view of our current ignorance about the neurology of semantics, it seems premature to push this analogy any further.

6.2.4.3.2 Extension and normalization

Turning to EXT, what is intriguing about it is how the expansion of the universe on which it is based recapitulates our discussion of the formal characteristics of normalization in Chapter 3, as illustrated in Fig. 3.19. Thus as a first approximation to a grounding of EXT we present 6.41:

6.41. Expansion and contraction of the universe in the sense of EXT are equivalent to expansion and contraction within the order topology on the Cartesian plane.

The one problem is that normalization traffics in both expansions *and* contractions, while EXT only traffics in expansion, since permitting contractions would rule out the numerals as possible quantifiers.

The curious thing is that numerals *are* different from quantifiers, see for instance Bartsch (1973), Jackendoff (1977), Verkuhl (1981), and Jong (1984). Numerals act more like adjectives, while quantifiers act more like determiners. For instance, numerals are serialized after quantifiers and determiners, (6.42a), and they can follow the copula, (6.42b):

6.42 a) every three questions; my many friends
b) My questions are {three / many / *every}.

Thus it appears justifiable to draw the contrast between the two categories as in (6.43):

6.43 a) A numeral allows expansions but not contractions in the order topology, i.e. it is sensitive to magnitude.
b) A quantifier allows expansions and contractions in the order topology, i.e. it is insensitive to magnitude.

A numeral by definition denotes a specific quantity, which remain in Q space as it grows but can be left out if it shrinks. A (true) quantifier, by the new definition of (6.43b), is not sensitive to magnitude in this way. It is normalized – expandable and contractible – making it usable even when its cardinality is not known exactly or is unimportant.

6.2.4.3.3 CONS and labeled lines

The idea underlying Conservativity is that the nominal input to a quantifier must be conserved in its output. This is a fundamental constraint on relational categories in language, yet a similar type of conservation of inputs is equally fundamental for all perceptual systems. Imagine what it would be like for tactile stimulation of your right index finger to be replaced by stimulation of your left index finger on the way to somatosensory cortex. Or for the part of your inner ear that is sensitive to frequencies of 40 Hz to be replaced by the response to frequencies of 400 Hz. Or for the cones that are sensitive to green wavelengths of light to be overwritten by the cones that respond to red wavelengths. A perceptual system that permitted such haphazard crossovers would be of little adaptive value to an animal attempting to survive and reproduce in the real world, since any action planned in response to such percepts would often be inappropriate.

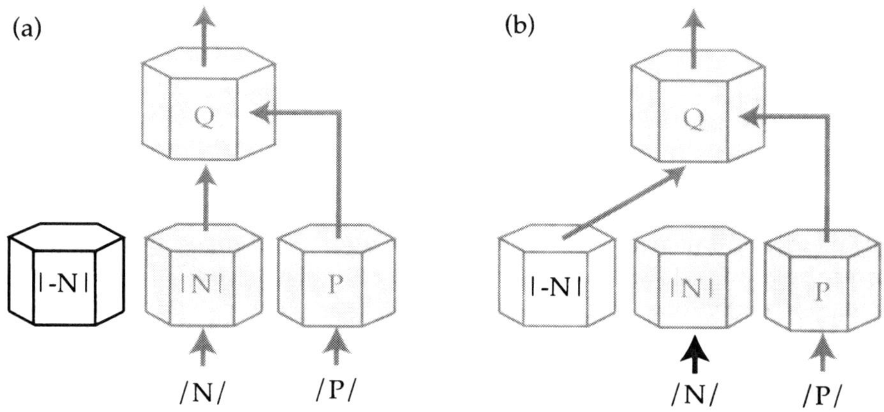

Figure 6.13. A labeled-line hypothesis of CONS. (a) Input to Q satisfies CONS; (b) input to Q violates CONS.

This analogy from perception suggests a neurological reduction of Conservativity to a similar mechanism. The mechanism is the theory that the pathways from sensory transducers to the corresponding areas of cortical representation do not intermix, so that the output of a particular neural transmission line is effectively labeled by its input location – whence the name, **labeled-line theory**, see for instance Hendry, Hsiao, and Brown (1999).

The linguistic realization of labeled-line theory builds on results from Chapter 5, in which it was shown how LVQ links a phonological and a semantic form. This linkage permits each form to label the other. For instance, in speech comprehension, the activation of a phonological form creates a labeled line to the corresponding semantic form, as depicted in Fig. 6.13. On the left, we see a correctly Conservative quantification, in which the nominal and predicate inputs are activated by their phonological counterparts and do not intermingle with other nearby nominals. On the right, we see a Conservativity-violating quantification, in which *-N* replaces the *N* input before reaching the quantifier, despite the fact that *-N* is not activated by any phonological material. Thus Conservativity can be reduced to an extremely general property of neurological organization. Note that labeled-line theory itself is presumably due to Hebbian learning strengthening the synapses that participate in a labeled line, but such speculation is beyond the reach of this monograph.

However, this labeled-line hypothesis of Conservativity is incomplete in one important respect, namely that there is an asymmetry between the nominal and predicate arguments. The nominal information can never be absent from the predicate, but the predicate information can be absent from the nominal – in fact, it presumably must be, otherwise there would be no informational point to uttering the quantification.

Q space has this asymmetry already built into it, since it occupies the eastern (positive in x) half-plane of Cartesian space, not the western (negative in x) half. Unfortunately, no principled reason was offered for this choice of location; only the empirical reason that it works best this way. We suspect that there *is* no principled reason; it is rather an arbitrary fact of language, one that must be learned. That is to say, nominal quantifiers are nominal precisely because they conserve their nominal argument; quantifiers of other categories would conserve some other argument. Such asymmetries are grounded in the representation of the corresponding spaces by assigning the conserved argument to (the positive half of) the x axis. Fig. 6.13 attempts to depict the nominal case in neuromimetic terms by making a short, straight connection between the N and Q minicolumns and longer, crooked connection between the P and Q minicolumns. Exactly what content this iconicity may have is a mystery to us at present.

6.2.5. The origin, presupposition failure, and non-correlation

The reader may have noticed that all of the quantifier patterns above include the origin, (0, 0), as a value for a quantifier. Yet, consider how a universe consisting only of the origin (0, 0) supports the logical quantifiers:

6.44 a) Every square circle is a two-dimensional figure.
b) Some square circles are two-dimensional figures.
c) Not every square circle is a two-dimensional figure.
d) No square circle is a two-dimensional figure.

6.45 a) Every square circle is filled in.
b) Some square circles are filled in.
c) Not every square circle is filled in.
d) No square circle is filled in.

In (6.44), at best the statement with *every* is judged true, though with the caveat "if there were any square circles". The others are false. In (6.45), at best the statement with *no* is judged true, with the caveat embodied in the continuation "... because there aren't any square circles!". The others are judged false. Thus intuitions from English support the exclusion of the origin as a possible quantifier value.

This oddness of sentences like those of (6.44) and (6.45) is studied as a branch of **presupposition failure**, for which there is a considerable body of literature; see Seuren (2000) for recent review. This topic is too far removed from the focus of this monograph to be taken up adequately here, so we must console ourselves with just the handful of basic comments that can be drawn from what has been said so far.

If a logical operator expresses correlation between its two arguments, it is bound to fail when there are no individuals upon which to estimate the correlation. Thus the fact that Q/COOR space is inherently structured with a discontinuity at $[0, y]^T$ automatically accounts for presupposition failure, with no further stipulation necessary. Yet for the sake of explicitness, let us call attention to this result.

In the next section, we discuss how Mostowski (1957) and van Benthem (1984) deal with trivial quantifiers. These authors formulate an explicit constraint against including the origin in quantifier denotations, reproduced in (6.46a). Our alternative is elaborated in (6.46b):

6.46 a) -0: Exclude (0, 0).

b) $-[0, y]^T$: Any presupposition of x fails at the x axis, i.e. an expression with such a denotation is radically false in the system of Seuren (2000) and Seuren et al. (2001).

Note that (6.46b) still allows the y argument to have a presupposition, which is a virtue whose bounty we cannot expound on here.

It can also be pointed out that normalization directly excludes the origin from consideration as input to a quantifier, due to the fact that the length of the vector $[0, 0]^T$ is zero, and it is arithmetically impossible to divide by zero – or it is possible, but the result is infinity. However, (6.46b) is more general, so we can retain it as the 'real' explanation.[34]

6.2.6. Triviality

As our final foray into GQ constraints, under the assumption that the values for a quantification consist of the positive integers, consider that *fewer than zero* is not satisfied by any argument denoting a positive integer, while *more than zero* is satisfied by all of them. These expressions consequently instantiate two new quantifiers, TAUTology and CONTRAdiction, respectively. Some means should be found to rule out such unintuitive meanings.

The general approach is to classify a quantifier as **trivial** if its denotation is either empty or universal for any choice of arguments, see Westerståhl, 1989, pp. 75-6 and 91, for review. At least three means of implementing this

[34] An additional argument can be designed from the Third Generation ontology taken up in Chapter 11. In particular, Gärdenfors' claim summarized in Sec. 11.5.2.1.1 that conceptual space defines all possible individuals from intersections of the various underlying quality dimensions suggests that an individual with no value on a quality dimension – such as the x axis of Q space – would not be a proper individual and so could not be a value for a quantification.

ALL	SOME	NALL	NO	TAUT	CONTRA
+	–	–	+	+	–
– +	– +	+ –	+ –	+ +	– –
– – +	– + +	+ + –	+ – –	+ + +	– – –
– – – +	– + + +	+ + + –	+ – – –	+ + + +	– – – –
– – – – +	– + + + +	+ + + + –	+ – – – –	+ + + + +	– – – – –

Figure 6.14. Accepting/rejecting regions of the logical quantifiers plus TAUT and CONTRA in the Tree of Numbers represented by +/–.

approach have been proposed. To understand them, let us introduce a new representational format, in which an accepted leaf in the Number Tree is labeled with '+' and a rejected leaf is labeled with '–'. The logical quantifiers, plus TAUTology and CONTRAdiction, are illustrated in this new format in Fig. 6.14. With the aid of this labeling scheme, trivial quantifiers can be ruled out by the constraints in (6.47):

6.47 a) NONTRIVIALITY: A quantifier is non-trivial on some universe, i.e., there is at least one + and one – in the Tree of Numbers.

b) ACTIVITY: A quantifier is non-trivial on every universe, i.e., there is at least one + and one – in the **top triangle** consisting of (0, 0), (1, 0) and (0, 1) in the Tree of Numbers.

c) VARIETY: For all M and all $A_1, \ldots, A_n \subseteq M$ such that $A_1 \cap \ldots \cap A_n \neq \emptyset$, there are $B_1, B_2 \subseteq M$ such that $Q_M\ A_1 \ldots A_n, B_1$ and $\neg Q_M A_1 \ldots A_n, B_2$, i.e., there is at least one + and one – on each diagonal, except (0, 0), in the Tree of Numbers.

(6.48) lists the three conditions in order of increasing strength, in order to uncover the implicit implicational hierarchy:

6.48. VARIETY ⇒ ACTIVITY ⇒ NONTRIVIALITY

Ideally, we would like natural language quantifiers to be characterized by the strongest one, that of VARIETY. However, there are exceptions like *four*, which are empty for a universe of less than four elements. It follows that the strongest claim that can be made as a universal of natural language is (6.49):

6.49. Simplex natural-language quantifiers satisfy NONTRIVIALITY.

Nevertheless, the four logical quantifiers do respect VARIETY, so we would like to know why this is so.

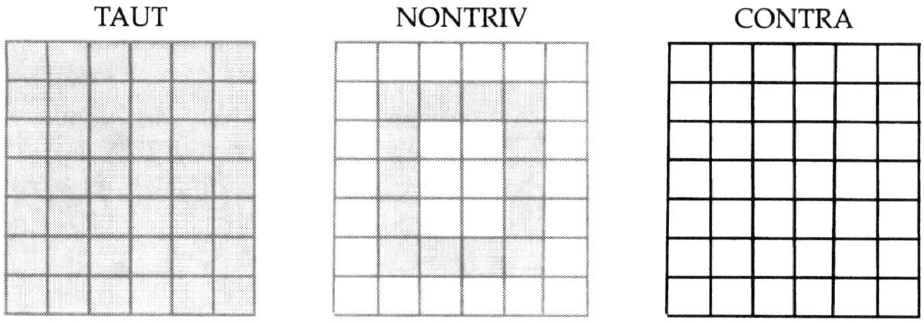

Figure 6.15. Object recognition on a retina as three quantifiers.

6.2.6.1. Triviality and object recognition

What is wrong with trivial quantification? From the perspective of neuromimetic learning rules, CONTRA and TAUT are not ill-formed, since both can be learned by the networks introduced in the preceding chapter. Yet if we examine how these patterns are learned, the results are rather strange. This strangeness is easiest to state for perceptron learning: the decision boundary induced for CONTRA and TAUT does not bisect Q space; the closest it will come is to approximate one cut point or the other. For cluster learning, the result is equally strange: either the whole of Q space is covered by the nodes learned, or none of it. In fact, cluster learning may terminate in an even more exaggerated result: one node can come to represent all of Q space – the case of TAUT – or no node at all can represent it – the case of CONTRA.

This latter observation suggests a principle of (semantic) object recognition that relies on an object having a non-uniform pattern of activation in a domain. By way of illustration, consider the recognition of three 'objects' on a rudimentary retina in Fig. 6.15. Each square on the seven-by-six grid indicates the receptive field of a retinal cell (rod) sensitive to achromatic color, i.e. shades of gray. A square that is darkened represents a rod that is on; otherwise, the rod is off. Under this convention, all the rods are on, on the left side, which recapitulates the status of the quantifier TAUT within Q space. Conversely, all the rods are off on the right side, which recapitulates the status of the quantifier CONTRA within Q space. Neither of these is recognizable as a visual object. It is only the pattern on the center retina, which is partially on and partially off, that is recognizable as a visual object. This is the one which recapitulates the status of the nontrivial quantifiers within Q space.

Inspired by this correspondence, we offer the following conjecture for NONTRIVIALITY:

6.50. A pattern that does not consist of at least one change of sign is not recognizable as an object to the relevant perceptual system.

This is the perceptual grounding of the family of triviality constraints in (6.47). By its action, TAUT and CONTRA are ruled out as possible natural language quantifiers – not because of any strictly quantificational effect, but rather because they are not available to the semantic system as potential objects.[35]

6.2.6.2. Continuity of non-triviality and logicality

The logical quantifiers satisfy the more stringent constraint of VARIETY in (6.47c), which requires there to be one + and one – on every diagonal of the Number Tree. Moreover, the + and – of the first diagonal is propagated down the Tree in a consistent manner, which suggests that this consistency may lie at the heart of logicality. This possibility has not been overlooked by logicians.

For instance, Westerståhl, 1989, pp. 98ff, uses the particular case of the logical quantifiers to review what it means to be a logical constant. He uncovers three fundamental desiderata, and one secondary one. The first two are well-formedness conditions on patterns in the Tree of Numbers, Nontriviality and Right Continuity. Nontriviality has already been seen; RIGHT CONT constrains how gaps in the Tree are filled in:

6.51. RIGHT CONT: Between two +'s on a diagonal there are only +'s.

The third desideratum, which can be traced in different versions back to Mostowski (1957) and van Benthem (1984), rules out any distinction between cardinal numbers. Mostowski's version can be called No Cardinality; Van Benthem's version is called Uniformity:

6.52 a) NO CARD: Q does not distinguish any pair of non-zero natural numbers.

b) UNIF: The sign of any point in the Tree determines the sign of its two immediate successors.

Both formulations are discussed in more detail below, but their common rationale is that such distinctions of cardinality "... belong to mathematics, not logic". Using these three desiderata, plus the additional exclusion of the

---

[35] This meta-constraint undoubtedly has an information-theoretic formalization in the fact that both CONTRA and TAUT are uninformative, but such speculation is more than can be taken up in this already rather eclectic treatise.

origin mentioned above with respect to presupposition failure, a simple proof demonstrates that the only quantifiers that can be defined are the logical ones.

Mostowski's version plays out in the following way. We begin by considering the top triangle minus (0, 0), which has four possible patterns. The first two, '++' and '– –', illustrated by TAUT and CONTRA in Fig. 6.14, satisfy RIGHT CONT and NO CARD. The way they satisfy NO CARD is interesting: the fact that '++' has + at the top left edge and + at the top right edge means that + must iterate all the way down both edges, since a Q cannot distinguish (1, 0) from, say, (2, 0) or (3, 0). Of course, both putative quantifiers fail NONTRIVIALITY, so they are discarded. The only remaining possibilities start with '+ –' or '– +' at the top and iterate them down either edge, as is the case of the logical quantifiers in Fig. 6.14. A chance for variation arises at the next diagonal, since the point between the two edges must be marked. Whatever the choice, it is iterated downward, and we get the four logical quantifiers. There are no other patterns consistent with the four conditions.

Van Benthem's version makes the iteration imposed by NO CARD much more explicit. In the Tree of Numbers, the cardinality property means that no point in the Tree is special: you always proceed downward in the same manner. The manner in question is to choose an arbitrary point (A, B) and go down by adding one element to $|A - B|$ or to $|A \cap B|$. The result should be **uniform**, that is, it should not vary according to the number of elements in these sets. This is the import of UNIF. The outcome is exactly as above: the trivial quantifiers are ruled out by NONTRIVIALITY, and only the logical quantifiers are ruled in.

6.2.6.3. Continuity and the order topology

These desiderata for logicality can also be reduced to the explanatory principles that we have been developing. RIGHT CONT is subsumed by the order topology on the unit arc, if the claim about monomorphemicity of coordinators in (5.29) is generalized to all functional categories:

6.53. Monomorphemic functional categories representable on the unit arc are convex subsets thereof.

Recall that the convexity of a subset requires that all of the points between *a* and *b* are also members of the category that includes *a* and *b*. Thus the stipulation of RIGHT CONT, that between any two +'s on a diagonal there are only +'s, follows naturally. Moreover, the topological definition is more general, in that it also accounts for the symmetrical case of –'s (between any two –'s on a diagonal there are only –'s), which is instantiated by NALL.

No Cardinality is directly accounted for by normalization, which, as we have taken pains to show, reduces all of the cardinalities of the leaves of the Number Tree – the magnitudes of unnormalized Q space – to the unit arc. Uniformity can also be subsumed under the magnitude neutralization effected by normalization, since it removes all of the cardinality information that would enable the identification of special, non-uniform leaves in the Tree.

6.2.7. Finite means for infinite domains

There still remains one GQ constraint which has not been attributed to any more general cognitive or perceptual process, namely FIN. FIN ensures that quantifiers can be represented in the Tree of Numbers by some finite sample of their extension and thus finesses the problem of universal quantification over infinite domains raised in Sec. 6.1.4. It is perhaps the easiest one of all to account for.

6.2.7.1. FIN, density, and approximation

In the light of the neuromimetic learning rules simulated for coordinators in the previous chapter, it should be abundantly clear that children can learn much from a small sample. In fact, given their limited memories and limited exposure to their ambient language, they may have to learn practically their entire first language from a small sample. Thus if a quantifier wishes to hold out any hope of being learned, it must realize a pattern this is instantiated within the first few diagonals of the Number Tree.

However, this focus on the initial steps of an infinite pattern is of no help to the adult logician or mathematician trying to grapple with the other, infinite, end of the Number Tree. Yet once again we can appeal to the tandem of normalization and the order topology to simplify the cognitive load.

On the one hand, to reach an infinite point in the unnormalized Cartesian plane, the successor function must be applied an infinite number of times. This computation has an obvious instantiation in terms of a finite-state automaton's traversal of the Tree of Numbers outlined at the beginning of Chapter 1. A finite-state automaton traverses the Tree outward, into the 'depth' of the quadrant and off the edge of the page.

In contrast, the traversal of the arc of the half-plane, whether by vector angle or vector normalization, proceeds by tracing the breadth of a certain span of the arc. Though this may not appear to require any kind of infinite sequence, this appearance is misleading: the arc is infinitely dense. There are, however, two important differences between density and extension, revolving around how infinite points are approximated and used in proportions.

In the arc, an infinite point is always near many finite points that can serve as approximations to it. It follows that if the arc supports a measure of distance, then infinite points cease to be a cognitive burden. And in fact, it does support a measure of distance, as given by the order topology of Chapter 3. Moreover, the quantization of a space performed by a competitive network suggests that an infinite point will fall with the Voronoi cell of the closest prototype and so be recognizable to the overall system.

Secondly, the fact that the absolute number of $|N|$ is discounted in favor of the proportion of $|N|$ to $|P|$ means that many potentially infinite calculations have simple solutions. Take the case of $|N| = |P| = \infty$. The proportion $\infty{:}\infty$ is obviously resolvable as $[0.71, 0.71]^T$ or cos = 0.71, which the cognitive system can (hopefully) recognize before it goes too far towards trying to count to ∞. The result is that the infinite extension of the Cartesian plane becomes tractable in the guise of the infinite density of the arc that spans it.

The stipulation of finiteness in GQ theory by means of the condition FIN in (6.30) can now be parsed into the two observations of (6.54):

6.54 a) FIN NUMBER: Only a finite number of points on the arc are considered.

b) FIN APPROX: Only finite points on the arc are considered; i.e. infinite points are approximated by nearby finite ones.

In the greater scheme of things, this is but one reflection of a more general tendency of neuromimetic systems to trade processing time, e.g. the steps in an automaton, for precision in the storage of activations, see Sontag, 1995, p. 122, for review and references, and Siegelman and Sontag (1995) for the original derivation of this result.

6.3. STRICT VS. LOOSE READINGS OF UNIVERSAL QUANTIFIERS

There is one final consideration of how people actually use quantifiers that needs to be raised. Based on a corpus of oral narratives, Labov (1984, 1985) identifies a strict and a loose interpretation of universal quantifiers. The strict interpretation conforms to the interpretation that we have taken for granted: "the quantifier is applied to a set to designate exhaustively all members of the set, with no exceptions" (Labov, 1985, p. 176). The loose interpretation "...is applied to designate the members of the set as a whole, but not necessarily exhaustively." (ibid.) Labov adduces the sentences in (6.55) to exemplify the two readings:

6.55 a) Now every one o' my kids turned back.

b) I left all my clothes down South.

In (6.55a), the universal quantifier is applied to a set of known size making the strict reading probable, while the pragmatic assumption that the speaker is wearing clothes at the moment of utterance rules out the strict reading of (6.55b), ruling in a looser reading in which maybe not quite all of the speaker's clothes were left down South. Labov marshals a broad range of utterances from his corpora to show that such loose interpretations are not at all uncommon. Labov concludes from these observations that ...

> ... the rules of logical inference taught in the schools are restricted in their application to public discourse, and we must

Table 6.1. Definitions of the logical quantifiers in Q space.

Q	Tree of Numbers	Correlation	Angle	Norm	Order features
ALL	$\|N \cap P\| = \|N\|$	$\rho(q) = 1$	$\angle q = 45°$	$N(q) = [.7, .7]^T$	[-blade, -down]
SOME	$\|N \cap P\| \neq 0$	$\rho(q) > 0$	$\angle q > 0°$	$N(q) > [1, 0]^T$	[+blade, -down]
NALL	$\|N \cap P\| \neq \|N\|$	$\rho(q) < 0$	$\angle q < 0°$	$N(q) < [1, 0]^T$	[+blade, +down]
NO	$\|N \cap P\| = 0$	$\rho(q) = -1$	$\angle q = -45°$	$N(q) = [.7, -.7]^T$	[-blade, +down]

> continue to ask whether or not these rules form the proper basis for the grammar of natural language. (Labov, 1985, p. 194)

Labov thus eschews the logical analysis of the constructions that have been examined in this and the last few chapters, though he does not offer any clear indications of how an 'anti-logical' alternative would work.

To our own way of thinking, both the logical and the anti-logical approaches are flawed in that, by failing to define the neurological processes that perform logical operations, it is not unexpected that they fail to delimit the proper extent of the phenomena in question. Under our approach, the fact that a universal node can capture nearby existential values falls out directly from the workings of a competitive network, namely from the fact that the prototype vector at $[0.71, 0.71]^T$ can capture nearby vectors in its Voronoi cell. Loose universal quantification consequently constitutes the core behavior of the logical quantifiers.

The strict interpretation is much harder to encode, given the limits of precision of the system. In fact, it may take an additional machinery not contemplated in GQ theory to enforce, as we shall see in the next chapter. Thus while we agree with Labov in essence, we do not go so far as to eject from semantics the insights of logical analysis – we just want to put them in their proper neuromimetic place.

6.4. SUMMARY

The main goal, and innovation, of this chapter is to derive a representation of the logical quantifiers that is compatible with our analysis of the logical coordinators from existing work on quantification, especially from Generalized Quantifier theory and its usage of the Tree of Numbers. Unfortunately, GQ theory only takes us only so far, because the procedure for converting sets to numbers relies on the linguistically implausible sets $N \cap P$ and $N - P$. We therefore turned to the more plausible sets N and P and would up with a quantificational version of COOR space that still satisfies the key constraints of QUANT, EXT, and CONS. We may therefore conclude that that the logical quantifiers share with the logical coordinators the characteristic of being convex subsets of the unit arc. Thus the definitions of the logical quantifiers are isomorphic to those of the logical operators and are reproduced in Table 6.1 from Table 3.5.

A secondary concern has been to ground the GQ constraints on neuromimetic principles. The basing of quantification on cardinality via QUANT suggests that semantic processing is separated into at least two streams, one of content and the other of location in some space, much like the ventral and dorsal pathways of the visual system. The insensitivity of quantification to expansion of the universe imposed by EXT is grounded in the order topology of Cartesian space. A distinction between (true) quantifiers and numerals can be grounded on this insight by including contraction: numerals are sensitive to contraction – they have a fixed magnitude – while true quantifiers are not – they have no fixed magnitude, reflecting their normalization. Conservation of the nominal argument via CONS defines nominal quantification and is grounded on the assignment of the nominal argument to the positive half of the x axis, though the exact details of this assignment are not known. The absence of trivial quantifications is grounded on the fact that such objects are unrecognizable in any cognitive domain, and presumably reduce to general principles of statistical pattern recognition as were introduced in Chapter 1. The continuity of sign found among the logical quantifiers is grounded on their status as convex subsets of Q space. Finally, the recognition of quantifier patterns even in a small, finite sample of Q space is a necessary condition for them to be learned by children. Their extension to infinite size is finessed by the quantization of Q space brought about a competitive network, and especially LVQ, as we will see in the next chapter.

Chapter 7

ANNs for quantifier learning and recognition

Chapter 5 performs most of the hard work of deciding which neuromimetic architectures and learning rules perform the best for the logical operators, taking the logical coordinators as a representative sample. The most parsimonious approach is to assume that these results hold over to the logical quantifiers. Given that the child should have access to more data points in learning the logical quantifiers, they present us with the opportunity to investigate the behavior of LVQ networks more closely.

7.1. LVQ FOR QUANTIFIER LEARNING AND RECOGNITION

The LVQ architecture of the logical coordinators generalizes directly to the logical quantifiers, with the single difference of the addition of extra neurons to the competitive layer in order to cover more accurately the larger data space of the quantifiers.

7.1.1. Perfect data, less than perfect data, and convex decision regions

Let us train an LVQ network to classify the patterns that describe the logical quantifiers. To do this in the most realistic manner possible, we would like to know what quantities children are exposed to the most or are best able to estimate. At the present, we know of no answers to these questions and so will just assume that children learn quantifier meanings from groups of two to n members, where n is some small integer, such as seven. Thus an ideal data set consists of the fifty-four points for all of the values of the logical quantifiers in the range $2 \leq |N| \leq 7$, henceforth Q7, which provides the background for Fig. 7.1a.

The number of L1 neurons for this data set should be greater than that of the corresponding coordinator L1, in view of the more densely populated data space. Intuitively, eight would be a good choice, with the goal of partitioning Q7 space into four positive and four negative subclasses. A glance at Fig. 7.1a shows that the asterisks marking the weights of the competitive neurons do indeed tessellate Q7 space in this manner. Their even spacing reflects the even statistical distribution of the Q7 points quite accurately.

Before taking up the contribution of L2, let us shift our perspective slightly to consider the effect on L1 of a less idealized data set. A more realistic data set is the subject of the LVQ simulation reported in Fig. 7.1b. Many values were deleted at random, so that not every logical quantification possible in Q7 is represented. Of course, we still do not know of any actual measure of the

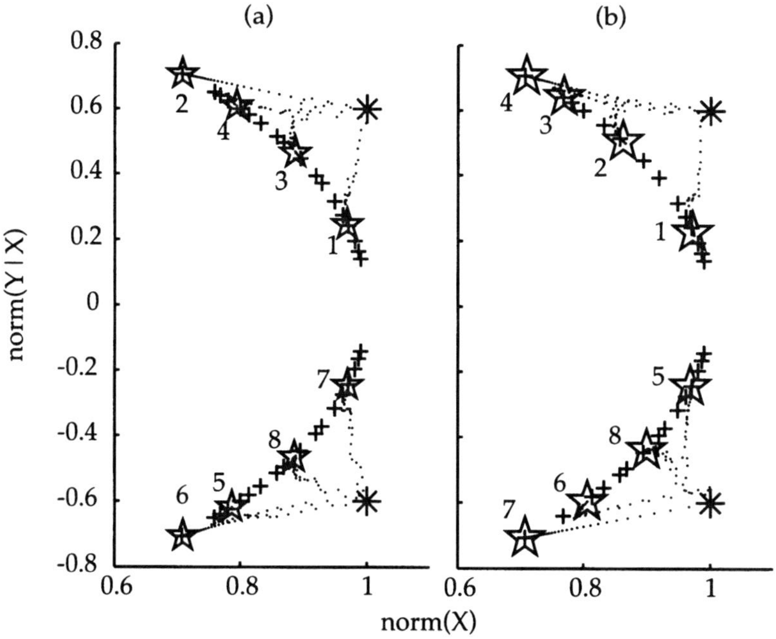

Figure 7.1. Supervised competitive learning of Q7: (a) idealized data (**p7.01_LVQQ_ideal.m**); (b) realistic data (**p7.02_LVQQ_real.m**). '+' = accepted point, '*' = initial weight, dot = intermediate weight, star = final weight.

frequency of these readings, so, in our current state of ignorance, it seems best to content ourselves with these rather artificial guesses as a guidepost for posterior research and. The visual result of this change is that the competitive neurons of Fig. 7.1b are no longer spaced evenly, especially in the positive quadrant.

Nevertheless, such data omission does not undermine the accuracy of the L1 network. This resilience is due to the fact that a competitive neuron learns a convex region of the input space and consequently 'closes the gap' left by incomplete data. That is to say, any point between two points accepted by a competitive neuron is also accepted. This ability of LVQ networks to extrapolate across missing data makes them an ideal mechanism for learning linguistic patterns, in the face of the well-known poverty of the linguistic stimulus that children are exposed to. The problem lies in the boundaries between receptive fields, which is best appreciated by a consideration of L2.

With respect to classificatory layer of the idealized data, Table 7.1a shows that the existential competitive neurons next to the universal neurons play a large role in the universal quantifiers, from 23% of ALL to 10% of NOR. Any

Table 7.1. Final W2 for Fig. 7.1.

	(a) Idealized data, Fig. 7.1a				*(b) Realistic data, Fig. 7.1b*			
L1	*ALL*	*SOME*	*NALL*	*NO*	*ALL*	*SOME*	*NALL*	*NO*
1	0	0.30	0	0	0	0.36	0	0
2	0.73	0.21	0	0	0.06	0.28	0	0
3	0.03	0.27	0	0	0.26	0.19	0	0
4	0.23	0.22	0	0	0.69	0.17	0	0
5	0	0	0.26	0.10	0	0	0.35	0
6	0	0	0.18	0.87	0	0	0.20	0.12
7	0	0	0.27	0.03	0	0	0.17	0.88
8	0	0	0.29	0	0	0	0.29	0

percentage of more than zero is inaccurate. Table 7.1b details the change in percentages that the 'corruption' of the ideal data brings about. The participation of neighboring existentials in the universals rises to 26% of ALL and 12% of NOR, making this simulation even less accurate than the previous one.

7.1.2. Weight decay and lateral inhibition

Much of the inaccuracy in the L2 weights springs from the retention of weights from the beginning of the simulation in the averages, despite the fact that these early weights are necessarily inaccurate. This drawback can be corrected by allowing the L2 weights to decay. The script **p7.03_LVQQ_decay.m** implements such weight decay by multiplying the updated L2 weights by a factor of 0.9. The graph of the resulting simulation is displayed in Fig. 7.2a, which does not look significantly different from Fig. 7.1b. More revealing are the L2 weights reproduced in Table 7.2a. The gradual erosion of the early weights allows the later and more accurate weights to prevail, so that the universal quantifiers are now correctly associated to the single competitive neuron on either edge of Q7 space.

A second way in which the path of quantifier learning can be made smoother is by allowing the L1 neurons to inhibit one another, as in the network drawn in Fig. 7.3. Whereas activity-dependent weight addition tends to correlate the active neuron with its input, activity-dependent weight subtraction among neurons tends to decorrelate the active neuron from all the other neurons. This propensity of lateral inhibition is evident in Fig. 7.2b, where one can readily appreciate the fact that the paths of competitive neurons now head straight for their final position in Q7 and are separate from one another through the simulation, without the clumping together and wandering about that characterized these same paths in the previous simulations.

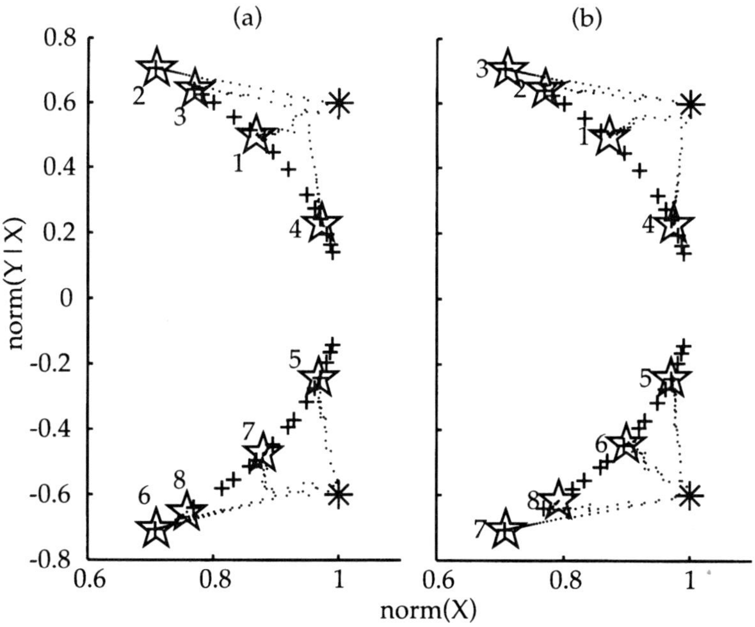

Figure 7.2. Supervised competitive learning of Q7: (a) W2 decay (**p7.04_LVQQ_decay.m**); (b) W1 lateral inhibition (**p7.05_LVQQ_latinh.m**). '+' = accepted point, '*' = initial weight, dot = intermediate weight, star = final weight.

7.1.3. Accuracy and generalization

The next step is to ascertain how well the network generalizes to quantifications on which it was not trained. However, this experiment we can do in our heads, rather than with the computer. The obvious *tierra incognita* in Q space as mapped in Fig. 7.2 is the gap between each universal point and the first existential point next to it. It is easy to image a quantification with a large $|N|$ whose $|P|$ is nearly as large, so that its value in Q space is nearly the maximum, but not quite. An example of such a quantification would be 49/50. The normalized value of this quantification is so close to $[0.71, 0.71]^T$ that it would be captured by the competitive neuron for ALL, yet it is not a valid value for ALL.

The solution that does not require any additional machinery is to simply claim that the network needs to be trained on a larger data sample, which would increase the resolution of the network and close any gaps. Unfortunately, there are at least two flaws in this argument. The philosophical one can be condensed into a single rhetorical question – how much is enough? For any size of training

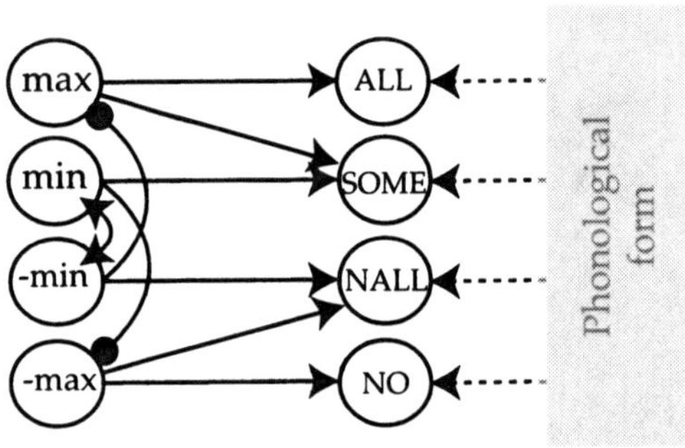

Figure 7.3. Pseudo-LVQ network for the logical quantifiers with L1 inhibition.

Table 7.2. Final W2 for Fig. 7.2.

	(a) Decay of W2, Fig. 7.2a				*(b) Lateral inhibition W1, Fig. 7.2b*			
L1	*ALL*	*SOME*	*NALL*	*NO*	*ALL*	*SOME*	*NALL*	*NO*
1	0	0.20	0	0	0	0.20	0	0
2	1.00	0.15	0	0	0	0.08	0	0
3	0	0.34	0	0	1.00	0.37	0	0
4	0	0.30	0	0	0	0.35	0	0
5	0	0	0.51	0	0	0	0.44	0
6	0	0	0.08	1.00	0	0	0.15	0
7	0	0	0.38	0	0	0	0.11	1.00
8	0	0	0.03	0	0	0	0.31	0

set, there will always be a number just a little larger that will fall into the gap alluded to above. A more empirical flaw has to do with the potential resolution of the network. There should be some maximal quantization of the data space beyond which biological competitive neurons cannot resolve the difference between two numbers, as was mentioned in Sec. 6.2.7.1. And in any event, there is a limited number of neurons that the brain can devote to any given network, so it is not to be expected that the competitive layer can continue growing without bounds as its brain is exposed to ever larger quantifications. One cannot help but conclude that some additional mechanism must be invoked.

On a final note, there is a sense in which the 'error' created by a universal node capturing a high existential simulates quite accurately the way in which people actually use the universals. It corresponds to Labov's "loose" universal quantification, see Sec. 6.3. The outstanding question is, then, how does the brain produce strict universal quantification?

7.2. STRICT UNIVERSAL QUANTIFICATION AS DECORRELATION

The way to achieve strict universal quantification with the mechanisms that have been introduced in this book is to ensure that values of one sign turn off the universal operator of the opposite sign. There are at least three means of effecting this outcome, which are discussed in the next three subsections. The first is to explicitly encode both polarities in the data structure, so that the universal nodes of a LVQ network will be turned off explicitly by an existential input. The second is to augment L2 so that the existential nodes of one sign inhibit the universal node of the opposite sign. This is a kind of antiphase inhibition, a notion reviewed in Chapter 1 in the context of early vision. The third mechanism is similar, being inspired on dendritic processing of selective attention.

One way of conceptualizing the resulting 'grammar' of logical quantification, and presumably all logical operations, is as sensitivity to exceptions. This is especially so for the universal operators: AND/ALL is turned off by a single negative exception, while NOR/NO is turned off by a single positive exception. This overall effect has been anticipated in work on dendritic computation, see Koch et al. (1983), Shepherd and Brayton (1987); Shepherd and Koch (1998), where it is noted that shunting inhibition can effectively veto information from more distal branches of a dendritic tree.

7.2.1. Three-dimensional data

The simplest means of enforcing strict universal quantification is to increase the data object from two to three dimensions, one for the input and two for the output. The idea is that each data point would have a slot for both its positive and its negative value. For instance, the positive existential value that we have been calling $[2, 1]^T$ would be augmented to $[2, 1, -1]^T$. The entire representation for $|N| = 2$, from maximal to minimal y, would be: $[2, 2, 0]^T$, $[2, 1, -1]^T$, to $[2, 0, -2]^T$. Given that each data point has a measure for both polarities, an existential measure of one polarity could inhibit the contrary universal node. A network diagram that includes the requisite inhibitory connections is that of Fig. 7.4. Each competitive neuron that responds to an intermediate measure inhibits the universal node of the opposite polarity in the supervised network.

There are two drawbacks to this proposal, one empirical and one theoretical. The obvious empirical drawback is that the two existential operators are collapsed in the data structure and in the network to such an extent that it seems impossible to disentangle them. The quick rejoinder is that natural language has only one monomorphemic existential operator in any event, OR/SOME, so maybe Fig. 7.4 is not that inaccurate after all. The rejoinder to the rejoinder is that Fig. 7.4 does not confer a privileged position to monomorphemic OR/SOME in detriment to polymorphemic NAND/NALL, but rather that it does not distinguish the two *at all*. It does seem to be an inescapable fact that the two should be distinguished.

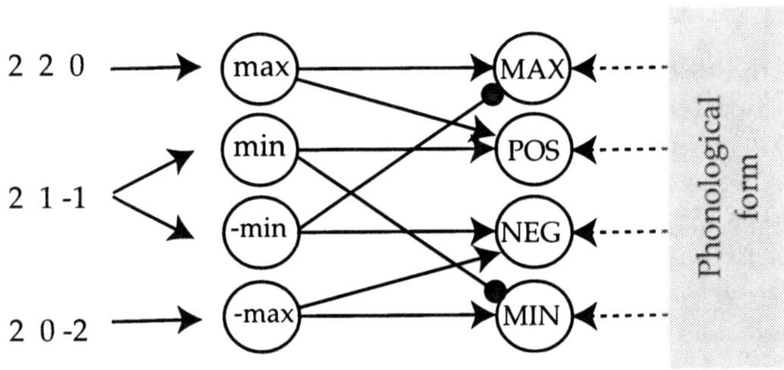

Figure 7.4. Pseudo-LVQ network for the logical operators with 3D input.

Turning to the theoretical problem, we have argued that any proposed semantic representation should match as closely as possible the observed linguistic form. Given that the morphosyntactic form of the logical coordinators and quantifiers at best evidences two argument positions, we are loath to augment it with a third. Or perhaps it would be more accurate to say postulating a third, unobserved argument should only be done as a last resort, after all other alternatives have been exhausted. Since there still remains a family of alternatives, they should be preferred to the innovation embodied in Fig. 7.4.

7.2.2. Antiphase complementation

The alternative is to augment the representational power of W2, so that it becomes the locus of the missing inhibitory connections. In particular, for input of order two, each intermediate node inhibits the maximal antiphase node, as depicted in Fig. 7.5a. However, order two is a special, and somewhat misleading, case of the more general layout illustrated in Fig. 7.5b. In logical operator space, any y value has a complement $1-|y|$, so that the y value and its complement add up to one. Fig. 7.5b indicates that a given y value, such as 2/3 should excite its complement at –1/3 and inhibit all other antiphase y values. The crucial outcome of these new antiphase connections is that they are reciprocal: if node 2/3 excites its complement –1/3, then –1/3 repays the favor by exciting its complement, which is again 2/3, but also inhibiting all of the other antiphase values. In this way, the antiphase connections build a feedback loop that tends to inhibit the universal nodes in favor of nearby existential nodes. This prevents high existential values from being captured by a nearby universal node.

The exact provenance of the antiphase connections is discussed in more detail in Chapter 8, since they are responsible for the fundamental inferences that are drawn from the assertion of a logical operator. In view of their being an important topic of the next chapter, we forego including them in our current

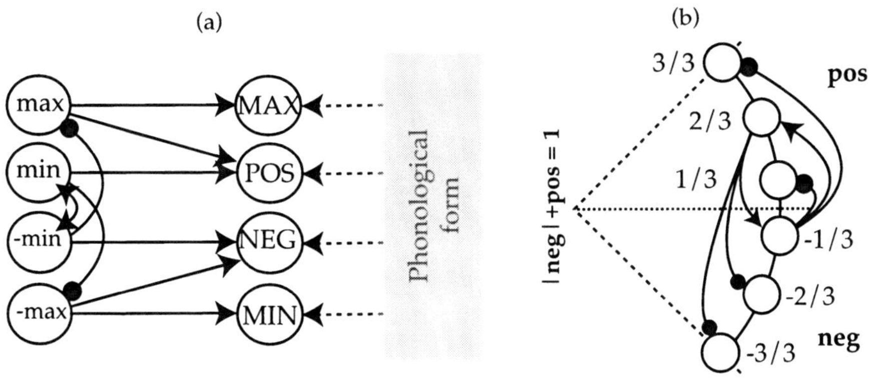

Figure 7.5. Sample antiphase connections in logical-operator space. (a) $|N| = 2$; (b) $|N| = 3$.

simulations. We should at least mention in passing, nevertheless, that these connections are similar in design to the self-organizing feature map of Kohonen (1982) and much posterior work, which is a kind of competitive network in which nearby neurons excite one another and inhibit neurons that are further away.

7.2.3. Selective attention

There is an alternative to be found among the neurological circuits introduced in Chapter 1, in particular the dendritic analysis of selective attention in Sec. 1.2.3.2.3. The simplest adaptation of this system takes advantage of the fact that, in the absence of selective attention, a V4 neuron performs an AND/ALL operation by responding to both visual stimuli. Thus if each visual stimulus is replaced with a subject-predicate symbol $p(n)$, the attentional connections necessary to assimilate the other three functors to the default AND mechanism are illustrated in Fig. 7.6, from which the inhibitory interneurons have been omitted for the sake of simplicity. OR and NAND share the same structure, in which one branch or the other (not shown) is selectively enhanced. If selective attention were applied to both branches simultaneously, each branch would presumably balance the other, and the result would be the inclusive reading of OR (equivalent to AND). NOR falls out from inhibiting both branches.

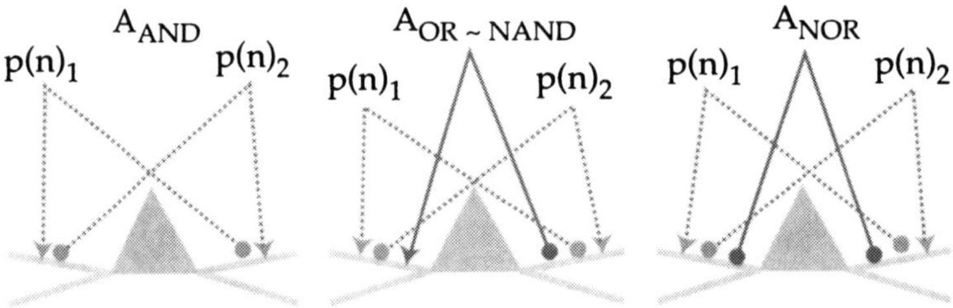

Figure 7.6. Dendritic computation of coordinators from unsigned values.

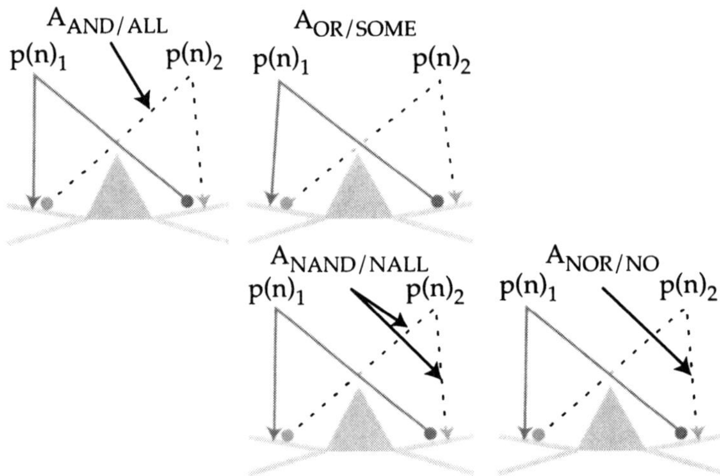

Figure 7.7. Dendritic computation of the logical operators from signed values.

This paradigm has the benefit of supporting the correlational interpretation of logical operations, since AND/ALL can be seen as marking the maximum correlation among inputs. However, it suffers from failing to distinguish OR from NAND – an unavoidable descriptive error. If there were no alternative, we would just have to make do with this system, but fortunately, there is an alternative.

The solution is to draw inspiration from early vision, and in particular the 'early' compartmentalization of photoreceptor responses into ON and OFF pools. The analogy to truth values is to compartmentalize them by polarity: the positive truth values constitute one "sub-stimulus"; the negative ones constitute another, as the theory of signed measures teaches us. Selective attention then has the role of enhancing one polarity over the other, as laid out in Fig. 7.7. Yet the two are not equal. Positive polarity somehow predominates over negative polarity, which is to say that the $+p(n)$neurons respond in a more efficient fashion than the $-p(n)$ neurons. The nature of this asymmetry is discussed in more detail in Chapter 8. This asymmetry is reflected in Fig. 7.7 by the solid lines emanating from $+p(n)$in contrast to the stippled lines emanating from $-p(n)$. Under this assumption, if the input is $-p(n)$ and $+p(n)$ to all of the neurons in the figure and the other connections are ignored, the left branch is excited and the right branch is slightly inhibited. We take that to mean that the neuron winds up accepting the input pattern of $-p(n)$ and $+p(n)$, and defines OR/SOME. Our strategy is to encode the other patterns by selectively enhancing one connection at the expense of the other.

AND/ALL implements this strategy by requiring that selective attention heighten the inhibition emanating from a negative truth value, effectively neutralizing the left branch and turning 'turning off' the neuron in the presence of $-p(n)$. Thus the neuron will fire optimally when there are no negatives in the input, and so accept only AND/ALL patterns. Conversely, for NOR/NO, selective attention heightens the excitation emanating from a negative input and so turns on the neuron in its presence, a positive value with still turn it off. Finally, NAND/NALL is the most complex, because it has to reverse the canonical response of OR/SOME. This entails enhancing both connections to the left branch, so that the neuron responds to a negative input despite the presence of an accompanying positive one.

7.2.4. Summary: AND-NOT logic

In this section, three mechanisms for the implementation of strict universal quantification have been discussed. We have rejected a direct encoding of the phenomenon into the data structure in favor of two indirect, connectivity-based approaches. The goal of both is to decorrelate universal from nearby existential values by making universal operators qualitatively different from existential operators of the same sign. The antiphase inhibition approach mimics mechanisms found in early vision, while the selective attention approach mimics mechanisms postulated for dendritic processing in 'middle' vision. Both of them provide the beginnings of an AND-NOT logic, in which an exception can override the response otherwise expected from the bulk of the data.

In the upcoming chapters we concentrate on antiphase inhibition, just because it is more consistent with, and easier to incorporate into, the rest of our analysis.

7.3. INVARIANT EXTRACTION IN L2

The one glaring flaw of our exposition so far is it does not address the fact that so many generations of logicians and linguists have analyzed the logical operators into two elements, an existential and a universal quantifier. It is easy enough to design a network that groups the convex subclasses of the

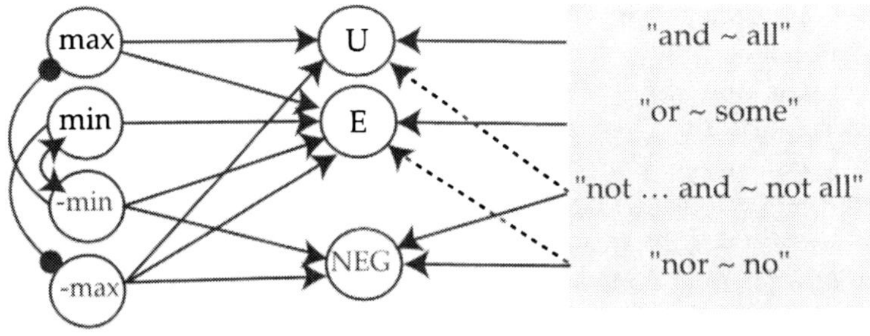

Figure 7.8. Pseudo-LVQ network for lexicalization of the logical operators.

competitive layer in these terms, plus one more for negation. Fig. 7.8 does so, building on the format of Fig. 7.5. For instance, the *–max* node activates both the single universal node *U* and the *NEG* node, to produce a semantic grouping "negative universal" that is realized phonologically as *nor/no*. While this outcome is factually correct, and it can be learned from the actual phonological forms, we do not know why this asymmetric semantic-phonological mapping is preferred over the symmetric four-item mapping that we have used up to now. To state it as precisely as possible, we do not know what evaluation metric the brain uses to choose the symmetric 4x4 network over the asymmetric 4x3 network.

Fortunately, we have already had reason to mention one neurological evaluation metric in the review of the visual system undertaken in Sec. 1.2.2.4. It is the recent hypothesis that the early visual system reduces the redundancy in the visual stream by having a given layer compress together correlated inputs.

While it is not immediately clear how to apply this metric to the semantic-phonological interface, one method is to count connections. In both networks, L1 and L2 are assumed to be completely connected, which is to say that each L1 neuron connects to each L2 neuron. In the 4x4 network, this connectivity produces a matrix of sixteen connections, of which only six are used – i.e. nonzero. These numbers allow us to calculate a **filling fraction** for the L1-L2 matrix, which is simply the number of nonzero connections divided by the total number of connections, 6/16 or 0.38, see Stepanyants et al. (2002). For the 4x3 network, eight connections are used out of a possible twelve, to give a filling fraction of 8/12 or 0.67. Thus the 4x3 network is almost twice as efficient as the 4x4 network. Note that the phonological connections are ignored because they have the same $1/n$ ratio in both cases.

The conclusion is that the 3x4 layout effectively compresses two kinds of the redundancy in logical-operator space and thus extracts two kinds of invariances. The first is the mirror-image redundancy around the x axis, for which L2 compresses together values that share the same magnitude but differ only in polarity and so extracts their invariance with respect to absolute value. The second is the redundancy in sign, for which L2 compresses together values that share negative polarity but differ only in magnitude and so extracts their invariance with respect to negative polarity, leaving positive polarity as the unmarked case.

The decrease in redundancy of the 3x4 layout is paid for by a massive increase in ambiguity. The activation of any L1 neuron excites both E and U, and the activation of any negative neuron excites NEG, so there are cases in which every neuron in L2 will be active. Of course, this ambiguity is managed in linguistic utterances by the scope of negation, so that, for instance, the activation of the L1 node [-max] can be realized as *Not all linguists are gourmets* or *Some linguists are not gourmets*. The combinatorial possibilities of wide or narrow scope of negation with respect to the other two operators add up to a system of six different realizations, of which only three are monomorphemic. The neuromimetic grounding of operator scope is taken up in the next chapter, where we introduce a network architecture that is more adaptable than the LVQ architecture at representing the bidirectional interplay of semantic and phonological information.

7.4. SUMMARY

Having laid out the analyses of the logical coordinators and quantifiers, it is now time to step back and take stock of the overall theory. The theory is comprised of five claims, laid out below.

7.1. The acquisition of logical coordinator/quantifier meanings is an instance of associative pattern classification.

This is the guiding tenet of what we have called pattern-classification semantics. In the particular case that concerns us here, a logical coordinator/quantifier denotation is associated to its phonological label.

7.2. Coordinator/quantifier patterns are better described as ellipsoids than planes.

This claim ties into the arguments for the convexity of natural properties. More specifically, we posit the following:

7.3. Coordinator/quantifier patterns are convex sets of measures of correlation on the unit arc.

We have developed a neurologically inspired means for learning these patterns:

7.4. LVQ is good first approximation to simulate the learning and recognition of the coordinator/quantifier patterns.

Table 7.3. Four-level analysis of the logical operators.

Level	*LOG OP*
Environmental (L0)	Talk about associations between two groups of entities
Computational (L1)	CLASS(CORR(X, Y)), where *X* and *Y* are the cardinality of a group
Algorithmic (L2)	LVQ
Implementational (L3)	Rate-based dynamical system

We say a "good first approximation" and not something stronger, for we will not take the time to augment LVQ to perform antiphase inhibition or redundancy reduction, but rather turn to other issues, of which there are still many.

Finally, not only have we shown how coordinators and quantifiers are similar, but also how they differ:

7.5. Coordinators differ from quantifiers in preserving magnitude information, though not necessarily in COOR space.

These five claims constitute the only analysis that we know of that sets forth a large literature on coordination and quantification in terms of a neurologically-inspired theory of cognition. The next chapter tests these results against an even larger literature on the inferences that can be drawn from the various logical operators.

As a more general summary, Table 7.3 organizes these conclusions within the four- or five-level analytical framework that was generalized from Marr's three-level framework in Sec. 1.4. The affordance that motivates the logical operators are associations between two groups of entities in the environment that people want to talk about. The function that computes this affordance classifies correlations between the cardinalities of the two groups. The algorithm that computes this function is that of Learning Vector Quantization. This algorithm is implemented in a rate-based, neurologically inspired, dynamical system, programmed in MATLAB.

Chapter 8

Inferences among logical operators

So far, the discussion has focused on neuromimetic learning of logical operator patterns. The only output from this process is a signal of recognition, generally some vectorial variation on a 1. But any realistic natural language system would do much more than just recognize a licit usage of a logical element – it would use this information in some way. One type of usage is extremely important to logicians and semanticists: it is the ability to infer some additional information from a legitimate assertion. This chapter takes up those relationships in which one logical element is inferred from another in the Square of Opposition, which are referred to here as "inferences of opposition". We first introduce the Square in terms of the logical quantifiers, then introduce a spreading-activation model that can account for them. The next task is to run IAC simulations of the three main types of oppositional inference, using our own implementations of the IAC algorithm in MATLAB.

8.1. INFERENCES AMONG LOGICAL OPERATORS

Imagine that you are told that Blinker is either a gof or a juppet, and then that Blinker is not a gof. If you are like most adolescents and adults to whom these statements have been posed, you infer that Blinker is a juppet, even though these propositions are undoubtedly unfamiliar to you. As Rader and Sloutsky, 2001, p. 838, conclude, the processing of sentences with logical coordinators cannot solely be a function of content and experience, but rather must depend on more general mechanisms; see also Braine, Reiser, and Rumain (1984) and Evans, Newstead, and Byrne (1993) for reviews.

There are two extended families of such general mechanisms, the semantic, which has its roots in model-theoretic deduction, and the syntactic, which has its roots in proof-theoretic deduction. According to the syntactic account, people extract and represent a logical structure of statements that include logical operators and reasoning consists of a manipulation of this structure. According to the semantic approach, people represent possibilities consistent with a statement, although these representations may deviate from logical prescriptions.

In this section, we briefly review one member of the semantic/model theoretic family. Our focus is drawn to the semantic family in preference to its syntactic contender because it is the one that is more compatible with the neuromimetic analysis assayed in the second and third sections of this chapter.

8.1.1. The Square of Opposition for quantifiers

Aristotelian logic[36] recognizes a series of inferences among the logical quantifiers that are abbreviated in a mnemonic shorthand which assigns the vowels *A, I, E,* and *O* to the four "forms" of quantified clauses, as in (8.1):

8.1 a) A: All F is G.
b) I: Some F is G.
c) E: No F is G.
d) O: Not all F is G / Some F is not G.

These abbreviatory vowels are drawn from the Latin verbs ***A****ff****I****rmo* 'I affirm' and *n****E****g****O*** 'I deny', as a mnemonic aid for the fact that *all/some* are affirmative quantifiers while *no/not all* are negative. However, this monograph finds it more perspicuous to use capitalized operator names for the single letters, though it must be confessed that the single letters have the advantage of not forcing a choice between *some ... not* and *not all* for *O*.

The inferential relationships among the forms can be divided into four groups:

8.2. Contradictories must have opposite truth values
a) ALL ≡ ¬NALL:
All ascetics are philosophers.
≡ It is not the case that not all ascetics are philosophers.
a') ¬ALL ≡ NALL:
Not all ascetics are philosophers.
b) SOME ≡ ¬NO:
Some girls are good archers.
≡ It is not the case that no girls are good archers.
b') ¬SOME ≡ NO:
It is not the case that some girls are good archers.
≡ No girls are good archers.

8.3. Contraries cannot both be true
ALL → ¬NO:
All politicians are liars, therefore, no politicians are not liars.

8.4. Subcontraries cannot both be false.
¬SOME → NALL:
Some logicians are not philosophers, therefore, not all logicians are philosophers.

---

[36] For further background and discussion, see Horn, 1989, p. 10ff; Parry and Hacker, 1991, Ch. 8; Seuren, 1984, pp. 574-581; Parsons (1999), and Levinson, 2000, pp. 77ff.

8.5. Superalterns (universals) entail subalterns (particulars)[37]

a) ALL → SOME:
All prudes are moral persons, therefore, some prudes are moral persons.

b) NO → NALL:
No prudes are moral persons, therefore, some prudes are not moral persons.

These binary relations can be organized graphically by placing each quantifier at one of the corners of a square, making a figure known as the Square of Opposition, seen below in Fig. 8.1.

Despite the symmetry of the square, there is an important difference in the kinds of negation that vertebrate it. The first four relationships set out pairs of opposites, so a judicious use of negation brings out their interdefinability. However, the negation operator in the formula of (8.2b, b′) actually covers up some interesting patterns in the linguistic realization of negation, which are brought out more clearly by the standard laws of quantified propositional logic.

The relationship between the first pair of contradictories translates transparently into the corresponding laws:

8.6 a) ALL ≡ ¬NALL: $\forall x\phi \leftrightarrow \sim\exists x{\sim}\phi$

b) ¬ALL ≡ NALL: $\sim\forall x\phi \leftrightarrow \exists x{\sim}\phi$

The relationship between the second pair of contradictories is contingent on recognizing that NO is equivalent to ALL¬:

8.7 a) SOME ≡ ¬NO (≡ ¬ALL¬): $\exists x\phi \leftrightarrow \sim\forall x{\sim}\phi$

b) ¬SOME ≡ NO (≡ ALL¬): $\sim\exists x\phi \leftrightarrow \forall x{\sim}\phi$

Löbner (1986) refers to the negation of a quantifier in these formula as **outer negation** and negation of the propositional variable ϕ as **inner negation**. While these two types of negation are homophonous in the data considered here, there are instances in which English distinguishes inner negation by a negative affix such as *un-* prefixed to a predicate:

---

[37] The usage of the word "particular" here hides a lengthy debate on whether the particulars have existential import or not, ably summarized in Parsons (1999). Our take on the debate is that the 0-value of trivalent logic takes care of any empty term, and the logical system devised in Seuren et al., 2001, tells us what can be inferred from it. As in the previous occasions when this topic has come up, we remit the reader to Seuren et al. for a more satisfactory treatment.

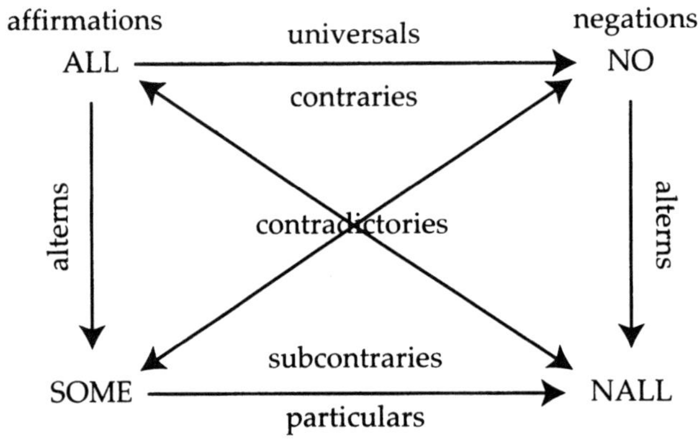

Figure 8.1. The Square of Opposition for the logical quantifiers.

8.8 a) Everybody was kind. ≡ Nobody was unkind. $\forall x\phi \leftrightarrow \sim\exists x\sim\phi$
b) Everybody was unkind. ≡ Nobody was kind. $\forall x\sim\phi \leftrightarrow \sim\exists x\phi$

In both examples, the negative adjective *unkind* translates logically into the inner negation $\sim\phi$, facilitated by the translation of NO as ¬SOME permitted by (8.9b).

8.1.2. A Square of Opposition for coordinators

The logical coordinators map into the Square in the way that we have come to expect from the preceding text. Plugging them into the quantificational Square of Opposition produces the coordinative Square of Fig. 8.2. The inferences that this figure summarizes are written out in (8.9):

8.9 a) AND → OR
a′) NOR → NAND
b) ¬OR ≡ NOR, hence ¬NOR ≡ OR. OR and NOR are contradictories.
b′) ¬AND ≡ NAND, hence ¬NAND ≡ AND. AND and NAND are contradictories.
c) AND → OR ≡ ¬NOR. AND and NOR are contraries. (Both cannot be true.)
c′) ¬OR ≡ NOR → NAND. OR and NAND are subcontraries. (Both cannot be false.)

The next few paragraphs examine whether they are accurate or not.

The inference from AND to OR is extremely well known in the logical literature as one of the fundamental rules of inference, that of conjunction

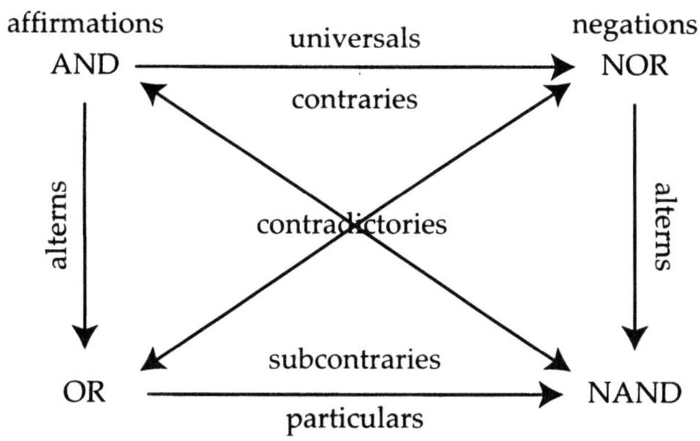

Figure 8.2. The Square of Opposition for the logical coordinators.

elimination or simplification. For instance, from the assumption in (8.10a) we are entitled to draw the inferences of (8.10b) or (8.10b′):

8.10 a) Socrates is mortal, and Plato is mortal.
b) ∴ Socrates is mortal.
b′) ∴ Plato is mortal.

These two conclusions are both potential denotations of OR, as seen in the following restatement of 8.10:

8.11. Socrates is mortal, or Plato is mortal.

In other words, conjunction elimination allows us to infer one of the disjuncts of OR from any of the conjuncts of AND.

The inference from NOR to NAND instantiates disjunction reduction of negatives, as exemplified in (8.12):

8.12 a) Neither Socrates nor Plato is immortal.
b) ∴ Socrates is not immortal.
b′) ∴ Plato is not immortal.

These two conclusions are both potential denotations of NAND, as seen in the following restatement of 8.12:

8.13. It is not the case that Socrates is immortal and that Plato is immortal.

In other words, disjunction elimination allows us to infer one of the conjuncts of NAND from any of the disjuncts of NOR.

Turning to the coordinators, the equivalencies established by negation between the contradictory coordinators have been taken for granted in the previous discussion, so let us bring them out here. If the focus of the matrix clause "it is not the case that" is placed on the subject of the embedded clauses, then it is clear that (8.9b), repeated here as (8.14a), is captured by the equivalencies between (8.14b, b′) and (8.14c, c′):

8.14 a) ¬OR ≡ NOR, hence ¬NOR ≡ OR. OR and NOR are contradictories.
b) It is not the case that either Elmer or Hortense smelled the custard.
b′) ≡ Neither Elmer nor Hortense smelled the custard.
c) It is not the case that neither Elmer nor Hortense smelled the custard.
c′) ≡ Either Elmer or Hortense smelled the custard.

For the second case, the invocation of NAND means that there is no one-to-one correspondence between morphemes to be set up in order to test the predicted inferences. (8.15) is the closest that we can come:

8.15 a) ¬AND ≡ NAND, hence ¬NAND ≡ AND. AND and NAND are contradictories.
b) It is not the case that Elmer and Hortense smelled the custard.
b′) ≡ Elmer and Hortense did not smell the custard.
c) Elmer and Hortense did not smell the custard.
c′) ≡ It is not the case that Elmer and Hortense smelled the custard.

One's intuitions about these sentences are in accord with the predictions of the corresponding contradictory quantifiers.

Granting conjunction elimination from AND to OR, inferences between the contrary coordinators are exactly as predicted.

8.16 a) AND → OR ≡ ¬NOR. AND and NOR are contraries. (Both cannot be true.)
b) Elmer and Hortense smelled the custard.
c) → Elmer or Hortense smelled the custard.
d) ≡ It is not the case that Elmer and Hortense smelled the custard.

The inferences between the subcontrary coordinators are more straightforward, and are also exactly as predicted:

8.17 a) ¬OR ≡ NOR → NAND. OR and NAND are subcontraries. (Both cannot be false.)
b) It is not the case that Elmer or Hortense smelled the custard.

c) ≡ Neither Elmer nor Hortense smelled the custard.
d) → Elmer and Hortense did not smell the custard.

Thus our hypothesis of a core similarity between quantification and coordination finds additional support in the parallel between the patterns of inference that they support.

8.1.3. Reasoning and cognitive psychology

Having laid out a regimentation of the inferences that we want to account for, the next step is to introduce the theoretical approaches in whose terms an account should be couched. As was mentioned in the introduction to this chapter, there are two: the syntactic/proof-theoretic family and the semantic or model-theoretic family. They are introduced in this order.[38]

8.1.3.1. Syntactic/proof-theoretic deduction

Following the overview in Rader and Sloutsky, 2001, p. 838, a syntactic/proof theory claims that an untrained reasoner converts discourse into a form defined by propositional logic, from which deductive inferences can be drawn, see Braine and O'Brien (1998) and Rips (1994). If the form matches one or more inference schemas stored in memory, the reasoner infers the conclusion licensed by the relevant schema(s). For instance, the example with which the chapter opened, *Blinker is either a gof or a juppet; Blinker is not a gof, therefore Blinker is a juppet*, matches a **disjunction-elimination schema** of the form *A or B; not-A, therefore B*.

Clearly, the logical schema should be abstracted regardless of the content of the statements. Several theorists have argued that such schemas are similar to grammatical frames in that they apply in an online, obligatory fashion whenever premises are present in working memory, see Braine et al. (1984), Lea (1995), and Lea et al. (1990). Automatic application of inference schemas implies that the abstraction of logical form should also be automatic.

Syntactic/proof theory makes several predictions about how deduction should work. The paramount one is that a reasoner must represent the logical syntax of each proposition in an argument, or else the appropriate inference schemas cannot be applied. To represent logical syntax, deep logical form must be extracted from surface linguistic form. Although errors may occur in extraction, these errors should not show systematic tendencies. Errors in reasoning should instead be a function of the number of inference schemas

---

[38] For further background on model-theoretic and proof-theoretic deduction, see the corresponding entries in the Routledge Encyclopedia of Philosophy, Hodges (1998) and Sieg (1998), though these articles do not address the difference between the two approaches. The syntactic/proof-theoretic family is exemplified in Chapter 11.

invoked, as well as the difficulty of each schema, see Braine et al. (1984) and Rips (1994).

However, experiments have shown that reasoners make errors that do not appear to spring from the complexity of the logical derivation. The easiest to explain in a handful of sentences is the fact that reasoning from conjunctive premises (universal coordination or AND) is associated with almost no errors, whereas reasoning from disjunctive premises (existential coordination or OR) is associated with many errors, see again Braine et al. (1984) and Rips (1994). Moreover, there is an unexpected systematicity to the latter: disjunctive errors seem to stem from a failure to represent all the possibilities consistent with the premises, see Evans et al. (1995), Johnson-Laird et al. (1992), and Klauer and Oberauer (1995). What is even more perplexing for the syntactic/proof-theoretic approach is that if a reasoner is provided with an external aid that allows possibilities to be represented visually, performance with disjunctive premises improves, as reported in Bauer and Johnson-Laird (1993), and Sloutsky and Goldvarg (1999). There are other predictions made by syntactic/proof-theoretic approaches to deduction that also fail to find empirical confirmation, for which the reader is remitted to the introductory paragraphs of Rader and Sloutsky (2001) for a recent summary.

8.1.3.2. Semantic/model-theoretic deduction and Mental Models

Continuing within the realm of experimental psychology, the most well-known alternative, that is, semantic or model-theoretic approach, is that of Mental Models, which can ultimately be traced back to Kenneth Craik's suggestion in 1943 that the mind constructs "small-scale models" of reality that it uses to anticipate events, according to Johnson-Laird (2001) and the Mental Models Website.[39] Mental Models are founded on the three assumptions listed in (8.18):

8.18 a) A mental model represents one possibility, capturing what is common to all the different ways in which the possibility may occur.
b) Mental models explicitly represent what is true according to the premises, but by default not what is false.
c) Deductive reasoning depends on mental models.

In the words of Johnson-Laird (2001), a mental model is iconic like a diagram, in that its parts correspond to the parts of what it represents, and its structure corresponds to the structure of the possibility. By way of illustration, the *gof or*

[39] See the book's website for the address.

juppet example calls for the two mental models set forth without parenthesis in (8.19) to represent the two premises:

8.19.	Blinker-gof	(¬Blinker-juppet)
	(¬Blinker-gof)	Blinker-juppet

The first line denotes a mental model of the possibility in which Blinker is a gof, and the second denotes a model of the possibility in which Blinker is a juppet. Note that (8.19) enters the false versions of the two models into their respective niches, in order to create a **fully explicit** model. This parenthetical material is introduced with the intention of reminding the reader that it is *not* there in the mental model – in accord with (8.18b), the default format is to omit false premises. It is only with the incorporation of the explicitly negated third model into the argument, *¬Blinker-gof,* that we get an indication of what is false. Its combination with (8.19) leads to the elimination of the model in the first line, leaving the second line, with the implicit negation now explicit, as the conclusion:

8.20. ¬Blinker-gof Blinker-juppet

With respect to why a reasoner should be more accurate at conjunctive rather than disjunctive deduction, consider the fact that the model for the conjunction of the premises used in our examples is the single line of (8.21):

8.21. Blinker-gof Blinker-juppet

The difference between this expression and the disjunction of (8.19) is that the latter is meant to express two entries in working memory, while (8.21) expresses but a single one. Thus the conjunctive format of (8.21) achieves a more efficient encoding in memory and consequently imposes less strain on limited storage resources. Moreover, Rader and Sloutsky (2001) argue that conjunctive efficiency imposes a bias on cognition in that conjunctive forms are recalled and recognized more accurately than other logical forms, and other logical forms are recalled and recognized as if they were conjunctions. In the terminology of dynamical systems theory introduced in Chapter 2 to analyze the Hodgkin-Huxley equations, we can say that the conjunctive bias acts as an attractor that draws other logical items into its format.

8.1.3.3. Modest vs. robust deduction?

The given and take between syntactic/proof-theoretic deduction and semantic/model-theoretic deduction is reminiscent of the debate between modest and robust theories of semantics introduced in Chapter 1. One can image the advocate of proof-theoretic deduction dismissing the empirical advantages of Mental Model theory as simply highlighting performance limitations of working memory that are orthogonal to deductive competence.

As the reader may have anticipated, our approach will partake of the semantic/model-theoretic approach, though not exactly in the form of Mental Models, but rather in the form of a computational paradigm that is more consistent with what is known, or at least suspected, about how the brain performs deduction.

8.2. SPREADING ACTIVATION GRAMMAR

We first return to Shastri's discussion of connectionism for guidance, and then look to recent work by Ray Jackendoff on the organization of grammar before proposing our own synthesis of all this material.

8.2.1. Shastri on connectionist reasoning

Shastri, 1990, p. 79ff, and (1991) argues that the scalarity of messages in a connectionist network precludes them from bearing symbolic content and so requires that all relevant distinctions, at every level of granularity, be built into the network. Moreover, in the absence of a distinct interpreter that mediates between the representation and the processes that operate on it, the pattern of connectivity, the weights on links, and the computational characteristics of nodes not only represent domain knowledge but also encode the retrieval and inferential processes that manipulate this knowledge. In other words, there is a strong coupling between the nature of representation, the nature of inference and the degree of efficiency with which inferences are carried out. Issues of implementation and performance cannot be separated from issues of representation and expressiveness.

Such considerations lead Shastri, 1991, p. 264, to state that connectionism supports the following metaphor for reasoning:

8.22. Assign a processing element to each unit of information and express each inferential dependency between pieces of information by explicitly linking the appropriate nodes.

More precisely, Shastri claims that the core features of connectionism impose the following constraints on knowledge representation and reasoning:

8.23 a) Inference is the spread of activation in a parallel network.
b) Messages lack symbolic content and are restricted to "who is talking, and how loudly?"
c) Representations make up for the lack of message content by being explicit, multifaceted and fine grained.
d) There is a strong coupling between a representation and the inferences that it can be expected to support.

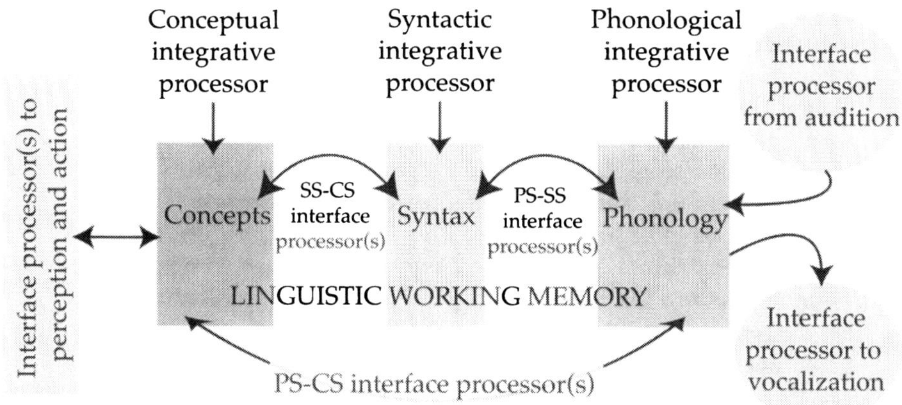

Figure 8.3. A parallel generative grammar implemented in the processing architecture of Jackendoff, 2002, Fig. 7.2.

e) Representations display this coupling by being vivid and directly mirroring the inferential structure of the domain.

f) Reasoning is evidential and/or probabilistic and can be modeled as constraint satisfaction and/or energy minimization.

g) Constraints may need to be placed on representations in order to support efficient reasoning.

The rest of this chapter illustrates many of these proposals by implementing inferences of opposition among quantifiers and coordinators as parallel spreading activation in a network.

8.2.2. Jackendoff (2002) on the organization of a grammar

But which way should activation spread? Here Jackendoff, 2002, points the way. In Part II, Jackendoff argues for a parallel constraint-based architecture of generative grammar, in contrast to the serial architectures that have been the standard assumption since Chomsky (1965). We unfortunately do not have the space here to delve into the particulars of Jackendoff's reasoning, so we must content ourselves with a short review of the global design presented in Fig. 8.3.

The core of the structure is the notion of a linguistic working memory divided into three buffers, represented by the contiguous boxes in the center of the diagram. Each box is labeled with its contents, drawn from the sub-modules of a generative grammar, with the single exception of the replacement of Semantics with Concepts. Mediating between each buffer is an interface processor. Likewise, mediating between the grammar module and the rest of the cognitive system are other interface processors.

In a nutshell, these components work together in the following way:

> In language production, the processor goes in the other direction, starting with an intended message in conceptual structure. The conceptual processor sends a cell to the lexicon: "Does anybody out there mean *this*?" And various candidates raise their hands: "Me!" and thereby become activated. But by virtue of becoming activated, they also activate their syntax and phonology, and thus establish partial structures in those domains and partial linking to the intended message. The phrasal constraints proliferate structure through syntax to phonology, until there is a complete phonological structure that can be sent off to the interface to the motor system, there to be pronounced. (Jackendoff, 2002, p. 201)

The preceding page of Jackendoff's text describes a more detailed derivation in the opposite direction, that of speech perception, and following pages sketch how certain grammatical phenomena are handled within the confines of the system.

Besides this neuro-friendly organization of grammar, perhaps the most relevant aspect of Jackendoff's model is that it is explicitly meant to marry linguistic competence to performance, or as Jackendoff himself says, "how stored pieces are used online to build combinatorial linguistic structures in working memory during speech perception and production." (ibid. p. 196) We will return to this topic below, but let us go ahead and introduce our own neuromimetic version of a generative grammar.

8.2.3. Spreading Activation Grammar

To simulate a range of grammatical phenomena, some model of an entire grammar needs to be devised. It should have at least the format of Fig. 8.4. There are at least four components: conceptual structure, semantics, morphosyntax, and phonology, much as in Jackendoff's model. All of the connections between components are excitatory at a standard weight of +1. Connections within components can be excitatory or inhibitory, but within pools of units only inhibitory connections are allowed, at a standard weight of –1. The standard weights can be altered to account for empirical asymmetries, but our first guess is always to set them at ±1. This makes analysis of the network much more transparent.

Most of the connections are self-evident, once one grasps their interpretation. Consider the two-way excitation between the semantic unit 'X' and the morphosyntactic unit [X]. This represents the intuition that 'X' is the meaning of [X] – you cannot activate one without activating the other. Likewise, /X/ is the phonological form of [X] – you cannot activate one without activating the other. There is consequently a path of activation connecting 'X', [X], and /X/. In this way, the model captures the structure of a lexical item as mutual activation

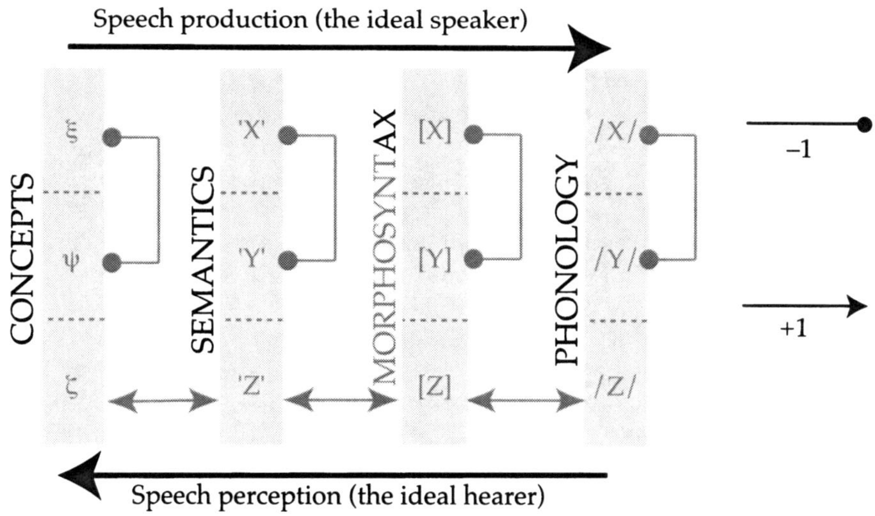

Figure 8.4. Spreading-activation grammar.

among elements in different components – echoing almost exactly what Jackendoff says in the quote above.

Conversely, consider the two-way inhibition between phonological units /Y/ and /Z/. This captures the intuition that these two units are mutually exclusive, which is to say that they compete for the occupancy of some slot in a specific context. Neurophysiologically, this corresponds to the notion of inhibition introduced in Sec. 1.2.2.1, which is integral to the competitive learning rule as explained in Chapter 5. It is the factor that imposes order within a component and is responsible for much of the self-organization of cognitive structure. The importance of this factor has not been recognized for linguistics, and it is what most distinguishes spreading-activation grammar from competing theories.

The existence of two-way inter-component excitation rules in the possibility of a flux of activation among components. This possibility is taken to model the ideal speaker-hearer. In particular, the flow of activation from side to side models the task of the ideal speaker to clothe a meaning in phonological form, while the flow of activation from bottom to top models the task of the ideal hearer to recover a meaning from a phonological form. In this way, spreading activation sets up an input-output mapping across the intervening units of the network.

Nevertheless, the only components of the network that are available for inspection to Jackendoff's extra-linguistic interface processors are its left and right edges. These processors can activate units in the conceptual component to see which units become active in the phonology, or vice versa, but the outcomes are always phonological or conceptual. To such a peripheral observer, the network is an impenetrable black box in which the activations of the semantic and morphosyntactic units are unseen. In neuromimetic parlance, the morphosyntactic component is said to consist of **hidden** units, see Sec. 5.3.3.1. Of course, the omniscient 'global' observer (the linguist sitting in her swivel chair) *can* see the activations of the semantic and morphosyntactic components, but one must distinguish the relative perspectives of the two observers. Given that hidden units are not available for local observation, one may argue that they should not be taken into consideration for the construction of the theoretical framework. In fact, schools of analysis affiliated with Cognitive Linguistics – see Chapter 11 – have argued for a direct phonological-conceptual mapping, without mediation by an intermediate syntactic level. This report hews to the traditional claim of generative grammar that such a hidden layers are indeed justified.

8.2.4. Interactive Competition and Activation

It was explained in Chapter 2 that the notions of neurological excitation and inhibition have simple mathematical interpretations. This mathematical tractability suggests that some clever soul could develop a computer program that would simulate a network of neurons and the spread of activation through it. We would therefore have a tool for the objective verification of 'pencil-and-paper' neurology, as well as a need to be extremely precise and explicit in our formulations, with the corresponding gains in replicability and falsifiability.

That some clever soul could develop a computer program for the simulation of spreading activation is indeed the case. The Interactive Competition and Activation (IAC) program of McClelland (1981, 1991), McClelland and Rumelhart (1981, 1989) and Rumelhart and McClelland (1981) is currently implemented as one of the constraint-satisfaction paradigms of PDP++, see O'Reilly and Munakata (2000) as well as the PDP++ Software Home Page.[40] However, since we wish to have more control over the network than is allowed in PDP++, and in order to do all of our programming in the same language, we have developed a MATLAB implementation of the IAC algorithm that is discussed in the upcoming sections.

Fig. 8.5 depicts the initial form of the IAC network that is used in this chapter. It can be understood as consolidating two LVQ networks into a single model of the logical operators, the model-to-phonology LVQ network studied in the previous chapters, plus a converse phonology-to-model network. The morphosyntactic component is omitted in favor of the phonological component,

[40] Again, see the book's website for the address.

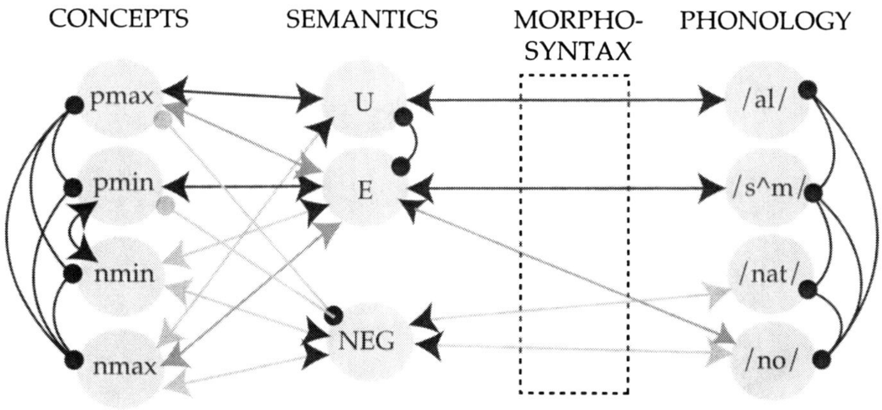

Figure 8.5. Initial IAC network for the logical operators, using the phonological form of the quantifiers. Note that the 'p' of *pmax* stands for 'positive', and the 'n' of *nmax* stands for 'negative'.

in keeping with the rest of the monograph. Starting at the left, the competitive neurons that map logical-operator space are reduced to four and labeled according to their place on the scale from the lowest value of $[0.71, -0.71]^T$ to the highest value of $[0.7, 0.71]^T$. The semantic layer contains the three operators, and the phonological layer consists of the phonological form of the logical quantifiers, plus a unit for negation.

The weights of the connections are indicated by the darkness of the line and its termination. The black elements are set to the highest weight for their kind, 0.35 for interlayer excitation and -0.35 for interlayer inhibition, and -0.25 for intralayer (i.e. lateral) inhibition. Lighter elements signal a reduction in these basic values. All of the excitatory connections from *E* to a dispreferred universal node are drawn in a lighter shade to indicate their reduction to 0.9. In addition, all of the excitatory connections emanating from *NEG* are drawn in an even lighter shade to indicate their reduction by half. This is to balance the negative with respect to the positive nodes, given that the negative nodes receive one additional excitatory input, namely *NEG*. Not taking this asymmetry into account wrecks havoc on the simulations, since any given negative node will receive a greater and slightly different pattern of excitation from the complementary positive node, leading to significant divergences in their behavior. A similar asymmetry between the positives and negatives arises in the inhibition of the positive end of the scale by *NEG*. Maintaining these connections at the standard value of -0.35 also leads to distortion of their expected behavior (especially *pmin*), so these connections are also reduced by half.

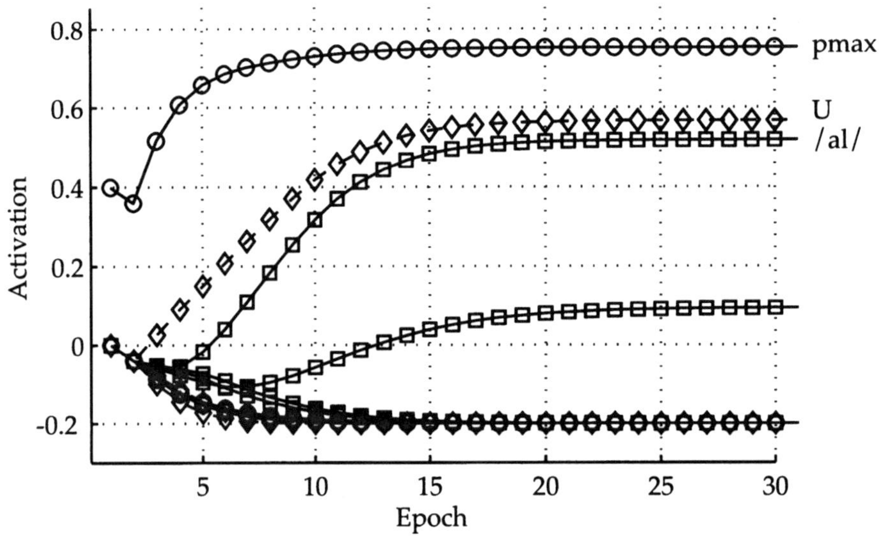

Figure 8.6. Response of basic IAC network to external input of 0.4 to *pmax*.

8.2.4.1. An example

To begin a simulation, external activation is applied to some unit or units, for instance, to neuron *pmax* in Fig. 8.5. Then activation is allowed to spread through the network for some number of epochs, say 30. Fig. 8.6. plots the network response to this initial state. The names at the far right of the plot label the neurons whose activation changes the most. Some activation is lost at each layer, so the gradual drop in the equilibrium value of the most active units indexes the path of activation through the network. We see that activation spreads from the scalar model, through semantic node *U*, to the phonological node */al/*. We interpret this to mean that */al/* recognizes *pmax*, which is correct and is tantamount to submitting $[1, 0]^T$ to the LVQ network of the previous chapter. The next few subsections sketch the mathematical calculations performed by an IAC simulation so that the reader may understand how this result is derived.

8.2.4.2. The calculation of input to a unit

Having illustrated what an IAC network does, let us pause for a few paragraphs to consider how it does it. Units in an IAC network change their activation based on a function that takes into account both the current activation of the unit and the net input to the unit from other units or from outside the network. The net input to a particular unit, say unit *i*, is the sum of the influences of all of the other units in the network plus any input from without.

The influence of some other unit, say unit *j*, on unit *i* is the product of the output of unit *j* times the strength or weight of the connection to unit *i* from unit *j*. This description can be condensed into the following equation:

8.24. $$net_i = \sum_j \left(output_j \times w_{ij}\right) + extinput_i$$

a) net_i = the net input to unit *i*
b) $output_j$ = the output from unit *j*
c) w_{ij} = the weight of the connection to unit *i* from unit *j*
d) $\sum_j \left(output_j \times w_{ij}\right)$ = the sum of weighted outputs from all units *j*, i.e. all the units connected to unit *i*
e) $extinput_i$ = the input to unit *i* from outside the network

A couple of clarifications are in order. In the IAC implementation, $output_j = [a_j]^+$, where a_j refers to the activation of unit *j*, and $[a_j]^+$ has the value a_j for all a_j greater than 0, and 0 otherwise.

8.2.4.3. The calculation of change in activation of a unit

Once the net input to a unit has been computed, the resulting change in its activation is given by (8.25):

8.25 a) If net_i is greater than 0: $\Delta a_i = (max - a_i)net_i - decay(a_i - rest)$.
b) Otherwise: $\Delta a_i = (a_i - min)net_i - decay(a_i - rest)$.

Here, Δa_i represents the change in activation of unit *i*, and there are four parameters, *max, min, rest*, and *decay*. In general, *max* is set to 1, $min \leq rest \leq 0$, and *decay* is set between 0 and 1. Finally, a_i is assumed to start, and to stay, within the interval [*min max*].

8.2.4.4. The evolution of change in activation of a network

The successive changes in the activation of a network leads to three broad outcomes, known as equilibrium, hysteresis, and resonance.

We say that the activation of unit *i* is at **equilibrium** when it stops changing, that is, when its change Δa_i equals zero. For positive activation, the mathematical value of equilibrium can be found by setting Δa_i to zero in (8.25a) and solving for a_i:

8.26 a) $0 = (max - a_i)net_i - decay(a_i - rest)$
b) $= (max)(net_i) + (rest)(decay) - a_i(net_i + decay)$

c) $a_i = (\text{max})(\text{net}_i) + (\text{rest})(\text{decay}) / \text{net}_i + \text{decay}$

Setting the parameters *max* to 1 and *rest* to 0 simplifies the final equation to the next one:

8.27. $a_i = \text{net}_i / \text{net}_i + \text{decay}$

Thus the activation of a unit stops changing when its value becomes equal to the ratio of the net input divided by the net input plus decay. Analogous results are obtained for negative net input, see McClelland & Rumelhart, 1989, p. 14.

This analysis of equilibrium depends on a fixed net input, but in a network, the net input to a unit changes as the units in its context respond to their input. The other two outcomes play on this interaction.

The first effect of variation in net input is to amplify differences between units in the same pool. Consider two units *a* and *b* in the same pool, the first of which is receiving more excitation from outside the pool that the second. If γ represents the strength of the inhibition each unit exerts on the other, then the net input to each unit is given by subtracting the other's output from its excitation:

8.28 a) $\text{net}_a = e_a - \gamma(\text{output}_b)$

b) $\text{net}_b = e_b - \gamma(\text{output}_a)$

As long as the activations stay positive, $output_i = a_i$, so the previous pair of equations yields the following pair:

8.29 a) $\text{net}_a = e_a - \gamma(a_b)$

b) $\text{net}_b = e_b - \gamma(a_a)$

The larger initial activation of unit *a* quickly overwhelms the smaller activation of unit *b*, so that the excitation of *b* is quickly suppressed. By the same token, the extinction of *b*'s activation quickly suppresses the inhibitory effect of *b* on *a*. Grossberg (1976) calls this process the "rich get richer" effect, and it is what produces competition in an IAC network: units with slight initial advantages in their external inputs amplify this advantage over their competitors.

There is a another, more extreme, way in which the "rich get richer" effect can be achieved in an IAC network, namely by forcing a unit to be on by giving it a large value for its external input. This value can inhibit all of the other competing neurons in the same pool, with the result that the forced unit 'wins' from the very beginning. Both the forced and the gradual winners are examples of the network behavior known as **hysteresis**: prior states of networks tend to put them into states that can delay or even block the effects of new inputs.

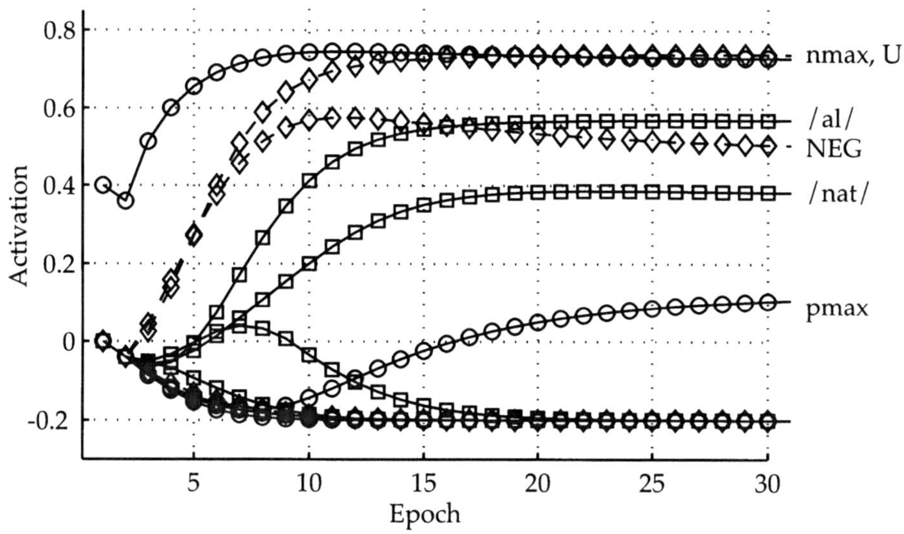

Figure 8.7. Response of basic IAC net to external input of 0.4 to *nmax*.

The second effect of variation in net input is to amplify activation between units in different pools. Consider two units *a* and *b* in different pools with mutually excitatory connections. Once one of them becomes active, it will tend to keep the other active. The network therefore winds up sustaining mutually excitatory interactions and so 'resonates', just as certain frequencies resonate in a sound chamber. Such resonance can be strong enough to overcome the decay parameter, see McClelland and Rumelhart, 1989, p. 16, for an example, so that a pattern of activation can be maintained even in the absence of continuing input.

As with hysteresis, a special case of resonance can be obtained in an IAC network by setting a unit's external input to 1. Forcing a unit to be on in this way serves to activate the other units that are fed by excitatory connections. The result is that units that are not activated directly by external input become active. In this way, a full pattern of activation can be retrieved from partial input, a phenomenon known as **pattern completion**.

8.2.5. Activation spreading from semantics to phonology

With this mathematical background, we can return to the topic that interests us, namely the spread of activation from semantics to phonology as a way of simulating a LVQ network in a more realistic forwards/backwards circuit. This is the unavoidable preliminary step to using IAC as a framework for the simulation of inferencing as the spread of activation from phonology to semantics.

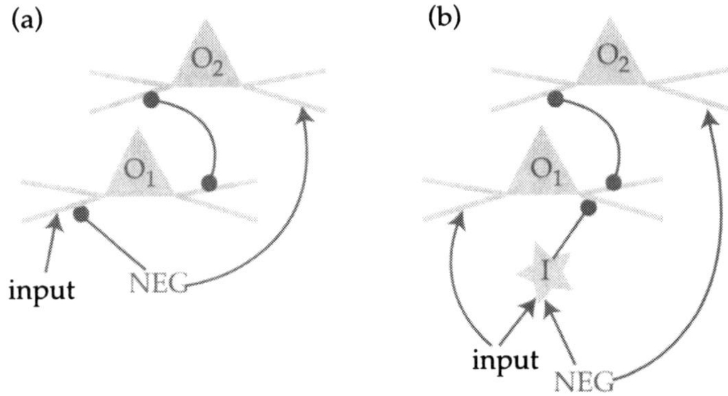

Figure 8.8. Two circuits for negation of O_1. (a) direct divisive inhibition (*input*/*NEG*) of O_1; (b) indirect subtractive inhibition of O_1, enabled by multiplicative excitation (*input* * *NEG*) of *I*.

8.2.5.1. The challenge of negation

Despite the success of the network diagrammed in Fig. 8.5 at deriving the correct response to *pmax* summarized in Fig. 8.6, it fails terribly on *nmax*, as plotted in Fig. 8.7. Three of the four phonological units become almost equally active, an outcome whose most charitable interpretation is as an inability to decide between */nat al/* and */s^m/*.

The reason for this failure stems from the passivity of *NEG*. While it correctly mediates the negative scale and its phonological realization, it needs to do more, namely to turn off the expected semantic operator and turn on its complement. This is not an operation that the standard form of the IAC algorithm lends itself to. However, the discussion of dendritic processing does supply enough tools to design a circuit or two that will perform the necessary changes. Two options are depicted in Fig. 8.8.

The option of Fig. 8.8a is the simpler, in that it can be implemented as divisive inhibition by *NEG* of the input that neuron O_1 receives, i.e. *input*/*NEG*. The greater the activation of *NEG*, the more it reduces the input to O_1. Since *NEG* simultaneously excites O_2, there should be a point at which O_2 comes to dominate O_1 through lateral inhibition and effectively turn it off, even though O_1 does not receive any direct input of its own. The one drawback of this circuit is that IAC neurons are never entirely inactive, so that the low background level of activity of *NEG* will greatly enhance the effect of the input through division by a small quantity.

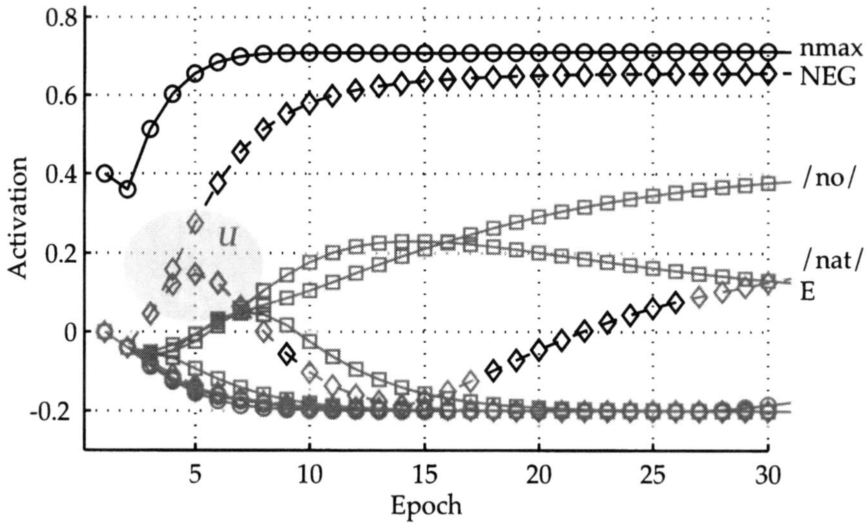

Figure 8.9. Response of enhanced IAC net to external input of 0.4 to *nmax*.

The alternative of Fig. 8.8b introduces one of the star-shaped inhibitory interneurons that are implicit in Fig. 8.8a in order to provide a locus for multiplicative excitation. In accord with our model of dendritic processing, the input to O_1 and the activity of *NEG* are multiplied together in order to drive the interneuron, which consequently only applies *effective* subtractive inhibition to O_1 when *input* and *NEG* are at their highest values. As in Fig. 8.8a, *NEG* also simultaneously excites O_2, so that there should be a point at which O_2 overcomes O_1 through their relation of mutual lateral inhibition, despite the lack of any direct input to O_2. This regime does not suffer from the excessive manipulation of the input pointed out for Fig. 8.8a, and it has the advantage of instantiating a learning rule for the afferent interneuron connections that has already been discussed, namely the Gaussian distance rule devised for spines in Chapter 5.

It is therefore incorporated into the implementation of the IAC network pictured in Fig. 8.5 through an *ad hoc* intermediary step in the calculation of the network activation which subtracts the product of the appropriate input and *NEG* from *U* and *E*. Fig. 8.5 is not redrawn here to reflect this augmentation, so it is up to the reader to bear in mind the appropriate amalgam of Fig. 8.5 and Fig. 8.8b during the upcoming discussions.

With this enhancement, the simulation of external input to *nmax* plotted in Fig. 8.7 produces the new plot of Fig. 8.9. Activation spreads from *nmax* to *NEG*

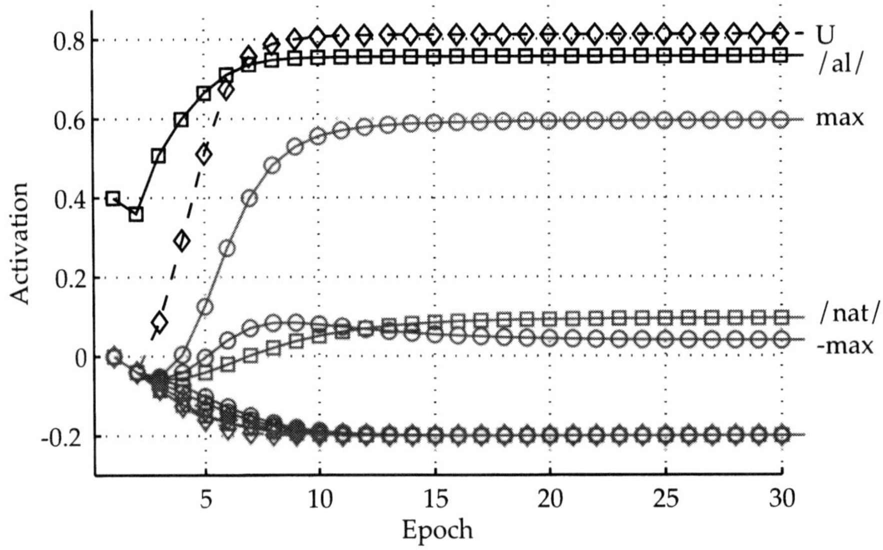

Figure 8.10. Response of enhanced IAC net to external input of 0.4 to */al/*.

and *E*, and then to */no/* and in an attenuated form to */nat/*. This is the correct derivation. What is difficult to discern amid the welter of falling units is the sharp rise and fall of *U* in the three to eight epoch period. This hump, highlighted by the shaded circle, demonstrates the effectiveness of the negation operation: *U* is initially excited by *nmax*, but its inhibitory interneuron is excited by the convergence of *nmax* and *NEG* so as to quickly shut it down.

8.2.6. Activation spreading from phonology to semantics

The spread of activation in the opposite direction, from phonology to semantics, is the cornerstone of our theory of inferencing, so let us first explain how it is accomplished in the enhanced implementation of Fig. 8.5. To demonstrate the existential operator, external activation is applied to */al/* and allowed to circulate for 30 epochs. The network reacts as depicted in Fig. 8.10. Activation spreads from */al/* to *U* to *max*, but given the connection from *U* to *nmax*, *nmax* and the units that it connects to, such as */nat/*, also become slightly active. This need not be considered an error if it is assumed that we are only conscious of activations that cross some threshold, such as 0.5. The other phonological nodes work just as well, as the reader will have the chance to appreciate in the upcoming discussion of inferencing.

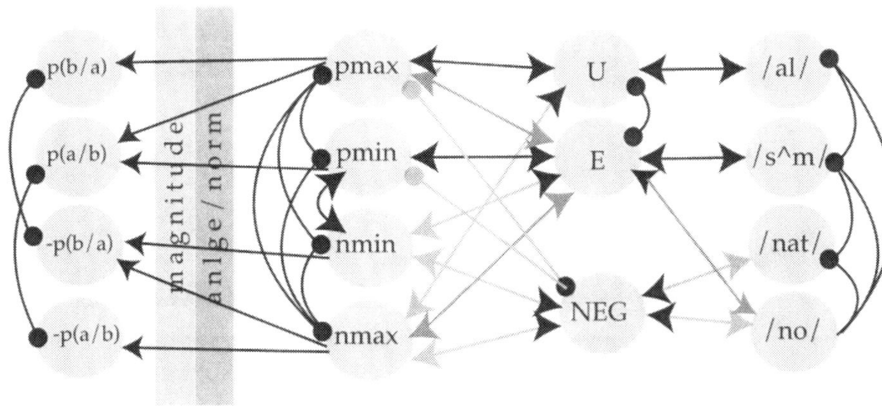

Figure 8.11. Beyond preprocessing in the IAC network.

8.2.7. Extending the network beyond the preprocessing module

In order to account for the inferences organized into the Square of Opposition, the IAC network needs to be extended to the right, past the initial phase of preprocessing that started us off with LVQ. The problem is that it was shown in Chapter 3 that the calculation of angle and norm discards the measure of magnitude, which prevents the derivation from proceeding any further in this direction. However, it was also mentioned in practically the same breath that some memory of the magnitude must be retained in any event. For the quantifiers, this memory trace takes the form of the NP argument, while for the coordinators, it takes the form of the coordinatees. Under the assumption that these arguments help to reconstruct the magnitude and therefore the original number of predicates, the IAC network should take on the form of Fig. 8.11, for the minimal input of two entities. The conversion of the scale into specific predications is accomplished by the block between these latter nodes and the scale nodes. The block is labeled to indicate its two subparts, the calculation of magnitude and the calculation of angle or norm.

An aspect of this augmentation of the basic IAC network that is crucial for the upcoming theory of inferencing is the fact that the choice of entity for any given predication is arbitrary, since this information is not preserved in the angle or norm calculations. Fig. 8.11 endeavors to represent this arbitrariness through the medium of the slash notation, so that *a/b* can be read as "*a* for the sake of illustration, but also *b* if all the other labels are also reversed". Thus the fact that Fig. 8.11 connects *pmin* to *p(a/b)* via excitation can be read as "*pmin* excites *p(a)*, or *p(b)* if all the entity labels are also reversed". Obviously, we would prefer for the network to implement this behavior directly, rather than

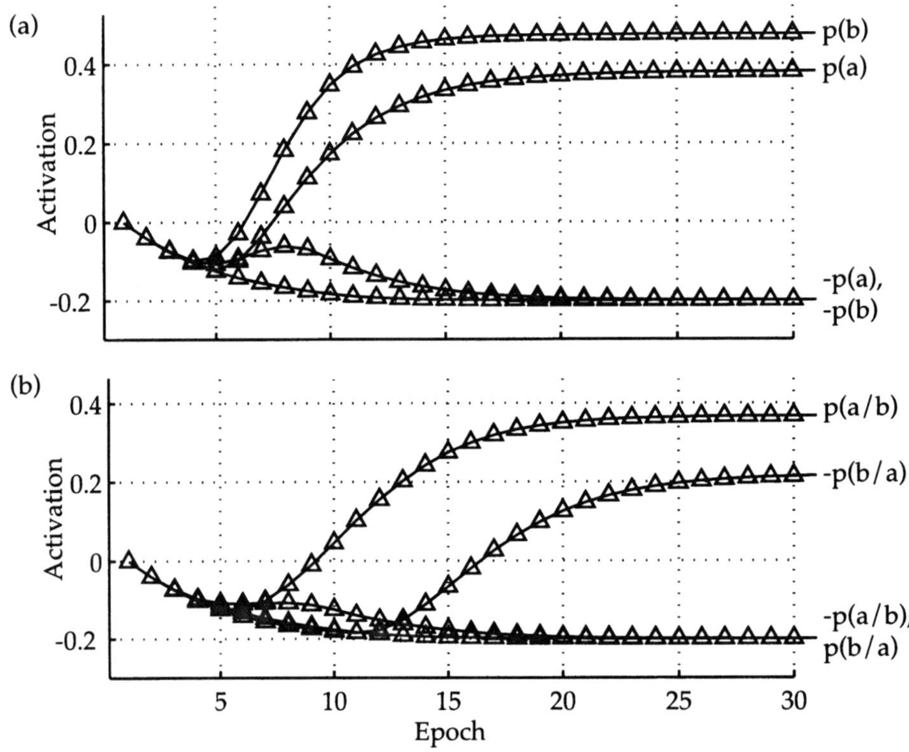

Figure 8.12. Spread of activation in the model for the positive subaltern inference. (a) from /*al*/; (b) from /*s^m*/

having to sneak it in via an *ad hoc* labeling convention, but for the purposes of the upcoming discussion the simple format of Fig. 8.11 is sufficient.

8.3. SPREADING ACTIVATION AND THE SQUARE OF OPPOSITION

This section uses the IAC network described above to simulate inferencing among the logical quantifiers. It takes up each of three kinds of inference described in the first section, and then demonstrates how exclusive SOME, XSOME, can be derived from SOME.

8.3.1. Subaltern oppositions

The subaltern oppositions are the ones that make up the vertical legs of the Square, namely (8.5a), ALL → SOME, and (8.5b), NO → NALL. They differ from the other four in not invoking negation. The crucial observation about this sort of inference is that a universal activates all of the relevant observations, any one of which satisfies the corresponding existential.

This claim finds support in the simulations reported in Fig. 8.12. In Fig. 8.12a, */al/* is once again activated, but only the spread of activation through the model is plotted. It is readily apparent that only the two positive predications are activated. Fig. 8.12b supplies the point of comparison by plotting the spread of activation through the model contingent on the external activation of */s^m/*. *P(a)* becomes active, as does its complement, *-p(b)*. Thus the two simulations concur on the final state of *p(a)*, while contradicting each other on the final sate of *p(b)*. A parallel result is achieved for */no/* and */nat al/*, but it is not reproduced here in the interest of saving space.

Fig. 8.12 would appear to provide a toehold on a neurologically-grounded theory of inference. The simplest first approximation is (8.30):

8.30. $\phi \rightarrow \gamma$ iff any node activated by ϕ corresponds to a node activated by γ at the relevant downstream layer *L*.

This formulation permits Fig. 8.12a (ALL) to imply Fig. 8.12b (SOME) through the fact that *p(a)* becomes active in both.

However, what if we had picked *p(b)* as the potentially corresponding node, a choice that "any" in (8.30) certainly allows? Here we must appeal to the fact that the selection of nodes that become active from SOME is arbitrary, a fact that serves as the motivation for the slash convention introduced in Fig. 8.11. This is the reason why Fig. 8.12b is labeled with the full notation for each node, in order to contrast it with Fig. 8.12a (ALL), for which the choice of label is irrelevant, since both nodes become active. Bearing this arbitrariness in mind, it can be understood that the activation of *p(b)* is a possibility implicit in the labeling of Fig. 8.12b, which allows the implication from ALL to SOME to go through. Conversely, it also prevents the implication from SOME to ALL from going through. This implication is blocked by the fact that no matter which predication is chosen as the active positive one in Fig. 8.12b (SOME), there is also an active negative predication which is never active in Fig. 8.12a (ALL). Thus the definition of implication (8.30) rules in all of the correct cases and rules out all of the incorrect ones. We know of no other theory that reduces implication to independently-motivated neural function.

8.3.2. Contradictory oppositions

The next step is to take up the contribution of negation. To begin, we remind the reader that the contradictories are ALL/NALL and NO/SOME, as defined in (8.2). To appreciate the import of contradictory opposition, the simulation of */s^m/* in Fig. 8.12b should be contrasted with a simulation of its contradictory, */no/*. Fig. 8.13 plots the requisite spread of activation from */no/* within the model. It is readily apparent that */s^m/* cannot imply */no/*, for the inverse reason that it cannot imply */al/*: the positive active node of Fig. 8.12b does not correspond to

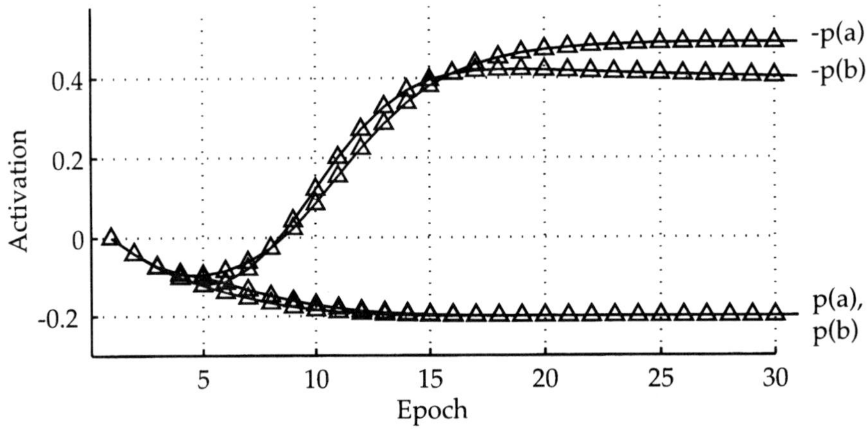

Figure 8.13. Spread of activation from /*no*/ within the model.

any active node of Fig. 8.13, because they are all negative. Yet if we accept this reasoning, we immediately confront a problem, because the implication from /*al*/ to /*s^m*/ also generalizes to the present case: any node activated by /*no*/ can correspond to some node activated by /*s^m*/; in particular, either active node in Fig. 8.13 corresponds to the active negative node in Fig. 8.12b. In other words, the current definition of implication allows NO to imply SOME.

A way to resolve this error comes to mind readily, but it may strain the reader's credulity, at least at first glance. It is to relativize implication to a hierarchy of potential correspondences. (8.31) states one approach:

8.31. $\phi \rightarrow \gamma$ iff any node activated by ϕ corresponds to *the most active* node activated by γ at the relevant downstream layer *L*.

The italicized addition "the most active" stipulates that the potential correspondence cannot skip the most active node of the target element. The most active node is a crucial datum for the existential quantifiers, because it signals the polarity of the quantifier. Clearly, without some indication of polarity, all manner of incorrect inferences will be licensed by the IAC system. This is the consideration that, it is hoped, will sway the reader to accept what otherwise could be considered a desperate attempt to save the framework under development from a fatal counterexample. Or two, since the other pair of contradictories, ALL and NALL, also enter into functionally parallel patterns of activation.

Since we can look at the internal state of the network, it is readily apparent how /*no*/ can stand in contradictory opposition to /*s^m*/, namely, through the mediation of *NEG*. This observation suggests a neurologically-grounded approximation to contradictory negation:

8.32. Contradictory negation reverses the polarity of the highest level of activation.

Of course, this observation is already built into the architecture of the IAC network, in the guise of *NEG*, so it may be concluded that our design of the *NEG* circuit has been validated empirically.

Finally, the usage of equivalence should also have some neurologically-grounded definition. The obvious one is downstream identity:

8.33. $\phi \equiv \gamma$ iff the activation of γ is identical to that of ϕ at the relevant downstream layer *L*.

We have seen this implicitly, in the sense that /*no*/ ≡ *NEG E*, and *E* is the operator that activates /*s^m*/ at L0. The next step is to see how this account generalizes to the phrasal negation of the (sub)contraries.

8.3.3. (Sub)contrary oppositions

The inference schema for (sub)contrary negation are repeated here:

8.34 a) ALL → ¬NO. ALL and NO are contraries. (Both cannot be true.)
b) ¬SOME → NALL. SOME and NALL are subcontraries. (Both cannot be false.)

The (sub)contraries can be derived from the other two inferences by the steps stated in (8.35):

8.35 a) ALL → SOME & SOME ≡ ¬NO
b) ¬SOME ≡ NO & NO → NALL

For the contraries, the correctness of the proof in (8.35a) can be grasped by visual inspection of the L0 plots for ALL and NO: the active nodes have opposite polarities. Moreover, the right-to-left inferences do not go through, because, even though SOME can be inferred from ¬NO, ALL cannot be inferred from SOME, as was argued above. It will be seen below that parallel results holds for SOME and NALL.

These observations suggest a neurologically-grounded approximation to contrary negation:

8.36. Contrary negation reverses the polarity of all nodes.

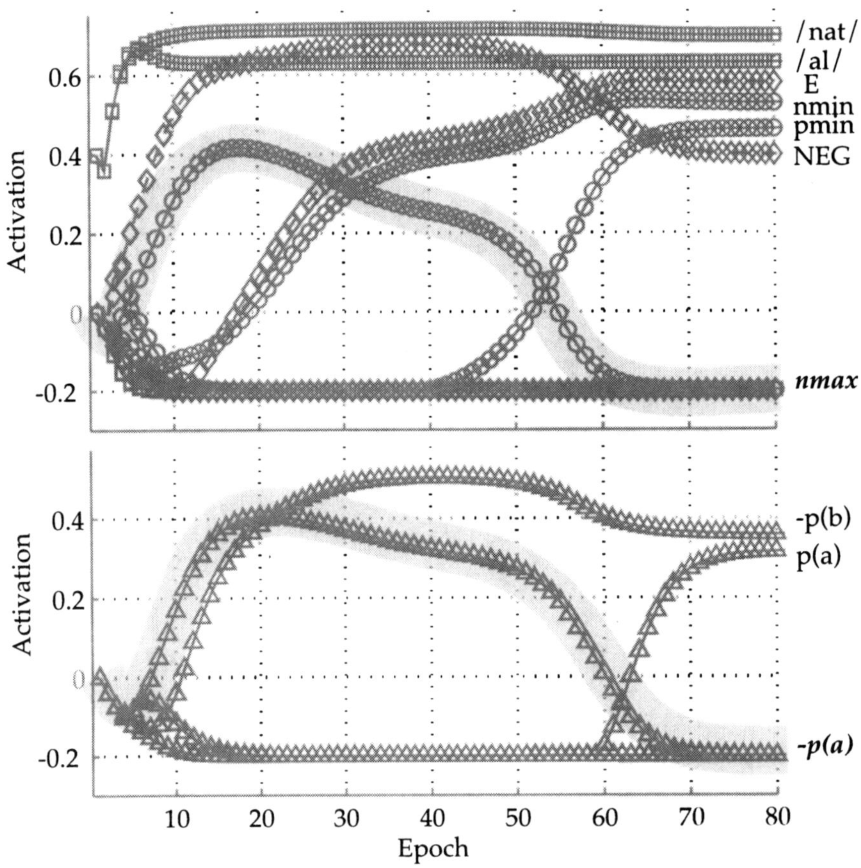

Figure 8.14. Spread of activation from */nat/ /al/*.

It will be demonstrated below that this conjecture holds for subcontraries as well.

8.4. NALL AND TEMPORAL LIMITS ON NATURAL OPERATORS

The exposition has so far studiously avoided any simulation of NALL. The missing simulation is now offered here, in Fig. 8.14. The explanation for not including it with the others should jump off the page at the reader: it takes more than twice as long to reach equilibrium, and then only after a complex dance of activations in which one set of nodes turns off as another set turns on.

The two nodes outlined in gray in Fig. 8.14 are the ones that turn off. *Nmax* is initially excited by */al/* via *U*, but then *U* is quickly overcome by *E* under the influence of *NEG*. The question is, why does *nmax* not fall in tandem with *U* and so clear the way for the full existential reading? The answer appears to be that *pmax* is also activated by *NEG E*, though it is the less preferred option. Thus it takes some time for the network dynamics underlying the preferred option of *nmin* to extinguish the potential option of *nmax* that is already active. That is, in the switch from *U* to *E*, the preferred (and only) option for *U* has to be reduced to the dispreferred option for *E*. This cannot be done directly, through excitation, but rather indirectly, through lateral inhibition from a competing node.

NO does not suffer from this problem of pernicious overlap because, in the switch from *E* to *U*, the dispreferred option for *E* is strengthened through direct excitation to the preferred option for *U*. On a final note, it should be pointed out that this asymmetry between NALL and NO is a network phenomenon and so is relatively insensitive to parameter settings.

The connectivity explanation given for the slow equilibrium of NALL exemplified in Fig. 8.14 suggests two neuromimetic hypotheses of lexicalization, or at least the negative incorporation that distinguishes the monomorphemic NO from the polymorphemic NALL. One is to simply claim that a potential lexical item must reach equilibrium by some maximum number of cycles, say thirty. The other is to cast lexicalization as a mechanism for economy of network effort. (8.37) is a first guess at what a principle of economy based on our limited data looks like:

8.37. Lexicalization of multiple items as a single morpheme is licensed if one item facilitates the activation of another. Lexicalization of multiple items as a single morpheme is blocked if one item reverses the activation of another.

"Facilitation" is seen in the case of NO as a monotonic increase in the excitation of an already excited element. The language of (8.37) leaves open the possibility of the inverse case, in which there is a monotonic increase in the inhibition of an already inhibited element. (8.37) specifically rules out the possibility of lexicalization of nonmonotonic cases in which the activation of an item winds up being neutralized or reversed.

8.4.1. Comparisons to other approaches

Horn, 1989, Chapter 4, wonders how it is that the surface morphology of a natural language like English can do without NAND/NALL, an apparently crucial ingredient of its semantics. The answer which he comes to is that, first, OR/SOME and NAND/NALL express the same information, which makes one of them redundant. Secondly, since negation itself is marked – by a morpheme – it is more economical to do without the version that it marks, NAND/NALL.

Levinson, 2000, p. 69ff., constructs a more elaborate account, based on the supposition that an implicated item can never be lexicalized, but his overall reasoning turns out to be parallel to Horn's: OR/SOME and NAND/NALL express the same information, but the latter is more difficult to process. Thus the grammar eschews its lexicalization.

While we are sympathetic with the overall thrust of Horn's and Levinson's argumentation, our neuromimetic model calls into question the explanatory depth of their principles. At the end of Chapter 3 we demonstrated indirectly that SOME and NALL do not bear an identical informational load. Given the greater population of negative facts about the world, it is expected that a negative operator should be less informative than any putative positive counterpart. This fact is encoded into the IAC network of this chapter through the lower weights of the relevant negative connections. The resulting asymmetry endows the network with a default preference for OR/SOME. As was demonstrated in the last simulation, the outcome is that NAND/NALL are indeed more difficult to process – to a degree that surprised even the author. This outcome motivated us to fix the locus of the lexicalization asymmetry not in redundancy *per se* but rather in mutual support or not among the items to be lexicalized. Mutual support or facilitative excitation means that the items in question are so correlated that it is 'easy' to conceive of them as a single item; mutual conflict means that the items in question are so anticorrelated that it is 'difficult' to conceive of them as a single item.

8.5. SUMMARY

This chapter has offered the neuromimetic account of the logical coordinators and quantifiers by offering a vision of how inferences of opposition among these elements can be effected in a neurologically-inspired manner.

Following McClelland and Rumelhart, 1989, 12ff, An interactive activation and competition or IAC network consists of a collection of processing units organized into pools by bidirectional connections of excitation or inhibition. Units in the same pool inhibit one another while units in different pools excite one another. The bidirectionality of intra-pool excitation gives rise to the competivity of an IAC network: the unit with the most activity within a pool tends to drive down the activity of all of the other units within the pool and thereby appear to 'win' in that pool. The bidirectionality of inter-pool excitation gives rise to the interactivity of an IAC network: processing in one pool can both influence and be influenced by processing in other pools.

We have reviewed a model of grammar in which activation spreads among all three major linguistic components, in order to trace how the excitation of a phonological form excites or inhibits other phonological forms. In this way, the inference from one logical clause to another can be simulated as the spread of activation from a phonological item to its semantic representation and then back to another phonological item.

The next chapter takes the analysis of inferences of opposition to a more detailed level by investigation of a class of constructions in which the subaltern inference from universals to particulars is systematically violated.

Chapter 9

The failure of subalternacy: reciprocity and center-oriented constructions

In this chapter, the symmetry between logical coordination and quantification is made even stronger by explaining why subaltern implications fail for both of them with symmetric predicates. In so doing, we will have the chance to design neuromimetic analyses for anaphora and certain spatial prepositions.

9.1. CONSTRUCTIONS WHICH BLOCK THE SUBALTERN IMPLICATION

Lakoff and Peters, 1969, p. 113, credit Curme (1931) with the observation that, despite the predictions of conjunction elimination, see Sec. 8.1.2, the pairs below are not synonymous:

9.1 a) The King and the Queen are an amiable pair.
a') ≠ *The King is an amiable pair, and the Queen is an amiable pair.
b) She mixed wine and oil together.
b') ≠ *She mixed wine together, and she mixed oil together.

Lakoff and Peters adduce a variety of other predicates which foil conjunction elimination. They fall into two main classes. The first includes the predicates in (9.1), which can be termed **collectives**. The second includes **symmetric** predicates such as *be similar* and *meet*, which often have reciprocal force:

9.2 a) Dylan and Simon are similar (to each other).
a') ≠ *Dylan is similar (to each other), and Simon is similar (to each other).
b) John and Mary met (each other) yesterday.
b') ≠ *John met (each other) yesterday, and Mary met (each other) yesterday.

Collective and symmetric predicates were taken to show that the Conjunction Reduction account of coordination, see Sec. 4.3.1, could not work.[41]

---

[41] See Oirsouw (1987) for a through review of the early generative treatments of coordination.

Given the evidence for a strong parallelism between coordination and quantification presented in the preceding chapters, we would expect ALL to pattern like AND in this unexpected failure of the subaltern inference. In fact, this is exactly what is found:

9.3 a) All my friends are an amiable bunch.
a') ≠ *Chris is an amiable bunch; Dana is an amiable bunch; etc.
b) She mixed all the ingredients together.
b') ≠ *She mixed wine together; she mixed oil together; etc.
a) All folk rockers are similar (to one another).
a') ≠ *Dylan is similar (to one another); Simon is similar (to one another), etc.
b) All the first year students met (one another) yesterday.
b') ≠ *Kim met (one another) yesterday; Lee met (one another) yesterday, etc.

There is no accepted term for this phenomenon that generalizes across the coordinational and quantificational domains, so we offer "failure of **subalternacy**" for the purposes of this monograph. The reader may recall from Sec. 8.1.1 and 8.1.2 that a superaltern corner of the Square of Opposition entails the subaltern corner directly beneath it. In more recent terminology, universals imply particulars of the same polarity. This chapter discusses those cases where this implication fails to obtain.

9.1.1. Classes of collectives and symmetric predicates

The predicates which block subalternacy can be separated into several notional subclasses, plus the reciprocal. Link (1983) lists the following eight:

9.4 a) comparison: be similar, alike, etc.
b) spatial comparison: be interspersed, perpendicular, parallel, etc.
c) group membership: be friends, classmates, etc.
d) group formation: come together, unite, combine, mix, etc.
e) group dissolution: separate, split up, scatter, etc.
f) group movement: spread out, surround, circle, etc.
g) group use: share, collaborate, etc.
h) other: outnumber, etc.

However, the extremely detailed analysis of some of the subclasses developed in the following pages points to a much more coherent and explanatory grouping into three main categories according to whether there is implicit movement towards or away from a center, or 'movement' in tandem. Link's notional subclasses can then be assigned to these three sorts in the following manner:

9.5 Centripetal constructions (motion towards a center)

a) group formation: come together, unite, combine, mix, etc.
b) group movement: surround, circle, etc.
c) others

9.6 Centrifugal constructions (motion away from a center)
a) group dissolution: disperse, separate, split up, scatter, fall apart, etc.
b) group movement: spread out, etc.
c) others

9.7 Tandem (symmetric) constructions ('motion' in tandem)
a) comparison: be similar, alike
b) spatial comparison: be interspersed, perpendicular, parallel
c) group membership: be friends, classmates;
d) group use: share, collaborate
e) conflict & conformity: fight, argue; agree
f) others

This categorization is meant just to provide an initial orientation for the reader. The third group, that of tandem predicates, is not discussed here, in order to allow space for a thorough treatment of the two center-oriented classes, the centripetals and the centrifugals, plus reciprocity.

9.2. RECIPROCITY

Many languages have a means of making almost any transitive verb symmetric, the reciprocal construction:

9.8 a) George and Martha hate each other.
b) ≠ *George hates each other, and Martha hates each other.

Link suggests that the reciprocal pronouns *each other/one another* turn transitive verbs into intransitive verbs which take collective subjects and perforce block subalternacy. In this section, the neuromimetic techniques that have been developed in previous chapters are called upon to implement a semantics of reciprocity that demonstrates that it is semantically transitive – except in the case of singular antecedents, for which no appropriate resolution of the pronoun can be found.

9.2.1. A logical/diagrammatic representation of reciprocity

In a thorough review of the literature on reciprocal constructions, Dalrymple, Kanazawa, Mchombo and Peters (1994, 1998) design a theory of reciprocity based on Langendoen's (1978) notion of **Strong Reciprocity** (SR), also called *each-the other* by Fiengo and Lasnik (1973), which is exemplified and defined in (9.9):

9.9 a) Willow School's fifth-graders know each other. (4)
b) $|A| \geq 2$, and $\forall x, y \in A\ [x \neq y \rightarrow xRy]$

In prose, every member of a set *A* with a least two members is related directly by relation *R* to every other member of *A*. For the example, this means that the sentence is false unless each Willow fifth-grader stands in the knowing relation to every other Willow fifth-grader. This notion of Strong Reciprocity implies four weaker notions which are briefly reviewed in the next paragraph.

In Intermediate Reciprocity (IR), Langendoen (1978), every member of *A* is related through *R* directly or indirectly to every other member of *A*. (9.10a) supplies an example, and (9.10b), the definition:

9.10 a) The telephone poles are spaced five hundred feet from each other.

b) $|A| \geq 2$, and $\forall x, y \in A\ [x \neq y \rightarrow$ for some $z_0, \ldots, z_m \in A\ [x = z_0 \wedge z_0Rz_1 \wedge \ldots \wedge z_{m-1}Rz_m \wedge z_m = y]]$

In One-way Weak Reciprocity (OWR), every member of *A* is related to some other member of *A* as the first argument of *R*. For instance, in (9.11a) it is required that each pirate stare at some other, but not that each pirate be stared at by another. (9.11b) states the definition:

9.11 a) "The captain!" said the pirates, string at each other in surprise.

b) $|A| \geq 2$, and $\forall x \in A\ \exists y \in A\ [x \neq y \wedge xRy]$

In Intermediate Alternative Reciprocity (IAR), all pairs of members of *A* must be related directly or indirectly via *R*. Dalrymple et al. (1998) adduce the example of the stones from which the National Cathedral in Washington D. C. is constructed. They are stacked in staggered, overlapping patterns like a brick wall, see (9.12a):

9.12 a) Instead, countless stones – each weighing an average of 300 pounds – are arranged on top of each other and are held in place by their own mass and the force of flying buttresses against the walls.

b) Mrs. Smith's third-grade students gave each other measles.

c) $|A| \geq 2$, and $\forall x, y \in A\ [x \neq y \rightarrow$ for some sequence $z_0, \ldots, z_m \in A\ [x = z_0 \wedge [z_0Rz_1 \vee z_1Rz_0] \wedge \ldots \wedge [z_{m-1}Rz_m \vee z_mRz_{m-1}] \wedge z_m = y]]$

As an additional instance, (9.12b) is true under IAR if each of Mrs. Smith's third-graders either gave measles to a classmate or got measles from a classmate. It is false if one of Mrs. Smith's third-graders neither gave measles to – nor got it from – a classmate. The final attested reading is dubbed Inclusive Alternative Ordering (IAO), Kanski (1987). Under IAO, every member of *A* is related to some other member of *A* as the first or the second argument of *R*, but not necessarily as both arguments. Dalrymple et al. (1998) offer the example of (9.13a).

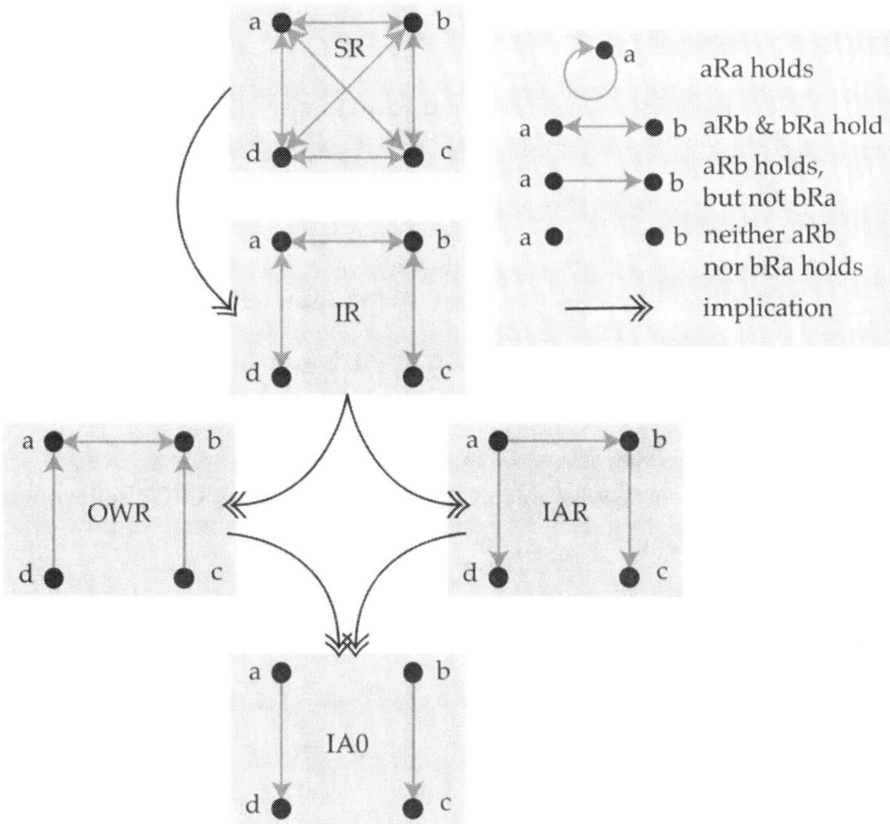

Figure 9.1. Reciprocal action for $A = 4$.

9.13 a) He and scores of other inmates slept on foot-wide wooden planks stack atop each other – like sardines in a can – in garage-sized holes in the ground.

b) $|A| \geq 2$, and $\forall x \in A\ \exists y \in A\ [x \neq y \wedge (xRy \vee yRz)]$

A simple image of the example consists of two side-by-side stacks of two planks – in Dalrymple et al.'s indexical scheme, this could be plank z_1 stacked on top of plank z_2, and plank z_3 stacked on top of plank z_4. Thus the odd indices are first arguments, the even indices are second arguments, and no plank bears both.

Dalrymple et al. reduce these five readings to the combination of five parameters. The first three concern how R should cover its domain A, and they are defined so as to exclude cases in which an individual bears *R* to itself, through the notation $X\backslash I$:

9.14 a) FUL(A, R)\I: each pair of individuals in *A* may be required to participate in the relation *R* directly, and $|A| \geq 2$.

b) LIN(A, R)\I: each pair of individuals in *A* may be required to participate in the relation R either directly or indirectly, and $|A| \geq 2$.

c) TOT(A, R)\I: Each single individual in *A* may be required to participate in the relation *R* with another one, and $|A| \geq 2$.

The final two parameters concern how *R* itself is organized; in particular, whether it includes its inverse or not:

9.15 a) R: *R* is the extension of *R*, i.e. *R* goes in both directions.

b) $R^{\vee}$: *R* is the extension of *R* minus its inverse, i.e. *R* only goes forward.

The five parameters cross-classify in six (3 x 2) ways, which correspond to the attested readings as listed in (9.16):

9.16 a) SR = FUL(A, R)\I.

b) IR = LIN(A, R)\I.

c) SAR = FUL(A, $R^{\vee}$)\I.

d) OWR = TOT(A, R)\I.

e) IAR = LIN(A, $R^{\vee}$)\I.

f) IAO = TOT (A, $R^{\vee}$)\I.

(9.16c) introduces a predicted reading, dubbed Strong Alternative Reciprocity, which is not attested.

Each family of parameters can be ordered from less to more inclusive, TOT < LIN < FUL, and $R^{\vee} < R$. Thus the five attested readings can be arranged into an inclusiveness or implicational hierarchy, which is drawn in Fig. 9.1 for a set *A* consisting of the four entities *a*, *b*, *c*, and *d*. The legend to the right side explicates the correspondence of arrows to instantiations of *R*, following the conventions of Langendoen (1978). This hierarchy enables us to reason about the relations among the various readings visually. For instance, SR implies the other four, since it stands at the top, which is the 'root' of all of the implicational arrows.

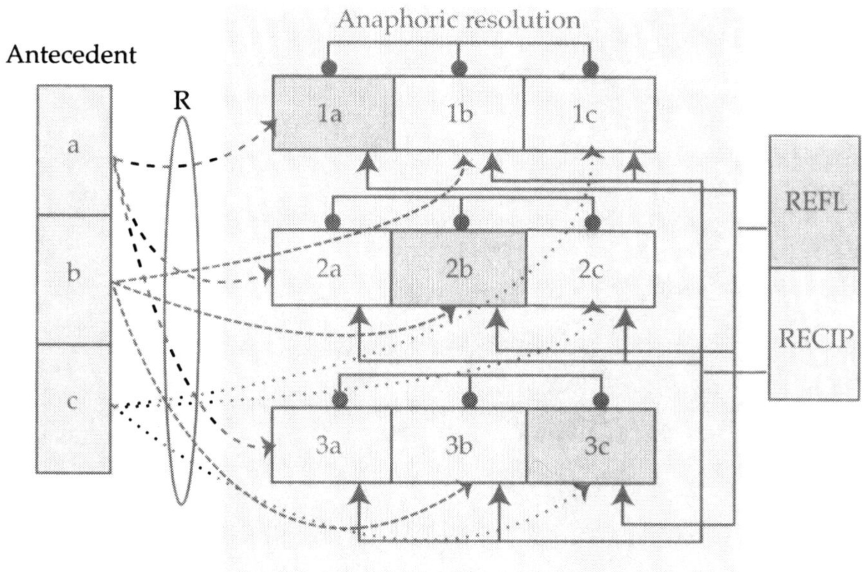

Figure 9.2. An IAC network for anaphora among three entities.

Table 9.1. Matrix of anaphora for a three-entity subject.

Subject	*Value of anaphor*			*Kind*
	Andy	*Betty*	*Cathy*	
Andy $[1, 0, 0]^T$	1	0	0	REFL
	0	1	1	SR
	0	1	0	IR
Betty $[0, 1, 0]^T$	0	1	0	REFL
	1	0	1	SR
	0	0	1	IR
Cathy $[0, 0, 1]^T$	0	0	1	REFL
	1	1	0	SR
	0	1	0	IR

This comes about because any configuration of arrows found in the lower four can be found in SR. The converse does not hold. For instance, IR lacks three double-headed arrows that SR has, so IR does not imply SR.

Dalrymple et al. go on to analyze several other aspects of reciprocal meaning, a review of which would take us too far afield from subalternacy. Let us therefore stop here and turn to a neuromimetic formalization of the basic facts.

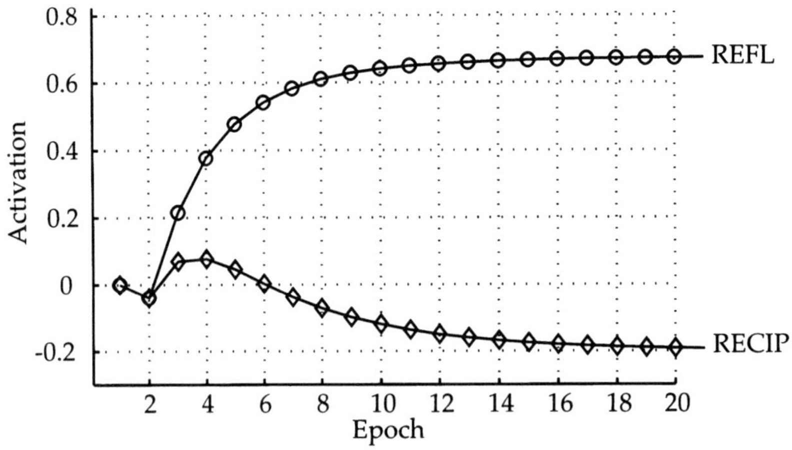

Figure 9.3. Response of anaphoric units in Fig. 9.2 to external input of 0.4 to REFL.

9.2.2. A distributed, *k*-bit encoding of anaphora

For the sake of concreteness, let us suppose that the letters *a*, *b*, and *c* in the preceding examples represent *Andy*, *Betty*, and *Carol*, respectively. For the sake of even more concreteness, this group can be converted into a numerical representation as a three-component vector whose components store the value for each entity in the order given above, i.e. $[a, b, c]^T$. The individual components refer to individual entities by taking 1 to mark an entity accepted by the pronoun as an antecedent and 0 to mark an entity rejected by the pronoun. The assumption is that a rejected entity is uncorrelated with the subject, not anti-correlated.

If *Andy* is chosen as subject, there are three sorts of values for a reflexive or reciprocal pronominal object. The value of a reflexive pronoun would be *Andy*; or $[1, 0, 0]^T$; the value of Strong Reciprocity would be *Betty* and *Cathy*; or $[0, 1, 1]^T$; and the value of Intermediate Reciprocity – according to Fig. 9.1 – would be *Betty*; $[0, 1, 0]^T$. These patterns are arranged in the shaded portion of Table 9.2. This is just one third of the story, however, for *Andy* is just one third of the subject *Andy, Betty, and Cathy*. The rest of Table 9.1 continues the patterns for *Andy* through *Betty and Cathy* in what is hoped is now an obvious way.

9.2.3. Anaphora in SAG

Now that we have a hypothesis of the contrast between reflexivity and reciprocity, we can turn to Spreading Activation Grammar for a model of how it

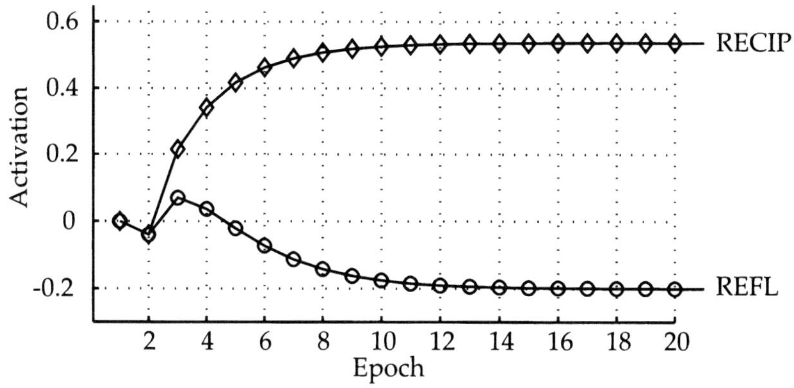

Figure 9.4. Response of anaphoric units in Fig. 9.2 to external input of 0.4 to RECIP.

works. Let us excise from SAG just the subnetwork of the conceptual component that models anaphora. For a subject consisting of three members, the architecture of anaphora is that of Fig. 9.2. We assume that the anaphora has already been resolved, which is to say that the reflexive/reciprocal pronoun has already found and accessed its antecedent. Each member of the antecedent pool on the left activates the corresponding unit among the anaphoric pools in the center. The two anaphoric morphemes on the left also send a pattern of activation into the anaphoric pools: the reflexive unit excites the unit of each anaphoric pool that is identical to the antecedent, while the reciprocal unit excites the others.

The IAC simulations in Fig. 9.3 and 9.4 confirm the accuracy of this design. In these plots of the network evolution, there is a clear demarcation between the active target reading and the inactive alternative. In Fig. 9.3, the reflexive units reach equilibrium at about 0.68, while the reciprocal units in Fig. 9.4 reach equilibrium at about 0.48. The reason for the lower value of the latter is due to the fact that more reciprocal units become active in a given pool and so send more lateral inhibition to one another.

9.2.4. Comments on the SAG analysis of anaphora

There are several questions that can be asked about this analysis of anaphora. Perhaps the main one is how it generalizes to more than five entities. It is certainly possible in principle to add as many units as are needed for the cardinality of the subject, but in practice we would expect to hit some upward bound on the number of neurons available for the anaphoric matrix. We have tried a number of formats that would lead to a normalized representation, and

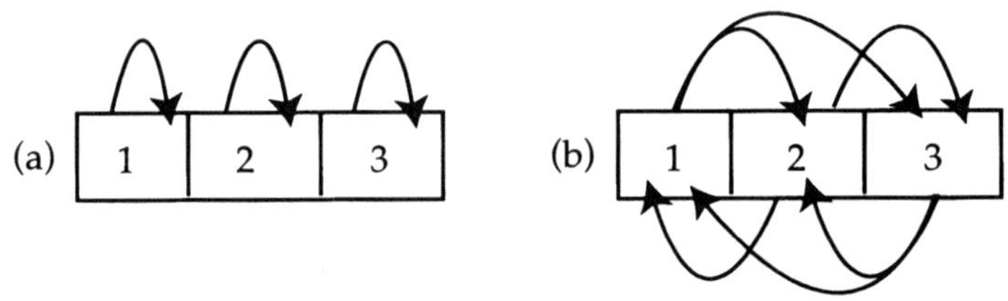

Figure 9.5. Data structure for anaphora. (a) reflexive; (b) strong reciprocal.

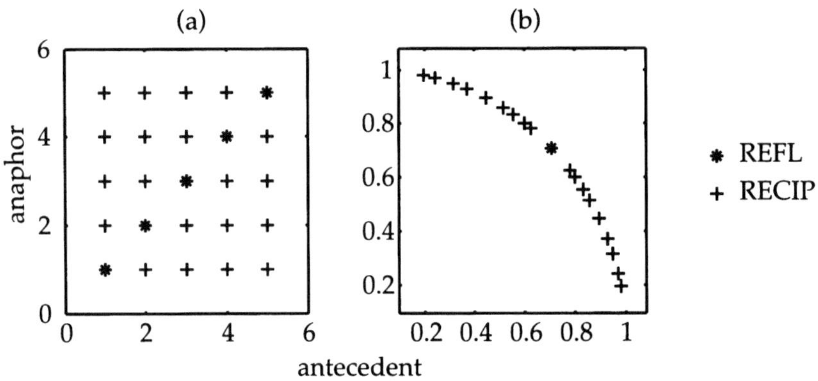

Figure 9.6. Anaphoric phase space. (a) sample; (b) normalization.

the next few paragraphs sketch the simplest format, and the one that provides a natural transition to the center-oriented constructions in the next section.

The idea is to build on the IAC network and represent an anaphoric binding as a cell in a linear array such as Fig. 9.5. The arrows in Fig. 9.5a show three links for Reflexivity among three members; the arrows in Fig. 9.5b show six links for Strong Reciprocity among three members. Each arrow can be represented by a vector $[x, y]^T$, standing for $[antecedent, anaphor]^T$. For instance, the reflexive link from 1 to itself is represented as $[1, 1]^T$, while the reciprocal link from 1 to 2 is represented as $[1, 2]^T$. The pairs for all links between 1 and 5 are graphed in Fig. 9.6a; this plot is normalized in Fig. 9.6b.

Fig. 9.6b defines three convex regions, one at the center point, $[0.71, 0.71]^T$, and the other two on either side. An LVQ network can learn these separate convex regions easily, though we will not offer a specific simulation. Moreover, given that all three regions are convex, it follows that Strong Reciprocity should be the default reciprocal reading. This is due to the pattern-completion ability of

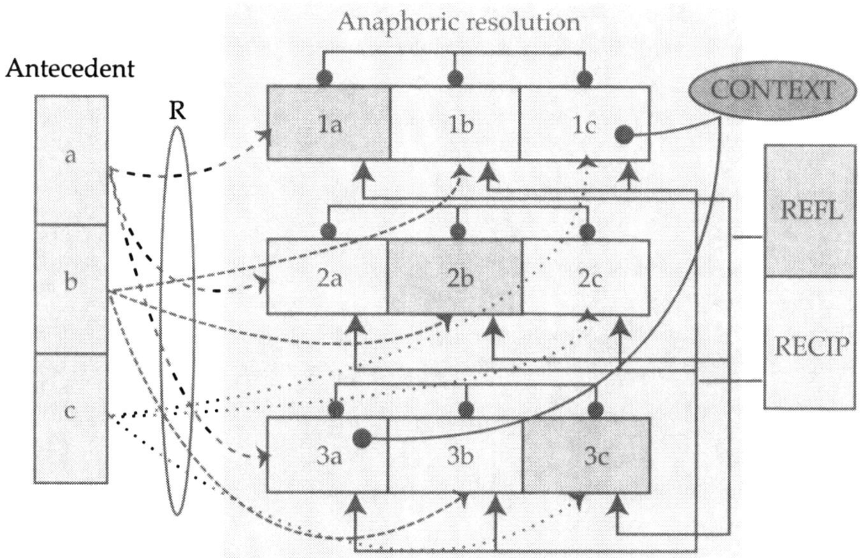

Figure 9.7. Adding contextual inhibition to Fig. 9.2 produces IR.

the neurons that map reciprocal phase space – the fact that they 'saturate' or close the gaps in anaphoric phase space, thereby producing the reading in which all the reciprocal links possible are made, the Strong Reciprocal.

Another interesting comment to be made about Fig. 9.6b is how reminiscent it is of the center/surround organization of early vision. The reflexive 'center' of Fig. 9.6b inhibits its reciprocal 'surround', and vice versa. We can only speculate that the general principles of statistical pattern recognition that were offered in Chapter 1 as the origin of the center/surround architecture of early vision are also what account for the ubiquity of anaphoric constructions in human language. In this respect, it is intriguing that the center of Fig. 9.6b is the point of correlation. Thus reflexivity appears to be reducible to the statistical notion of correlation between subject and object, while reciprocity reduces to partial correlation, which supports our thesis of the statistical nature of natural language semantics.

9.2.5. The contextual elimination of reciprocal links

Strong Reciprocity can be weakened to produce the lower readings on the hierarchy in Fig. 9.1 by selectively inhibiting units in the anaphoric pools. For instance, Intermediate Reciprocity can be modeled by inhibiting the anaphoric units of the 'missing' connection between *a* and *c* of Fig. 9.1. The source of this inhibition is an additional pool CONTEXT, which sends enough inhibition to

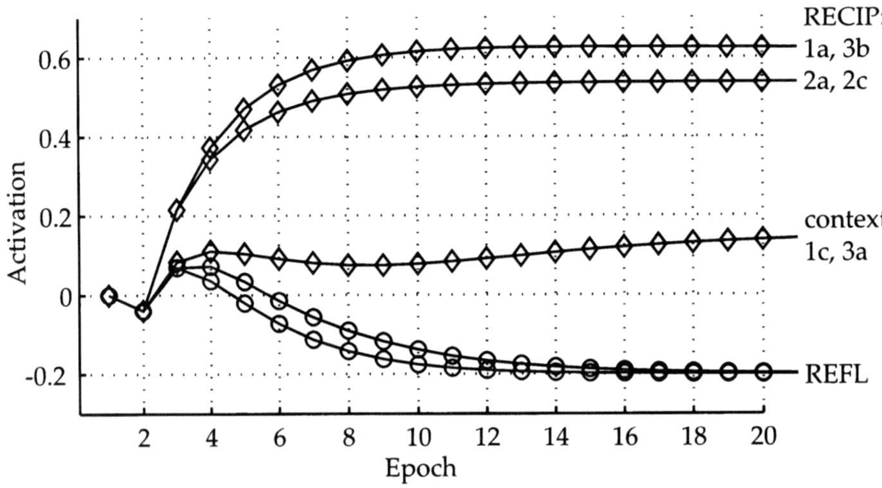

Figure 9.8. Response of anaphoric units in Fig. 9.7 net to external input of 0.4 to RECIP.

overcome the excitation impinging from RECIP. Fig. 9.7 overlays the necessary units and connections onto the format of Fig. 9.2, and Fig. 9.8 introduces a plot of the network dynamics that result from this alteration. The new curve labeled "context" shows that the group of reciprocal units inhibited by the contextual node in Fig. 9.7 has fallen to a state of low activation. The remaining highly active units represent the target IR reading, as the labels on the right margin indicate.

Clearly, we can continue turning off units through contextual inhibition. For instance, chain $a \leftrightarrow b \leftrightarrow c$ or IR can be reduced to chain $a \rightarrow b \rightarrow c$ of IAR by inhibition of the backward links *2a* and *3b*. However, there is some minimum point of elimination beyond with the pattern of activation becomes ungrammatical as a reciprocal meaning. Judging by IAO in Fig. 9.1, this lower bound is crossed when one of the entities ceases to participate in the relation as either subject or object – an observation that Dalyrmple et al. also make.

Presumably there is some feedback from the conceptual network to the semantic network that either keeps RECIP active or deactivates it, but the nature of this feedback is not clear to us. We therefore leave its elucidation for some other venue and turn to the things that we can explain.

9.2.6. The failure of reciprocal subalternacy

In principle, it is child's play to point out the reason that the subaltern inference fails with reciprocals. All that must be done is to turn on a single

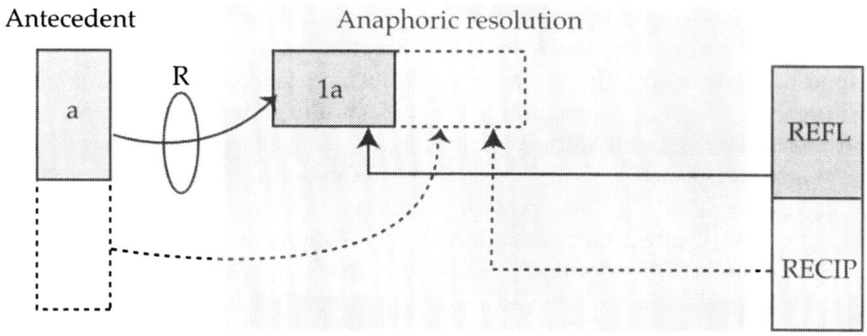

Figure 9.9. The reduction of Fig. 9.2 to a single antecedent.

antecedent unit in Fig. 9.2 and see what pattern the network settles on. In practice, this simple experiment gives unhelpful results, because Fig. 9.2 is 'hard-wired' to expect a three-member antecedent. Fig. 9.9 corrects this drawback by depicting the layout of a network whose antecedent consists of but a single member. The reason for the failure of the subaltern inference becomes glaringly obvious – there are no additional antecedents for the reciprocal node to activate, so that only the reflexive supplies a potential pattern of activation. We therefore conclude that there is no need to stipulate that reciprocals are limited to at least two antecedents, or any other mechanism that specifically rules out the subaltern inference. Instead, it follows naturally from the architecture of reciprocity in neuromimetic networks.

9.2.7. Reflexives and reciprocals pattern together

The matrix layout of Fig. 9.6a that serves as the basis for anaphoric phase space in Fig. 9.6b implies an intimate relationship between linguistic reflexivity and reciprocity, since the two fill complementary subspaces of the overall space. Such a tight relationship has been postulated for the syntax of these expressions at least since the Binding Theory of Chomsky (1981), in that both reflexives and reciprocals must be coindexed to a c-commanding antecedent in their bounding category, usually the smallest noun phrase or clause that contains them. If their antecedent lies outside of this category, it is ungrammatical:

9.17 a) Mary$_i$ does not know that John saw {her$_i$ / *herself$_i$}.
b) Mary$_i$ and Rachel$_j$ do not know that John saw {them$_{ij}$ / *each other$_{ij}$}.

As the contrast in grammaticality shows, a pronoun drawn from the non-anaphoric series *he, her, them,* etc. can easily be bound to an antecedent outside of its bounding category, but such is not the case for pronouns drawn from the anaphoric series, *himself, herself, themselves,* etc. and *each other/one another*. Of course, (and unfortunately), the matrix layout does not explain this contrast, but at least it predicts that reflexives and reciprocals should pattern together for any process that attempts to resolve their antecedents.

What is less well-known about these constructions is that they also pattern together morphologically, at least in some languages. For instance, Slavic, Baltic, Romance, and non-English Germanic languages share a set of etymologically related reflexes of IndoEuropean **s-* that mark both reflexive and reciprocal meanings, see Knjazev (1998) for examples. By way of illustration, consider the sentence in (9.18), which illustrates the usage of this reflex in Spanish, namely the clitic pronoun *se*:

9.18. Juan y María se rascaron ({a sí mismos / el uno al otro}).
John and Mary SE scratch-3pPAST (to self same / the one to-the other
John and Mary scratched {themselves / each other}.

In the first line, the material in curly brackets are tonic pronouns that disambiguate *se* as either reflexive, *a sí mismos*, or reciprocal, *el uno al otro*. Yet the parentheses around the curly brackets indicate that the material between them is optional. That is to say, *se* is ambiguous between a reflexive and a reciprocal reading without the parenthetic material, though a specific context may bias the interpretation towards one or the other. Knjazev (1998) goes on to point out that such a morphological identity may also be found some Finno-Ugric, Nilo-Saharan, Austronesian, and Australian aboriginal languages. The analysis introduced in the preceding pages leads us to expect this formal identity, given that these constructions are drawn from the same semantic space.

9.3. CENTER-ORIENTED CONSTRUCTIONS

At the beginning of this chapter, four categories were postulated for the constructions that block subalternacy: reciprocals, and centrifugal, centripetal, and tandem constructions. In the upcoming section, it is argued that the two center-oriented constructions are covert reciprocals, so the analysis of reciprocity sketched in the previous section generalizes naturally to them.

9.3.1. Initial characterization and paths

The content of centrifugal and centripetal constructions can be diagrammed as in Fig. 9.10. What these diagrams strive to depict is the movement of some entities towards or away from some center. These entities move in tandem, each one with respect to all the others, which is represented by the equality in length of the arrows in the diagrams. Fig. 9.10a depicts the Strong Reciprocal graph for a subject with four members. Fig. 9.10b is an idealization of the centrifugal situation in which all four move towards each other, and Fig. 9.10c likewise

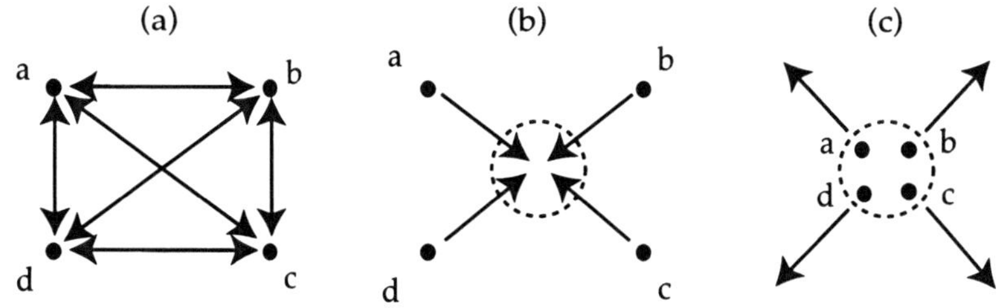

Figure 9.10. Schemata for (a) reciprocal, (b) centrifugal, and (c) centripetal constructions for four entities.

idealizes the centripetal situation in which all four move away from each other. It is our claim that all center-oriented constructions can be paraphrased in such a way as to bring out this reciprocity. (9.19) demonstrates this claim for the illustrative verbs *gather* (centrifugal) and *disperse* (centripetal):

9.19 a) The yaks gathered in the meadow.
a′) ≈ The yaks moved towards one another until they came together.
b) The yaks dispersed after grazing the meadow.
b′) ≈ The yaks moved away from one another until they came apart.

At the level of idealization which is assumed here, the sort of movement seen in both constructions is the same. The only difference lies in its orientation with respect to the center: towards or away from it.

It will help to have a bit of terminology to talk about such movements. Let us refer to a complex of prepositional phrases that include at least an origin and a goal as a **path**, in particular, as a $\textbf{path}_2$. (9.20) gives two sample paths_2:

9.20 a) Mary walked from the lodge to the forest.
b) Mary walked out of the lodge, down the path, and into the forest.

Center-oriented constructions differ with respect to the kind of paths that they license. Predicates like *gather* and *disperse* permit a path_2, while predicates like *collide* and *separate* generally only permit one of the endpoints of a path_2. Centrifugal *collide* permits the center where the motion converges, a goal. Centripetal *separate* permits the center where the motion originates, a source. For simplicity, call either one a $\textbf{path}_1$, though if need be we can specify a source or ablative path, $\textbf{path}_{So/Ab}$, a goal or allative path, $\textbf{path}_{Go/Al}$ and a 'with' or comitative path, $\textbf{path}_{Com}$.

9.3.2. Centrifugal constructions

Centrifugal constructions come from the semantic field of "x_1 and x_2 move towards each other until they came together". They can be divided into **intersectives** (verbs of overlap) and **congregatives** (verbs of gathering), and resultative *together*. Intersectives and congregatives can be distinguished by the minimum number of entities that come together: for intersectives, the smallest number is two, while for congregatives, the smallest number is about four or five. The three subclasses are further characterized in the next subsections.

9.3.2.1. Verbs of intersection

It takes at least two conterminous entities to constitute an instance of intersection. Verbs that express such a relation can be recognized by several properties that are illustrated by *collide* in (9.21a) below. They can be paraphrased with *come together*, (9.21b), or more distantly *move towards*, (9.21c). The monodic form allows the optional usage of comitative *with* and a reciprocal pronoun, (9.21d), which is synonymous with a dyadic conjunctive paraphrase of the reciprocal subevents, (9.21e). The translation of these verbs into Spanish requires the usage of the reflexive or reciprocal clitic pronoun *se*, (9.21f):

9.21 a) John and Mary collided after the show.
b) John and Mary came together after the show.
c) John and Mary moved towards each other after the show.
d) John and Mary collided with each other after the show.
e) John collided with Mary and Mary collided with John after the show.
f) Juan y María se chocaron (uno con el otro) después del espectáculo.
Juan and María SE collide-3pPAST (one with the other) after the show.

The centrifugal argument permits modification by *all*, (9.22a). If singular, it can at best be a collective term, (9.22b), and cannot be quantified by *every* or *each*, (9.22c):

9.22 a) All the musicians collided (with one another) after the show.
b) The {?group/*fog/*guitarist} collided after the show.
c) *{Each/Every} musician collided after the show.

The intersectives license the comitative alternation, in which a member of the subject is realized as the object of *with*, see (9.23a). Intersective *with* does not permit the continuation of the motion with a goal phrase, in either the x_1 *and* x_2 version, (9.23b), or the x_1 ... *with* x_2 version, (9.23b′):

9.23 a) John collided with Mary after the show.

b) *John and Mary collided with each other into the crowd after the show.
b′) *John collided with Mary into the crowd after the show.

We next identify several notional subclasses of intersectives, which fall into the two general categories of partial and total intersection.

Intersection at a single point or region is physical contact:

9.24 a) The two lines contact *(each other) at point P.
a′) ≈ Line A contacts line B, and Line B contacts line A.
b) The two lines cross (each other) at point P.
c) The two lines intersect (with each other) at point P.
d) The two conferences overlap ((with) each other) on Friday.

This leads to a host of forms of touching between humans:

9.25 a) John and Mary touched (each other).
a′) John touched Mary, and Mary touched John.
b) John and Mary {cuddled/embraced/hugged/kissed} (each other) for several minutes.
c) John and Mary {copulated/made out} (with each other) for several minutes.

Total intersection in the same space is coalescence:

9.26 a) The two lakes coalesce (with each other) every spring.
a′) ≈ A coalesces with B, and B coalesces with A every spring.
b) Mary combined oxygen and hydrogen (with each other) in the test tube.
c) The two halves melded (with each other) easily.
d) The two scents mingled (with each other) in the morning breeze.
e) Delta and TWA will merge (with each other) by next summer.

There are transitive verbs that have the same properties:

9.27 a) Mary illegally commingled the two funds (with each other).
b) Mary consolidated her two companies (with each other) to cut costs.
c) Mary tried to mix oil and water (with each other) in the test tube.
c′) cf. The oil and water did not mix (with each other) in the test tube.

Note that at least *mix* participates in the causative/inchoative alternation, in which the object of a transitive verb alternates with the subject of an intransitive verb. Continuing in the transitive vein, contact that creates a new whole is connection:

9.28 a) Mary connected the two pieces (to each other).
a′) ≈ Mary connected A to B and B to A.
b) Mary attached the two pieces (to each other).
b) Mary coupled the two pieces (to each other).
c) Mary joined the two pieces (to each other).
c) Mary linked the two pieces (to each other).
d) Mary {fused / welded / glued / soldered / nailed / etc.} the two pieces (to each other).

Note that connection has the idiosyncrasy of licensing allative *to* instead of comitative *with*. Finally, there are abstract usages of intersection in the social domain, which can be called association:[42]

9.29 a) John and Mary rarely associate (with each other).
a′) ≈ John rarely associates with Mary, and Mary rarely associates with John.
b) John and Mary rarely consort (with each other).
c) John and Mary rarely meet (with each other).
d) John and Mary rarely contact *(each other) by cell phone.
e) John and Mary allied (with each other) in the face of a common enemy.
f) John and Mary affiliated (with each other) to pool their talents.
g) John and Mary united (with each other) behind the President.

There is no physical contact in these situations, though it is not unreasonable to suppose that the subjects come into social proximity.

9.3.2.2. Resultative together

There is a centrifugal sense of *together* that patterns with the intersectives in being licensed by nominals whose minimum cardinality is two. It is the sense that Lasersohn (1995) refers to as the "assembly reading":

9.30 a) assembly: John put the bicycle together.
b) ≈ John put the parts of the bicycle with one another

---

[42] The semantic field of cooperation, e.g. *cooperate, collaborate, collude, concur,* is included among the tandem predicates of sharing, (9.7d), which are not dealt with in this monograph. The semantic field of association differs from sharing in that association has a sense of coming together that sharing lacks.

c) ≈ John put one part of the bicycle with another

However, in this case there is a significant contrast in meaning between *together* and its comitative paraphrase: the comitative *put with one another* only has the reading of putting into spatial proximity, whereas *put together* means not only to put parts into spatial proximity, but also for them to stay that way, as a new whole. This stronger reading is found with many other verbs in construction with *together*:

9.31 a) Mary finally pulled the deal together.
b) John's death brought the family together.
c) Mary pushed the campfire back together.
d) John glued the pieces of the plate together.
e) Mary held the company together until it could be sold.

This extra iota of meaning appears to be causative, since the comitative act can be understood as causing the parts of the whole to retain their proximity on their own:

9.32 a) assembly: John put the bicycle together.
b) ≈ John put the parts of the bicycle with one another in a way that caused them to stay with one another, presumably because they fit with one another as a whole

This semantico-syntactic configuration is currently known as a resultative construction.[43] Few other resultative constructions are centrifugal; the only one that we are aware of is *add up* and its synonyms:

9.33. The corks in the bottle add up (with one another) to fifteen.

These meanings are very similar to those of the verbs of congregation.

9.3.2.3. Verbs of congregation

It takes several proximate entities to constitute an instance of congregation. Verbs that express such a relation can be recognized by several properties that are illustrated by *gather* in (9.34a) below. They can be paraphrased with *come together*, (9.34b), or more distantly *move towards*, (9.34c). The monodic form is not felicitous with the optional usage of comitative *with* and a reciprocal pronoun, (9.34d), nor is the putatively synonymous conjunctive paraphrase of the

---

43 See Carrier and Randall (1992), Napoli (1992), Neelman and Weerman (1993), Rapoport (1993), and Li (1995) for discussion of resultatives.

reciprocal subevents grammatical, (9.34e). The translation of these verbs into Spanish requires the usage of the reflexive or reciprocal clitic pronoun *se*, (9.34f):

9.34 a) The yaks gathered after sunrise.
b) The yaks came together after sunrise.
c) The yaks moved towards each other after sunrise.
d) ?? The yaks gathered with one another after sunrise.
e) *Yickety Yak gathered with Yackety Yak and Yucky Yak, Yackety Yak gathered with Yickety Yak and Yucky Yak, and Yucky Yak gathered with Yickety Yak and Yackety Yak.
f) Los yakes se juntaron (unos con otros) en el valle.
the yaks SE gather-3pPAST (one with the other) in the valley.

The centrifugal argument permits modification by *all*, (9.35a). If singular, it can collective and perhaps mass, (9.35b), but never singular count. It cannot be quantified by *every* or *each*, (9.35c):

9.35 a) All the yaks gathered after sunrise.
b) The {group/??fog/*guitarist} gathered after sunrise.
c) *{Each/Every} yak gathered after sunrise.

The congregatives do permit a full $path_2$: a goal phrase, which denotes the center around which the entities congregate and a source:

9.36. The yaks gathered around the newborn calf from the far corners of the valley.

Other such intransitive verbs are adduced in (9.37). Note how they are uncomfortable with a comitative phrase, but easily bear an allative phrase.

9.37 a) The particles accumulated (?with one another) on the cathode.
b) The protesters converged (??with one another) on City Hall.
c) The guests crowded (?with one another) around the bar.
d) The yaks congregate (?with one another) around the trough.
e) The guests massed (?with one another) around the liquor cabinet.
f) The bees clustered (?with one another) around the queen.

Congregation that creates a new whole is amalgamation:

9.38 a) The pebbles amalgamated (?with one another) into a solid ball.
b) The sticky seeds agglutinated (?with one another) into a solid ball.
c) The sticky seeds agglomerated (?with one another) into a solid ball.

Like connection, these verbs license an allative phrase, *into* instead of *to*.

Table 9.2. Summary of centrifugal predicates

	Property	*Intersectives* 'collide'	*Congregatives* 'gather'
a)	means *move together*	yes	yes
b)	translates with *se*	yes	yes
d)	*all X*	yes	yes
d)	collective X	yes	yes
e)	singular count X	*	*
f)	*each/every* X	*	*
g)	mass X	*	??
h)	cardinality of X	2+	4/5+
i)	preposition + reciprocal	with/to	??with
j)	comitative alteration	yes	???
k)	licenses $path_2$	*	yes

A few congregatives participate in the causative/inchoative alternation, in which the object of a transitive verb alternates with the subject of an intransitive:

9.39 a) Mary amassed volumes of data (*with one another) in the archive.
a') Volumes of data are amassing (*with one another) in the archive.
b) Mary collected dust bunnies (*with one another) under the bed.
b') Dust bunnies collected (*with one another) under the bed.
c) Mary pooled her samples (?with one another) in a bucket.
c') Mary's samples pooled (??with one another) in a bucket.

Some other verbs such as *surround* show this behavior with plural subjects, (9.40a), but it would be inaccurate to classify *surround* as an exclusively congregative verb, since it permits singular count subjects, (9.40b):

9.40 a) The yaks surrounded the new-born calf.
b) A fence surrounds the park.

It is more accurate to claim that one of the by-products of the special spatial relation of *surround* is a congregative reading.

9.3.2.4. Summary of centrifugal constructions

The properties of the two sorts of verbs are summarized in Table 9.2 so as to bring out their similarities, Table 9.2a-g, and their differences, Table 9.2i-k. Resultative *together* is excluded in order to concentrate on the verb classes, though it is implicit in Table 9.2a.

9.3.3. Centripetal constructions

Centripetal constructions come from the semantic field of "x_1 and x_2 move away from each other". They can be divided into predicates of separation and of dispersion, and resultative *apart*. These three subclasses can be recognized by several properties that parallel those of centrifugal verbs. **Separatives** and **dispersives** can be distinguished by the minimum number of entities that come apart: for separatives, the smallest number is two, while for dispersives, the smallest number is about four or five. Resultative *apart* patterns with the separatives in allowing the cardinality of its centripetal argument to be at least two. The three subclasses are further characterized in the next subsections.

9.3.3.1. Verbs of separation

It takes at least two conterminous entities to constitute an instance of separation. Verbs that express such a relation can be recognized by several properties that are illustrated by separate in (9.41a) below. They can be paraphrased with *move apart*, (9.41b), or *move away from*, (9.41c). The monodic form allows the optional usage of ablative *from* rather than comitative *with* and a reciprocal pronoun, (9.41d), which is synonymous with a dyadic conjunctive paraphrase of the reciprocal subevents, (9.41e). The translation of these verbs into Spanish requires the usage of the reflexive/reciprocal clitic pronoun *se*, (9.41f):

9.41 a) John and Mary separated after the show.
b) John and Mary moved apart after the show.
c) John and Mary moved away from one another after the show.
d) John and Mary separated from each other after the show.
e) John separated from Mary after the show, and Mary separated from John after the show.
f) Juan y María se separaron (uno del otro) después del espectáculo.
Juan and María SE separate-3pPAST (one from of-the other) after the show.

The centripetal argument permits modification by *all*, (9.42a). If singular, it can only be a collective term, (9.42b), and cannot be quantified by *every* or *each*, (9.42c):

9.42 a) All the musicians separated (from one another) after the show.
b) The {group/*fog/*guitarist} separated after the show.
c) *{Each/Every} musician separated after the show.

Separatives license the ablative alternation, in which one member of the subject is realized as the object of *from*, (9.43a). Separative *from* does not permit the continuation of the motion with a goal phrase, in either the monodic or the dyadic version, (9.43b, b'):

9.43 a) John separated from Mary after the show.
b) *John and Mary separated from each other into the crowd after the show.
b') *John separated from Mary into the crowd after the show.

We next identify several notional subclasses of separatives, which fall into the two general categories of partial and total separation.

Most centripetal verbs come from the semantic field of separation. A few participate in the causative-inchoative alteration, such as *separate*:

9.44 a) The cook separated the white and the yolk (from each other).
a′) ≈ The cook separated the white from the yolk and the yolk from the white.
b) The white and the yolk separated (from each other).
b′) ≈ The white separated from the yolk, and the yolk separated from the white.

Other examples are:

9.45 a) This lever disengages the clutch and the drive shaft (from each other).
a') ≈ This lever disengages the clutch from the drive shaft and the drive shaft from the clutch.
b) The sailor disentangled the two lines (from each other).
b') ≈ The sailor disentangled A from B and B from A.

The rest are exclusively transitive, (9.46), or intransitive, (9.47):

9.46 a) The engineer detached the two cars (from each other).
a′) ≈ The engineer detached A from B and B from A.
b) The electrician disconnected the two wires (from each other).
c) The cook divided the white and the yolk (from each other).
d) The government segregated the two castes (from each other).
e) The surgeon severed Chang and Eng (from each other).
f) Mary unbuckled the two straps (from each other).
g) Mary untied the two lines (from each other).
h) The electrician untwisted the two wires (from each other).

9.47 a) John and Mary disassociated (?from each other) after the accident.
a′) ≈ John disassociated from Mary, and Mary disassociated from John.
b) John and Mary parted (?from each other).

It is interesting to note the presence of a large proportion of instances of the reversative prefixes *dis-* and *un-* among the verbs of separation. The unprefixed versions are centrifugal, which suggests that centrifugality is the unmarked member of the opposition, while centripetality is marked.

The notion of separation in space is extended to separation in kind for *distinguish* and *differentiate*:

9.48 a) Few entomologists can distinguish the males and the females of this species (from each other).
a') ≈ Few entomologists can distinguish the males from the females and / or the females from the males.
b) Botanists differentiate the two species (from each other) by diet.
b') ≈ Botanists differentiate A from B and B from A by diet.

This is an instance of a recurring type of analogy from the spatial to a more abstract domain, which was found among the intersectives in the guise of association.

9.3.3.2. Verbs of extraction and the ablative alternation

One of the most insightful facts about the verbs of separation comes from noticing what is not included among them. One broad semantic field is particularly informative, a field which for lack of a better term shall be referred to as **predicates of extraction**. It encompasses those predicates that license the usage of *from* that is synonymous with *out of*. (9.49) gives two examples:

9.49 a) Mary accidentally dislodged a rock from the cliff face.
b) The advancing troops freed their comrades from the prison camp.

These verbs overlap with the verbs of separation in having a paraphrase of the form *move X from Y*, e.g.:

9.50 a) Mary accidentally moved a rock from the cliff face.
b) The advancing troops moved their comrades from the prison camp.

However, they are crucially different in not permitting a monodic alternative:

9.51 a) *Mary accidentally dislodged a rock and the cliff face (from each other).
b) *The advancing troops freed their comrades and the prison camp (from each other).

This ungrammaticality can be traced to the ungrammaticality of the conjunctive subevents into which (9.52) would have to be analyzed:

9.52 a) *Mary accidentally dislodged a rock from the cliff face and the cliff face from a rock.

b) *The advancing troops freed their comrades from the prison camp and the prison camp from their comrades.

In particular, the second conjunct is not a viable usage of either verb.

This violation comes about in the following way. It can be claimed that extractive predicates instantiate the container-contained schema of Langacker (1986), which requires the trajector to be contained within the landmark. For extractive predicates, the trajector moves out of the landmark, or more precisely, the subject of *from* moves out of its container, the object of *from*. In (9.49), a rock moves out of the cliff face that contained it, and the comrades move out of the prison camp that contained them. (9.53) labels the two members of the relation as *Cr*, container, and *Cd*, contained:

9.53 a) Mary accidentally dislodged [$_{Cd}$ a rock] [$_{Cr}$ from the cliff face]

b) The advancing troops freed [$_{Cd}$ their comrades] [$_{Cr}$ from the prison camp]

The converse spatial relationship between these objects is nonsensical: a rock cannot contain a cliff face, nor can a group of comrades contain a prison camp. It follows that the two data sets above that try to affirm these spatial relations are ungrammatical, to wit, (9.51) and (9.52). This is why the verbs of extraction lack a monodic alternative.

Given that the verbs of separation have the centripetal usage reviewed in the previous subsection, they cannot instantiate the container-contained schema, see (9.41), repeated below in (9.54a, a'):

9.54 a) John and Mary separated (from each other) after the show.

a') ≈ John separated from Mary and Mary separated from John after the show.

b) John separated from Mary after the show.

However, the asymmetry of the ablative alternate in (9.54b) does suggest a container-contained analysis, and this asymmetry stands out much more clearly in other usages of the separatives:

9.55 a) The engineer detached [$_{Cd}$ the fender] from [$_{Cr}$ the car].

a') ≠ ??The mechanic detached the fender and the car (from each other).

b) The surgeon cut [$_{Cd}$ the wart] (away) from [$_{Cr}$ my knee].

b') ≠ ??The surgeon cut the wart and my knee (from each other).

Thus it is incorrect to label the separatives as lexically centripetal. It is more accurate to say that they are lexically ablative, and ablatives can become centripetal when the trajectory between the trajector and landmark is symmetrical. The verbs of extraction are also lexically ablative, but they are also lexically marked to instantiate the container-contained schema and so can never be centripetal.

9.3.3.3. Resultative apart

The centripetal converse of resultative *together* is *apart*. If resultative *together* denotes assembly, then resultative *apart* denotes disassembly:

9.56 a) John took the bicycle apart.
a') ≈ John took the parts of the bicycle away from one another
b) Political differences drove John and Mary apart.
b') ≈ Political differences drove John and Mary away from each other.

Again, there is a significant contrast in meaning between *apart* and its ablative paraphrase: the paraphrase *take the parts away from one another* only has the reading of taking them out of spatial proximity, whereas *take apart* means not only to take the parts out of spatial proximity, but also for them to stay that way, putting an end to the whole. This stronger reading is found with many other verbs in construction with *apart*:

9.57 a) Mary finally pulled the deal apart.
b) John's death tore the family apart.
c) Mary pushed the campfire apart.
d) John pried the pieces of the plate apart.
e) Mary split the company apart so that it could be sold.

As before, this extra iota of meaning appears to be causative, since the ablative act can be understood as causing the parts of the whole to loose their proximity on their own:

9.58 a) John took the bicycle apart.
b) ≈ John took the parts of the bicycle away from one another in a way that caused them to stay away from one another, presumably because they ceased to fit with one another as a whole

This is yet another manifestation of the resultative construction.

9.3.3.4. Verbs of dispersion

There are handful of verbs illustrated by *disperse* in (9.59) below that have paraphrases with *move apart* and *move away from* that qualify them as centripetal predicates, see (9. 59c, d). In contrast to the separatives, they reject an ablative argument, see (9.59d), and they do not license a conjunctive paraphrase of the reciprocal subevents, (9.59e). Nevertheless, they do translate into Spanish with the reflexive / reciprocal clitic pronoun *se*, (9.59f):

9.59 a) The yaks dispersed after sunrise.
b) The yaks moved apart after sunrise.
c) The yaks moved away from one another after sunrise.
d) * The yaks dispersed from one another after sunrise.
e) *Yickety Yak dispersed away from Yackety Yak and Yucky Yak, Yackety Yak dispersed away from Yickety Yak and Yucky Yak in the valley, and Yucky Yak dispersed away from Yickety Yak and Yak Yackety.
f) Los yakes se dispersaron después del amanecer.
the yaks SE disperse-3pPAST after of-the sunrise

The centripetal argument can be modified by *all*, (9.60a), and if singular, it can only be a collective or mass term, (9.60b), and cannot be quantified by *each* or *every*, (9.60c):

9.60 a) All the yaks dispersed after sunrise.
b) The {herd / fog / *yak} dispersed after sunrise.
c) *{Each / Every} yak dispersed after sunrise.

The dispersives do not license the ablative alternation, (9.61a), but they do license a path$_2$, (9.61b):

9.61 a) *The male yaks dispersed from the female yaks after sunrise.
b) The yaks dispersed into the forest from their feeding ground.

The other verbs that fit this pattern are exemplified in (9.62):

9.62 a) The troops scattered after the first burst of gunfire.
b) The players spread out across the field.
c) The party-goers disbanded once the beer ran out.
d) The particles diffused into the surrounding solution.

Again, the presence of the prefix *dis-* suggests that the dispersives are marked with respect to the congregatives.

9.3.3.5. Summary of centripetal constructions

The properties of all four sorts of verbs are organized into Table 9.3 on the pattern of Table 9.2. Table 9.2a, b highlight the differences between centripetal and centrifugal predicates, while Table 9.3c-g highlight their similarities. Table 9.3h-l present the properties that distinguish intersectives and separatives from congregatives and dispersives. Having elaborated a basic description of the

Table 9.3. Summary of centripetal predicates, and comparison to centrifugals.

	Property	*Centrifugal*		*Centripetal*	
		Inter-sectives 'collide'	*Congre-gatives 'gather'*	*Separ-atives 'separate'*	*Disper-sives 'disperse'*
a)	means *move ...*	*together*	*together*	*apart*	*apart*
b)	*dis-, un-*	no	no	yes	yes
c)	translates with *se*	yes	yes	yes	yes
d)	*all X*	yes	yes	yes	yes
e)	collective *X*	yes	yes	yes	yes
f)	singular count *X*	*	*	*	*
g)	*each/every X*	*	*	*	*
h)	mass *X*	*	??	*	yes
i)	cardinality of *X*	2+	4/5+	2+	4/5+
j)	preposition + reciprocal	with/to	??with	from	??from
k)	*with/from* alteration	with	???with	from	??from
l)	licenses $path_2$	*	yes	*	yes

centrifugal-centripetal contrast, let us now turn to deepening our understanding of it.

9.4. CENTER-ORIENTED CONSTRUCTIONS AS PATHS

It is now time to round out the discussion of spatial prepositions and center-oriented constructions with a formalization. As in the case of logical coordination and quantification, the first and perhaps most important step is to decide on a numerical representation of the denotation of the linguistic expressions in question. For the sake of perspicuity, we will make several simplifying assumptions. The first assumption is to restrain the model to one-dimensional space. The second assumption is to only permit five positions in this space, which are named by the five real numbers 1, 0.5, 0, -0.5, and -1. This is a rather loaded assumption, in that it imposes the perspective of three-valued logic onto the array of positions. The rationale behind it is our claim that in any movement of some entity *J* to or from some landmark *M*, there is an intermediate point at which it is hard to affirm conclusively that *J* is at *M* or that *J* is not at *M*. In the language of correlation, if *J* is at *M*, the two are spatially correlate (hence a 1 at the beginning of the path). Conversely, if *J* is not at *M*, the two are spatially anticorrelated (hence a -1 at the end of the path). If you cannot tell whether *J* is at *M* or not, the two are spatially uncorrelated. The third assumption is that movement in this space is a change from one of the five positions to another. The fourth assumption is that movement is smooth or connected, i.e. it does not skip intermediate positions. The final assumption is to limit the number of entities that occupy a position in this space to four.

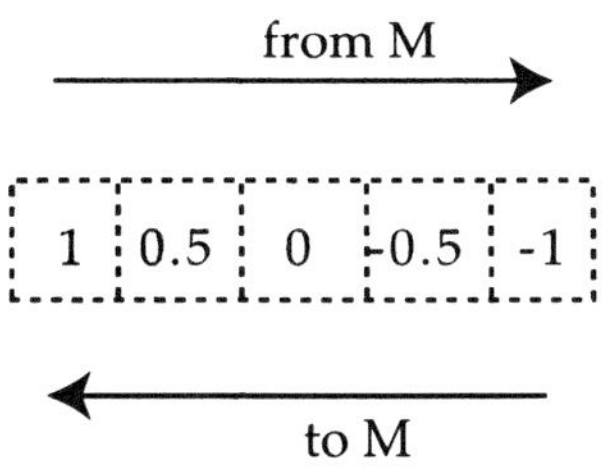

Figure 9.11. The data structure of paths.

Table 9.4. Spatial location of *J* and *M* for FROM, WITH, and TO.

	'J (went) from M'		*'J (stayed) with M'*		*'J (went) to M'*	
time	*loc(M)*	*loc(J)*	*loc(M)*	*loc(J)*	*loc(M)*	*loc(J)*
t_1	1	1	1	1	1	-1
t_2	1	0.5	1	1	1	-0.5
t_3	1	0	1	1	1	0
t_4	1	-0.5	1	1	1	0.5
t_5	1	-1	1	1	1	1

Having made these assumptions, the ablative/allative contrast can be modeled by convergence on one point from the other four, or dispersal from one point to the other four. Fig. 9.11 diagrams the main patterns of change, in which the central point is 1. Another way of understanding this representation is to imagine that the cells in Fig. 9.11 correspond to potential positions of John and Mary, in the manner organized into Table 9.4. In the ablative pattern of *from,* John starts at the same location as Mary, namely 1, and ends at any anticorrelated position. Conversely, in the allative pattern of *to,* John starts at any anticorrelated position and ends up spatially correlated with Mary at position 1. A third option has been added in the middle, one in which John retains his original position with respect to Mary. This describes the comitative pattern of *with.*

Fig. 9.12 plots the three different patterns in a space defined by the locations of Mary and John. Fig. 9.12a effortlessly wins the prize as the worst diagram in the book. Since the ablative, allative, and comitative paths are routed through the same points in space, they pile up one on top of the other to produce an unintelligible jumble of symbols. Fig. 9.12b unpeels the three patterns by adding a third dimension, that of direction. TO is labeled with the +1 direction, towards the center; FROM is labeled with the -1 direction, away from the center; and WITH lacks motion altogether and so bears the label of 0. Note that by assigning the negative value to the FROM direction, it is to be understood as marked, i.e.

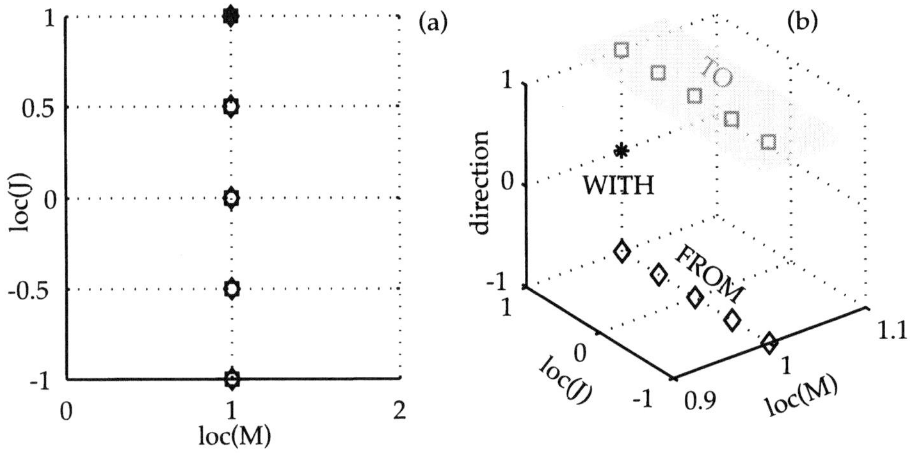

Figure 9.12. Phase space for prepositions in Table 9.4. (a) location of *M* x location of *J*; (b) location of *M* x location of *J* x direction.

anticorrelated, with respect to the TO direction. This appears to be the proper account of the facts, but we do not know why.

It is now straightforward to turn the dimensions of Fig. 9.12b into the arguments of a definition for the corresponding English prepositions. Let us postulate an abstract three-place predicate PATH(x, y, z), with the first two arguments defining a path between two entities given by the *x* and *y* axes of Fig. 9.12b, and the third argument consisting of an indication of direction. The prepositions *to*, *from*, and *with* can then be defined as in (9.63):

9.63 a) to(x, y) = PATH(y, x, 1).
b) from(x, y) = PATH(y, x, -1).
c) with(x, y) = PATH(y, x, 0).

This provides the background for our account of center-oriented constructions.

9.4.1. Covert reciprocity

Limiting our sights to the intransitive center-oriented verbs and assuming that they are distinguished from one another by a binary feature, a first approximation to their meanings is given by (9.64), where x_1 and x_2 are members of the location of the subject X, *loc(X)*, and are not the same entity, and the *V* operator finds the value of the *X* constituent that bears the feature *{+F}*:

9.64 a) V(X, [+centrifugal]) = ∀i, -i ∈ X (PATH(i, -i, 1) & PATH(-i, i, 1)).

b) $V(X, [+centripetal]) = \forall i, -i \in X$ (PATH(i, -i, -1) & PATH(-i, i, -1)).
c) $V(X [+tandem]) = \forall i, -i \in X$ (PATH(i, -i, 0) & PATH(-i, i, 0)).

The markedness of the anticorrelated direction -1 accounts for the usage of *dis-* and *un-* as prefixes to the centripetal verbs.

There is one way in which (9.64) is wildly at odds with the results of the previous subsection. It is that centrifugal verbs invariably choose *with* over *to*, despite having a value of 1 – not 0 – for their third argument in (9.64a). The reason for this unexpected choice of preposition can be attributed to the fact that an instance of TO ends with the trajector arriving at its goal, which is to say, the point at which both are spatially correlated. It appears that most centrifugal verbs focus on this final state and choose a preposition accordingly. It may be assumed that this 'weakening' of the expected morphosyntactic marking of the verbal subclass is facilitated by the unmarked nature of the centrifugals – and indeed may count as another bit of evidence in favor of their unmarked status.

9.4.2. The failure of center-oriented subalternacy

From the perspective of this chapter, the crucial stipulation in (9.64) is that the first two arguments of each path are reversed, so that the verbs are lexically reciprocal. In terms of Dalrymple et al.'s typology, they qualify as instances of Strong Reciprocity, since each member of the subject set stands in a relation *R* of moving towards or away from every other member of the set. Yet English does not normally realize a center-oriented object with a reciprocal pronoun. However, English does not realize many other anaphoric pronouns, either. For instance, many verbs of grooming require reflexive pronouns in Spanish that are optional in English. (9.65) gives but two examples:

9.65 a) Juan se afeitó.
John SE shave-3sPAST ~ "John shaved (himself)."
b) María se baño.
MARY SE bathe-3sPAST ~ "Mary bathed (herself)."

This leads us to conclude that the optionality of reciprocal pronouns with center-oriented constructions may be a peculiarity of English's.

Given the interpretational equality of center-oriented constructions with reciprocals, the reason for the failure of the subaltern inference with the former now becomes clear: a singular antecedent does not supply an alternative entity for the implicit reciprocal to refer to, so the semantic process of antecedent resolution fails.

9.4.3. Path$_2$ and gestalt locations

Perhaps the most puzzling property of the center-oriented predicates is how they cross-classify with respect to paths$_1$ and paths$_2$. Intersectives and separatives license a $path_1$ but not a $path_2$, while congregatives and dispersives behave in the opposite fashion.

It is rather difficult to pin down a reason for the contrast in collocation with $paths_2$. There are only two clues. One is that the congregatives-dispersives seem to distinguish themselves from the intersectives-separatives by being the verbal analog of a collective noun, that is, a sort of collectivizing predicate. The other is that intersection-separation can take place anywhere that (at least) two individuals intersect-separate, while congregation-dispersion seems to only take place at a given location, though this location may only be given rather vaguely in the context.

For lack of an alternative, we attempt to bring these two hunches together by writing a location into the definitions and reorganizing the PATH predicates so that the members of X move with respect to it. If *LOC* is a set of locations in the format of Fig. 9.12b, then ...

9.66 a) $V(X, [+congregative]) =$
$\exists i \in LOC\ \forall j, -j \in X\ (PATH(j, -j, 1)\ \&\ PATH(-j, j, 1)\ \&\ PATH(i, j, 1)).$
b) $V(X, [+dispersive]) =$
$\exists i \in LOC\ \forall j, -j \in X\ (PATH(j, -j, -1)\ \&\ PATH(-j, j, -1)\ \&\ PATH(i, j, -1)).$

In prose, all members of X move towards/away from one another and towards/away from the location *i*. To our visual imagination, the only schemata that can satisfy both requirements are center oriented.

The existential quantification of *i* invokes a location that is salient in the context. It therefore becomes the 'hook' on which to hang a $path_2$ – and in fact can constitute one of its end-points. In the absence of this location, an intersective-separative asserts no location other than that that which is given by the path that its subject members traverse.

Nothing that has been conjectured so far constrains the cardinality of the congregative-dispersive subject to the fuzzy range of four or five. In order to accommodate this last datum, it may be claimed that the location *i* is itself given by the locations of some members of the congregative-dispersive subject X. Since these entities are moving, the group location synthesized from their individual locations will not become apparent until some critical mass is reached. That is to say, the group location is a perceptual gestalt – a location that pops out from the background only when its elements are in the proper configuration, see for instance Palmer and Rock (1994) and Hochberg (1998) for recent discussion. Uncovering the exact principles of organization that enable us to perceive such gestalts would take us far beyond the goals of this chapter, so let us assume that *LOC* in (9.66) is actually the appropriate gestalt function of X, say *LOC(X)*, and leave a deeper investigation for another venue. The result is that the cardinality of the congregatives-dispersives need not be stipulated explicitly, since it follows from the perceptual requirement for several members

Table 9.5. General summary of center-oriented constructions.

Direction/ LOC(X)	*1*	*-1*
0	Intersective resultative *together*	Separative resultative *apart*
1	Congregative	Dispersive

of *X* to stand together in order to define a location that all members of *X* can be oriented towards.

9.4.4. The comitative/ablative alternation

Having established an analysis for $paths_2$, the reason for the restriction of $paths_1$ to the intersectives-separatives has to do with defining the comitative-ablative alternation as a single PATH:

9.67 a) $V(X, [+comitative]) = \exists x_1, x_2 \in X\ (PATH(x_1, x_2, 1))$.
b) $V(X, [+ablative]) = \exists x_1, x_2 \in X\ (PATH(x_1, x_2, -1))$.

These definitions can instantiate the same class of verbs as those of the intersectives-separatives due to the PATH predicate that they all share. However, the congregatives-dispersives are distinguished by reference to some 'center' *LOC(X)* which is not mentioned in (9.67), so they cannot be shoe-horned into this more restricted definition that omits the crucial specification of a center. It follows that the congregatives-dispersives cannot enter into the comitative-ablative alternation, which is correct.

It thus turns out that the four sorts of center-oriented constructions are distinguished by opposite settings of two parameters. The first is the setting for the *z* argument of direction in the constitutive PATHS, either '1' (allative) or '-1' (ablative). The second is the presence or absence of a location defined from the center-oriented subject, *LOC(X)*. These parameters are used to regiment the four sorts of center-oriented constructions as in Table 9.5.

9.5. SUMMARY

This chapter began with the observation of a number of predicates that prevented the subaltern implication from universals to particulars and followed up with an neuromimetic analysis of the main group, that of reciprocals, and a logical analysis of the center-oriented constructions. The failure of the subaltern inference with single-individual antecedents follows directly from the architecture of the network that resolves the reference of a reciprocal pronoun. In such a network, a single individual does not supply an alternative individual on which to base a reciprocal reading, so the network fails to satisfy reciprocal processing. Along the way, several discoveries were made. It emerged that reflexives and reciprocals share the same spatial representation, so that it is not unexpected that they pattern together syntactically and morphologically. Center-oriented constructions were shown to rely on the semantics of basic spatial prepositions, and to be distinguished by the parameters of direction of motion and existence or not of a 'center' to which the motion of the individuals is oriented.

Chapter 10

Networks of real neurons

This chapter takes a closer look at real neural networks, namely those for language and episodic memory. Given all of the speculation that we have indulged in about how logical coordinators, logical quantifiers, and collective predicates should be represented neurologically, it is about time that we looked at the areas of the brain that are held to be responsible for them. Unfortunately, the results of this investigation will be disheartening. Current techniques do not have the resolution to reveal how the human brain deals with such fine-grained aspects of language. On a more positive note, much more is known about episodic memory, and we will weave it into an analysis of how the two arguments of the logical operators are bound together by correlation.

10.1. NEUROLINGUISTIC NETWORKS

10.1.1. A brief introduction to the localization of language

Dronkers, Pinker and Damasio, 2000, p. 1174, state the problem quite succinctly:

> The lack of a homologue to language in other species precludes the attempt to model language in animals, and our understanding of the neural basis of language must be pieced together from other sources. By far the most important source has been the study of language disorders known as **aphasias**, which are caused by focal brain lesions that result, most frequently, from stoke or head injury.

The specific areas of the cerebral cortex responsible for linguistic functions were initially identified from post-mortem analyses of restricted lesions that correlated with a linguistic deficit during the patient's lifetime.

10.1.1.1. Broca's aphasia and Broca's region

Post-mortem studies of patients with slow or non-fluent speech but unimpaired comprehension enabled Paul Broca to identify the posterior third of the left inferior frontal gyrus as the seat of speech production, see Broca (1861), and Schiller, 1992, pp. 186-7, and Ryalls and Lecours (1996) for historical perspective. A patient with damage to this – Broca's – region displays a propensity for speech which varies from complete muteness to a slow, deliberate delivery characterized by impaired articulation, flat intonation, and a simplified grammar. This latter characteristic is quite striking, as the following quote from Kandel, 1995, p. 640, makes clear:

> Patients express nouns in the singular and verbs in the infinitive or participle, and often eliminate articles, adjectives, and adverbs altogether. For example, instead of saying "I saw some large gray cats", a patient with Broca's aphasia might say "see gray cat". These omissions are even more dramatic in more complex sentences. Here we can see the second characteristic of this defect: a breakdown in syntax. Consider the sentence: "Ladies and gentlemen, you are now invited into the dining room." A patient with Broca's aphasia may only be able to say "Ladies, men, room." When asked his occupation, a mailman with Broca's aphasia said "Mail ... Mail ... M ...."

To give the reader more of the flavor of this disorder, consider the following excerpt drawn from Brookshire, 2003, p. 153, of a patient attempting to describe a famous picture used in aphasia diagnosis called "The Cookie Theft", see the book's website for pointers to reproductions:

10.1. "uh ... mother and dad ... no ... mother ... dishes ... uh ... runnin[g] over ... water ... and floor ... and they ... uh ... wipin[g] dis[h]es ... and ... uh ... two kids ... uh ... stool ... and cookie ... cookie jar ... uh ... cabinet and stool ... uh ... tippin[g] over ... and ... uh ... bad ... and somebody ... gonna get hurt."

This utterance is remarkable in that it appears to be constructed almost entirely by juxtaposition of isolated words. It is practically devoid of the markers of hierarchical grammatical relationships that bind together normal English – with the recurrent exception of *and*. Such grammatical simplification is known as **agrammatism** or **telegraphic speech**. Note that not only does it involve the omission of function words, but it also involves distortion of word order. Damasio, 1992, p. 533, cites the attempt of a Broca's aphasiac to express *I will go home tomorrow* coming out as *Go I home tomorrow*.

A series of studies in the late 1970's demonstrated that Broca's patients also suffer from an impairment in comprehension, especially in those patients suffering from agrammatism, e.g. Caramuzza and Zurif (1976) and Berndt and Caramuzza (1980); see Damasio (1992) and Linebarger, 1998, p. 160, for further review. Agrammatics experience more difficulty in understanding sentences that are reversible, passive, and have object gaps than sentences that are nonreversible, active, and have subject gaps. The following examples are reproduced from Linebarger, 1998, pp. 160-1:

10.2 a) The clown chased the violinist. [Meaningful if reversed to *The violinist chased the clown.*]

b) The boy ate the apple. [Nonsensical if reversed to *The apple ate the boy.*]

10.3 a) The cop shot the robber. [Active]

b) The robber was shot by the cop. [Passive]

10.4 a) It was the cop who __ shot the robber. [Gap in subject position of embedded clause]

b) It was the robber who the cop shot __. [Gap in object position of embedded clause]

It is the (b) sentences that challenge the comprehension of agrammatic Broca's patients. Given that passive sentences are analyzed on a par with object-position gaps in intuition-based theories of syntax, the pattern of errors in examples (10.2-4) has attracted considerable interest from linguists working outside the neurolinguistic community, see for instance the relevant chapters of Visch-Brink and Bastiaanse (1998) and Bastiaanse and Grodzinsky (2000).

10.1.1.2. Wernicke's aphasia and Wernicke's region

A decade after Broca's discovery, post-mortem studies of patients with fluent speech but limited understanding lead Carl Wernicke to claim that speech comprehension was located at a different site, one in the posterior portion of the left superior temporal lobe, see Wernicke (1861). However, not only do patients with a lesion in this – Wernicke's – region have extreme difficulty understanding the speech produced by others, they also have difficulty selecting phonemes or entire words with which to express their own meaning, producing errors known as **paraphasias**. Errors in the selection of phonemes include addition, omission, or change in position. For instance, Damasio, 1992, p. 535, cites *trable* for *table* and *pymarid* for *pyramid*. Clearly, the more such **phonemic paraphasias** accumulate in a word, the harder it is to understand it, to the extent that the intended word may become unidentifiable. This is the point of **neologism**, illustrated in another of Damasio's examples by the utterance of *hipidomateous* for *hippopotamus*. The following are more extensive samples drawn from Brookshire, 2003, p. 155:

10.5 a) Clinician: "Tell me where you live."
Patient: "Well, it's a meender place and it has two ... two of them. For dreaming and pinding after supper. And up and down. Four of down and three of up ..."

b) Clinician: "What's the weather like today?"
Patient: "Fully under the jimjam and on the altigrabber."

Note that the function words, especially the logical coordinator *and*, are preserved in both cases.

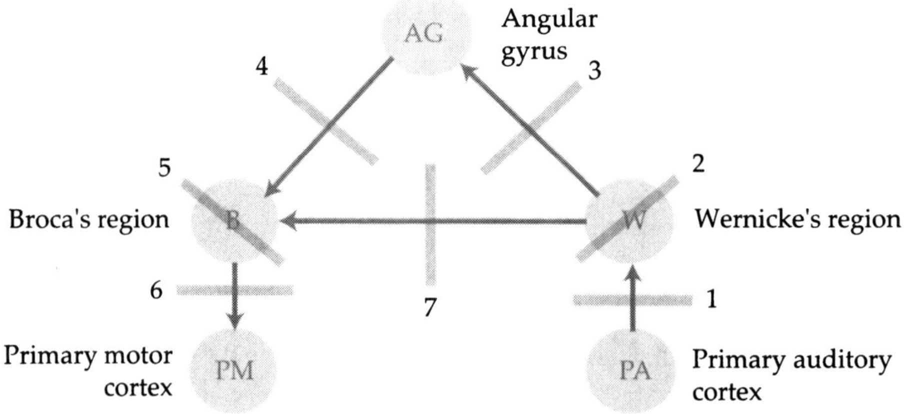

Figure 10.1. The Wernicke-Lichtheim-Geschwind boxological model.

A patient with damage to Wernicke's region may also fail to select the proper words with which to convey her ideas, though this deficit can be compensated for by the usage of paraphrases. Such **semantic paraphasias** are often quite simple, such as relying on generic terms like *thing* or *stuff* to stand in for the more specific words that do not spring to mind. Other times, they become quite elaborate. Kandel, 1995, p. 640, cites the example of a Wernicke's patient who was asked where he lived and answered "I came there before here and returned there." Such an overabundance of speech is referred to as **logorrhea**, or even more colorfully as **press of speech**.

10.1.1.3. Other regions

Early research also turned up other regions of the cerebral cortex that mediate spoken language in a secondary fashion. One clear candidate for such a ancillary role is primary auditory cortex, since a lesion in this region impairs the perception of any sound, not just spoken language. Another ancillary region is primary motor cortex, since a lesion in this region can produce **dysarthria**, an inability to articulate speech which also encompasses stuttering and stammering. Early researchers also made allowance for a conceptual contribution to spoken language, whose presence was indicated by impairments in the production of content words. Such impairments, known as **anomia**, were found mainly accompanying lesions in the angular gyrus, but also in other regions of association cortex.

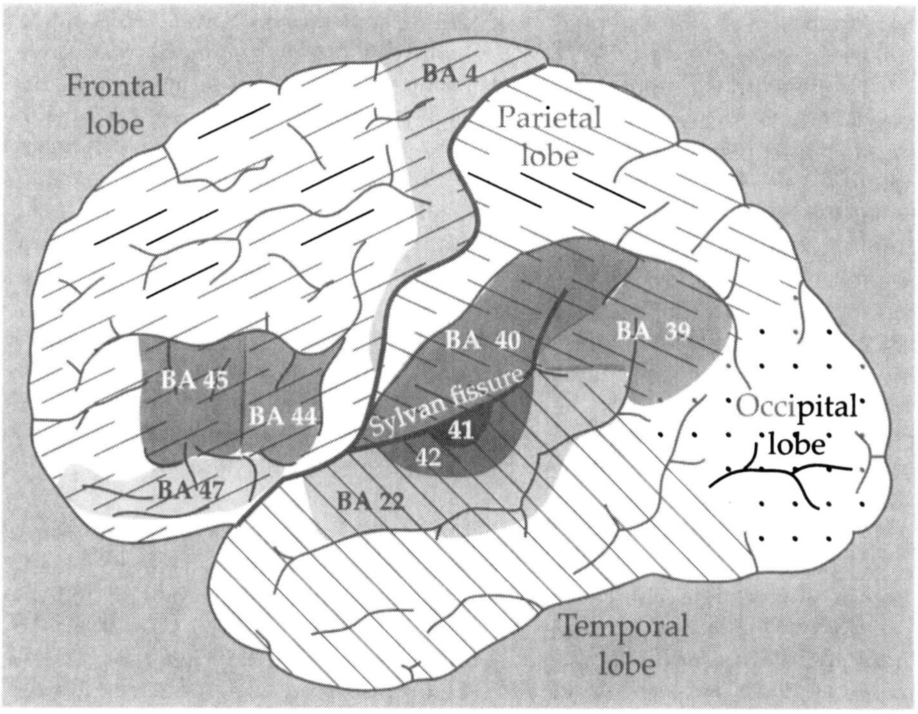

Figure 10.2. Left lateral view of a brain showing cerebral lobes and Brodmann's areas implicated in the Wernicke-Lichtheim-Geschwind model of language. Note that BA 44 is also known as the pars opercularis and BA 47 as the pars orbitalis.

10.1.1.4. The Wernicke-Lichtheim-Geschwind boxological model

From such observations, a model of the neurological organization of human linguistic ability was formulated as early as the second half of the nineteenth century. Wernicke (1874) presents the germ of this model. It organizes speech into three cerebral centers: (i) an acoustic center that is in charge of the perception of speech, located in Wernicke's region, (ii) a concept (Ger. 'Begriff') center which is assumed to store the meaning of words and to be involved in their encoding and decoding, and (iii) a motor center which is responsible for the articulation of speech, located in Broca's region. Each center is assumed to work independently and in serial, which is to say that a given center processes an incoming stimulus and then sends the result on to the next one. The flow of information from speech perception to speech production in this system is depicted by arrows pointing from one center to the next in Fig.

10.1. Potential sites of lesions that would disrupt the flow of information are depicted by gray bars cutting across the pathway or center in question.

It is interesting to point out that this system makes a prediction that was subsequently confirmed. Wernicke himself postulated that some pathway should connect his speech comprehension region to Broca's speech production region, which would permit the repetition of speech without necessarily understanding it. He speculated that a lesion in this pathway would create another type of aphasia, even though both the production and the comprehension regions were intact and functioning. This is the lesion represented by bar 7 in Fig. 10.1. Lichtheim (1885) reported the existence of such a speech impairment due to lesions of the **arcuate fasciculus**, thereby confirming Wernicke's prediction. Patients with a lesion in this bundle of white matter that connects Wernicke's and Broca's regions show **conduction aphasia**, which is characterized by fluent speech and normal comprehension but a diminished ability to repeat speech and frequent substitution of incorrect words or sounds for correct ones.

Finally, Geschwind (1965, 1970, 1972) presents a more recent elaboration of the Wernicke- Lichtheim model, so that his name is often attached to it along with one of the other two. We have added the modifier "boxological" in order to emphasize its status as a flow chart, in which each node is a 'black box' whose internal computation is unknown or unexamined. Thus most of the processing down by such a model is accomplished through its connections, which readily lends it to implementation in the connectionist computational paradigm, about which more is said below.

10.1.1.5. Cytoarchitecture and Brodmann's areas

The delimitation of Broca's and Wernicke's regions from other nearby regions was initially accomplished by examination of the cellular structure of the cortex through a light microscope. By mapping subtle differences in the thickness of the layers and the density of neurons in each layer, one very observant anatomist, Korbinian Brodmann, identified 50 or so areas in the cerebrum. In 1909 he published a book that remains the only comprehensive work on the subject. A map from this book of the various areas of the cortex seen from the left side has gained wide currency. Fig. 10.2 gives an approximation to Brodmann's brain map, with the areas relevant to speech highlighted. A glance at the full map[44] reveals that the numbering scheme does not trace a continuous path around the cerebrum. Areas 1, 2, and 3 are in the postcentral gyrus, all nicely adjacent to each other. Area 4 is in the precentral gyrus, right next to area 3. Area 5, however, is in the parietal lobe posterior to the

[44] See for instance Bailey and Von Bonen, 1951, p. 190, or the book's website for color versions of Brodmann's maps.

Table 10.1. Five centers implicated in the Wernicke-Lichtheim-Geschwind model of language, adapted from Pulvermüller and Preissl, 1994, p. 77.

Center	Results of lesions, numbered as in Fig. 10.1	Cortical region
Acoustic	(1) hearing impairment	Bilateral superior temporal gyri - BA 41, 42 (primary auditory cortex)
Speech comprehension	(2) Wernicke's or sensory or receptive aphasia: • severe deficit in comprehension • slight deficit in production (paraphasia, neologism, logorrhea)	Left posterior superior temporal lobe - BA 22 (Wernicke's region)
Concept	(3, 4) anomia	Angular gyrus - BA 39/40 - and other association cortices
Speech production	(5) Broca's or motor or expressive aphasia: • slight deficit in comprehension • severe deficit in production • agrammatism	Left posterior inferior frontal lobe - BA 44, 45, 47 Pars opercularis (Broca's region)
Motor	(6) dysarthria	Bilateral sensorimotor cortex-BA 4

postcentral gyrus, and area 6 is anterior to area 4! Likewise, area 7 is back in the parietal lobe, while area 8 is up front in the frontal lobe. It is not known what possessed Brodmann to number so erratically, but he did such a thorough job that his scheme has befuddled neuroanatomy students for over a hundred years.

10.1.1.6. Cytoarchitecture and post-mortem observations

It soon began to be appreciated that Brodmann's cytoarchitectonic areas house specific cognitive functions. For instance, some combination of Brodmann areas (BAs) 44, 45, and 47 house Broca's region. Uylings et al., 1999, p. 323, present a brief survey of studies that localize it to BA 44, BA44 and 45, or BA 44, 45, and 47. As Uylings et al. say, the reasons for this disagreement are partly historical and partly due to different understandings of the impairments encompassed by Broca's aphasia. Table 10.1 combines cytoarchitectonic information with the lesion information to give a general synopsis of the Wernicke-Lichtheim-Geschwind model. All of the linguistically relevant Brodmann areas lie within the first gyrus surrounding the Sylvian fissure, a region known as perisylvian cortex.

10.1.1.7. A lop-sided view of language

The reader may have noticed that so far the discussion has focussed on the left hemisphere of the cerebrum to the exclusion of the right. This focus is due to the fact that, since the late 1860's, aphasia was only found in patients with damage to the left hemisphere. This set the stage for a century-long lop-sided view of language. As Myers, 1999, p. 1 puts it,

> For the next 100 years, neurologists, speech-language pathologists, psychologists, and linguists studied aphasia and, later, apraxia, of speech, for clues about how humans process language, the nature of language breakdown, and the best means of remediating these breakdowns. Language *was* communication, and the left hemisphere (LH) was where it all happened. The LH appeared to house that which distinguishes us in a fundamental way from other living creatures; we could communicate using speech, and they could not. So powerful was this notion that the LH become to be known as the 'dominant' hemisphere. Little regard was given to the right hemisphere (RH), which was rather ignominiously dubbed the *silent* or *minor* hemisphere.

This lop-sided perspective of the classical model began to change in the 1960's with the development of a radical surgical procedure to control epilepsy.

10.1.1.8. The advent of commissurotomy

An epileptic seizure derives much of its disruptive power from the way in which it spreads from one hemisphere to the other via the corpus callosum. The **corpus callosum** is a bundle of about two million fibers that tie the two hemispheres together, see Aboitz, Scheibel, Fisher and Zaidel (1992) and LaMantia and Rakic (1990). Accounting for about 2% to 3% of all cortical fibers, these connections are established in a **homotopic** fashion, which is to say that points on one hemisphere are connected to points on the other in a roughly mirror-symmetric fashion, see Innocenti (1986) and Pandya and Seltzer (1986). The contralateral neurons that these fibers synapse on cover a fairly small cortical region, see Hartenstein and Innocenti (1981) and Innocenti (1986).

Commissurotomy is an operation that severs the corpus callosum. Such a radical intervention as the severance of this capacious pathway was devised in the hopes of preventing an epileptic seizure from spreading from one hemisphere to the other. It had the unexpected consequence of supplying a pool of patients with "split brains", in which one hemisphere could be investigated without interference from the other. The results of research on this new group of patients confirmed earlier suspicions about the importance of the right hemisphere for visual perception and pointed to a new role for it in decoding simple linguistic forms, see Myers, 1999, p. 2, for references.

10.1.1.9. Experimental 'commissurotomy'

In the 1970's, the development of two techniques that were more attuned to the experimentalist's lab than the surgeon's operating room permitted further insight into the differential contribution of the two hemispheres. The techniques are **dichotic listening**, which is described below, and **hemifield tachistoscopy**, a technique for dividing a viewer's visual field into two halves, which is not discussed here.

10.1.1.9.1. Dichotic listening

As Ivry and Robertson, 1998, p. 25, put it, dichotic listening ...

> ... was developed to mimic processing demands in the natural world, where sensory overload is common. Consider the cocktail party or, more appropriate for today, the wine-tasting party. We may attempt to speak with one individual, but the speaker's voice is intermixed with a multitude of incoming auditory signals: conversations going on about us, music from the compact disc player, the clatter of plates being filled at the buffet table, the children watching a video in the next room. Despite this cacophony of sound, we are quite proficient at focusing on the relevant signal—the words being spoken by our conversational partner.

Dichotic listening explores this ability through the presentation of two simultaneous messages, one to each ear.

In the seminal study of Kimura (1961a), the stimuli were digits, presented so that one digit was heard in the left ear at the same time as a second digit was heard in the right ear. Kimura found that people were much more likely to report having heard the stimuli presented to the right ear, an effect dubbed the **right-ear advantage**. Kimura's usage of dichotic listening to confirm the lateralization of language to the left hemisphere in normal subjects was soon substantiated by studies with lesioned subjects. Kimura (1961b) showed that patients with left temporal lobe lesions performed worse at the task than did patients with right temporal lobe lesions. In addition, split-brain patients showed a considerable right ear advantage in a study using words as stimuli. They succeeded in recognizing words presented to the right ear, but performed no better than chance for words presented to the left ear, see Milner, Taylor, and Sperry (1968) and Sparks and Geschwind (1968).

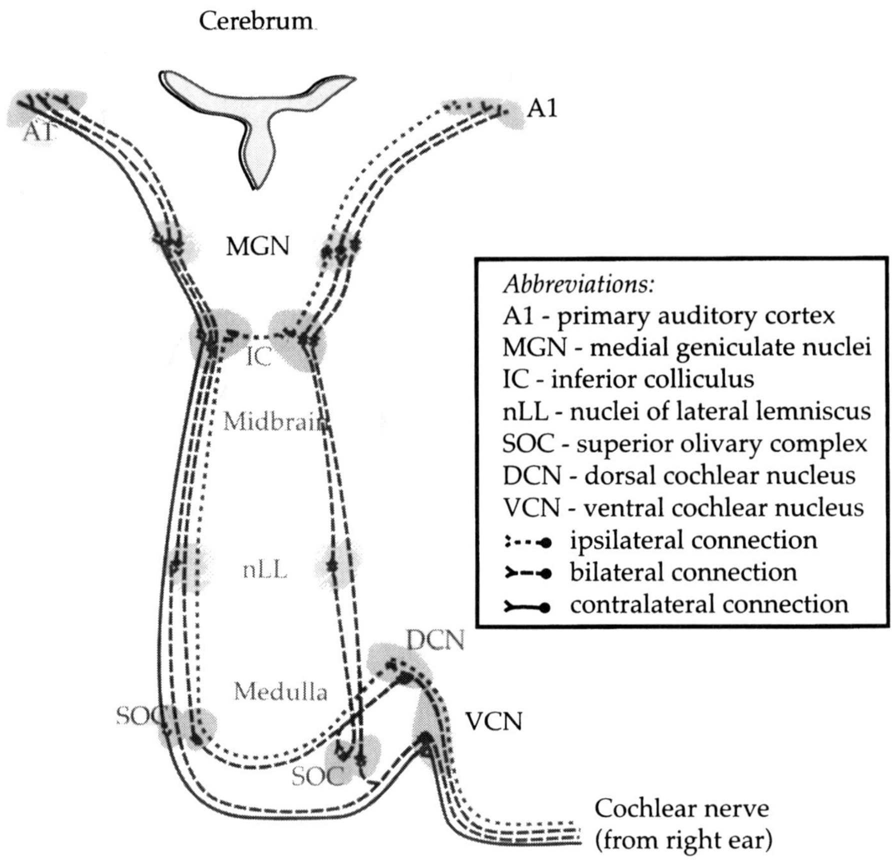

Figure 10.3. Simplified ascending auditory pathways. Comparable to Purves et al., 1997, Fig. 12.11; Bear, Connors, and Paradiso, 1996, Fig. 11.17; and Henkel, 1997, Fig. 20-10.

10.1.1.9.2. An aside on the right-ear advantage

The reader may find it surprising that the right ear wins out over the left for linguistic stimuli, after we have accepted at face value the Lichtheim-Geschwind claim that the left hemisphere is dominant for language. The reason for this apparent contradiction lies in the fact that each ear sends its output signal to both auditory cortices. The example input in this figure enters from the right ear and divides into four sample streams. Using the terms introduced in Sec. 1.2.1 for vision, an ipsilateral stream goes to the auditory cortex on the same side of the head as the ear in question, a contralateral stream goes to the auditory cortex on the opposite side, and a bilateral stream goes to both sides. Tracing the various connections in Fig. 10.3 reveals that each primary auditory cortex receives a signal from the right ear, and a symmetrical pathway also departs the left ear.

This double wiring permits each auditory cortex to compare the input that it receives from both ears. The advantage of one ear over the other comes from each primary auditory cortex weighing its contralateral input more than its ipsilateral input, see for instance Henkel, 1997, p. 298, for a brief explanation. Thus the right-ear advantage reflects the underlying domination of the contralateral left hemisphere for processing linguistic stimuli.

10.1.1.9.3. A left-ear advantage for prosody under dichotic listening

Nevertheless, the right-ear advantage for linguistic stimuli is not absolute. Blumstein and Cooper (1974) used a dichotic-listening task to demonstrate the importance of the right hemisphere for perceiving intonation contours. The stimuli in their first experiment were four three-word sentences: a declarative ("It has come."), an interrogative ("Has it come?"), an imperative ("Hal, come here!"), and a conditional ("If he came,..."). The sentences were filtered so that no frequency above 510 Hz was heard. The words were unintelligible in filtered form, but with training the subjects learned to identify the pitch contour. In each trial of the experiment, subjects were first presented dichotically with a pair of the four stimuli and then with a probe to both ears that either matched one member of the dichotic pair or was different from both. Subjects were more accurate when the probe matched the dichotic stimulus that had been presented to the left ear, indicating right-hemisphere dominance.

An additional piece of evidence for the left-ear advantage of prosody comes from the observation that recognition of individual voices is also more accurate for stimuli presented to the left ear, see Kreiman and Van Lancker, (1988); and Van Lancker, Kreiman, and Cummings (1989). Individual differences between speakers are clearly manifest in pitch contours.

10.1.1.10. Pop-culture lateralization and beyond

As the usage of dichotic listening and other techniques such as hemifield tachistoscopy, see footnote 45, deepened and broadened, the understanding of the role of the two hemispheres began to evolve from functional specialization to more abstract considerations of "styles" or "strategies" for the processing of information. The specialization of the left hemisphere for language was generalized to a strategy for analytic, linear, and rule-like processing in any domain. The right hemisphere took on a complementary strategy for holistic, nonlinear, and parallel processing. Moreover, the popular press began to filter this research out to the public at large. To quote Myers, 1999, p. 3, again,

> ... the RH became the seat of artistic capacity and creativity, less bound by rules, more fluid and flexible, and more adept at managing novel input than the LH. Allusions to the two sides of the brain appeared in all sorts of social contexts. People talked about whether they were more "right-brained" or "left-brained". A popular automobile ad detailed the car's features in printed text on the left page, and pictured it speeding up a twisting mountain road on the right page with the words, "A car of the left side of your brain, and a car for the right side of your brain" written across the two-page layout.[45]

In this context, I cannot resist relaying a quote from Hopper (1998), a book dedicated to the teaching of college study skills, that adopts this rigid classification of people by their 'dominant' hemisphere:

> The left side of the brain processes information in a linear manner. It process from part to whole. It takes pieces, lines them up, and arranges them in a logical order; then it draws conclusions. The right brain, however, processes from whole to part, holistically. It starts with the answer. It sees the big picture first, not the details. If you are right-brained, you may have difficulty following a lecture unless you are given the big picture first. Do you now see why it is absolutely necessary for a right-brained person to read an assigned chapter or background information before a lecture or to survey a chapter before reading? If an instructor doesn't consistently give an overview before he or she begins a lecture, you may need to ask at the end of class what the next lecture will be and how you can prepare for it. If you are predominantly right-brained, you may also have trouble outlining (you've probably written many papers first and outlined them latter because an outline was required). You're the student who needs to know why you

---

[45] In view of the right-ear/left-hemisphere effect uncovered by dichotic listening, the careful reader may wonder whether there is a corresponding effect in vision. Indeed there is, and it is revealed by hemifield tachistoscopy, mentioned in passing above. Thus the layout of the ad described by Myers is neurophysiologically incorrect, in that the text should be on the right side and the picture on the left, for the eyes to channel the correct information to the contralateral hemisphere intended by the copy writers.

> are doing something. Left-brained students would do well to exercise their right-brain in such a manner.

Would that it were so simple.

The assumptions about hemispheric specialization conveyed in Hopper's quote eventually began to be seen as narrow, overly precise, and oversimplified. As Banich and Heller, 1998, p. 1, say:

> The topic fell from grace during the reign of dichotomania, when it seemed imperative to determine for every imaginable mental function whether it was better performed by the left hemisphere or by the right. As the mountain of research results grew, the inconsistencies also increased, until to some observers, the whole endeavor seemed a bit futile.

For instance, laterality effects were found to vary with one's familiarity with the experimental task, type of stimulus material, and other factors that contradicted any fixed division between left and right, see Zaidel (1985) for review.

Thus by the mid 1980's, two kinds of insights had been gained. On the one hand, the significance of nonverbal processing in general, and the contribution of the RH to intellectual functioning in particular, had been recognized. The RH would never again be labeled a minor functionary in the cognitive enterprise, and a crucial facet of this revision was an ever-broadening appreciation of the linguistic abilities of the RH. On the other hand, the brain was seen as a much more dynamic organ than Wernicke or Lichtheim could have conceived of. Allow us a long quote from Banich and Heller, 1998, p. 1, who summarize this point with more authority than we have:

> Researchers have come to appreciate that the seemingly contradictory nature of the findings from the past 20 years were saying something very essential about hemispheric asymmetries, namely, that they are more dynamic than static, more process-oriented than representation-specific. It is not that the left hemisphere processes verbal information and the right hemisphere processes spatial information. Rather, the left hemisphere may be better conceptualized as being specialized for processing information in a piecemeal, analytic, and sequential manner, which just happens to be a good method for processing verbal information, and the right hemisphere may be better conceptualized as being specialized for processing information in an integrative and holistic manner, which just happens to be ideal for processing spatial information. These complementary processing modes allow the hemispheres to adapt their processing styles in a dynamic

manner depending on the nature, context, and demands of the task.

We return to further details on the lateralization of language in Section 10.1.2.1, so let us now return to our brief history of neuroscience.

10.1.1.11. Neuroimaging

The last quarter century has seen the invention and deployment of several technologies that allow us to see with our own eyes how the brain works.

10.1.1.11.1. CT and PET

In 1971 Robert S. Ledley invented the computerized (axial) tomographic scanner, known more colloquially today as the CT or CAT scanner. This device rotates 180° perpendicular to a patient's body, sending out a pencil-thin X-ray beam at 160 different points. Crystals positioned at the opposite points of the beam detect and record the absorption rates of the varying thickness of tissue and bone, creating a cross-sectional "slice" of the body. The slices are relayed to a computer, which mathematically stitches them together to form a three-dimensional image of the body on the computer screen, see Brookshire, 2003, p. 82 or the book's website for sample images. CT scans offer clear views of any part of the anatomy, including soft organ tissues, making it invaluable for diagnostic studies of internal bodily structures, such as the detection of tumors or cerebrovascular accidents that cause aphasia.

Michael Phelps, Edward Hoffman and colleagues at Washington University achieved the first image of a human from positron emission tomography (PET) in 1974, after several years of development, see Nutt (2002). To produce a PET scan, a patient is administered a solution of a metabolically-active substance, such as glucose, tagged with a positron-emitting isotope. The substance eventually makes its way to the brain and concentrates in areas of high metabolism and blood flow, which are presumably triggered by increased neural activity. The positrons emitted by the isotopes are collected by detectors arrayed around the patients' body and converted into signals which are amplified and sent to a computer for construction of an image, see Brookshire, 2003, p. 84, or the book's website for sample images.

PET differs from CT in that it uses the body's basic biochemistry to produce images. The positron-emitting isotope is chosen from elements that the body already uses, such as carbon, nitrogen, oxygen, and fluorine. By relying on normal metabolism, PET is able to show a biochemical change even in diseases such as Alzheimer's in which there is no gross structural abnormality.

10.1.1.11.2. MRI and fMRI

In 1977, a team lead by Raymond Damadian produced the first image of the interior of the human body with a prototype device using nuclear magnetic resonance, based on ideas that Damadian had pioneered over the preceding six years (Damadian 1971), though the technology itself had been developed during World War II to probe the composition of various substances.

Damadian's device uses liquid helium to supercool magnets in the walls of a cylindrical chamber. A subject is introduced into the chamber and so exposed to a powerful magnetic field. This magnetic field has a particular effect on the nuclei of hydrogen atoms in the water which all cells contain that forms the basis of the imaging technique.

All atoms spin on their axes. Nuclei have a positive electronic charge, and any spinning charged particle will act as a magnet with north and south poles located on the axis of spin. The spin-axes of the nuclei in the subject line up with the chamber's field, with the north poles of the nuclei pointing in the 'southward' direction of the field. Then a radio pulse is broadcast toward the subject. The pulse causes the axes of the nuclei to tilt with respect to the chamber's magnetic field, and as it wears off, the axes gradually return to their resting position (within the magnetic field). As they do so, each nucleus becomes a miniature radio transmitter, giving out a characteristic pulse that changes over time, depending on the local microenvironment surrounding it. For example, hydrogen nuclei in fats have a different microenvironment than do those in water, and thus transmit different pulses. Due to such contrasts, different tissues transmit different radio signals. These radio transmissions can be coordinated by a computer into an image, see Gregg (2002), based on Horowitz (1995). This method is known as magnetic resonance imaging (MRI), and it can be used to scan the human body safely and accurately. The reader is referred to Brookshire, 2003, p. 83, or the book's website for sample images.

An elaboration of MRI called functional magnetic resonance imaging (fMRI) has become the dominant technique for the study of the functional organization of the human brain during cognitive, perceptual, sensory, and motor tasks. As Gregg (2002) explains it, the principle of fMRI imaging is to take a series of images in quick succession and then to analyze them statistically for differences. For example, in the blood-oxygen-level dependent (BOLD) method introduced by Ogawa et al. (1990), the fact that hemoglobin and deoxyhemoglobin are magnetically different is exploited. Hemoglobin shows up better on MRI images than deoxyhemoglobin, which is to say that oxygenated blood shows up better then blood whose oxygen has been depleted by neural metabolism. This has been exploited in the following type of procedure: a series of baseline images are taken of the brain region of interest when the subject is at rest. The subject then performs a task, and a second series is taken. The first set of images is subtracted from the second, and the areas that are most visible in the resulting image are presumed to have been activated by the task.

One of the most vexing issues surrounding fMRI is that until recently, it was not known exactly what component of neural metabolism was responsible for the BOLD signal, see Russo (2000) for an overview. There are two possibilities. One we have already discussed in considerable detail, namely the action potential. The other is known as the **local field potential**. These are the more slowly varying electrical potentials that arise from the input to, and integrative processes within neurons, and especially within dendrites.

It now appears that Logothetis et al. (2001) have taken the first steps towards resolving this issue by combining fMRI with microelectrode recordings such as those used by Hodgkin and Huxley to measure the action potential, though using many more microelectrodes. As reviewed in Raichle (2001), Logothetis et al. were able to distinguish between action potentials and local field potentials, and identify the local field potential as the major determinant of the fMRI signal. In other words, the activation of an area of the brain seen in fMRI predominantly reflects synaptic input to that area and the accompanying changes in dendritic processing, rather than output from the area.

Raichle (2001) puts this result together with recent insights into brain metabolism to forge an understanding of the BOLD signal as concomitant on the recycling of the excitatory neurotransmitter glutamate. Glutamate recycling draws energy from the break-down of glucose but does so anaerobically – without consuming oxygen. Thus blood flow increases in an active area in order to deliver more glucose, but since glucose metabolism does not consume oxygen, the proportion of oxygen in the blood does not fall. This local excess of oxygen accounts for the blood-oxygen-level dependent (BOLD) signal.

Despite this success, there is still a long way to go to relate recordings from individual neurons with the resolution of MRI. The smallest unit of an MRI image is an imaging **voxel**, which is the three-dimensional element determined by the two-dimensional MRI pixel size, which varies from 1mm x 1mm to 4mm x 4mm, and the slice thickness, which varies from 2 to 5mm. Consequently, in a 1 mm^3 voxel, one might typically find 10^5 neurons and 10^8 synapses, whereas current techniques for multiunit recordings can only implant and monitor microelectrodes up to two orders of magnitude, from 20 to 100, see Menon (2001).

10.1.1.11.3. Results for language

The new perspective afforded by the more recent methodologies has, on the whole, confirmed the Wernicke-Lichtheim-Geschwind boxology of language, as laid out by Price (2001) in a recent review. What innovations there are tend to involve small-scale refinements of the overall architecture. For instance, Hickok and collaborators, Hickok (2000, 2001) and Hickok and Poeppel (2000), find that the picture of speech perception is in need of revision, in that it appears to be more bilaterally organized than is accepted in the classical model, and Wernicke's region plays a larger role in speech production. Nevertheless, they do not call into question the overall layout as set forth in Table 10.1, and I have not found any references that do.

Having said this, we should point out that there is a major challenge to the classical model circulating in a variety of sub-disciplines that are slightly removed from neurolinguistics proper. The challenge has to do with whether the Wernicke-Lichtheim-Geschwind flow diagram is specialized for language, or whether it is specialized for certain computations that happen to be useful for language processing, but could be used for other domains as well. We have already broached this controversy in Sec. 1.3.4, where a distinction between strong and weak modularity is drawn. We will lift up the carpet to expose another corner of this controversy in Section 10.1.2.4 below.

10.1.1.12. Computational modeling

One new methodology has been omitted from this review – the one that is utilized in this monograph, namely computational modeling. Nadeau (2000) provides a convenient recent review of computational models of language, with emphasis on the parallel distributed processing or **PDP** approach inaugurated in McClelland, Rumelhart, and the PDP Research Group (1986) and Rumelhart, McClelland, and the PDP Research Group (1986). The main algorithms of this approach are backpropagation of error, which was briefly discussed in Chapter 5, and spreading activation, such as the IAC model of Chapter 8. Such models have a variety of properties that make them "brain-like" and so helped them to gain popularity among cognitive scientists in the late 1980's and early 1990's.

Oddly enough, most of models that Nadeau reviews have to do with reading/writing and not speech perception/production. One PDP-based model which does is that of Pulvermüller and Preissl (1994), whose design is essentially that of an IAC network, and so is very similar to our Spreading Activation Grammar in Chapter 8.

To design a simulation that actually articulates or understands speech adds considerable complexity to the problem, though the gains in descriptive and explanatory adequacy would seem to be worth the effort. One of the few groups to take up this challenge is Frank Guenther's lab, which has produced the DIVA model of speech production, see Guenther (1994, 1995a, b); and Guenther, Hampson, and Johnson (1998). DIVA accounts for a variety of experimental data, from kinematic analyses of articulator movements to functional imaging studies of the human brain, so it has the closest fit to the Wernicke-Lichtheim-Geschwind model of language that is covered in this monograph.

Yet even in DIVA, every neuron computes the same function, so the computational ability of all of the linguistic models arises mainly from the weights of its connections. For this reason, such paradigms are often known collectively as **connectionism**. Connectionistic computation is the simplest stop beyond the boxological model of language of Wernicke-Lichtheim-Geschwind, in which the nodes do not compute any function at all. We know of no models in which neurons in different layers compute different functions, which is more consonant with how real brains work. After all, the whole basis for the identification of Brodmann's areas has to do with discerning histological differences among neuronal populations. If this histological differentiation did not contribute to any functional differentiation, it would be much less taxing of the brain's resources to not create it in the first place.

To bring it all together briefly, there are a handful of neuromimetic computational implementations of language, which have been used to simulate a handful of observations about normal and deficient linguistic processing. Unfortunately, none of this body of research addresses the semantic operations performed by function words, which is a gap that this monograph takes the first steps towards filling.

10.1.2. Localization of the logical operators

Much of the attractiveness of the Wernicke-Lichtheim-Geschwind boxological model of language comes from its grounding in simple considerations of the sensorimotor organization of speech. Yet this strength has the consequence of relegating the non-sensorimotor aspects of speech, such as semantics, to some hidden cubbyhole in the brain. Fortunately, this handicap has not prevented neurophysiological investigation into non-phonological linguistic processing, though the results are much less firm than for the phonological aspects of language.

10.1.2.1. Word comprehension and the lateralization of lexical processing

Studies of the comprehension of individual words are usually undertaken in a research design known as semantic priming. **Semantic priming** is the process through which a subject responds more quickly or accurately to a target word if it is preceded by a semantically related word than if it is preceded by a semantically unrelated word. For instance, *dog* is more easily recognized if it is preceded by *cat* than if it is preceded by *cap*. This increase in performance is taken to indicate that some facet of meaning is accessed in the exposure to the initial word that is shared with the target word and so helps to accelerate its recognition.

Following the summary of Beeman and Chiarello, 1998, p. 4, the consensus in the literature is that both hemispheres have access to similar mental dictionaries, which is to say that they appear to not differ in semantic knowledge. Where they differ is in how this common knowledge is activated. In general, the LH primes closely related meanings and a single interpretation for a target word very quickly, whereas the RH primes more loosely related meanings and multiple interpretations for a target word and maintains this facilitation over a longer period.

Two experimental procedures that reveal this contrast are known as **direct** vs. **summation priming**. Direct priming is the simplest application of the priming paradigm: a target word is facilitated by a strongly related prime word, such as *scissors* by *cut*. Summation priming, in contrast, is the facilitation of a target word by three prime words that are only loosely related to it, such as *launch* by *shuttle*, *ground*, and *space*. Experiments have shown that direct

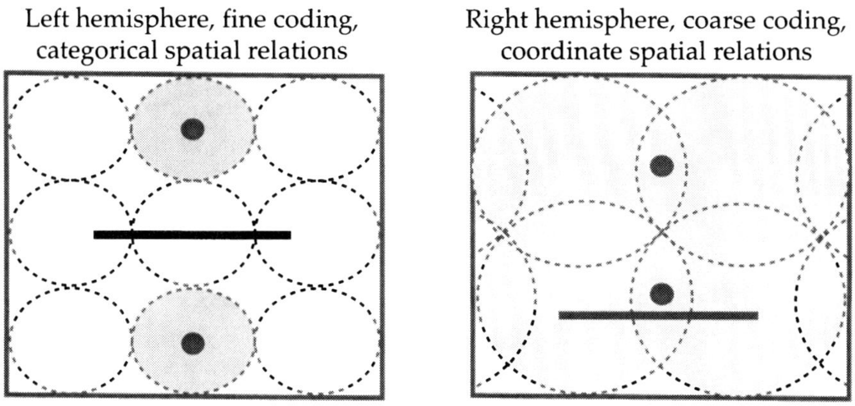

Figure 10.4. Hemispheric difference in spatial coding. On the left, each dot is above or below the bar; on the right, one dot is near and the other is far from the bar. Darker shading indicates higher activation of a receptive field by a dot.

priming is more robust in the left hemisphere, while summation priming is more robust in the right, see Beeman et al. (1994).

Another way in which the processing advantage of the LH comes to the fore can be seen when subjects pay attention to the meaning relation between words, and when suppression of alternate meanings or selection of a single best response from many choices is required. For instance, a subject may be presented with a noun such as *hammer* and be asked to supply a verb, giving the response *(to) pound*. Such testing, which evokes a single response, is sometimes called **convergent** semantic processing, see Myers, 1999, pp. 92ff. Conversely, one way of bringing the processing advantage of the RH to the fore is to continue the experiment and ask the subject to supply yet another verb, resulting in a response such as *(to) throw*. Such testing, which evokes many responses, is sometimes called **divergent** semantic processing. The conclusion drawn from these observations is that the RH is adept at facilitation for words that share few semantic features, e.g. *arm/nose*, and for the less frequent meaning of an ambiguous word, e.g. *bank/river*. Moreover, this happens at time spans beyond which LH priming is observed; see references in Beeman and Chiarello, 1998, p. 5.

The challenge is to reduce this asymmetry to some simpler, independent factor that could hold over a variety of experimental paradigms. One of the earliest proposals was designed for spatial representations by Stephen Kosslyn and his associates. They have argued that the two hemispheres are specialized for solving different computational problems, see Kosslyn et al.

Figure 10.5. Meanings activated from an utterance of the word "pig". Starting at the top center and going clockwise, the images are of a cow, a piggy bank, a shopping cart (for 'food'), and a farm.

(1989), Kosslyn, 1994, pp. 192ff, Jacobs and Kosslyn (1994), and Ivry and Robertson, 1998, pp. 111ff. The left hemisphere is claimed to specialize for **categorical** spatial relationships such as *above/below*, *on/off*, or *left/right*, which describe discrete, qualitatively different positions in space. In contrast, the right hemisphere is claimed to specialize for **metric** or **coordinate** spatial relationships such as *near/far*, which describe dense, quantitatively different positions in space.

This distinction, in turn, is claimed to reduce to the receptive field size of the neurons found in the visual centers of either hemisphere. Fig. 10.4 organizes all of this information together into a single contrast. Left hemisphere visual neurons are claimed to have small, non-overlapping receptive fields, and right hemisphere visual neurons, large, overlapping receptive fields. The nine non-overlapping receptive fields of the **finely coded** LH idealization index nine patches of space (one for each neuron) in the square. This permits strong activation of just a few neurons and delineation of large areas of space, such as those above or below the bar. In contrast, the eight overlapping receptive fields in the **coarsely coded** RH idealization index at least eighteen patches of space (one for each area of overlap). This permits weak activation of many neurons which collaborate to localize points in space

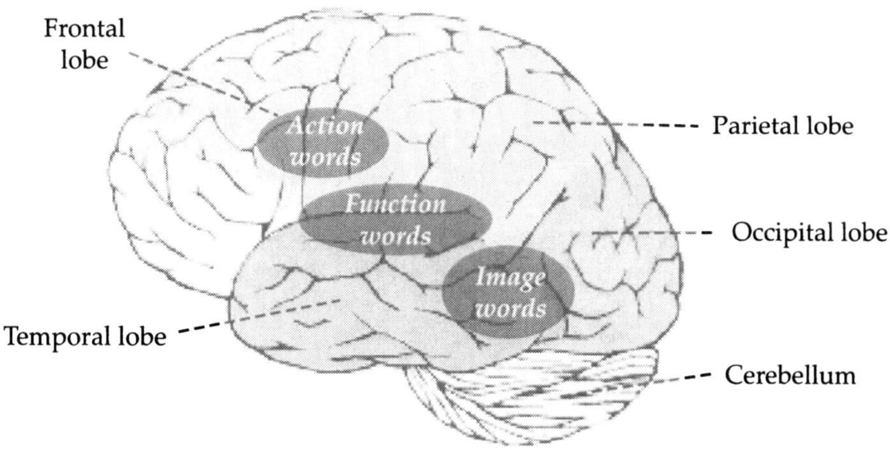

Figure 10.6. Cerebral localization of content words with strong associations for actions and for images, along with function words.

precisely. It is the precise spatial localization of coarse coding that subserves the discernment of gradation denoted by coordinate relations.

With respect to language, the coarse/fine distinction carries over almost intact to semantic processing. As Myers, 1999, p. 95 puts its, fine semantic coding is seen in the LH in its ability to rapidly select the most familiar or dominant meaning of a word while suppressing other less closely related meanings. The RH, in contrast, slowly produces multiple meanings that are weakly activated and weakly associated. Fig. 10.5 introduces an example, based on the utterance of the word "pig". On the left, the image representing the dominant meaning is activated strongly; on the right, images representing several associated meanings are activated weakly.

10.1.2.2. Where are content words stored?

From these general studies of lexical lateralization have emerged two main schools of thought of the localization of lexical entries for content words. One is that they are stored by syntactic category, so that there would be an area for nouns, another for verbs, another for adjectives, and so on. The other is that they are stored in terms of semantic properties. For instance, words with a strong visual component would be stored in occipito-temporal areas, while words with a strong action component would be stored in frontal areas responsible for motor programming.

The advent of noninvasive brain-imaging techniques has afforded us the means to test these predictions. A recent case in point is Pulvermüller, Mohr and Schleichert (1999), which extends previous research using event-related brain potentials (ERPs) to discover what zones of the brain are activated upon hearing German nouns with strong associations (i) for images and (ii) for actions, and (iii) verbs with strong associations for actions. German was chosen as the source language of the stimuli due to its consistent morphological distinction between nouns and verbs.

The result is that brain responses to visual nouns and action verbs differed at central and occipital sites, and a very similar difference was found for the comparison of visual vs. action nouns. No reliable differences were found between action verbs and action-related nouns. The diagram in Fig. 10.6 gives a very rough idea of the difference in localization of the two word classes. This result thus supports the semantic hypothesis of word storage over the syntactic hypothesis outlined above. More generally, it supports an associative theory of word learning, in which the perisylvian sensorimotor representation of a word is activated together with a distinct cortical topography corresponding to its semantic class, such as the frontal areas involved in motor programming for action words or the occipito-temporal areas involved in visual perceptions for imagable words.

10.1.2.3. Where are function words stored?

In contrast to words with concrete and easily-imagable meaning, function words such as pronouns, auxiliary verbs, conjunctions, and articles primarily serve a grammatical purpose. Under Pulvermüller's theory of associative cell assemblies, it follows that the meaning of function words cannot be explained based on objects or actions that they refer to, but rather must be a more complex function of their use, learned in highly variable linguistic and non-linguistic contexts. Evidently, the correlation between the occurrence of a particular function word and a particular sensory stimulus or motor action is low. Therefore, there is no reason why the perisylvian assembly representing the phonological form of a function word should incorporate additional neurons. If this is correct, neural assemblies representing function words are predicted to congregate in the perisylvan region to the exclusion of the extra-perisylvan zones that are observed with content words.

Pulvermüller, 1999, §5.1, summarizes several converging lines of research that support this prediction. To give the reader a taste of the data, let us consider but a single example. It is predicted that a perisylvian lesion will destroy a large percentage of neurons included in function word representations, but only a smaller part of the representations of content words. This prediction practically defines agrammatism, in which patients have difficulty producing function words and show deficits in language comprehension that can be explained by the assumption that they have a selective deficit in processing these lexical items, see Caramazza and Berndt (1985), Pick (1913), and Pulvermüller (1995a). Crucially, lesions within the entirety of the perisylvian region can be the cause of agrammatic aphasia, see Vanier and Caplan (1990).

10.1.2.4. Function-word operations and anterior/posterior computation

Thus we are lead to the conclusion that the cortical representation of function words, and especially the logical operators, is concentrated in the perisylvian region, but this conclusion leads to a paradox, or at least an indeterminacy. Given that the perisylvian region is also the seat of phonological representation, the fact that function-word cell assemblies aggregate there suggests that they have no more content than a phonological representation. Yet several decades of research in truth-conditional and model-theoretic semantics have demonstrated that function words perform many subtle and often complex operations on semantic representations. Where in the brain are these operations taking place?

A rather involved hint is found in chapter ten of Deacon (1997). Deacon reviews several reasons why the Wernicke-Lichtheim-Geschwind allocation of certain language functions to certain cortical areas may be misguided. He feels that the current weight of evidence supports an alternative theory of cerebral organization in which cortical areas specialize for computations. In this view, a specific cortical area is specialized not for a particular linguistic function, but rather for a particular computational operation that can implement the linguistic function in question. In the context of the evolution of speech, Deacon, 2000, p. 280, puts it the starkest terms:

> One way to describe this is to say that there are no intrinsically pre-specified language circuits in the brain, only connection patterns that have been biased in unique ways so that they are slightly better suited to the unique demands imposed by language.

It is the cooperation of a myriad of such cortical circuits or computational centers that enables humans to learn a language and communicate with it.

In contrast to what this monograph has taken pains to do, Deacon does not actually propose specific algorithms, preferring to state the computations that he has in mind informally. He draws considerable inspiration from a dichotomy introduced by Roman Jakobson, Jakobson (1956), which identifies two poles of linguistic relationship, syntagmatic and paradigmatic. Relationships between words of different parts of speech are syntagmatic, while relationships between words of the same part of speech are paradigmatic. For instance, the fact that a quantifier in English modifies a noun is syntagmatic, while the similarity of quantifiers to one another with respect to their ability to quantify nouns is paradigmatic. Jakobson himself attributed syntagmatic relations to anterior cerebral cortex – witnessed in the disruption brought about by Broca's aphasia – and paradigmatic relations to posterior cerebral cortex – witnessed in the disruption brought about by Wernicke's aphasia.

Deacon extends Jakobson's dichotomy to computation by characterizing syntagmatic relations as requiring a shift of attention to alternate features, in contradistinction to paradigmatic relations, which require the recognition of common features. The invocation of attention in syntagmatic computation implicates frontal cortex, with the implicit conclusion being that Broca's region is located where it is located in order to partake of the attentional mechanisms that are the provenance of the frontal lobe. Likewise, the invocation of recognition in paradigmatic computation implicates sensory or associational cortex, with the implicit conclusion being that Wernicke's region is located where it is located in order to partake of the associational mechanisms that are the provenance of the parietal and occipital lobes.

10.1.2.4.1. Goertzel's dual network model

We cannot leave this topic without mentioning one more theoretician who, in grappling with the computational abilities of the brain has come to conclusions that are eerily similar to those of Deacon, despite a considerable difference in starting point and goals. In a series of books, Ben Goertzel, Goertzel (1993a, 1993b, 1994, 1997), proposes deriving a myriad of facts about human cognition from mathematical principles of chaos and self-organization. Prominent among his assumptions is that the memory is organized into two networks, a hierarchical one and a "heterarchical" – non-hierarchical – one, see Goertzel, 1993a, p. 170ff; Goertzel, 1994, p. 32ff; Goertzel, 1997, p. 22ff.

Goertzel finds evidence for a hierarchical network in what he calls command-structured perceptuomotor control. The perceptual aspect is familiar to us from the description of the visual system in Chapter 1, in which initial layers extract small features from the visual image which are combined into larger features by following layers. The motor aspect has the same layered organization processing monotonically increasing patterns, but producing motion. As Goertzel, 1994, p. 28 puts it,

> ... say level 1 represents muscle movements, level 2 represents simple combinations of muscle movements, level 3 represents medium-complexity combinations of muscle movements, and level 4 represents complex combinations of movements such as raising an arm or kicking a ball. Then when a level 4 process gives an instruction to raise an arm, it gives instructions to its subservient level 3 processes, which then give instructions to their subservient level 2 processes, which give instructions to level 1 processes, which finally instruct the muscles on what to do in order to kick the ball.

Goertzel refers to this as a multilevel control hierarchy, finding antecedents for it in the subsumption architecture of robotics, see Brooks (1989).

A heterarchical network is simply an associative memory, of which we have seen several in our simulations of the logical operators. In such memories, items are organized by their similarity, in that more similar items are stored more closely together than less similar items. In our simulations, logical operators are ultimately stored in terms of their similarity on a scale of correlation between 1 and -1.

It is tempting to map Goertzel's dual network to the Jakobson-Deacon division of cortical labor, so that anterior cortex has the structure of a hierarchical network, while posterior cortex has the structure of a heterarchical network.

10.1.2.5. Some evidence for the weak modularity of language circuits

While Deacon's proposal may appear to be little more than a change in name, it does suggest an empirically-verifiable difference between the classical and his own approach. If the computation performed by a particular cortical area is general enough, it can be pressed into service for both linguistic and nonlinguistic purposes. That is to say, Deacon's hypothesis predicts that one could find non-linguistic operations being performed in the putative language module of left perisylvan cortex. It should be emphasized that this prediction is made by the current author; neither Deacon (1997) nor Deacon (2000) mentions it. In Sec. 1.3.4 we coined the term "weak modularity" for the possibility that the linguistic system could be structured by algorithms that are not specialized for language but rather are shared throughout the brain.

Perhaps the strongest piece of evidence for Deacon's 'prediction' comes from Greenfield (1991), which proposes a parallel in the developmental complexity of speech and object manipulation. In studying object manipulation among children aged 11-36 months, Greenfield observes that an increase in their ability to combine objects mirrors an increase in their ability to combine phonemes to produce speech. Drawing on evidence from neurology, neuropsychology, and animal studies, Greenfield argues that the two processes are built upon a common neurological foundation in Broca's region, which divides into separate specialized areas as development progresses. Using a hierarchical representation of the object-manipulation and language data, Reilly (2002) puts Greenfield's observations on firmer conceptual ground by devising a computational simulation in which a network's ability to learn linguistic patterns is facilitated by prior exposure to object-manipulation patterns. Moreover, the network gradually resolves itself into two sub-networks, each specialized for one task or the other. These results spur Reilly to conclude that the linguistic specialization of Broca's region could be entirely due to learning rather than genetic programming.

Several other recent papers have uncovered nonlinguistic operations performed in Broca's and Wernicke's regions. Binkofski et al. (2000) found that Brodmann's area 44 was activated when subjects imagined movement from a third party's perspective, where the movement was either of the subject's own limbs or a target. In the former case, the left hemisphere BA 44 was activated, and in the latter, the right hemisphere BA 44. Binkofski et al. couch these results within the context of the recent discovery of so-called 'mirror neurons' in inferior frontal cortex of non-human primates, suggesting that BA 44 is the human analog of these cells.[46] As such, it houses representations concerned with movement control or identification – forelimb movement in Binkofski et al.'s experiments. To bring this proposal into congruence with the established role of the left-hemisphere BA 44 in speech, Binkofski et al. propose that it subserves the recognition of abstract or higher-order motor behavior that is relevant for communication.

The mention of forelimb movement and communication in the same context brings to mind sign language, so perhaps it comes as less of a surprise that Petitto et al. (2000) discovered with the aid of fMRI that a major portion of LH Wernicke's region becomes activated when deaf subjects observe syllable-sized units of sign language. It came as a complete surprise to Petitto et al., however, for the activated portion of Wernicke's region was thought to be unimodal, that is, specialized for a single modality, namely the sound of speech – as one would expect given its contiguity to auditory cortex.

Finally, Maess et al. (2001) studied how a group of non-musicians react to a sequence of five musical chords. Some of the chords followed the rules of Western classical music, while others contained a chord toward the end of the progression that was 'wrong'. Using MEG, Maess et al. discovered that the brain responses to the rule-violating sequences were different from the responses to the rule-obeying sequences and that signal of violation was localized to BA 44, bilaterally.

Taken together, these three studies point to the cortical interpenetration of speech, signing, and music. They cast doubt on the hypothesis of cortical specialization for function, and are consistent with the Jakobson-Deacon-Goertzel alternative of cortical specialization for computation. Of course, stating with any degree of precision what the computation performed in Broca's or Wernicke's regions is, still remains a daunting task. Taking up just the more-discussed case of Broca's region, in Deacon's terms we would say that it is specialized for some computation which processes hierarchical relations. Reilly (2002), following Van Essen, Anderson and Olshausen (1994), sees his own model as a way to store abstract motor schema, which are presumably organized hierarchically. Binkofski et al. (2000) propose something similar for their own results, which is that BA 44 recognizes higher-order motor behavior involved in communication, whether vocal or manual. Finally, Maess et al.'s (2001) observations on music cognition can be integrated with these antecedents by claiming that the five-chord stimuli are understood as Western music by imposing a hierarchical structure on them, which consists of certain

---

[46] First reported in Gallese et al. (1996) and Rizzolatti (1996), these neurons discharge when a monkey observes another individual performing an action; see Gallese and Goldman (1998) and Rizzolatti and Arbib (1998) for review, as well as the recent volume dedicated to the topic, Stamenov and Gallese (2002).

Table 10.2. Summary of processing classification: lateral by longitudinal

	Left (fine coding)	*Right (coarse coding)*
Anterior (attentional)	syntax, semantics	discourse
Posterior (associative)	segmental phonology, lexicon	prosodic phonology

expectations about what chord should be found at which location in a sequence. All of this fits together nicely with the uncontroversial hierarchical structure of linguistic grammar. And ultimately, it all depends on the ability of attention to select some items – those that are grouped together at a given level of a hierarchy – and exclude others – those at all other levels of the hierarchy, or even outside of it, and on the ability of attention to be guided by expectations about what item should be found at which position within a hierarchy.

10.1.2.6. Where does this put the logical operators?

With this background, we can make an educated guess about the cortical localization of the computations performed by the logical operators. On the one hand, they invoke a minimal syntagmatic structure, since a logical coordinator must apply to some set of elements and a logical quantifier must apply to a nominal. On the other hand, they also have a minimal paradigmatic structure, in the sense that the elements that the logical operators apply to are all similar to one another: they are all drawn from the same category for the coordinators and are all nominal entities for the quantifiers. However, this similarity effect may be an epiphenomenon of an anterior mechanism (selective attention) activating a portion of posterior associative memory. Given its associative format, any items selected by attention from posterior memory will unavoidably be associated, which is to say, similar. Thus there is a slight advantage to attributing the logical operators to the anterior, attentional processing of Broca's region.

10.1.2.7. BA 44 vs. BA 47

This claim is not as definitive as it may seen, since Broca's region consists of two or three Brodmann's areas, namely 44, 45, and 47. It turns out that a recent fMRI experiment reported in Dapretto and Bookheimer (1999) supplies a clue as to which of these three areas may ultimately be responsible. Dapretto and Bookheimer asked subjects to judge whether two sentences had the same meaning. In the critical contrasts, the two sentences meant the same but differed in syntax or semantics. The syntactic condition presented subjects with different versions of the same sentence, such as "The pool is behind the gate" and "Behind the gate is the pool". The semantic condition presented subjects with synonyms, as in "The car is in the garage" and "The auto is in the garage". By subtracting the activation patterns in these two conditions, the authors pinpointed areas in the left inferior frontal gyrus selectively related to syntax, BA 44, and to semantics, BA 47. Of course, the recognition of synonyms and the logical operations are undoubtedly rather different cortically, but this is as far as current research takes us.

10.1.3. Summary

Much of the content of this first three-quarters of the chapter can be summarized in the Table 10.2, which classifies the lateral axis of the brain against the longitudinal axis. It is more of a guidepost to work that needs to be done than a self-assured summary of what has been discovered. It is therefore time to turn to a topic about which we can be more confident, namely memory, and especially episodic memory as housed in the hippocampus.

10.2. FROM LEARNING TO MEMORY

Up to now, our models of associative pattern recognition rely on vector-theoretic representations with little to nothing in the way of internal structure. In Sec. 1.3.3, it was mentioned that such representations are challenged at representing structured information, such as the hierarchical relations that have been found to be irreplaceable in the analysis of natural language. This section sketches one way in which the hippocampus may represent structured information, due to Shastri. Along the way, it finally cashes out our promise to show how correlation is encoded neurologically, and in particular how negatives are marked with respect to positives. The account follows from a meditation on how memory works, with which we begin.

10.2.1. Types of memory

If learning is the process by which new information about the world is acquired, **memorization** is the process by which what is learned is retained with the possibility of drawing on it later. Psychological research has generated an enormous number of theories of memory, but following Sternberg, 1996, p. 157ff, two stand out as being simple enough to inform neuromimetic modeling.

10.2.1.1. Quantitative memory: memory storage

Atkinson and Schiffrin (1968) propose a metaphor for memory articulated into three sequential stores: sensory memory, short-term memory, and long-term memory. Sensory memory is the initial repository of perception. It holds sensory impressions extremely accurately, but is easy to erase and fades quickly, and we have little or no introspective access to it. Short-term memory holds a handful of items in storage for seconds up to a couple of minutes. Finally, long-term memory is the introspectively-available store of knowledge that lasts for minutes or an entire lifetime.

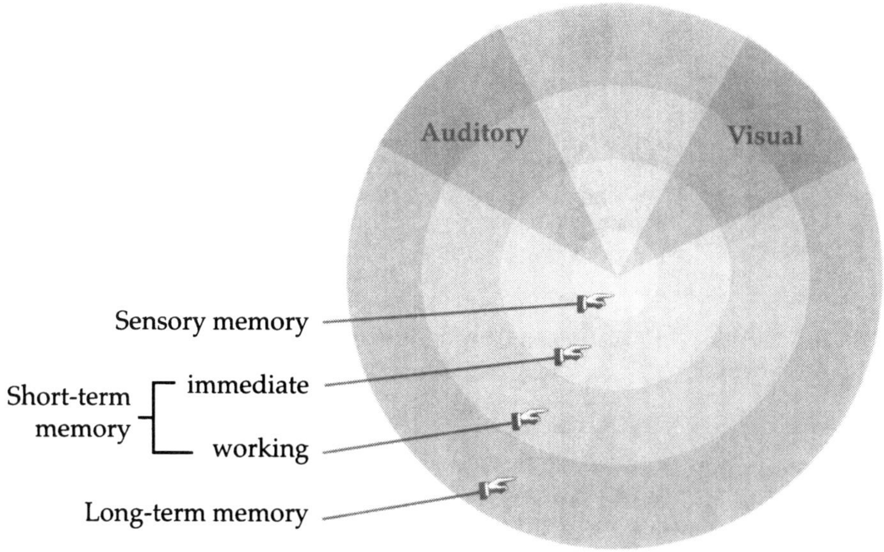

Figure 10.7. Slices through memory rings along the lines of single senses.

More recent refinements of this model expand the role of short-term memory by teasing out two further subdivisions, immediate memory and working memory, see Squire and Kandel, 1999, chapter 5. Immediate memory refers to what can be held actively in mind beginning the moment that information is received. It is the information that forms the focus of current attention and that occupies the current stream of thought. Its capacity is extremely limited – limited to the famous 7±2 items of Miller (1956) – and persists for less than 30 seconds. This 30-second cap can be overcome by rehearsing the items in question through conscious repetition. Such rehearsal is different enough to deserve its own name, that of working memory. Short-term memory also englobes any other memory processing before the establishment of the new information as a stable long-term memory. Thus the global organization of memory is not a series of discrete stores, but rather a series of overlapping stores, which can be visualized more appropriately as the nested concentric circles in Fig. 10.7. The darker the circle, the longer it holds an item in storage.

There is one way in which this diagrammatic metaphor is potentially misleading. It is that the different memories appear to occupy different cerebral spaces. While working memory may be segregated and localized in prefrontal cortex, it is believed that the other memory processes are grouped together. That is to say that long-term memories are thought to be stored in the same set of structures that perceive, process, and analyze what is to be

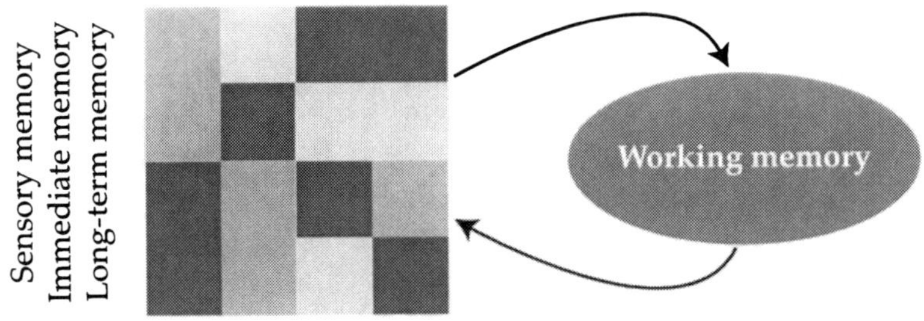

Figure 10.8. Grid of units for a single modality, orchestrated by working memory.

remembered; see Squire and Kandel, 1999, p. 88. Thus the proper metaphor may be more akin to a grid of memory units, orchestrated by working memory, and with each unit potentially in a different memory state. Fig. 10.8 depicts this metaphor, where, as before, increasing darkness represents increasing time in storage. This grouping of memory units in different states together in the same neighborhood can be seen as an efficient way of using what has already been learned as a foundation for new memories, in that similar temporary memories should become permanent together.

10.2.1.2. Qualitative memory: declarative vs. non-declarative

This quantitative theory of memory goes hand in hand with a qualitative theory of memory which classifies memories by their content. It has been elaborated most extensively by Larry Squire and colleagues, see Squire and Zola-Morgan (1988), Squire and Knowlton (1995) and Squire and Kandel (1999), as well as the summaries in Sternberg, 1996, p. 168-9, Beggs et al. (1999) and Gluck and Myers, 2001, pp. 21ff. The taxonomy in Fig. 10.9 summarizes the major claims and observations. The two major sorts of content are termed **declarative** and **non-declarative**. A simple rule of thumb for distinguishing them is that declarative memory is the knowledge that something has happened, while non-declarative memory is the knowledge of how something happens.

Declarative memories are known explicitly or knowable from introspection. They are sub-classified according to their temporal indexation. **Episodic memories** are indexed to a specific time referent and are most robust for events drawn from one's own life, e.g. "What did you have for lunch yesterday?" or "Who was the first person you saw this morning?". **Semantic memories**, in contrast, constitute a cache of facts that are not indexed by a specific time referent, such as vocabulary items or general knowledge of the world, e.g. "What do you usually have for lunch?" or "What is the name of the first person you saw this morning?". Note that this usage of the term "semantic" is

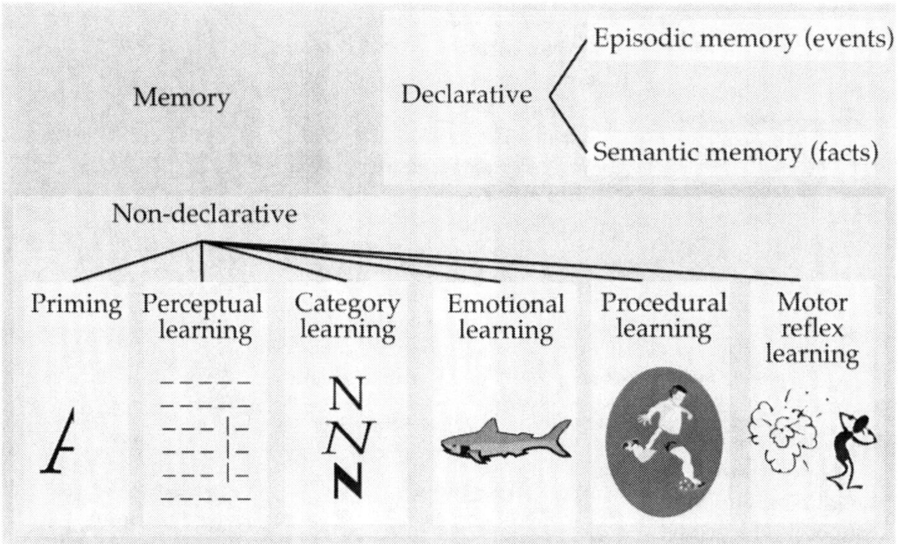

Figure 10.9. Taxonomy of qualitative memory.

far more circumscribed than its standard usage in linguistics, a difference that can lead to confusion. Linguistic semantics deals with representations that are presumably subserved by *both* sorts of declarative memory.

Non-declarative memories are known implicitly or unconsciously and are opaque to introspection. They are derived from at least the six processes illustrated in Fig. 10.9. Of these, it is category and procedural learning that interest us most here. The memories produced by **category learning** embody the accrual of knowledge about a category from interaction with specific instances. I cannot improve on Squire and Kandel's, 1999, p. 183, description:

> If we see a display of cups or chairs, for example, we may remember specific cups and individual chairs, bur we can also learn about the category of "cupness" and "chairness". Our knowledge of what a cup is, or a chair, is not something taught to us by instruction. Cups and chairs cannot be reduced to an unambiguous set of rules. Rather, our concepts of cup and chair are build up gradually as the result of numerous encounters with different kinds of cups and various kinds of chairs. When one thinks about it, much of our knowledge about the world is in the form of categories.

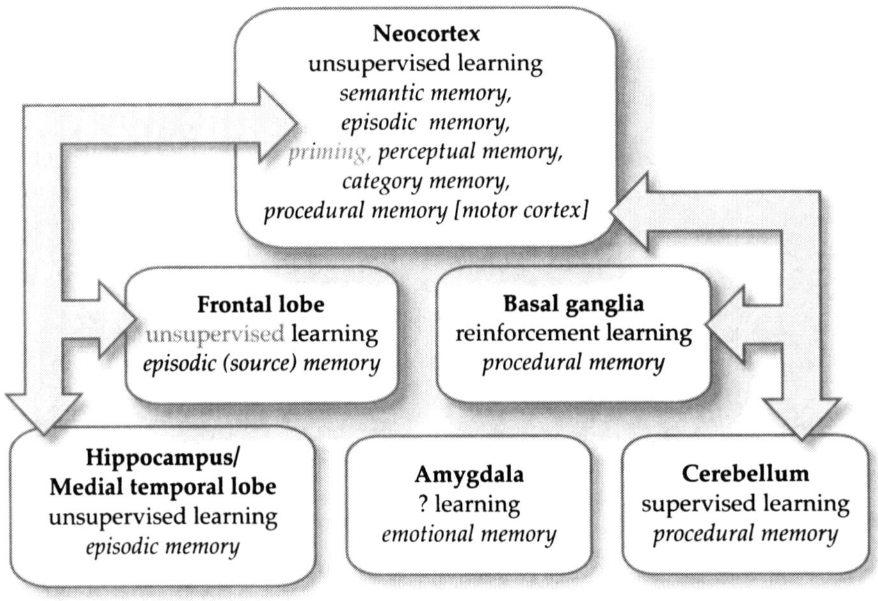

Figure 10.10. Cerebral structures for learning and memory.

The memories that result from **procedural learning** are motor skills or habits. Such memories are acquired incrementally by practicing a particular routine, such as tying a shoelace or playing the violin.

10.2.1.3. Synthesis

The relevance of quantitative and qualitative memory to our concerns is that a given cerebral system implements a specific learning rule to learn a specific sort of information and store it for a specific amount of time. This implies a three-way classification of cerebral structures in terms of how they learn (learning rule), what they learn (qualitative memory), and how long they retain it (quantitative memory). Fig 10.10 provides a graphic summary of the principal components and their interconnections. The right-side pathway is none other than the three-way classification of learning rules by cerebral structure proposed by Doya, reviewed in Sec. 5.2. The contextualization of this pathway within the broader theory of learning, memory, and brain function laid out in Fig 10.10 localizes it as the route of procedural memory. It must be pointed out, nonetheless, that it has only recently become apparent that the cerebellum plays a more general role in cognition than Fig 10.10 allows, though what this role may be remains unclear.

The new pathway that emerges on the left is the one that interests us the most. Fig 10.10 localizes it as the route of episodic memory, along which much of the information for learning the logical operators travels. Let us take up the contributions of the three components as a separate unit.

10.2.2. A network for episodic memory

The computational relationship between the hippocampus and neocortex has been explored intensively in a school of thought founded on McClelland, McNaughton and O'Reilly (1995) and refined several times since then; see O'Reilly, Norman and McClelland (1998), O'Reilly and Rudy (2000, in press), and O'Reilly and Munakata (2000). Its principles have been applied across several species (rats, monkeys and humans) to a wide range of learning and memory phenomena, such as impaired and preserved learning capacities with hippocampal lesions in conditioning, habituation, contextual learning, recognition memory, recall, and retrograde amnesia.

By way of introduction, in a recent review, O'Reilly and Rudy (in press) deduces the hippocampal/neocortical division of labor in memory from first principles of neural computation. In the following paragraphs, we adapt O'Reilly's exposition to our own interests.

Imagine seeing a new gadget and hearing its name, such as "iPod". The visual and auditory pathways end in separate regions of the brain, so episodic memory must bind together the image of an iPod and the sound of its name so that you can associate the two and thereby learn a new word. This binding must be very fast, because you can learn a new word after only one presentation of the image and sound. In neurocomputational terms, a speedy network is one in which weights change in large increments. Furthermore, episodic memory must not be confused by images or sounds that are similar to the target ones, such as any other mp3 player/portable hard drive or words that share the morpheme "pod", such as "seed pod". That is to say, representations in episodic memory which share parts must not interfere with one another. In neurocomputational terms, a network which prevents interference is one which encodes input patterns orthogonally.

There are several reasons to posit the hippocampus as the neural structure that implements these desiderata. The hippocampus has been a magnet for research on memory ever since Scoville and Milner (1957) described how the patient H.M. had lost all ability to form episodic memories after the removal of the inner surface of his temporal lobe as a last-resort treatment for his epilepsy. Posterior and now-classic studies confirmed that humans with severe damage to their hippocampus do not form any long-term memories of episodes after the time of damage, though memories of episodes from earlier periods remain intact, see O'Keefe and Nadel (1978), and Squire and Zola-Morgan (1991). A second startling property of the hippocampus is that it has a regular structure that is laid bare by standard cellular staining techniques, so much so that diagrams from Ramón y Cajal (1911) can still be used to trace its connectivity. As is discussed below, the connectivity of the hippocampus suggests that it separates its input patterns orthogonally. Equally intriguing is the fact that long-term potentiation, the rapid and sustained increase in synaptic efficacy following a brief but potent stimulus introduced in Sec. 2.3.3, was first observed in the mammalian hippocampus by Bliss and Lomo (1973). It has since been observed in many other pathways, but the most detailed observations still come from the hippocampus, see Koch, 1999, pp. 317ff. LTP can proceed at a fast or a slow pace, so it is feasible for the hippocampus to implement a very fast variety. Finally, the hippocampus receives input from practically every other area of the cerebrum, so that it is ideally placed to perform the multimodal binding function mentioned above. In this way, the conceptual argument that O'Reilly advances for a fast-learning, pattern-separating component to episodic memory finds a plausible biological substrate in independently-arrived at observations of the hippocampus.

Nevertheless, if all bindings were kept separate, there will be no way to generalize across them to novel situations. Yet humans do indeed generalize to novel situations, for instance, more sophisticated versions of the initial iPod can still be called "iPods". Episodic memory must therefore be able to slowly integrate representations of overlapping elements so as to abstract away from their specifics and encode the overall statistical structure of the environment, which enables generalization to novel situations. Such a slow-learning, pattern-combining mechanism has already been seen in the competitive networks designed for the logical coordinators and quantifiers in the preceding chapters. O'Reilly asserts that these characteristics define the architecture/representational format of neocortex as a whole.

To summarize briefly, the system for episodic memory espoused by O'Reilly and his collaborators synthesizes a large number of observations to propose that the hippocampus exists in part to provide a memory system which can learn arbitrary information rapidly without suffering undue amounts of interference. This memory system sits on top of, and works in conjunction with, the neocortex, which learns slowly over many experiences, producing integrative representations of the relevant statistical features of the environment.

The frontal cortex way-station included in Fig 10.10 as part of episodic memory is not incorporated into this model, though it is not hard to make a conceptual argument that accurately predicts its role. So far, we have described a fast component that binds disparate incoming percepts into a whole, and a slow component that integrates these particular observations into general schema in long-term memory. It is not difficult to conceive of an intermediate component that keeps some particular binding of incoming percepts at the center of attention for posterior processing, such as to decide on a particular course of action to be taken in response. This is just the function that Usher and Cohen (1999) attribute to frontal cortex, though on the basis of more direct neurophysiological recordings, functional imaging, and cognitive-

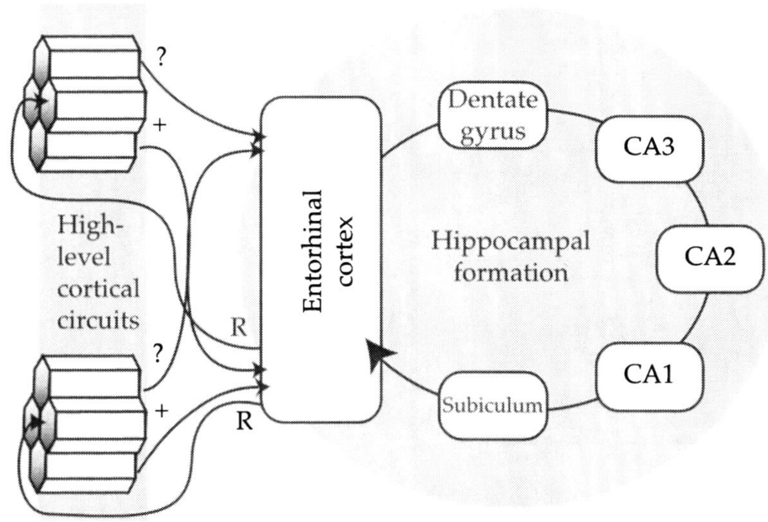

Figure 10.11. Schematic view of cortico-hippocampal interactions in SMRITI. Comparable to Shastri, 2001, Fig. 1, 2, and Rolls and Treves, 1998, Fig. 6.1. Parahippocampal gyrus, perirhinal cortex, and many back-projections excluded.

psychological experimentation. However, we do not pursue the contribution of frontal cortex any further, since the cortico-hippocampal loop provides an sufficient neural substrate for the points that we wish to make.

10.2.2.1. The hippocampal formation and Shastri's SMRITI

This fortuitous convergence of an isolatable cognitive function, a plausible biophysical substrate, and a traceable circuit diagram has proved fertile ground for neuromimetic simulation of the hippocampus. Marr (1971) was the first to propose that it performs autoassociation, and posterior research has revised and expanded on his account, see Rolls and Treves, 1998, chapter 6, and Gluck and Myers, 2001, pp. 117, for recent discussion.

Perhaps the most recent, complete, and ambitious model is that of Shastri (1999, 2001). In this model, christened by the acronym SMRITI (System for the Memorization of Relational Instances from Transient Impulses), the hippocampus binds together entities and their role labels in a way that is very familiar to linguists. For instance, Shastri proposes that the entities and roles in the clause "John give Mary a book in the library on Tuesday" are bound together as in the representation of (10.6):

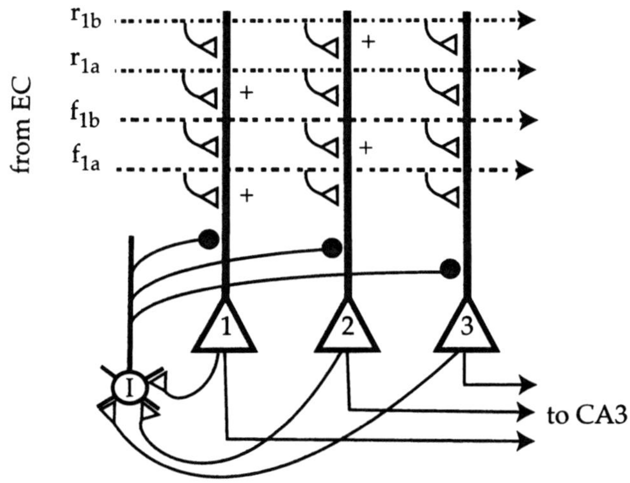

Figure 10.12. Schematic dentate gyrus. Comparable to Shastri, 2001, Fig. 11.

10.6. (GIVE: ⟨giver = John⟩, ⟨recipient = Mary⟩, ⟨give-object = a book⟩, ⟨location = the library⟩, ⟨temporal-location = Tuesday⟩).

The angled brackets enclose a role label and an entity bound together by virtue of the equals sign. The labels themselves are familiar to linguists since Fillmore's Case Grammar, see Fillmore (1968, 1976) and the posterior notion of thematic roles in Jackendoff (1972) and Chomsky (1981).

At a second, more inclusive level, all of the role-entity bindings are bound together within parenthesis by the label *GIVE*, in order to distinguish them from role-entity bindings pertaining to other events. Thus not only does (10.6) draw on Case Grammar and its descendents, but it also appears to be a notational variant of Event Semantics, see Davidson (1967) and Parsons (1990) among many others. (10.6) is translated into a pseudo-event-semantic representation in (10.7), in which the existential quantification implicit in (10.6) is made explicit and used to index each binding in (10.6):

10.7. $\exists e\ [\text{GIVE}_e$: $\langle \text{giver} = \text{John}\rangle_e$, $\langle \text{recipient} = \text{Mary}\rangle_e$, $\langle \text{give-object} = \text{a book}\rangle_e$, $\langle \text{location} = \text{library}\rangle_e$, $\langle \text{temporal-location} = \text{Tuesday}\rangle_e]$.

Where SMRITI stands out from these linguistic antecedents is in the role of the hippocampal formation in effecting that bindings.

SMRITI consists of two principal stages, a cortical stage and a hippocampal stage, which are ordered into a loop on the basis of their anatomical connections as in Fig. 10.11. High-level cortex sends three kinds of signals to the hippocampal formation, two event signals and a query signal. The event signals + and – assert the occurrence or non-occurrence of an event so that the hippocampal formation will memorize it. The query signal ? asks whether the hippocampal formation has memorized the associated event or not. The hippocampal formation propagates three kinds of signals in response. A reply *R* to an event signal indicates the extent of memorization of the event. A reply *R* to the query signal indicates whether an event being held in the hippocampus matches the input event (+) or not (–).

10.2.2.2. Long-term memory, the hippocampus, and COORs

10.2.2.2.1. The dentate gyrus

The dentate gyrus is where Shastri postulates that the various roles and entities are bound together into a whole. This is the area that interests us most for the logical operators, since numbers of role-entity bindings are the stuff of COOR and Q space. Shastri attributes to the dentate gyrus a structure such as that of Fig. 10.12, in which input from entorhinal cortex propagates across the three dentate granule cells in a regular fashion, modulated by a far smaller number of inhibitory interneurons. The synapses represented by triangles undergo long term potentiation and so are recruited to detect the binding of a correlated role and entity.

It is more in accord with the terminology of Chapter 5 to say that these synapses experience Hebbian learning of correlated inputs, as is modeled in the following simulation. Let the two different entity-role pairs $\langle e_1, r_1 \rangle$ and $\langle e_2, r_2 \rangle$ depicted as inputs in Fig. 10.12, be $[1, 0, 1, 0]^T$ and $[0, 1, 0, 1]^T$, respectively, which represent activation of the input fibers e_1 and r_1 at one moment and e_2 and r_2 at the next. The initial and final weights of a three-neuron network trained on these inputs using the Hebbian learning rule are summarized in Table 10.3. In comparison to the randomly-chosen initial weights, the final weights have adjusted themselves so that the first neuron, n_1, has the highest weights for components (columns) e_1 and r_1, the second neuron has negative weights for all components, and the third neuron has the highest weights for components e_2 and r_2. The upshot of these weights is that running the inputs through them returns the results on the right side of Table 10.3. The first neuron responds best to the first input, while the third neuron responds best to the second input. The second neuron does not respond to either. Thus the network has indeed learned to detect the two role-entity correlations.

Table 10.3. Initial and final weights of hetereoassociative Hebbian network. Test of final weights on each input, without/with threshold = 1. (**p10.01_HA_DG.m**)

		e_1	e_2	r_1	r_2	*[1, 0, 1, 0]*	*[0, 1, 0, 1]*
Initial weights	n_1	0.84	0.69	0.70	0.96		
	n_2	0.57	0.62	0.55	0.52		
	n_3	0.37	0.79	0.44	0.88		
Final weights	n_1	0.68	0.32	0.66	0.37	1.35/1	0.69/0
	n_2	-0.03	-0.32	-0.03	-0.33	-0.07/0	-0.65/0
	n_3	-0.32	0.63	-0.31	0.65	-0.63/0	1.28/1

Before declaring victory, a few clarifications are in order. The main one is that this specialization of one neuron for each input is not the only outcome. It may happen that no neuron reaches the threshold for responding, or that only one reaches threshold. This is not an error, for it simply means that the patch of dentate gyrus modeled here would not have learned these particular inputs, but some other patch surely would have. What is more worrisome is that sometimes a single neuron responds best to *both* inputs. This result should be considered an error, since the notion of a (reliable) binding detector implies a biuniqueness limitation: one detector per binding, and one binding per detector. Otherwise, the detectors will signal ambiguously. However, Shastri does allow for optional long term depression on those neurons which have already learned a binding in such a way that the synapses that are not potentiated are depressed. This additional mechanism would prevent the binding of more than one role-entity pair per neuron, but it is not included in the simulation for the sake of simplicity.

Finally, the crucial contribution of the interneuron should not be overlooked. If the interneuron is excluded by setting its weight to 0, there is a tendency for the same neuron to capture both inputs time after time and so signal ambiguously. It is the enhancement of contrasts imposed by the inhibitory interneuron that keeps this unwanted effect at bay.

10.2.2.2.2. An integrate-and-fire alternative

While the previous simulation does provide a linguistically-interesting illustration of Hebbian learning, it does not do justice to the depth of Shastri's model. Shastri's model uses integrate-and-fire neurons to implement binding as correlation in time. It would make our own model more falsifiable, and neurologically plausible, to do likewise. Thus the following simulations rely on spike timing rather than spike rate to provide the correlations that are the basis of Hebbian learning. This alteration makes for a much more imposing picture.

As a first approximation, consider Fig. 10.13. The four entorhinal cortex neurons along the bottom are parameterized so as to spike in unison. The

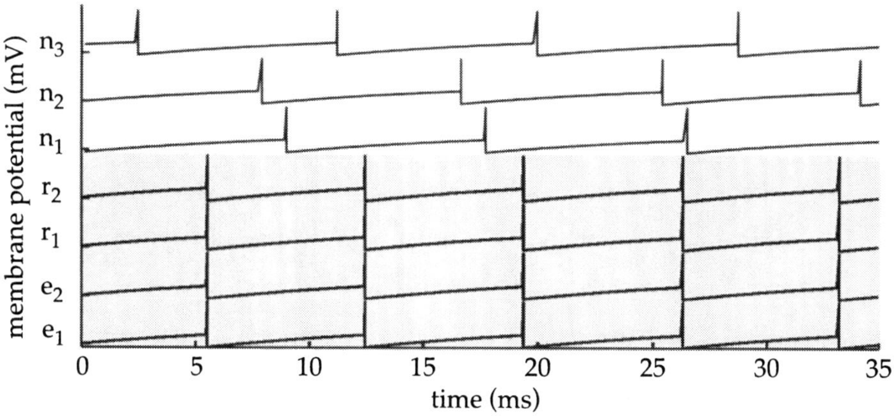

Figure 10.13. Integrate-and-fire model of dentate gyrus. (**p10.02_IF_DG.m**)

Table 10.4. Initial and final weights for Fig. 10.13.

(no operator)		e_1	e_2	r_1	r_2
Final weights	n_1	0.01	0.09	0.09	0.09
	n_2	0.05	0.03	0.02	0.02
	n_3	0.07	0.03	0.08	0.02
Final weights	n_1	0.01	0.09	0.09	0.09
	n_2	0.05	0.03	0.02	0.02
	n_3	0.07	0.03	0.08	0.02

subscripts of the input channels indicate the event or entity-role pairs that can be correlated, $\langle e_1, r_1\rangle$ and $\langle e_2, r_2\rangle$. The consequences of this behavior are explained in more detail in the next sub-section.

The three dentate gyrus neurons labeled n_n across the top do not spike in unison, and while they have the same period, it is slightly longer than that of the entorhinal neurons. It is hoped that the reader can appreciate that none of the dentate gyrus neurons are synchronized with the entorhinal neurons. Moreover, the system does not learn. The weights at the beginning of the simulation are the same as the weights at the end, as reproduced in Table 10.4.

10.2.2.2.3. The dentate gyrus and coordinator meanings

All of this is background for a hypothesis of how coordinator meanings would be represented by Shastri's implementation of episodic memory in the hippocampus. Though Shastri only discusses singular terms, his model is

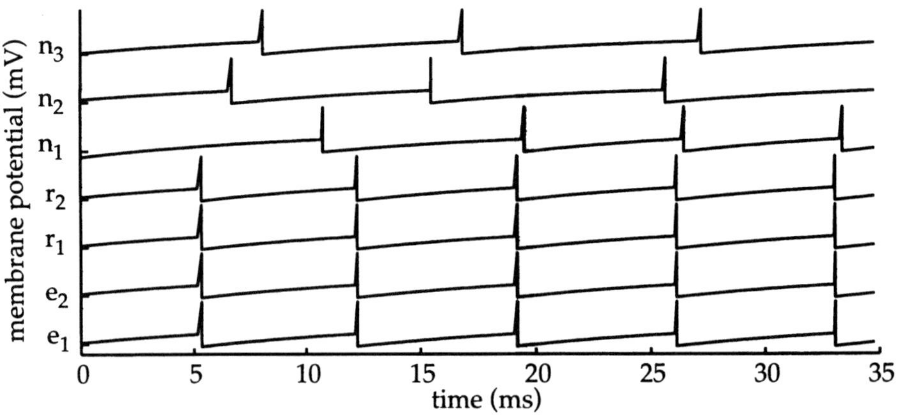

Figure 10.14. I&F dentate gyrus AND; window = 1 ms. (**p10.03_IF_and.m**)

Table 10.5. Initial and final weights for dentate gyrus neurons of Fig. 10.14.

AND		e_1	e_2	r_1	r_2
Initial weights	n_1	0.01	0.06	0.002	0.09
	n_2	0.02	0.03	0.07	0.05
	n_3	0.02	0.02	0.05	0.04
Final weights	n_1	1.64	1.67	1.63	1.69
	n_2	0.02	0.03	0.07	0.05
	n_3	0.02	0.02	0.05	0.04

specific enough for us to able to make a reasonable guess about how non-singular terms such as multiple coordinatees would be processed.

The most obvious first guess is to simply list the coordinatees. This is conveyed by the simulation of AND in Fig. 10.14. It is designed so that enough synaptic input is supplied to a dentate gyrus neuron to trigger learning only when the entity and relational spikes coincide within a given temporal window, here, one of 1 ms. In practice this design imposes the constraint that a dentate gyrus neuron can learn only *after* the appropriate entorhinal neurons have spiked.

In the simulation depicted in Fig. 10.14, dentate gyrus neuron n_1 synchronizes with the entorhinal neurons on its second spike. This can be appreciated in Fig. 10.14 by the fact that n_1 shortens its period enough to 'catch up' with the four entorhinal neurons. The initial and final weights of the simulation are reproduced in Table 10.5, where the shading highlights the change in weights of the synchronized dentate gyrus neuron from the initial to

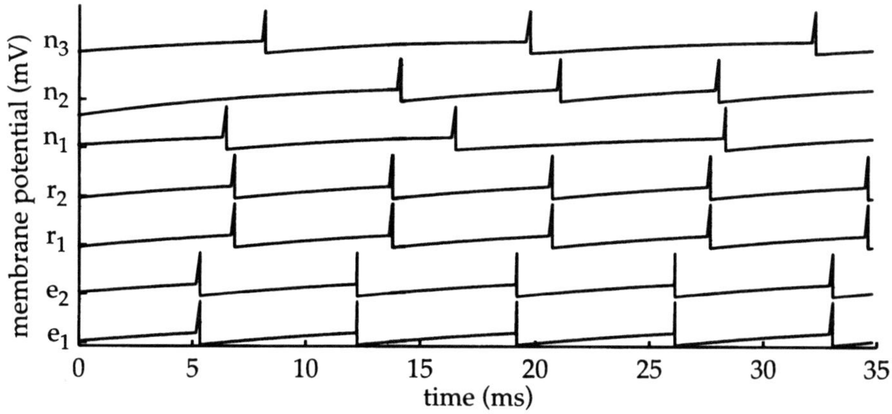

Figure 10.15. I&F dentate gyrus NOR; window = 1 ms. (**p10.04_IF _nor.m**)

Table 10.6. Initial and final weights for Fig. 10.15.

NOR		e_1	e_2	r_1	r_2
Initial weights	n_1	0.03	0.07	0.02	0.09
	n_2	0.02	0.03	0.07	0.09
	n_3	0.02	0.05	0.04	0.06
Final weights	n_1	0.03	0.07	0.02	0.09
	n_2	0.02	0.03	1.68	1.69
	n_3	0.02	0.05	0.04	0.06

the final epoch. The weights of the other two neurons do not evolve. It is important to emphasize that the dentate gyrus weights are initialized to small random values; it is only the Hebbian learning that the dentate gyrus neurons undergo that makes them sensitive to the temporal correlations in the role-entity pairs. This representation of AND as temporal synchronization of all of the coordinate sub-events thus implements quite directly the conjecture vertebrating this book that coordinators measure correlation, and in particular the conjecture that AND measures the highest degree of correlation.

Conversely, the obvious tack to pursue for NOR is to peg it to the point of anticorrelation in the entity cycle. Unfortunately, the integrate-and-fire model that we are using is too coarsely-grained to find such a point easily, so we decided to arbitrarily designate a moment about 2.5 ms after the entity spikes as the point of anticorrelation. It is here that entorhinal neurons spike for anticorrelated roles, as illustrated for r_1 and r_2 of NOR in Fig. 10.15. A dentate gyrus can learn an association with any neuron spiking at this point, if

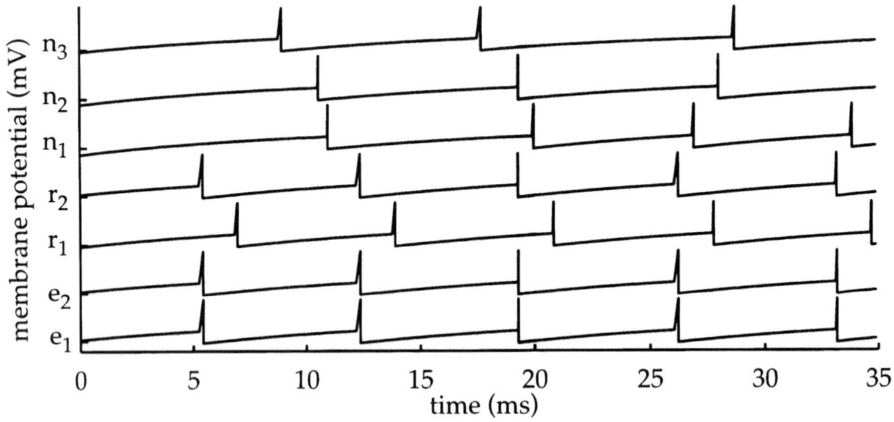

Figure 10.16. I&F dentate gyrus OR; window = 1 ms. (**p10.05_IF_or.m**)

Table 10.7. Initial and final weights for Fig. 10.16.

OR		e_1	e_2	r_1	r_2
Initial weights	n_1	0.02	0.03	0.06	0.04
	n_2	0.05	0.07	0.07	0.07
	n_3	0.09	0.04	0.04	0.08
Final weights	n_1	1.64	1.65	0.05	1.66
	n_2	0.65	0.66	0.06	0.66
	n_3	0.09	0.04	0.04	0.08

it is restricted to the appropriate temporal window. Such is the case for n_2 in Fig. 10.15, which happens to spike at the very end of the window for anticorrelation. Its weights are strengthened, and by its first spike, it has synchronized with the entorhinal neurons.

The final weights in Table 10.6 confirm that it is indeed n_2 that has learned to detect the anticorrelation in the entorhinal spike trains, but notice that there is a crucial difference from AND: the pattern learned does not include the entity neurons, for their weights have not changed from their initial low random assignment. In fact, with the simple assumptions made here, it is not possible to learn from the entity spike trains without making NOR indistinguishable from NAND.

We initially thought that this result was an error in the algorithm; after some reflection, we are no longer so sure. This result clearly identifies the negative operators as not just symmetrical to the positive operators, changing correlation for anticorrelation, but actually different and more effortful, in

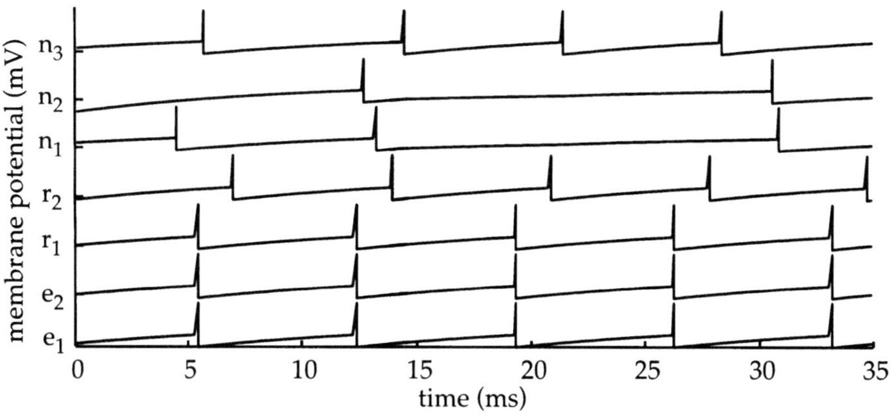

Figure 10.17. I&F DG NAND; window = 2.5 ms. (**p10.06_IF _nand.m**)

Table 10.8. Initial and final weights for Fig. 10.17.

NAND		e_1	e_2	r_1	r_2
Initial weights	n_1	0.07	0.05	0.02	0.07
	n_2	0.07	0.04	0.07	0.05
	n_3	0.01	0.04	0.07	0.06
Final weights	n_1	0.07	0.05	0.02	0.07
	n_2	0.07	0.04	0.07	0.05
	n_3	0.01	0.03	0.05	1.67

that it takes some additional mechanism to inscribe the entity representation in the overall pattern of the negative operations – a mechanism that we will not pursue here. Since generations of philosophers, logicians, linguists, and psychologists have commented on the 'difficulty' or markedness of negation, see the up-coming discussion, we take the fact that our simple hippocampal assumptions cannot entirely rule in what we have posited to be the representation of negative operations to be an indication that they are on the right track, though obviously in need of considerable refinement. So let us return to the positive operators, where our assumptions fare better.

OR can be assimilated to this scheme, if the temporal window of AND is allowed to *exclude* the cycles of anticorrelated relations. Thus in Fig. 10.16, role spike train r_1 is anticorrelated with its entity spike train e_1. The temporal window of Fig. 10.16 is too narrow to include it, so it is effectively omitted from the pattern learned by n_1. This indeed is our assumption about the representation of OR/SOME patterns initially broached in Chapter 3, so it is reassuring that it can be encoded so straight-forwardly in our stripped-down model of the hippocampus.

Given these results, what predictions can be made about NAND? It should be an OR based on anticorrelation, which is to say that at least one anticorrelated role is learned by a dentate gyrus neuron. This is what we find in the simulation depicted in Fig. 10.17 and the accompanying list of weights in Table 10.8. On its second spike, n_2 falls into the window of anticorrelation around role r_1, and a glance at the increase in the final weight of r_1 reveals that n_2 has learned to recognize this pattern.

10.2.2.3. Discussion

The previous simulations reveal that negative relations of the kind hypothesized for NOR and NAND are very difficult, if not impossible, for the simple system based on Shastri's model of the dentate gyrus to learn. This implies that some additional mechanism must be responsible for the complete representation of negativity. Shastri himself is not unaware of the challenges posed by negation for his model, but he finesses them by formalizing negative facts as equivalent to positive facts, but not occurring. For instance, a hypothetical negative binding $-\langle e_1, r_1\rangle$ is meant to state that the relation $\langle e_1, r_1\rangle$ does not occur; the positive version that we have been using says that relation $\langle e_1, r_1\rangle$ does occur.

This formalization of negation as equivalent to zero affirmation is familiar from first order logic and can be traced all the way back to the Stoics, which Horn, 1989, chapter 1, does. It enables Shastri's hippocampal model to fit neatly into familiar first-order theories of reasoning, but consider the challenge it raises for a theory of binding as phase locking. For a negative such as NOR, to maintain parallelism with the positives, each entity neuron must be phase-locked with ... *zero* relational neurons! How can something phase-lock with nothing? Well, it cannot, and this unavoidable fact stopped us in our tracks for a considerable period of time.

However, the hypothesis of negation championed all the way from the Stoics to Shastri flies in the face of the linguistic, developmental, and psychological evidence that negation is not the absence of affirmation, but rather is subordinate to affirmation – an equally ancient theory founded in Aristotle, also summarized in Horn, 1989, chapters 1, and one that is more consonant with our own argumentation in Chapter 3.

The synchronizational account developed here partakes more of an Asymmetricist perspective than it does of a Symmetricist perspective – to use Horn's nomenclature – since negative operators wind up taking more processing time and resources than positive coordinators. This is in complete consonance with reaction-time experiments performed in the 1970's that Horn, 1989, chapter 3, reviews, and so provides a neuromimetic grounding for – if not an explanation for – the oft-remarked markedness of negation with respect to affirmation, see again Horn, 1989, chapter 3, for abundant references. Unfortunately, the neuromimetic source of the temporal asymmetry between positive and negative coordinators is not revealed within the dentate gyrus, but rather is already present in its input in the guise of the time lag between spike trains for affirmative and negative relations. Thus our reduction of negative markedness and processing difficulty to the Hebbian learning of binding detectors within the dentate gyrus is still more of a promise than a proven fact.

10.3. SUMMARY

We hope the reader now has a clearer picture of two of the principal contentions of this book. The first is that in our current ignorance of the how the brain processes language, one of the few explanatory methodologies that we have to draw on is neuromimetic modeling based on non-language neurology, with the hope that someday imaging technologies will catch up and provide the data needed to validate the models. By our short review of the neuroscience of language, do not mean to disparage what has already been learned, but rather show how much still remains to be learned.

The second contention is that the logical operators express correlation, which is fundamental neurological process. We have demonstrated on the basis of simple, though well supported, simulations how the dentate gyrus can bind together through correlation the various roles and entities that we have postulated for logical coordination.

The concluding task of this monograph is to synthesize all that we have postulated into the larger perspective of cognitive science – in fact, in the most recent of three generations of cognitive scientists.

Chapter 11

Three generations of Cognitive Science

This chapter concludes our exploration of the explanatory adequacy of the analyses of logical coordination, quantification, and collectivity undertaken in the previous chapters. The reader may recall that the core of the monograph is the analysis of logical operatorhood laid forth in Chapter 3 and substantiated in Chapters 4 through 7. The expansion of this core analysis to inferences of opposition and the failure of subalternacy in the following two chapters is intended to demonstrate the descriptive adequacy of the core analysis by generalizing it to phenomena for which it was not designed.

The final step is to demonstrate its explanatory adequacy. As stated in (1.15), for Chomsky ...

> a linguistic theory that aims for explanatory adequacy is concerned with the internal structure of the acquisition device; that is, it aims to provide a principled basis, independent of any particular language, for the selection of the descriptively adequate grammar of each language.

From the neuromimetic perspective, we claimed in (1.18) that the following is more appropriate:

> An explanatorily adequate model gives an output in a way that is consistent with the abilities of the human mind / brain.

In other words, the claim is that explanatory adequacy in the current context is an issue to be decided on general principles of cognitive organization, even if the analysis whose adequacy is at stake is of a linguistic phenomenon.

The challenge of this claim is how to execute it in the current welter of views on the shape of cognitive theory. To help us keep our bearing in this confusing sea of assumptions, counter-assumptions, and counter-counter-assumptions, we turn to George Lakoff's identification of two 'generations' of cognitive scientists in Lakoff (1987, 1988) – though it is unclear to us whether anyone outside of linguistics would actually recognize these two generations as Lakoff paints them. Given the precedent of a First and Second Generation, it is natural to conceptualize those that build on both of them – among whom we count ourselves – as the Third Generation.

11.1. GEN I: THE DISEMBODIED AND UNIMAGINATIVE MIND

Lakoff's First Generation of Cognitive Science coalesced in the mid-1970's from ideas in classical artificial intelligence, information processing, psychology, and generative linguistics. The philosophical underpinnings of this approach go by several names, of which the most prominent are realist semantics, the sentential paradigm of cognitive science, and symbolism. They are inspired on, if not designed after, first order logic. First order logic is so well-known that we hesitate to rehash its tenets here, but doing so will ensure that we all begin at the same place.

11.1.1. A first order grammar

11.1.1.1. First order logic

First order logic assigns a meaning to a sentence in some language *L* by assigning a truth value in a model to the formula expressed by the sentence. If the formula is an unanalyzable proposition, the system is known as propositional logic. If the formula is analyzable into a subject-predicate pair, the system is known as predicate logic. Propositional logic has less structure than predicate logic and so is simpler to use as an example of how logic works, but the status of the predicates of predicate logic is crucial to the rejection of the Cognitive Science of the Disembodied and Unimaginative Mind, so it behooves us to sketch the more complex structure.

A logical language *L* is an ordered pair of objects $\langle U, [\![\cdot]\!]\rangle$. *U* is any nonempty set, called the universe of the model, which contains the values of the individual variables of *L*. $[\![\cdot]\!]$ is a function which assigns to the variables in *L* their **denotation** in *U*. Two modes of assignment can be distinguished, syntactic and semantic. They are explained and illustrated in the following subsections.

11.1.1.2. First order syntax

The **syntax** of *L* is a set of rules for concatenating *L*'s symbols. These rules can be classified into two kinds, **rules of formation**, including a **vocabulary**, and **rules of inference**. (11.1) and (11.2) exemplify the former, and (11.4) exemplifies three of the latter:

11.1. Vocabulary

a)	apple	= CN
b)	ruby	= CN
c)	emerald	= CN
d)	my	= POSS
e)	your	= POSS
f)	is_red	= VP
g)	is_green	= VP
h)	every	= Q
i)	some	= Q

j)	no	= Q
k)	and	= CONJ
l)	or	= CONJ
m)	neither ... nor	= CONJ

11.2. Rules of formation

a) S → NP VP
b) NP → POSS CN
c) NP → Q CN
d) S → S CONJ S
e) Only that which can be generated by the Vocabulary plus clauses (11.2a-d) in a finite number of steps is a well-formed formula.

The rules of formation stipulate those symbols and concatenations of symbols that correspond to specific interpretable objects. More exactly, the Vocabulary in (11.1) stipulates those symbols that belong to syntactic categories, and the rules in (11.2) stipulate those concatenations of syntactic categories that belong to complex syntactic categories. (11.3) gives a sample:

11.3 a) My apple is_green. [by 11.2b, 11.2a]
b) Your ruby is_red. [by 11.2b, 11.2a]
c) My apple is_green and your ruby is_red. [by 11.2b, 11.2a, 11.2d]

These rules can be as complex as the natural language that they analyze.

There is also a set of rules of inference, a few of which are illustrated in (11.4):

11.4. Rules of inference

a) From S_1 *and* S_2 you may infer S_1.
b) From S_1 *and* S_2 you may infer S_2.
c) From S_1 and S_2 you may infer S_1 *and* S_2 .

The rules of inference stipulate those concatenations of symbols that correspond to premises and the conclusions that can be drawn from them. In the examples of (11.4), the object of *from* marks the premise, and the object of *infer* marks the conclusion. The rules of inference themselves can be concatenated into lists called **proofs**. For instance, (11.3) constructs some sentences for us, so we can plug them into the variables in (11.4) to build a toy proof:

11.5. a) From My apple is_green and your ruby is_red, you may infer:
b) *My apple is_green* [by 11.4c], and
c) Your ruby is_red [by 11.4c]

Of course, proofs can be more much complex, and useful. Fenstad (1998) characterizes the standard logical approach to reasoning and understanding as a search for ever more sophisticated proof procedures. Given that proof procedures are syntactic, we may ask why the approach is not to search for ever more sophisticated semantic procedures. It is to this question that we now turn.

11.1.1.3. First order semantics

The **semantics** of the language assigns the objects defined by the syntax a denotation in a universe and then evaluates the truth of this denotation. The standard **universe** is an ordered n-tuple *U* of the form (11.6):

11.6. $U = \langle A, R_1, \ldots, R_n \rangle$

Here *A* is a non-empty set of entities, and $R_1, \ldots, R_n$ are a number *n* of relations over *A*. Note that it is often convenient to single out one-place relations as **properties**. These semantic objects, plus a **truth value**, true (1) or false (0), are assigned to the syntactic objects of the previous subsection by the interpretation function $[\![\cdot]\!]$.[47] For our sample vocabulary and formation rules, some sample assignments are as in (11.7):

11.7. Vocabulary

a)	$[\![\text{apple}]\!]$	$= \{a_1, a_2\}$
b)	$[\![\text{ruby}]\!]$	$= \{r_1, r_2\}$
c)	$[\![\text{emerald}]\!]$	$= \{e_1, e_2\}$
d)	$[\![\text{my}]\!]$	$= \{x: x \in [\![\text{CN}]\!]$ & x belongs to me$\} = \{a_1, r_1, e_1\}$
e)	$[\![\text{your}]\!]$	$= \{x: x \in [\![\text{CN}]\!]$ & x belongs to you$\} = \{a_2, r_2, e_2\}$
f)	$[\![\text{is\_red}]\!]$	$= 1$ iff $[\![\text{NP}]\!] \in \{a_1, r_1, r_2\}$
g)	$[\![\text{is\_green}]\!]$	$= 1$ iff $[\![\text{NP}]\!] \in \{a_2, e_1, e_2\}$
h)	$[\![\text{every}]\!]$	$= 1$ iff $\forall x[x \in [\![\text{CN}]\!] \rightarrow x \in [\![\text{VP}]\!]]$
i)	$[\![\text{some}]\!]$	$= 1$ iff $\exists x[x \in [\![\text{CN}]\!] \;\&\; x \in [\![\text{VP}]\!]]$
j)	$[\![\text{no}]\!]$	$= 1$ iff $\neg\exists x[x \in [\![\text{CN}]\!] \;\&\; x \in [\![\text{VP}]\!]]$
k)	$[\![\text{and}]\!]$	$= 1$ iff $[\![S_1]\!] = 1$ and $[\![S_2]\!] = 1$
l)	$[\![\text{or}]\!]$	$= 1$ iff $[\![S_1]\!] = 1$ or $[\![S_2]\!] = 1$
m)	$[\![\text{neither ... nor}]\!]$	$= 1$ iff $[\![S_1]\!] = 0$ and $[\![S_2]\!] = 0$

[47] $[\![\cdot]\!]$ gloms the truth valuation together with the set-theoretic interpretations for the sake of brevity. It would be more perspicuous to do the set-theoretic interpretation first and then the truth valuation, so that the truth value assignments are separated out as **truth conditions**, but it would also be much more prolix.

11.8. Formation rules

a) [NP] = [POSS] ([CN])

b) [S] = [CONJ] ([S], [S])

Some sample interpretations in this model are given in (11.9):

11.9. a) [every ruby is red] = 1

b) [my emerald is green] = 1

c) [every apple is red] = 0

Moreover, the proof in (11.9) is both syntactically and semantically well-formed.

The model sketched above only lets us consider interpretations in one scenario at a time. However, there are many occasions when we would like to consider several alternative scenarios at a time. For instance, we may want to add to the scenario sketched in (11.7) another one in which both apples are red. To do so, the interpretations must be relativized to a given state of affairs, known technically as a **possible world**. This can be achieved by indexing the denotations according to the world. To continue with the apple example, we can add $world_2$ to what was already stipulated about $world_1$:

11.10. a) $[\text{is\_red}]_{w1} = 1$ iff $[NP] \in \{a_1, r_1, r_2\}$

b) $[\text{is\_red}]_{w2} = 1$ iff $[NP] \in \{a_1, a_2, r_1, r_2\}$

This now changes the interpretation of (11.9c) in $world_2$ to true:

11.11. a) $[\text{every apple is red}]_{w1} = 0$

b) $[\text{every apple is red}]_{w2} = 1$

All other readings stay the same.

11.1.2. More on the ontology

Entities and their relations are the only two types of objects permitted in *U* and are referred to as its **ontology**. Given that the relations are all defined at the level of the universe, they cannot be reduced to any more fundamental elements. That is to say, the relations do not have any more internal structure than that of a list:

> Any *n*-ary relation *R* on *A* can be represented in the form of two lists. One is a list of basic or atomic positive facts, i.e. a list of all $Ra_1 \ldots a_n$ such that $(a_1 \ldots a_n)$ belongs to the relation *R*; and a supplementary list of basic or atomic negative facts, *not*-$Rb_1 \ldots b_n$, for $(b_1 \ldots b_n)$ not belonging to the relation *R*. (Fenstad, 1998)

It follows that a given entity either does or does not have a particular property – since it is listed either on the positive list *R* or the negative list *not-R*. There are no fuzzy boundaries.

Nor are there any vertical divisions. All of the objects of the universe have the same ontological status. There are no structurally defined hierarchies, so *U* is 'flat'. Despite the vast literature on extensions to first order logic, these extensions are 'tame', in Fenstad's words – they all rest on the ontology of lists.

A third inference that can be drawn from the atomic status of relations in *U* is that they are 'objective': since they come predefined as part of the universe, they cannot vary from observer to observer or speaker to speaker. This inference is often understood to mean that the entities of the universe form objectively existing categories based on their shared (objective) relations, what Lakoff, 1987, p. 161, calls the Doctrine of Objective Categories.

This inference can be taken one step further by allowing entities to be categorized by having one or more relations in common. Under such a definition, categorization inherits the objectivity of the relations that make it up. Since membership in a relation is all or nothing, the grouping of relations into categories becomes all or nothing as well. This leads to the Aristotelian division of properties into necessary and sufficient for categorization: a property is **necessary** for categorization if its lack means that the entity *cannot* belong to the category, and a property is **sufficient** for categorization if its possession means that the entity *must* belong to the category. For instance, *having hair* is sufficient to categorize an animal as a mammal, but it is not necessary, since dolphins lack it.

11.1.3. Classical categorization and semantic features

The first formal framework that linguists hit upon for the analysis of necessary and sufficient categorization is that of componential semantic analysis, which Hjelmslev and Jakobson developed from Trubetzkoy's (1969[1939]) theory of componential phonological analysis.[48] Chomsky (1965) introduced the formalism of abbreviating such notions with semantic features, which take the form [±F] for some property *F*. The +F and –F collocations abbreviate the notions of 'being F' and 'not being F', respectively, which enforces the constraint of the excluded middle: an entity cannot be both *F* and not *F*. For instance, an equilateral triangle can be defined by four features: [+closed], [+three sided], [+equal length], and [+line segments]. These features combine by conjunction into larger feature matrices, such as [+F, –G], which is understood to mean that the entity in question is *F* and not *G*.

---

[48] See Lyons, 1977, pp. 318-9, for a brief history of componential analysis.

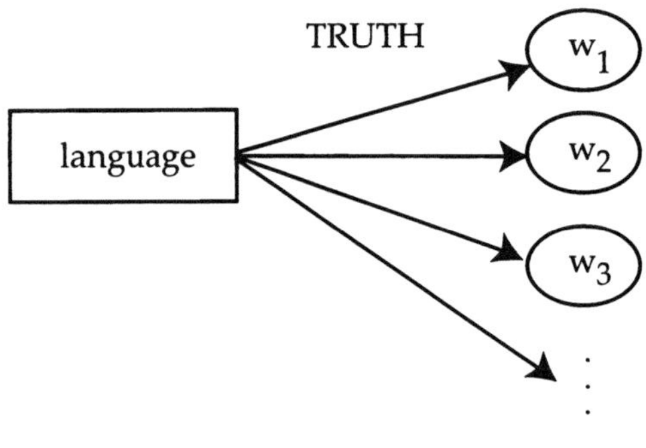

Figure 11.1. Relationship between language, truth, and possible worlds.

11.1.4. Objectivist metaphysics

Johnson, 1987, pp. xxii ff. summarizes the notion of meaning implemented in first order predicate logic as so:

> Meaning is an abstract relation between symbolic representations (either words or mental representations) and objective (i.e. mind-independent) reality. These symbols get their meaning solely by virtue of their capacity to correspond to things, properties, and relations existing objectively "in the world".

Gärdenfors (1996) schematizes this process as in Fig. 11.1. Lakoff (1987) finds that this notion of meaning rests on an "objectivist metaphysics" which Lakoff (1988) distills into the following nine doctrines:

11.12 a) The world consists of entities with fixed properties and relations holding among them at any instant.
b) The entities of the world are divided up naturally into categories called *natural kinds* .
c) All properties are primitive or complex.
d) There are rational relations that hold objectively among the entities and categories of the world.
e) Meaning is based on reference and truth.
f) Truth consists in the correspondence between symbols and states of affairs in the world.
g) There is an "objectively correct" way to associate symbols with things in the world.

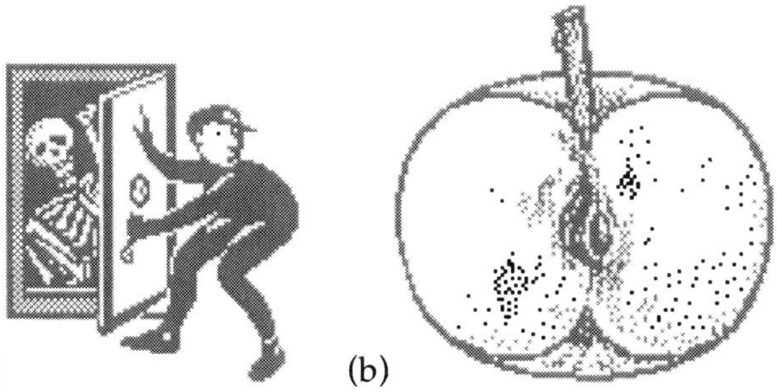

(a) (b)

Figure 11.2. (a) The skeleton is in the safe. (b) The seed is in the apple.

h) Conceptual categories are designated by sets characterized by necessary and sufficient conditions on the properties of their members.

i) A complex concept is defined by a collection of necessary and sufficient conditions on less complex (and, ultimately, primitive) concepts.

Gärdenfors (1996) summarizes these claims is a slightly more concise fashion:

11.13 a) Semantic elements are symbols that can be concatenated according to some system of rules.

b) Cognitive models are independent of perception.

c) Meaning is truth conditions in possible worlds.

d) Cognitive models are propositional. Metaphoric and metonymic operations are exceptional.

e) Syntax can be described independently of semantics.

f) Concepts follow the Aristotelian paradigm based on necessary and sufficient conditions.

The rest of this chapter is devoted to the ways in which human cognition departs from this model, but first, let us examine a linguistic example that is near and dear to the hearts of GenII-ers, namely that of spatial prepositions, especially English *in*.

11.1.5. An example: the spatial usage of *in*

At first glance, the denotation of *in* would appear to be so simple as to be trivial. Vandeloise (1991) reviews the following "logical" denotations of *in*:

11.14 x in y =

a) x is located internal to y with the constraint that x is smaller than y. (Cooper, 1968)

b) x is enclosed or contained either in a two-dimensional or in a three-dimensional place y. (Leech, 1970[1969])

c) locative(interior y). (Bennett, 1968)

d) referent x is in a relatum y if: part (x, z) & Incl (z, y). (Miller and Johnson-Laird, 1976)

Vandeloise recognizes that these definitions apply satisfactorily to the data for which they were designed, such as the sentences in the caption to Fig. 10.2 with are meant to describe the corresponding pictures. However, there are situations for which they are inadequate, which we will see presently.

11.2. REACTIONS TO GEN I

Reactions to this first generation of assumptions about Cognitive Science have been numerous and disparate – in fact, too numerous and disparate to review in the depth that they deserve here. This section therefore contents itself with the barest mention of some of the drawbacks that have been raised. The objections will take on a more vivid form as we work our way through the Second and Third Generations.

11.2.1. Problems with classical categorization

Classical or Aristotelian categorization has been subject to considerable criticism, see Smith and Medin (1981) and Murphy (2002) for review. Suffice it to say that the work of Rosch and colleagues, e.g. Rosch (1973, 1975, 1978) and Mervis and Rosch (1981), precipitated a re-evaluation of categorization theory as it was realized that many categories do not have identifiable necessary and sufficient features.

11.2.2. Problems with set-theory as an ontology

The listing ontology posited by Fenstad for the First Generation is captured naturally by set theory. Yet Smith (1994) reviews three arguments for the inadequacy of any ontology that depends on sets of points as a primitive. Their force depends on Smith's assertion that a fundamental cognitive faculty is the recognition of boundary-continuum structure. Smith, 1994, p. 10, draws a simple illustration of a boundary-continuum structure from Lewin (1936):

> We begin, following Lewin, with the opposition *thing* (intuitively, a closed connected entity) and *region* (intuitively, a space within which things are free to move about). As Lewin points out, what is a thing from one psychological perspective may be a region from another: 'A hut in the mountain has the character of a thing as long as one is trying to reach it from a

> distance. As soon as one goes in, it serves as a region in which one can move about.' (Lewin, 1936, p. 116) We then define the notion of a *boundary zone z* between two regions *m* and *n*, as the region, foreign to *m* and *n*, which has to be crossed in passing from one to the other. The whole $m + n + z$ is then *connected* in the topological sense. (ibid., p. 121)

Lewin's "boundary zone" is thus for Smith a boundary between two continuums, the exterior of the hut and its interior.

With this introduction, we can now paraphrase Smith's, 1994, p. 9-10, three objections to point-set ontologies:

11.15 a) The experienced continuum is not isomorphic to any real-number structure.

b) The experienced continuum is not organized out of particles or atoms, but rather in such a way that the wholes, including the medium of space, come before the parts which these wholes might contain and which might be distinguished on various levels within them.

c) No basic level of *Urelemente* can be isolated from the experienced continuum by means of which all higher structure can be simulated by successively higher types.

Thus the best that treatments of the experienced continuum as structured sets of points can hope for is a certain amount of descriptive adequacy, but never any explanatory adequacy.

11.2.3. Problems with symbolicism

Sec. 1.3.3 has already introduced several considerations of natural computation that call into question symbolic representations of cognition, and they are reviewed in the light of our results in Sec. 11.5.3. One that is most apropos for our current discussion is where symbols come from. As Cangelosi and Harnad (2000) put it, "Just as the values of the tokens in a currency system cannot be based on still further tokens of currency in the system, on pain of infinite regress -- needing instead to be grounded in something like a gold standard or some other material resource that has face-value -- so the meanings of the tokens in a symbol system cannot be based on just further symbol-tokens in the system."

11.3. GEN II: THE EMBODIED AND IMAGINATIVE MIND

The reaction to the First Generation of Cognitive Science that Lakoff names the Second Generation of Cognitive Science, that of the Embodied and Imaginative Mind, is characterized by Gärdenfors, 1996, pp. 162ff, in terms of six slogans:

11.16 a) Meaning is *conceptualization* in a cognitive model (not truth conditions in possible worlds).

b) Cognitive models are mainly *perceptually* determined (meaning is not independent of perception).

c) Semantic elements are based on *spatial* or *topological objects* (not symbols that can be concatenated according to some system of rules);

d) Cognitive models are primarily *image-schematic* (not propositional). Image schemas are transformed by *metaphoric* and *metonymic* operations.

e) *Semantics* is primary to syntax and partly determines it (syntax cannot be described independently of semantics).

f) Concepts show *prototype* effects (instead of following the Aristotelian paradigm based on necessary and sufficient conditions).

These are partially synthesized from Lakoff's own review of the debate between logical and cognitive semantics. In this section, the Second Generation is summarized only in sufficient detail to point out enough drawbacks to provide a springboard to the innovations of the Third Generation, on which we wish to dwell at length.

11.3.1. Prototype-based categorization

The alternative that sprang from Rosch's elucidation of the problems with classical categorization came to be articulated by the claim that a category is distributed around its most typical member, called the prototype, see Posner and Keele (1968, 1970). For instance, when asked to assign an item to a category, a subject responds with the category possessing the most similar prototype, see Reed (1972).

11.3.2. Image-schematic semantics

Prototype theory provided the motivation for a non-logical representational format based on **image schemata**, so analyses formulated with the aid of these objects can be referred to as **image-schematic semantics**. By way of illustration, consider Langacker's (1986) claim that linguistic representations have six "dimensions of imagery":

11.17 a) Every linguistic predication imposes a **profile** on a **domain**.

b) Every linguistic expression has a level of specificity.

c) Every linguistic predication has a scope or scale which it must include. A smaller scope will not permit the predication to be evaluated.

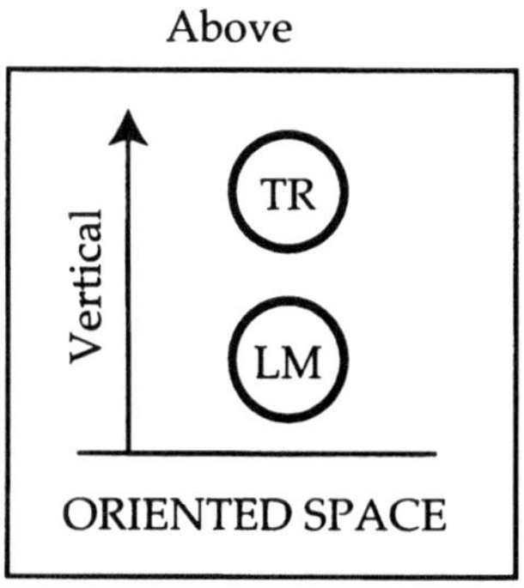

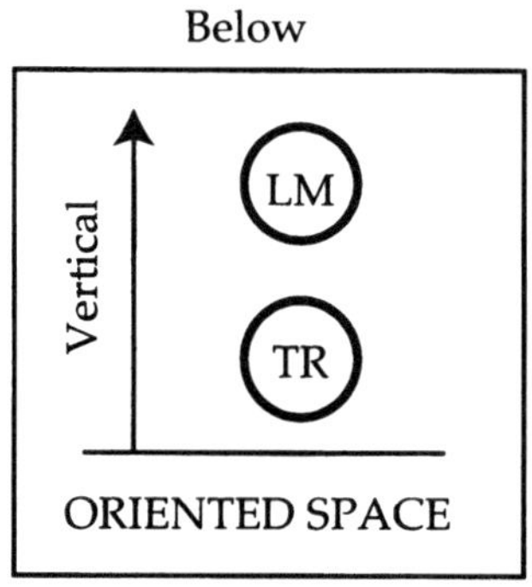

Figure 11.3. *Above* vs. *below*

d) Every linguistic predication assigns a degree of salience to its substructures. There are several variations of this desiderata, but the most important concerns the asymmetry between an entity X which is evaluated against a reference point Y. Intuitively, the X or **trajector** participant is tracked against the background provided by the Y or **landmark** participant. (Langacker, 1987, pp. 231-2)

e) Every linguistic predication is construed relative to background assumptions and expectations.

f) Every linguistic predication is construed relative to a perspective of orientation, assumed vantage point, directionality, and objectivity.

An example of the synthesis of these six dimensions into a single unit is provided by the representations of the English prepositions *above* and *below,* in Fig. 11.3, based on Langacker, 1987, Fig. 6.4. The domain in both cases is given by a box with a label across the bottom which describes its content, "ORIENTED SPACE" in this case. The profile is the rest of the material in the box, which is what the linguistic expressions designate. The level of specificity of Fig. 11.3 is rather broad: it focuses only on the level which most succinctly relates the participants in *above/below*. The participants themselves can be taken to much greater levels of specificity. The scope of *above/below* is rather imprecise, but if the boxes in Fig. 11.3 were shrunk to exclude one of the circles, *above/below* could not be evaluated appropriately. The trajector (TR) and landmark (LM) participants of Fig. 11.3 are inverses of each other. For instance, in *the clouds are above the airplane,* 'the clouds' corresponds to the trajector, which *above* situates higher on a scale of verticality than the landmark 'the airplane'. In contrast, in the inverse predication of *the clouds are below the airplane, below* situates the same trajector 'the clouds' lower on a scale of verticality with respect to the same landmark of 'the airplane'. The background of construal does not have a

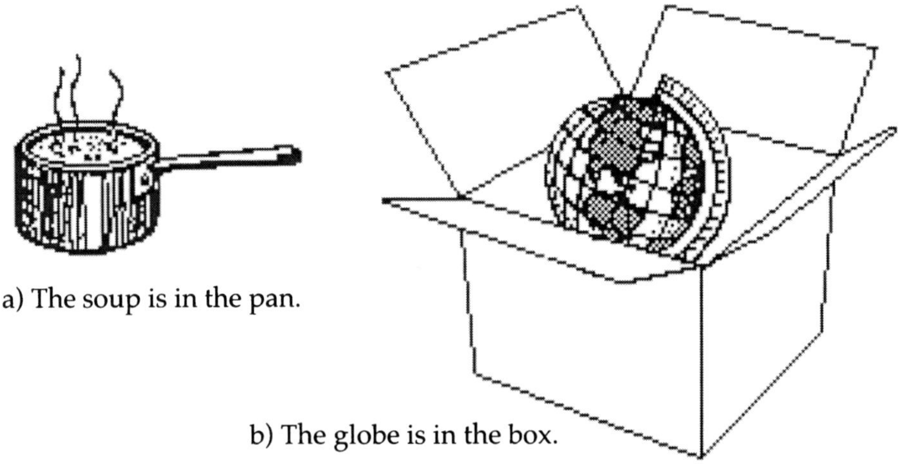

a) The soup is in the pan.

b) The globe is in the box.

Figure 11.4. Problematic usages of *in*.

particular effect on Fig. 11.3, but the perspective of construal does: Fig. 11.3 is meant to be understood as a view from the side, along the axis of verticality given by the upward arrow.

11.3.3. Image-schemata and spatial *in*

Returning to the particular example of the spatial usage of the English preposition *in*, where Vandeloise finds fault with the objective definitions of (11.14) is in images such as those of Fig. 11.4. On the one hand, neither the pan nor the box is covered, so their interiors extend upwards to the ends of the universe. On the other hand, even if the box in Fig. 11.4b were closed, the globe protrudes slightly above its top, so it is not entirely within the enclosure. Thus the globe is not a subset of, or included in, the interior of the box. A similar argument can be made for the stereotypical images associated with the following sentences:

11.18 a) The egg is in the eggcup.
b) The tree is in the ground/planter.
c) The thread is in the needle.
d) The straw is in the glass.
e) The dog is in the doghouse.
f) The fish is in my hand.
g) The wire is in the pliers.

Vandeloise, in line with the Second Generation philosophy, takes this descriptive failure to stem from the reliance of logical approaches on necessary

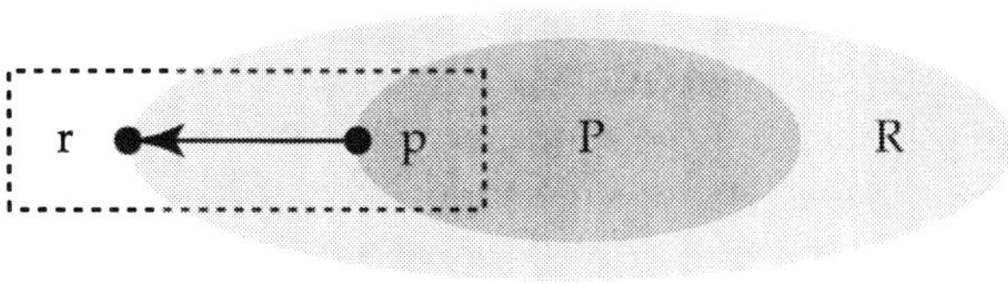

Figure 11.5. Langacker's conception of quantification.

and sufficient conditions when natural language actually relies on fuzzier categories of prototypes and family resemblances.

With respect to the case of French *dans* "in", Vandeloise, 1991, p. 255, proposes that it instantiates the **container/contained** prototype, which is characterized as follows:

11.19 a) The contained object moves towards the container but not the reverse.
b) The container controls the position of the contained object and not the reverse.
c) The contained object is included, at least partially, in the container or in the convex closure of its containing parts.

Since English *in* appears to behave in much the same way, Vandeloise's hypothesis can be stated semi-formally in the following way:

11.20. in(Contained, Container)

All of the problematic cases instantiate this fuzzier notion. In the globe and box example Fig. 11.4, the globe qualifies as the contained object and the box as the container, because (i) the globe stereotypically moves into the box, and not vice versa, cf. (11.19a); (ii) the box controls the position of the globe, and not vice versa, cf. (11.19b); and (iii) the globe is included at least partially in the box, and not vice versa, cf. (11.19c).

11.3.4. Image-schematic quantification

We know of no Second Generation analysis of the empirical issues dealt with in this book other than the theory of quantification adumbrated in Langacker, 1991, pp. 107ff. Langacker refers to *all, most, some* and *no* as "relative quantifiers". They are relative in that the nominal is profiled as a mass *P* that is a proportion of a reference mass *R*. Langacker illustrates this concept with a diagram like Fig. 10.5. In this diagram, *P* and *R* are conceptualized as spatially continuous entities with boundaries. *P* is superimposed on *R* and matched against it to see how close their boundaries come to coinciding. This matching is executed by taking a "slice" of the figure – the dashed box – and estimating the distance between the boundary points *p* and *r*. The quantifier *all* specifies coincidence or identity of the two points, while *most* locates *p* in the vicinity of *r*.

There is one notable point of agreement between Langacker's and our own theory of quantification. It is that both treat quantification as the calculation of a proportion between measurements of two objects. Where they differ is in how the measurements are stated. Langacker avoids stating them in terms of numbers, despite the fact that the measurements he has in mind could quite easily be so stated. It follows that if one were to recast Langacker's framework in a format explicit enough to support neurologically-inspired simulations of learning and retrieval, all of these numbers would have to be calculated. The pattern-classification approach, in contrast, states its measurements quite explicitly in a numerical format, one based on the Tree of Numbers, which lends itself readily to simulation and thereby testing of the theory. It should consequently be preferred on the meta-theoretical grounds of greater explicitness and broader cognitive plausibility.

There are also two ways in which Langacker's approach is inaccurate. The first is that it treats all quantifiers as mass quantifiers, despite the obvious distinction between mass and count quantifiers in English, such as *much* and *many*. To rule in the count quantifiers, some auxiliary translation must be made from the mass to the count domain. Such an extra step is not needed in the pattern classification approach, once it is realized the numbers actually refer more generally to measured entities, taken either as groups of count individuals or extents of mass individuals, though there is no space to develop this notion adequately here. The second problem of Langacker's account is how it would deal with infinite quantifications, such as *every even number*. Langacker would presumably agree with us that a human brain does not have the resources to find a proportion between two infinite masses, so some allowance must be made for a finite solution. The pattern-classification approach achieves this explicitly, through quantization of the unit arc. It is likely that if Langacker's account is modified to address these two problems, it will turn out to be little more than a notational variant of the pattern-classification approach.

11.4. REACTIONS TO GEN II

Image-schematic semantics has not diffused very far beyond its linguist adherents, so it has received little in the way of disinterested evaluation. What evaluation that it has received from Cognitive Science and Philosophy has not been positive.

11.4.1. The math phobia of image-schematic semantics

At a general level, Wildgen, 1994, p. 40, evaluates the entire Second Generation enterprise rather negatively:

11.21 a) The imaginistic language used is neither systematic nor conclusive and rests only on an intuitive perception of a possible relation between pictorial schemata and linguistic expressions.

b) Those imaginistic representations which are intuitively plausible cover only a small sub-field of the lexicon and of basic syntax.

c) There is no theoretical account of how the images may be constructed; they are mere illustrations based on a set of vaguely defined conventions.

d) The enormous possibilities of space-oriented modeling using geometry, topology, differential topology and other mathematical models, which have dealt with similar conceptual problems (partially since antiquity) are systematically ignored.

e) The epistemological claim that grammar must be independent of mathematical techniques is incompatible with the integration of standard techniques used in generative grammar, as these are based on algebraic concepts and not on 'natural' categories.

In the absence of appropriate formal theories drawn from space-oriented modeling or similar fields, as Smith, 1996, p. 298, says, "the work of cognitive linguists will remain subject to the charge that it has not gone beyond the stage of narrative evidence-gathering."

By way of illustration, Smith, 1996, p. 298, cites Talmy (1983) on the preposition *in*, which is a precursor of Vandeloise's analysis of *in* and Langacker's more general development of image schema. Smith points out that *in* is neutral with respect to several fundamental mathematical properties:

11.22 a)	magnitude neutrality:	in a thimble, in a volcano
b)	shape neutrality:	in a well, in a trench
c)	closure neutrality:	in a bowl, in a ball
d)	discontinuity neutrality:	in a bell-jar, in a bird cage

The task of defining the precise nature of the transformation that maps one *in*-structure into another under these conditions – or absence of conditions – is a difficult one. Smith cites Wildgen, 1994, p. 32, in this respect: "The quasi-formal symbols in Talmy's descriptions come from algebra, geometry, topology, and vector calculus, but the mathematical properties of these concepts are neither exploited nor respected." It is undoubtedly due to such "absence of appropriate formal theories" that Second Generation ideas have not been applied to natural language processing, with the honorable exception of Regier (1995).

Finally, we would like to add that the rejection of any mathematical ontology for language by the second Generation is really a rejection of set theory as an appropriate ontology, as was adumbrated above. But there are ways of defining ontologies that are not based on sets. Our discussion of mereotopology below will touch on one way of doing so.

11.4.2. Problems with prototype-based categorization

Despite its improvement over classical categorization, experimental psychologists soon noticed that prototype-based categorization suffered from its own defects, see Smith and Medin (1981), Medin and Ross (1989), Nosofsky (1992), Ashby, 1992, pp. 450-1, Ross and Spalding, 1994, pp. 124-5, and Ross and Makin, 1999, pp. 211ff. There are three main problems.

Categorization appears to make use of knowledge that goes beyond the prototype and into the particular instances that were used to learn the prototype. For example, if a stimulus is very similar to an item that was presented during the synthesis of the prototype, this stimulus will be categorized more easily than another stimulus that is just as similar to the prototype, but not similar to any of the instances used to learn the prototype, see Whittlesea (1987).

Moreover, categorization appears to make use of knowledge that goes beyond the prototypical values for a feature used in categorization and into particular values, and even combinations of values, that were used to learn the prototype. On the one hand, subjects know the range of values that a feature might have and can use this knowledge to make categorization decisions, see Walker (1975) and Rips (1989). On the other hand, subjects know that some properties go together – the example that is usually quoted is that small birds often sing, while big birds rarely do – and can use this information to aid in categorization – a singing bird is probably not a big one, see Medin, Altom, Edelson and Freko (1982) and Malt and Smith (1984).

Finally, categorization varies as a function of context, but prototypes do not, see Roth and Shoben (1983). The oft-quoted example is that a robin is a more typical bird than a turkey, but in the specific context of "The holiday bird looked delicious", a turkey becomes much more typical than a robin. As currently constituted, prototype models have no means of expressing this variation.

11.4.3. The biological plausibility of the Second Generation

Finally, one should not overlook the fact that 'embodied' cognitive science is actually 'embrained'. That is, some allowance must be made for how the brain processes the primitives of image-schematic semantics. To the best of my knowledge, there has been no attempt within the Second Generation to draw inspiration from the neurophysiological critiques of symbolicism sketched in Sections 1.3.3 and 11.5.3. This is unfortunate, for it means that those that take these critiques to heart jump directly from the First Generation to the Third, with nary a glance at the Second.

11.5. GEN III: THE IMAGED AND SIMULATED BRAIN

Despite the heated debate over sets, schema, and such, the First and Second Generations agree on two fundamental propositions, which concern methodology and the essence of cognition. Methodologically, they both rely on native-speaker intuitions, at least for the linguistic phenomena that they deal with. With respect to cognition, they rely on its 'macrostructure'. O'Reilly and Munakata, 2000, p. 14, say it best:

> Our introspections into the nature of our own cognition tend to emphasize the "conscious" aspects (because this is by definition what we are aware of), which appear to be *serial* (one thought at a time) and *focused* on a subset of things occurring inside and outside the brain. This fact undoubtedly contributed to the popularity of the standard serial computer model for understanding human cognition...

Though the final sentence casts an oblique glance at the First Generation, our reading of the Second Generation does not reveal it to be in substantial disagreement. For instance, though an image schema for a morpheme is a gestalt that is grasped as a whole or in parallel, its form is culled from the linguist's conscious (introspective) understanding of the morpheme in question.

11.5.1. The microstructure of cognition

Neither 'introspectionism' nor 'macrostructurism' are necessarily wrong, but their usefulness to a practicing cognitive scientist depends on how much importance should be allotted to what they omit. And it turns out that what they omit could be considerable. Let us finish O'Reilly and Munakata's thought:

> We argue that these conscious aspects of human cognition are the proverbial "tip of the iceberg" floating above the waterline, while the great mass of cognition that makes all of this possible floats below, relatively inaccessible to our conscious introspection. ... Attempts to understand cognition by only focusing on what's "above water" may be difficult, because all the underwater stuff is necessary to keep the tip above water in the first place – otherwise, the whole thing will just sink!

O'Reilly and Munakata call the alternative "underwater stuff" the **microstructure** of cognition, following Rumelhart et al. (1986).

If its microstructure is fundamental to understanding how cognition works on the macro scale, yet it is inaccessible to introspection, the practicing cognitive scientist is in a methodological bind. The only sources of information are the imaging technologies described in Chapter 10 and the single cell recordings used to such effect by Hubel and Wiesel in their investigation of V1, summarized in Chapter 1. Yet the former do not yet have the scale of resolution to answer the kinds of linguistic questions that are asked in this monograph, and the later are too dangerous to perform on humans except in extreme cases (pursuant to commissurotomy). So the practicing cognitive scientist is stuck. The only way to get loose is to develop a theory of cognitive microstructure from single cell recordings of non-humans and to try to verify the predictions it makes against some aspect of human behavior, or introspection. This is the project of computational neuroscience, begun by Hodgkin and Huxley and whose latest installment is sitting in your hand, or on your computer screen.

The commitments of the Third Generation should be getting clearer: imaging for data collection and computational simulation for analysis and explanation. Certainly for linguistics, we are not there yet. The following is what we can do in the mean time.

11.5.2. Mereotopology during the transition

The formal vacuum left by the Second Generation of Cognitive Science has attracted the attention of several researchers, among them, Barry Smith, Peter Gärdenfors, and Jens Erik Fenstad. Smith is at the center of a group working on the combination of topology and mereology into a hybrid framework called mereotopology. Gärdenfors is developing a theory of conceptual space inspired on Stalnaker's (1981) logical space, to which is added the notions of metric and topological structure. Fenstad (1998) combines aspects of both of these frameworks, as well as notions drawn from the theory of dynamical systems.

This section reviews conceptual spaces and mereotopology, with an eye towards accentuating those aspects that are most compatible with the ANN approach of the monograph, as well as demonstrating how they improve the First and Second Generation theories of cognitive science. The next chapter attempts to synthesize all of these considerations into a coherent whole.

11.5.2.1. Gärdenfors' conceptual spaces

Gärdenfors' theory of Conceptual Spaces has been elaborated in a series of publications which culminate in Gärdenfors (2000). The next few subsections limit themselves to pointing out the principle assumptions of his approach, which were already cited in Chapter 3.

11.5.2.1.1. Conceptual Spaces

A conceptual space is composed of a number of quality dimensions. Some examples of such dimensions are arrayed in Table 11.1. These dimensions are taken to be prelinguistic in the sense that humans, and other animals, can think about the qualities of objects, for instance, when planning an action, without presuming a language in which these thoughts are expressed. The reason for separating a cognitive structure into dimensions is that this separation expresses the incommensurability of dimensions: the assumption that an individual can be assigned one property independently of another, when the two lie on different dimensions. Moreover, many inferences can be drawn from a quality dimension due to its metric and topological structure, as we will see in upcoming paragraphs.

Table 11.1. Sample quality dimensions, biased towards metric system.

Name	*Units*	*Type of scale*
length	meter	linear non-negative real numbers
hue	(degree)	circular non-negative real numbers
pitch	Hertz	logarithmic non-negative real numbers
time	second	linear real numbers
sex	(none)	discrete and polar

Table 11.2. Sample conceptual spaces.

Name	*Dimensions*	*Type of scale*
Euclidean space	length × width × height	linear × linear × linear
color space	hue × brightness × saturation	circular × linear × linear
vowel space	$F_1 \times F_2$	logarithmic × logarithmic

Quality dimensions are plotted against one another to compose a conceptual space, three examples of which are laid out in Table 11.2. More technically, a conceptual space S is composed of a number of quality dimensions, $D_{1, \ldots,} D_n$. A point in S is a vector $s = \{d_1, \ldots, d_n\}$ with one component for each dimension, i.e. where each d_i is an element of D_i. An interpretation of a language L consists in a mapping of the components of L onto a conceptual space S.

As a first step, a **location function** – a term borrowed from Stalnaker, 1981, p. 347 – maps individual constants onto points in S. In this way, each individual gets a specific color, spatial position, weight, temperature, and so forth. If an individual is assigned a partial vector, this means that not all of the properties of the individual are known or have been determined. The incommensurability of dimensions means that each point in S represents a 'possible individual', that is, a possible assignment of properties to individuals. Moreover, each possible individual will always have an internally consistent set of properties, inherited from the underlying dimensions. For instance, given that blue and yellow are disjoint properties in color space, no individual can be both. There is no need for meaning postulates to exclude such contradictory properties. Thus a possible individual is a cognitive notion that need not have any form of reference in the external world.

A second step is for the predicates of L to be assigned regions in the conceptual space. Such a predicate is satisfied by an individual just in case the location function locates the individual at one of the points included in the region assigned to the predicate.

11.5.2.1.2. Properties in conceptual space

From Sec. 3.4.5, the reader may recall Criterion P, reproduced here:

11.23. Criterion P: A natural property is a convex region of a domain in a conceptual space.

The reader may further recall from Sec. 4.5 that a convex region is characterized by the fact that for every pair s_i, s_2 of points in the region, all points between s_i ands_2 are also in the region. This definition presumes that the notion of 'between' is meaningful for the relevant dimensions – a rather weak assumption about the topological structure of dimensions. Notably, it does not presume either the concept of an individual or the concept of a possible world.

Most meanings expressed by simplex words in natural languages are natural properties. For instance, all color terms in natural languages express natural properties with respect to the psychological representation of the three color dimensions. Intensional properties like *tall* and *former* are secondary properties, defined over a reference class given by some (primary) property.

The underlying dimensions endow natural properties with their logical characteristics. For instance, if time is isomorphic to the real number line, then *earlier* will automatically be transitive, asymmetric and acyclical. There is no need for meaning postulates or any other mechanism to guarantee that natural properties have their expected characteristics. Moreover, *any* natural property that is isomorphic to the real line will have a corresponding comparative, e.g. 'long' ~ 'longer', 'bright' ~ 'brighter', etc. Gärdenfors finds this to be more accurate than defining such properties as sets of ordered pairs.

11.5.2.1.3.Prototypes and Voronoi tesselation

Prototype effects are to be expected in conceptual space, given that properties are defined as convex regions. Some subregions of a convexity are more central than others, and these most central exemplars constitute the prototype. Moreover, 'the prototype' does not necessarily have to be among the existing members of the category. It can just as well be one of the possible individuals that is constructed from the quality dimension but not instantiated in the real world. In this case, it would be a partial vector, with some number of real-world dimensions left undetermined. For instance, the prototypical bird would have a specification for the prototypical bird shape, but no specification for age or color.

Gärdenfors goes on to argue in the converse direction, namely that the postulation of prototypes leads to the conclusion that natural properties have the structure of convex regions. The idea is that if we start from the prototypes $p_1, \ldots, p_n$ for a set of related categories, then these individuals should be the central points of the categories that they represent. A non-central individual p can be related to a central one p_i by stipulating that p belongs to the same category as the closest prototype p_i. This rule generates a partitioning of the space into convex regions, a process we are already familiar with from the discussion of Voronoi tessellation in Chapter 3.

11.5.2.1.4. Conclusions

Gärdenfors' theory of conceptual spaces has many points of overlap with the LVQ analysis of the linguistic phenomena discussed in this monograph, despite its lesser reliance on biological plausibility and neuromimetic simulation. Gärdenfors (2000) in particular can be read as a helpful companion to this book that fills out many of the philosophical issues that are raised but not addressed here.

11.5.2.2. Smith's and Eschenbach's mereotopology

Another congenial approach is developed in the unfolding mereotopological framework championed by Barry Smith, Carola Eschenbach, and their collaborators. Thus, despite our admiration for conceptual spaces, the formal methods that tie our neuromimetic approach to larger issues in cognitive science are drawn from mereotopology.

11.5.2.2.1. Mereology + topology

Topology is the branch of ontology that attempts to elucidate the concepts of 'region', 'interior', 'exterior', 'boundary', 'integrity', 'continuity', 'separation', 'surface', 'point', 'neighborhood', 'nearness', and 'closure', among others. Standard topology defines these concepts in terms of set theory, but such set-theoretic atomism leads to several foundational problems as a theory for human cognition.

In order to attain explanatory adequacy, Smith proposes to use mereology as a basis for articulating topological structures without points and without requiring topological entities to be sets. Mereology is the branch of ontology that attempts to elucidate the concepts of x being a part of y and its conceptual relatives 'overlap', 'discreteness' and 'sum', without giving rise to any concept of integrity or what it means to be a whole. The hybrid analytic framework that results from basing topology on mereology is called **mereotopology.**[49] There are several versions. Smith's (1996) version combines the mereological primitive of parthood or constituency and the topological primitive of an interior part, while Eschenbach's (1994) version combines the mereological primitive of discreteness and the topological primitive of a region. I find Eschenbach's framework to be more tractable for the limited purposes of this report and so briefly review its major definitions and axioms in order to give the reader a feel for mereotopology and a basis for understanding how it relates to LVQ.

11.5.2.2.2. Mereotopological notions of Eschenbach (1994)

The mereological primitives of Eschenbach's framework are reproduced below in their prose versions:

[49] See Smith (1994) and Eschenbach, 1994, p. 64, for references to historical antecedents, dating back at least to Husserl.

D1. x is a **part** of y iff x is discrete from everything y is discrete from.
D2. x is a **proper part** of y iff x is a part of y, and y is not a part of x.
D3. x and y **overlap** iff they have a common part.
D4. x is the **sum** of some entities iff x is discrete from exactly those entities which are discrete from each of them.
D5. x is the **product** of some entities iff x is the sum of all their common parts.
D6. x is an **atom** iff it has no proper part.
D7. The (mereological) **complement** of x is the sum of all entities discrete from x.

The axioms are:

A1. x and y are discrete iff x and y do not overlap.
A2. If x is a part of y and y is a part of x, then x and y are identical.
A3. For any entities, their sum exists.

The primitive notion is that of **region**:

D9. x is a **topological entity** iff x is part of a region.
D10. The **topological universe** is the sum of all regions.
D11. Regions x and y are **internally connected** iff they have a common part which is a region.
D12. Regions x and y are **externally connected** iff they overlap but are not internally connected.
D13. Region x is **open** iff it is not externally connected to any region.
D14. The **interior** of x is the sum of the open regions which are part of x.
D15. y is **adherent** to x iff every open region which overlaps y overlaps x.
D16. The **closure** of x is the sum of the open regions which are part of x.
D17. The topological entity x is **closed** iff it is identical to its closure.
D18. The **topological complement** of x is the sum of all topological entities discrete from x.
D19. The **boundary** of x is the product of its closure and the closure of its topological complement.
D20. The topological entities x and y are **separated** iff x is discrete from the closure of y and y is discrete from the closure of x.
D21. The topological entity x is **self-connected** iff it is not the sum of two separated topological entities.
D22. y is an **inner part** of the topological entity x iff y is part of an open region z which is part of x.
D23. If x is a topological entity, then y is a **dangling part** of x iff y is a proper part of x, x has a topological complement z, and y is an inner part of the closure of z.

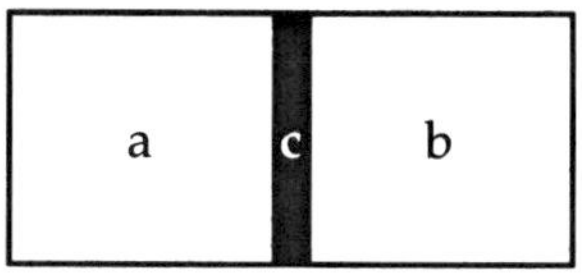

Figure 11.6. Mereotopological example.

D24. A topological space is **grounded on closed entities** iff every region is the sum of the closed entities which are part of it.

The axioms are:

A4. Every region has an open region as a part.
A5. For any regions their sum is a region.
A6. The product of any two overlapping open regions is an open region.

The following example helps to illustrate the definitions.

Let a, b, c be three mutually discrete entities and a, b, $a+b$, $a+c$, $b+c$, $a+b+c$ be the regions, as in Fig. 11.6. Instantiations of some mereotopological concepts found in Fig. 11.6 are listed in (11.24):

11.24 a)	topological universe:	$a+b+c$
b)	topological entities:	any part of $a+b+c$
c)	internally connected regions:	$a+c$ & $a+b$
d)	externally connected regions:	$a+c$ & $b+c$
e)	open regions:	a, b, $a+b$, $a+b+c$
f)	closed entities:	c, $a+c$, $b+c$, $a+b+c$
g)	interior of $a+c$:	a
	interior of $b+c$:	b
	interior of $a+b+c$:	$a+b+c$
h)	c is adherent to:	a, b, c, $a+b$, $a+c$, $b+c$, $a+b+c$
i)	adherent to c:	c
j)	closure of a:	$a+c$
	closure of b:	$b+c$
	closure of $a+b$:	$a+b+c$
k)	c is the boundary of:	a, b, c, $a+b$, $a+c$, $b+c$
l)	separate:	a, b
m)	self-connected:	$a+b+c$

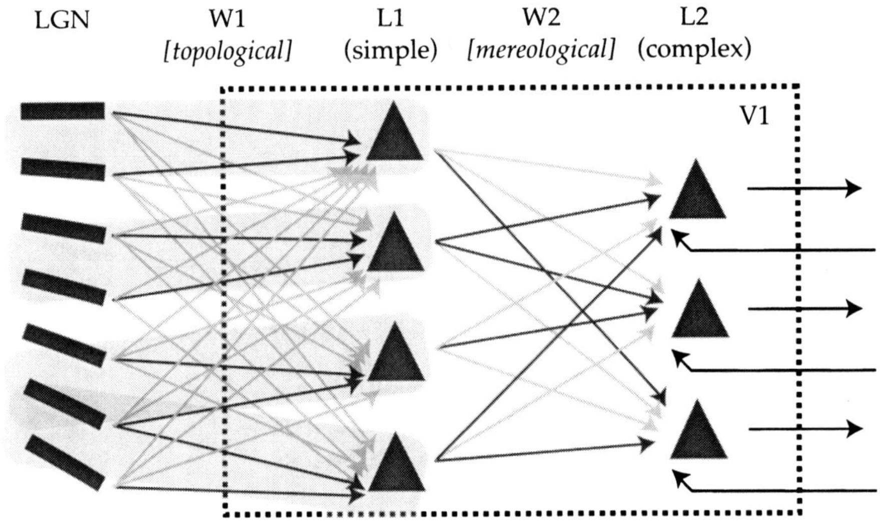

Figure 11.7. Mereotopological organization of V1.

Only a handful of these notions are needed for the mereotopological elaboration of LVQ, but it is important for the reader to get a glimpse of mereotopology in its natural habitat first.

11.5.2.2.3.LVQ mereotopology

Our hypothesis is simple: the competitive layer of a LVQ network encodes topological objects and the linear layer encodes topological regions. This hypothesis has already been met in Fig. 1.34 , which is reproduced here as Fig. 11.7. Mereotopology provides the tools to redefine these primitives more exactly:

D1′. *x* is a **topological entity** iff *x* is encoded by a competitive neuron.
D2′. *x* is a **topological region** iff *x* is encoded by a linear neuron.
D3′. The **topological universe** of network N is the collection of linear neurons of N.

The other topological definitions and axioms should retain their original form and so need not be repeated here. The mereological part-of relation is defined by the feedforward connections from the competitive layer C to the linear layer *L*:

D4′. *x* is a **part** of *y* iff *x* feeds *y*.
D3′. *x* and *y* **overlap** iff they have a common feeder.

The other mereological definitions also should retain their original form and so need not be repeated.

Given these reductions of pivotal mereotopological primitives to network relationships, we can entertain the prospect of training an LVQ network on a particular object as a way of analyzing the object mereotopologically. Such simulations are not conducted here, however, in order to not loose sight of the linguistic phenomena that are the empirical focus of the monograph.

11.5.3. Intelligent computation and LVQ mereotopology

From the Third Generation perspective, what it means to be intelligent is to learn to recognize patterns in the environment and respond to them in some way. Recognizing patterns means to parse the environment into spaces defined by varying degrees of shared similarity. Responding to them means associating certain responses to certain subregions of the space. With respect to the linguistic data of this monograph, this has meant parsing the environment into coordinator, quantifier, and prepositional vector spaces and associating certain phonological representations to certain subregions of these spaces. This is the essence of the function computed by learning vector quantization.

But note that this model of intelligent linguistic behavior recapitulates the theory of the relationship between topology and mereology found in mereotopology: topology parses a given domain into a continuum of topological entities sharing some degree of similarity, and mereology uses the part-of relation to combine these entities into larger topological regions. This suggests that the two frameworks can be mapped to each other.

The rest of this section tests these ideas against the review from the end of Sec. 1.3 where we summarized the ideas of Amit, Touretzky, Eliasmith, and Shastri on the criteria for designing cognitive architecture and intelligent computation, with some additional considerations from Haykin, 1994, pp. 415ff. and O'Reilly and Munakata, 2000, pp. 14ff. Along the way, we will incorporate the discussion from Nadeau (2000) and Bullinaria (2000) on ways in which PDP models capture important properties of cerebral organization or human performance. Nadeau concentrates on the present-/past-tense pattern associator of Rumelhart and McClelland (1986), which has given rise to a veritable cottage industry in the simulation of the acquisition of the English past tense and other cases of morphology with regular and irregular realizations, see Bullinaria (1997), Seidenberg and Hoeffner (1998), O'Reilly and Munakata, 2000, pp. 350ff, and Oaksford (2001). Bullinaria concentrates on models of reading: his own, Bullinaria (1997), and that of Seidenberg and McClelland (1989). Given Nadeau's stronger interest in linguistically relevant models, we will draw more frequently on his exposition.

11.5.3.1. Neural plausibility

As Amit, 1989, p. 6, says, the elements composing a cognitive model should not be physiologically outlandish. The basic LVQ model developed in Chapters

5 and 7 attempts to satisfy this criterion by using a local Hebbian learning algorithm, that of competitive learning, implemented by a spike-rate model of the neuron. The IAC model of inferencing developed in Chapter 8 adopts the LVQ network layout to bidirectional processing via spreading activation, again implemented by a spike-rate model. However, we have also had occasion to confect a spike-timing model to simulate the binding together of subject-predicate pairs in the hippocampus, and to sketch a novel dendritic learning algorithm. All of these proposals have been sandwiched between micro- and macro-level descriptions of neurophysiology. Chapter 2 explains where spikes and learning rules come from at the micro-level, Sec. 5.2 situates neocortex in contrast to the basal ganglia and cerebellum in the distribution of learning rules, and Sections 10.1.2.4-5 further subdivide neocortex into anterior and posterior processing zones and try to place the logical operators in one or the other.

Neuromimetic considerations also permeate the representations offered for the various linguistic phenomena that the text has touched upon. Chapter 3 proposes a signed measure/trivalent logic for the logical operators which is ultimately motivated by the hypothesis that logical operators detect coincidences, that is, they register correlations between two linguistic categories. Chapter 3 attempts to drive home this message by showing how signed measures cut across several mathematical domains, thereby unifying a number of disparate traditions and providing a wealth of tools for neuromimetic linguistic analysis. The confluence of these tools is the vector-theoretic understanding of mereotopology developed in the various simulations of Chapters 4, 6, 9, and 11, though considerable work remains to be done before this framework reaches the level of precision and explicitness that a real mathematician or logician would require.

Amit and Shastri draw up several architectural criteria for neural plausibility that can be examined fruitfully from the perspective of this monograph's claims and results. The LVQ network processes its input in parallel in that a given input is exposed to all of the neurons at the next level. Of course, these computations are performed for the most part on a serial, von Neumann computer, so no simulation is literally parallel, but until parallel computers become more readily available, this is the best we can do. A decision on what to do with an input is reached by competition among the neurons, without guidance from any central controller. Each neuron computes an out-going activation based on the magnitudes of its in-coming activations and transmits it to all of the neurons to which it is connected. The message communicated by these exchanges of activation is a scalar number without internal structure. The connections along which the neurons compete are numerous and given in advance, usually as all-to-all connectivity.

11.5.3.1.1.Interactivity

LVQ networks are designed hierarchically, so that the intermediate layer has to balance its response to ambiguous, small-scale stimuli active in the lower layer against large-scale contextual information active in the higher layer. It thus performs a kind of parallel bidirectional constraint satisfaction that is rampant in real nervous systems.

In Sec. 1.2.4.2, top-down supervision of V1 is conceptualized as Bayesian error correction, a proposal that should be extended to the phonological supervision of our LVQ networks. Unfortunately, we have not had the time to do so and must defer the topic to another venue.

11.5.3.1.2.Cross-domain generality

Finally, LVQ satisfies Touretzky's and Eliasmith's criterion of cross-domain generality, since there is nothing in its componential structure that limits it to, say, sight, taste, sound, touch, smell, or language.

11.5.3.2. Self-organization

Of course, a major goal of all this biology is to be able to learn from experience and through learning improve performance, see Haykin, 1994, p. 352. What makes this task even more difficult is the assumption that there is no teacher to supply an error signal in the acquisition of the logical operators. Competitive learning is the algorithm that we have promoted to solve this problem. While it may seem like magic, the way in which it works was anticipated as early as Turing (1952), where Turing says:

11.25. Global order can arise from local interactions.

The global order of a competitive network is a Voronoi tesselation of the input space, which produces many of the intelligent aspects of LVQ mereotopology. In this way, a LVQ network is free of an exogenous observer that assigns meaning to the output – the meaning of the output is given entirely by the internal mapping from the input, and usually is not even apparent and so must be discovered by careful testing. It may be concluded that Voronoi tessellation goes a long way towards meeting Amit's desideratum of freeing the cognitive system from reliance on homunculi. Such homunculi are often postulated surreptitiously in symbolicist systems in the absence of any external link from symbols to their referents in order to learn the system and interpret it to itself.

In the next subsections, we tease apart some of the reasons why Voronoi tesselation works the way it does, drawing on Haykin's, 1994, pp. 415ff., explanation of the three cognitively interesting properties of Voronoi tesselation effected by a competitive network.[50]

---

[50] Haykin actually refers to an elaboration of a competitive network called the self-organizing feature map, see Kohonen (1997), but the difference for our concerns is minimal.

11.5.3.2.1.Density matching and statistical sensitivity

The one Haykin calls **density matching** helps to explain the desideratum of statistical sensitivity developed by Touretzky and Eliasmith. A network Φ that performs density matching encodes statistics of the input distribution into its structure:

11.26. Regions in the input space *X* from which sample vectors are drawn with a high probability of occurrence are mapped onto larger domains of the output space *A*, and therefore with better resolution than regions in *X* from which sample vectors are drawn with a low probability of occurrence.

In LVQ terms, density mapping is reflected by the fact that more competitive neurons are assigned to those regions of the input space that are populated by more inputs. This is one way of saying that the greater the probability of finding an input in a particular region, the greater the probability of finding a competitive neuron nearby. Density matching thus constitutes one way of enabling an intelligent cognitive system to evolve through its experience, by learning more about frequent phenomena than infrequent phenomena.

11.5.3.2.2.Approximation of the input space

A second property of competitive networks is **approximation of the input space**, by which Haykin means the following:

11.27. A network Φ, represented by the set of synaptic weight vectors $\{w_j \mid j = 1, 2, ..., n\}$ in the output space *A*, provides a good approximation to the input space *X*.

Such approximation has two cognitive advantages. One speaks to the symbol-grounding problem and the other to self-organization.

Mentioned briefly in Sec. 11.2.3, and to be discusses somewhat obliquely in Sec. 11.5.3.6, the symbol-grounding problem asks how higher cognitive functions know what they refer to. Under the LVQ analysis of semantics as an exercise in pattern classification which begins with an approximation of the input space, there is always a path that can be traced backwards from the phonological output units to the conceptual input units to show how a sound-meaning symbol is ultimately grounded in whatever caused the conceptual input.

The second advantage is that approximation of the input space is the first step in self-organization, which is to say, learning from experience. This step is just for the system to create an internal representation of the input that it must process. Unsupervised competitive learning accomplishes this quite accurately.

11.5.3.2.3. Topological ordering and associativity

LVQ also takes the next step, which is to turn the good approximation into a **topological ordering**, by which Haykin means the following:

> A competitive network Φ computed by the [competitive] learning rule is topologically ordered in the sense that the spatial location of a neuron in the lattice corresponds to a particular domain or feature of input patterns.

Topological ordering creates a faithful representation of the important features that characterize the input vectors. Moreover, it implements Amit's criterion of associativity, the collapsing of similar inputs into a prototype. Of course, in a competitive network there are as many prototypes as there are neurons.

11.5.3.2.4. Implicit rule learning

Another brain-like property of LVQ is implicit rule learning. Networks learn common patterns – a central tendency in a data sample – and so appear to the eyes of a human observer to have learned a rule – and perhaps they have, though this is an issue of some contention, see Pinker (2000).

11.5.3.2.5. Emergent behavior

To summarize all of four these properties in a single phrase, we may say that the learning of an input-output mapping in a LVQ network is an instance of emergent behavior, in the general sense that the human experimenter does not encode it into the network, and in Amit's more specific sense that it seems rather unlikely that a network could ever settle on any useful mapping all by itself – this is the magic of LVQ and its sister neural-network algorithms!

11.5.3.3. Flexibility of reasoning

There still remain a fistful of other intelligent and brain-like properties of LVQ, which can be organized around the central topic of flexible reasoning. This 'miscellaneous' group includes resistance to degradation, generalization to novel inputs, pattern completion, and potential for abstraction.

11.5.3.3.1. Graceful degradation

Another brain-like property of LVQ networks lies in their resilience in the face of damage that mimics cerebral lesions. This robustness follows from distributed representation in that a given unit only stores a small and partially redundant fraction of the total knowledge of a network, so that its failure does not seriously compromise the rest of the network. Gradually increasing the number of failing units gradually disrupts the accuracy of the network, with the result that its performance degrades in a manner that is invariably called "graceful".

11.5.3.3.2. Content-addressability

The various schemes for representing information discussed above do so on the basis of the properties of the information to be stored, creating a kind of memory whose content is addressed directly by an external query. A digital computer, in stark contrast, stores information among various labeled cubbyholes which have no relation to one another. Its content can only be accessed by first finding the label of the cubbyhole holding the information, which may require a search of the entire memory. The content-addressable status of an LVQ network creates associative memory, which is to say that a query to a LVQ network will activate a range of items based on their similarity and frequency. Querying one item tends to activate all similar items, and querying all items tends to activate the most frequent ones to a greater degree.

Nadeau attributes associative memory to distributive encoding, since similar items share features and so are encoded by the same units and therefore tend to have similar activation patterns. However, Bullinaria, 2000, p. 223ff, offers an account based on statistical learning. The similarity among items in a training sample facilitates the detection by the network of the input-output mapping to be learned for them – creating an often-observed regularity effect. In a similar vein, frequent items are also easier to learn because the network has more exposure to them – creating an equally often-observed frequency effect. Hence both similar and frequent items have strong connections because they provide more instances for a network to learn from.

Mereotopology expands on Bullinaria's account of the regularity effect by positing that layers are organized topologically by some measure of similarity, so that regularity is part and parcel of their structure.

11.5.3.3.3.Pattern completion

One obvious aspect of intelligent behavior is the ability to deal with partial or incomplete information. In a pattern-recognition paradigm, the main way of implementing this ability is by retrieving a complete pattern from some fragmentary or distorted input. The LVQ networks designed in the previous chapters do not show this ability in any obvious way, due to the extreme simplicity of the patterns that they were trained on. It is difficult to imagine how to design a fragmentary or distorted input for AND or ALL that could test the pattern-completion ability of the coordinator or quantifier networks. However, the IAC networks for logical opposition display a kind of pattern completion, in the sense that they link logical elements into larger groups so that the excitation of one leads to the activation of similar elements and the inhibition of dissimilar ones.

However, a different way of understanding pattern completion lies in an LVQ network's ability to draw inferences. A query to the net activates some pattern of associations, which may wind up activating some coherent sub-pattern, which counts as an inference drawn from partial data, namely the query. Of course, this sub-pattern may be incorrect – the network has no internal way of knowing – so that its response counts as a confabulation. As Nadeau, 2000, p. 326, quite eloquently frames it, "the blurring of the distinction between veridical recall and confabulation or plausible reconstruction seems to be characteristic of human memory".

11.5.3.3.4. Generalization to novel inputs

Another sign of intelligent behavior is the ability to generalize to novel inputs. Given that an LVQ network learns how to classify its input patterns accurately from only a small sample of the input space, it follows that it will automatically extend a learned pattern to unfamiliar inputs that fall into the proper receptive field. In fact, one of the ways that the accuracy of a LVQ network was tested in the preceding chapters was by watching how it responded to novel input.

11.5.3.3.5. Potential for abstraction

Finally, it would appear that all of our simulations instantiate Amit's desideratum of a potential for abstraction. To refresh the reader's memory, this requirement is that "the model should operate similarly on a variety of inputs that are not simply associated in form but are classified together only for the purposes of the particular operation". Given that our LVQ networks find patterns in inputs that are grouped together solely for the purposes of the simulation, it follows that the architectures themselves have a tremendous potential for abstraction. In fact, this is one of the reasons that these architectures were investigated in the first place.

However, there is another way of understanding this desideratum that is less complimentary. Our networks are invariably fed a data set carefully designed so that the items are all of a similar form that is understandable to the network, so its ability for abstraction is only apparent in the broad applicability of the algorithm mentioned in the previous paragraph, not in any particular trained network. This may however merely reflect the fact that current experimental goals are overly narrow, rather than constituting a fatal flaw in the global cognitive plausibility of LVQ.

11.5.3.4. Structured relationships

Finally, the ability of LVQ networks to deal with structured information is rather limited, which is one reason why Sec. 10.2.2 is devoted to demonstrating how the hippocampus can bind together roles and their arguments in a way that begins to meet this desiderata. Pulvermüller (2002) treats this issue in considerably more detail, and goes a long way to answering the criticism of V1-like systems formulated in Jackendoff, 2002, p. 80.

11.5.3.5. Exemplar-based categorization

Given that we have reviewed evidence for rejecting both classical and prototype-based categorization, whatever is left standing should be taken to be the theory of categorization for the Third Generation. This third theory is known as exemplar-based categorization, which is briefly reviewed in the next few paragraphs.

The essential claim of exemplar-based models is that there is no single privileged representative of a category to which a stimulus is compared; instead, a stimulus is compared to every representative that the subject has been exposed to, see Brooks (1978, 1987), Medin and Schaffer (1978), Estes (1986), Hintzman (1986), Nosofsky (1986), and considerable research since then.

Several advantages are claimed for exemplar models. For instance, there may be an adaptive advantage to storing exemplars rather than creating abstractions. The creation of an abstraction requires that the organism be prescient as to what information will be required at a later time for survival. By contrast, the retention of detailed information about particular instances allows the organism to generate flexible abstractions on-line which may have been unanticipated when the category was first acquired. This is not to suggest that exemplar models do not allow abstraction to occur. They do, but it occurs on-line in the service of some particular task, rather than at the time of storage, see Barsalou (1990). This is why Palmeri can claim that exemplar models are superior to prototype models: they account for all the data that prototype models do, plus some.

From our LVQ perspective, a competitive neuron acts like a 'small' prototype, accounting for just a few exemplars, and competitive neurons can be grouped together mereologically at the linear layer in order to create flexible abstractions on-line. LVQ would therefore appear to be a good candidate to formalize many of the properties claimed for exemplar models, though pursuing this suggestion would consume another volume all in itself.

11.5.3.6. LVQ and the evolution of language

To round out our defense of LVQ within the unfolding Third Generation framework, let us look at a topic that has not been touched upon so far in these pages, namely the evolution of language. There is a story about the evolution of language implicit in LVQ, but we have to take a round-about path to get to it. The path passes through Cangelosi and Harnad (2001), where a simulation is reported which combines backpropagation of error with a genetic algorithm to evolve a community of virtual organisms that learn to forage for 'mushrooms' and communicate what they find to one another. While giving a full account of this simulation reaches far beyond the goals of this monograph, what fascinates us about it enough to mention it at all is the fact that the foragers learn which mushrooms to eat from their sensorimotor experience with them, but they learn much more rapidly if they are aided by the calls of fellow foragers.

Cangelosi and Harnad contrast learning a category by "sensorimotor toil" to learning a category by "symbolic theft", that is, cued by communication. The enhancement afforded by symbolic theft is conceptualized as learned categorical perception, i.e. the compression of within-category distances and the expansion of between-category distances, see Andrews et al. (1998).

They then tell the following story about the evolution of language – well, actually a Martian anthropologist tells the story, but you have to read the paper:

> All other species on this planet get their categories by toil alone, either cumulative, evolutionary toil or individual lifetime toil: Individuals encounter things, [they] must learn by trial and error what to do with what, and to do so, they must form internal representations that reliably sort things into their proper categories. In the process of doing so, they keep learning to see the world differently, detecting the invariants, compressing the similarities and enhancing the differences that allow them to sort things the way they need to be sorted, guided by feedback from the consequences of sorting adaptively and maladaptively (as in the mushroom world).

That's how it proceeded on our planet until one species discovered a better way:

> First acquire an entry-level set of categories the honest way, like everyone else, but then assign them arbitrary names. ... Once the entry-level categories had accompanying names, the whole world of combinatory possibilities opened up and a lively trade in new categories could begin...

While LVQ may not appear to have anything to do with backpropagation and genetic algorithms, it has everything to do with Cangelosi and Harnad's understanding of their results. "Sensorimotor toil" is the perceptual/conceptual input to our LVQ networks, encoded in the competitive layer. "Symbolic theft" is the top-down influence of a phonological form, encoded into the linear layer and perhaps ultimately modeled by Bayesian error correction. The compression of within-category distances and the expansion of between-category distances is practically a definition of Voronoi tesselation, so what Cangelosi and Harnad call categorical perception just follows from the correct functioning of a competitive network. The fact that its categorization is enhanced by additional guidance from an external information source, such as verbal names, rounds out the definition of LVQ. The story of linguistic evolution turns out to be much simpler and more perspicuous when told through the eyes of an LVQ network.

11.6. SUMMARY

What this chapter endeavors to do in the meantime is organize the recent history of cognitive science in terms that should be accessible to a linguist, for which Lakoff's identification of two 'generations' of cognitive scientists has proven to be conceptually invaluable – though it is unclear to us whether anyone outside of linguistics would actually recognize these two generations. Given the precedent of the two generations recognized by Lakoff, it is natural to conceptualize those that disagree with both of them as the Third Generation.

So far, Generation III has been characterized distinctively. On the one hand, there are those who fault the First and Second Generations for paying scant attention to the biological plausibility of their representations. The claim implicit in their criticism is that biological implausibility robs the theories in question of any hope of explanatory adequacy. This is a burning issue for this monograph, for we wish to demonstrate the explanatory adequacy can only be attained by studying neurology. On the other hand, there are those who fault the First and Second Generations on philosophical grounds – in particular for assuming incorrect or inadequate ontologies. We have given name and address to this latter group: Gärdenfors (2000) for conceptual space and Eschenbach (1994) and Smith (1994) for mereotopology.

The crucial step is to put the neurophysiologists and the philosophers together in a common framework that presents the Third Generation more as a coherent theory than a collection of tendencies. This is attempted in the last few subsections, where it is argued that something like the LVQ network structure is the linchpin that holds neurophysiology and mereotopology together.

References

Aboitz, F. et al., 1992, Fiber composition of the human corpus callosum, Brain Research, 598, 143-153.

Adams, D. and S. Zeki, 2001, Functional organization of macaque V3 for stereoscopic depth, Journal of Neurophysiology, 86, 2195-2203.

Albright, T.D. and G.R. Stoner, 2002, Contextual influences on visual processing, Annual Review of Neuroscience, 25, 339-379.

Allen, J. and M.S. Seidenberg, 1999, The Emergence of Language: The Emergence of Grammaticality in Connectionist Networks, Ed. MacWhinney, B. Erlbaum, Mahwah, NJ, pp. 115-152.

Amari, S. and M.A. Arbib, 1977, Systems Neuroscience: Competition and cooperation in neural nets, Ed. Metzer, J. Academic Press, San Diego, pp. 119-165.

Ames, A., 1997, Mitochondria and Free Radicals in Neurodegenerative Disease: Energy requirements of brain functions: When is energy limiting?, Eds. Beal, M.F., N. Howell and I. Bodis-Wollner John Wiley, New York, pp. 12-27.

Amit, D.J., 1989, Modeling Brain Function. The world of attractor neural networks. Cambridge University Press, Cambridge.

Andrews, J., K. Livingston and S. Harnad, 1998, Categorical perception effects induced by category learning, Journal of Experimental Psychology: Human Learning and Cognition, 24, 732-753.

Applebaum, D., 1996, Probability and Information. An integrated approach. Cambridge University Press, New York.

Arbib, M.A., 1993, Allen Newell, Unified Theories of Cognition, Artificial Intelligence, 59, 265-283.

Arbib, M.A., 1995, The Handbook of Brain Theory and Neural Networks: Part I: Background, Ed. Arbib, M. The MIT Press, Cambridge, USA & London, England, pp. 3-25.

Archangeli, D., 1997, Optimality Theory. An overview: Optimality Theory: an introduction to linguistics in the 1990s, Eds. Archangeli, D. and D.T. Langendoen Blackwell Publishers, Malden, Mass., pp. 1-32.

Archie, K.A. and B.W. Mel, 2000, An intradendritic model for computation of binocular disparity, Nature Neuroscience, 3, 54-63.

Archie, K.A. and B.W. Mel, 2001, Advances in Neural Information Processing Systems 13: Dendritic Compartmentalization Could Underlie Competition and Attentional Biasing of Simultaneous Visual Stimuli, Eds. Leen, T.K., T.G. Dietterich and V. Tresp The MIT Press, Cambridge, USA, Ashby, F.G., 1992, Multidimensional Models of Perception and Cognition:

Multidimensional models of categorization, Ed. Ashby, F.G. L. Erlbaum, Hillsdale, N.J., pp. 449-484.

Atick, J.J. and A.N. Redlich, 1990, Towards a theory of early visual processing, Neural Computation, 2, 308-320.

Atkinson, R.C. and R.M. Shiffrin, 1968, The Psychology of Learning and Motivation: Advances in Research and Theory. Vol. 2.: Human memory: A proposed system and its control processes, Eds. Spence, K.W. and J.T. Spence Academic Press, New York, pp. 89-195.

Atlas, J.D. and S.C. Levinson, 1981, Radical Pragmatics: It-clefts, Informativeness, and Logical Form: Radical Pragmatics (Revised Standard Version), Ed. Cole, P. Academic Press, New York, NY, pp. 1-62.

Bailey, P. and G. Von Bonin, 1951, The Isocortex of Man. University of Illinois Press, Urbana, Ill..

Balkenius, C., 1999, Are There Dimensions in the Brain? Lund University Cognitive Science, Lund, Sweden.

Ballard, D.H., 1997, An Introduction to Natural Computation. MIT Press, Cambridge, Mass. & London.

Banich, M.T. and W. Heller, 1998, Evolving perspectives on lateralization of function, Current Directions in Psychological Science, 7, 1-2.

Bar-Hillel, Y. and R. Carnap, 1952, An outline of a theory of semantic information. MIT Technical Report 247.

Bar-Lev, Z. and A. Palacas, 1980, Semantic command over pragmatic priority, Lingua, 51, 137-146.

Barlow, H.B., 1972, Single units and sensation: a neuron doctrine for perceptual psychology?, Perception, 1, 371-394.

Barlow, H.B., 1996, Banishing the Homunculus, Eds. Knill, D.C. and W. Richards Cambridge University Press, New York, pp. 425-450.

Barsalou, L.W., 1990, Content and Process Specificity in the Effects of Prior Experiences: Advances in social cognition, Vol. 3: On the indistinguishability of exemplar memory and abstraction in category representation, Eds. Srull, T.K. and J. Robert S. Wyer Lawrence Erlbaum Associates, Hillsdale, N.J., pp. 61-88.

Bartsch, R. and T. Vennemann, 1972, Semantic Structures. Athenäum Verlag, Frankfurt/Main, Germany.

Bartsch, R., 1973, Syntax and Semantics, vol. 2: The syntax and semantics of number and numerals, Ed. Kimball, J.P. Seminar Press, Inc., New York, NY, pp. 51-93.

Barwise, J. and R. Cooper, 1981, Generalized quantifiers and natural language, Linguistics and Philosophy, 4, 159-219.

Bastiaanse, R. & Grodzinsky, Y. (Eds.) (2000) Whurr Publishers, London.

Bates, E., 1994, Modularity, domain specificity and the development of language, Discussions in Neuroscience, 10, 1/2, 136-149.

Bauer, M.I. and P.N. Johnson-Laird, 1993, How diagrams can improve reasoning, Psychological Science, 4, 372-378.

Bayes, T., 1764, An essay towards solving a problem in the doctrine of chances, Philosophical Transactions of the Royal Society of London, 53, 370-418.

Bear, M.F., B.W. Connors and M.A. Paradiso, 1996, Neuroscience: Exploring the brain. Williams & Wilkins, Baltimore.

Becker, S. and M.D. Plumbley, 1996, Unsupervised Neural Network Learning Procedures for Feature Extraction and Classification, Applied Intelligence, 6, 185-203.

Becker, S., 1995, The Handbook of Brain Theory and Neural Networks: Unsupervised learning with global objective functions, Ed. Arbib, M. The MIT Press, Cambridge, USA & London, England, pp. 997-1000.

Beeman, M.J. and C. Chiarello, 1998, Complementary left and right hemisphere language comprehension, Current Directions in Psychological Science, 7, 2-8.

Beeman, M.J. et al., 1994, Summation priming and coarse semantic coding in the right hemisphere, Journal of Cognitive Neuroscience, 6, 26-45.

Beggs, J.M. et al., 1999, Fundamental Neuroscience: Leaning and memory: Basic mechanisms, Eds. Zigmond, M.J. F.E. Bloom, S.C. Landis, J.L. Roberts, L.R. Squire Academic Press, Inc., San Diego, pp. 1411-1454.

Bennett, D.G., 1968, English Prepositions: A Stratificational Semantics. Longman, London.

Benthem, J.v., 1984, Questions about quantifiers, Journal of Symbolic Logic, 49, 443-466.

Benthem, J.v., 1986, Studies in Discourse Representation Theory and the Theory of Generalized Quantifiers: Semantic Automata, Eds. Groenendijk, J.A.G., D.d. Jongh and M.B.J. Stokhof Foris Publications, Dordrecht, Holland, pp. 1-27.

Ben-Yishai, R., L. Bar-Or and H. Sompolinsky, 1995, Theory of orientation tuning in visual cortex, Proceedings of the National Academy of Sciences of the United States of America, 92, 3844-3848.

Berndt, R.S. and A. Caramazza, 1980, A redefinition of the syndrome of Broca's aphasia: implications for a neuropsychological model of language, Applied Psycholinguistics, 1, 225-278.

Bickhard, M. and L. Terveen, 1995, Foundational Issues in Artificial Intelligence and Cognitive Science. North-Holland, Amsterdam, The Netherlands.

Billard, A. and M.A. Arbib, 2002, Mirror Neurons and the Evolution of Brain and Language: Mirror neurons and the neural basis for learning by imitation: Computational modeling, Eds. Stamenov, M.I. and V. Gallese John Benjamins Publishing Co., pp. 343-352.

Binkofski, F. et al., 2000, Broca's region subserves imagery of motion: A combined cytoarchitectonic and fMRI study, Human Brain Mapping, 11.4, 273-285.

Bishop, C., 1995, Neural Networks for Pattern Recognition. Clarendon Press, Oxford.

Blakemore, D. and R. Carston, 1999, The pragmatics of and conjunctions: The non-narrative cases, UCL Working Papers in Linguistics, 11,

Blakemore, D., 1987, Semantic Constraints on Relevance. Basil Blackwell,

Bliss, T.V.P. and T. Lomo, 1973, Long-lasting potentiation of synaptic transmission in the dentate areas of the anaesthetized rabbit following stimulation of the perforant path, Journal of Physiology, 232, 331-356.

Bloomfield, S.A., J.E. Hamos and S.M. Sherman, 1987, Passive cable properties and morphological correlates of neurones in the lateral geniculate nucleus of the cat, Journal of Physiology, 383, 653-692.

Blumstein, S. and W.E. Cooper, 1974, Hemispheric processing of intonation contours, Cortex, 10, 146-158.

Bochenski, J.M., 1959, A précis of mathematical logic. D. Reidel Pub. Co., Dordrecht, Holland.

Botha, R.P., 1989, Challenging Chomsky. The Generative Garden Game. Blackwell, Oxford.

Bower, J.M., 1997, Is the cerebellum sensory for motor's sake, or motor for sensory's sake: the view from the whisker of a rat, Progress in Brain Research, 114, 463-496.

Braine, M.D.S. and D.P. O'Brien, 1998, Mental Logic. Erlbaum, Mahwah, NJ.

Braine, M.D.S., B.J. Reiser and B. Rumain, 1984, Some empirical justification for a theory of natural propositional logic, The psychology of learning and motivation. Vol. 18, 317-371.

Broca, P., 1861, Remarques sur le siège de la faculté du langage articulé, suivies d'une observation d'aphemie (pert de la parole), Bulletin Société Anatomique de Paris, 36, 350-357.

Brooks, L., 1978, Cognition and Categorization: Nonanalytic concept formation and memory for instances, Eds. Rosch, E. and B.B. Lloyd L. Erlbaum Associates, Hillsdale, N.J., pp. 170-216.

Brooks, L., 1987, Concepts and Conceptual Development: Ecological and Intellectual Factors in Categorization: Decentralized control of categorization: the role of prior processing episodes, Ed. Neisser, U. Cambridge University Press, New York, pp. 141-174.

Brookshire, R.H., 2003, An Introduction to Neurogenic Communication Disorders, 6e. Mosby Year Book, St. Louis.

Browman, C.P. and L. Goldstein, 1986, Towards an articulatory phonology, Phonology Yearbook, 3, 219-252.

Browman, C.P. and L. Goldstein, 1989, Articulatory gestures as phonological units, Phonology, 6, 201-251.

Browman, C.P. and L. Goldstein, 1990, Gestural specification using dynamically-defined articulatory structures, Journal of Phonetics, 18, 299-320.

Browman, C.P. and L. Goldstein, 1990, Representation and reality: Physical systems and phonological structure, Journal of Phonetics, 18, 411-424.

Browman, C.P. and L. Goldstein, 1992, Articulatory phonology: An overview, Phonetica, 49, 155-180.

Bullinaria, J.A., 1997, Modelling Reading, Spelling and Past Tense Learning with Artificial Neural Networks, Brain and Language, 59, 236-266.

Bullinaria, J.A., 2000, Information Theory and the Brain: Free gifts from connectionist modelling, Eds. Baddeley, R., P. Hancock and P. Földiák Cambridge University Press, Cambridge, U.K. & New York, pp. 221-240.

Bush, P.C. and T. Sejnowski, 1993, Reduced compartmental models of neocortical pyramidal cells, Journal of Neuroscience Methods, 46, 159-166.

Bush, P.C. and T. Sejnowski, 1995, The Cortical Neuron: Models of cortical neurons, Eds. Gutnick, M.J. and I. Mody Oxford University Press, New York, pp. 174-189.

Cangelosi, A. and S. Harnad, 2001, The Adaptive Advantage of Symbolic Theft Over Sensorimotor Toil: Grounding Language in

Caramazza, A. and R.S. Berndt, 1985, Agrammatism: A multicomponent deficit view of agrammatic Broca's aphasia, Ed. Kean, M.L. Academic Press, Orlando, Caramazza, A. and E. Zurif, 1976, Dissociation of algorithmic and heuristic processes in language comprehension: Evidence from aphasia, Brain and Language, 3, 572-582.

Chance, F.S. and L.F. Abbott, 2000, Divisive Inhibition in Recurrent Networks, Network, 11, 119-129.

Chance, F.S., S.B. Nelson and L.F. Abbott, 1999, Complex Cells as Cortically Amplified Simple Cells, Nature Neuroscience, 2, 277-282.

Chater, N. and M. Oaksford, 1999, The Probability Heuristics Model of Syllogistic Reasoning, Cognitive Psychology, 38, 191-258.

Chierchia, G. and S. McConnell-Ginet, 1990, Meaning and Grammar. An Introduction to Semantics. The MIT Press, Cambridge, Mass..

Chomsky, N., 1957, Syntactic Structures. Mouton, The Hague.

Chomsky, N., 1964, Current Issues in Linguistic Theory. Mouton, The Hague.

Chomsky, N., 1965, Aspects of the Theory of Syntax. MIT Press, Cambridge, Massachusetts.

Chomsky, N., 1980, Rules and Representations. Columbia University Press, New York.

Chomsky, N., 1981, Lectures on Government and Binding. Foris Publications, Dordrecht.

Chomsky, N., 1986, Knowledge of Language: Its Nature, Origin and Use. Praeger, New York.

Chomsky, N., 1988, Language and Problems of Knowledge: The Managua Lectures. The MIT Press, Cambridge, Massachusetts.

Chomsky, N., 1991, The Chomskyan Turn: Linguistics and Adjacent Fields: A Personal View, Ed. Kasher, A. Blackwell, Cambridge, Mass., USA, Chomsky, N., 1993, Language and Thought. Moyer Bell, Wakefield, R.I..

Churchland, P.S. and T. Sejnowski, 1990, Neural representation and neural computation, Mind and cognition : a reader, 224-251.

Churchland, P.S. and T.J. Sejnowski, 1992, The Computational Brain. The MIT Press, Cambridge, Mass..

Cleland, B.G., M.W. Dubin and W.R. Levick, 1971, Simultaneous recording of input and output of lateral geniculate neurones, Nat. New Biol, 231, 191-921.

Cohen, L.J., 1971, Pragmatics of Natural Languages: Some Remarks on Grice's Views about the Logical Particles of Natural Language, Ed. Bar-Hillel, Y. D. Reidel Publishing Co., Dordrecht, Holland, pp. 50-68.

Connor, J.A., D. Walter and R. McKown, 1977, Neural repetitive firing: Modifications of the Hodgkin-Huxley axon suggested by experimental results from crustacean axons, Biophysical Journal, 18, 81-102.

Connors, B.W. and M.J. Gutnick, 1990, Intrinsic firing patterns of diverse neocortical neurons, Trends in Neurosciences, 13, 99-104.

Cooper, G.S., 1968, A Semantic Analysis of English Locative Prepositions. Report 1587. Bolt, Beranek and Newman,

Coslett, H.B., 2000, Handbook of Neuropsychology, 2nd Edition. Vol. 3: Language and Aphasia: Language and attention, Ed. Berndt, R.S. Elsevier Science, Amsterdam, New York & Oxford, Cotterill, R., 1998, Enchanted Looms: conscious networks in brains and computers. Cambridge University Press, Cambridge, U.K. & New York.

Cover, T.M. and J.A. Thomas, 1991, Elements of Information Theory. Wiley, New York.

Craik, K., 1943, The Nature of Explanation,

Cronin, J., 1987, Mathematical Aspects of Hodgkin-Huxley Neural Theory. Cambridge University Press, Cambridge, U.K..

Cummins, R., 1983, The nature of psychological explanation. MIT Press, Cambridge, MA.

Curme, G., 1931, A Grammar of English Syntax. D. C. Heath & Company, Boston.

Dalrymple, M. et al., 1994, Proceedings from Semantics and Linguistic Theory IV: What do reciprocals mean?, Eds. Harvey, M. and L. Santelmann Dept. of Modern Languages and Linguistics, Cornell University, Ithaca, New York, pp. 61-79.

Dalrymple, M. et al., 1998, Reciprocal expressions and the concept of reciprocity, Linguistics and Philosophy, 21, 159-210.

Damadian, R., 1971, Tumor detection by nuclear magnetic resonance, Science, 19:171, 1151-113 .

Damasio, A.R., 1992, Aphasia, The New England Journal of Medicine, 326.8, 531-539.

Dan, Y., J.J. Atick and R.C. Reid, 1996, Efficient coding of natural scenes in the lateral geniculate nucleus: experimental test of a computational theory, Journal of Neuroscience, 16, 3351-3362.

Dapretto, M. and S.Y. Bookheimer, 1999, Form and content: dissociating syntax and semantics in sentence comprehension, Neuron, 24, 427-432.

Davidson, D., 1967, The Logic of Decision and Action: The Logical Form of Action Sentences, Ed. Rescher, N. University of Pittsburgh Press, Pittsburgh, pp. 81-95.

Dayan, P. and L.F. Abbott, 2001, Theoretical Neuroscience. Computational and Mathematical Modeling of Neural Systems. MIT Press,

Dayan, P. et al., 1995, The Helmholtz machine, Neural Computation, 7, 889-904.

De Yoe, F.A. and D.C. Van Essen, 1988, Concurrent processing streams in monkey visual cortex, Trends in Neurosciences, 11, 219-226.

Deacon, T.W., 1997, The Symbolic Species: The Co-Evolution of Language and the Brain. W. W. Norton & Co, New York.

Deacon, T.W., 2000, Evolutionary perspectives on language and brain plasticity, Journal of Communication Disorders, 33, 273-290.

Deco, G. and D. Obradovic, 1996, An Information Theoretic Approach to Neural Computing. Springer-Verlag, New York.

Delcomyn, F., 1998, Foundations of Neurobiology. W. H. Freeman and Co., New York, NY.

Dennett, D., 1987, The Intentional Stance. MIT Press, Cambridge, MA.

Desimone, R. and L.G. Ungerleider, 1989, Handbook of Neurophysiology, vol. 2: Neural mechanisms of visual processing monkeys, Eds. Boller, F. and J. Grafman Elsevier, Amsterdam, pp. 267-299.

Desimone, R. and S.J. Schein, 1987, Visual properties of neurons in area V4 of the macaque: sensitivity to stimulus form, Journal of Neurophysiology, 57, 835-868.

Desimone, R., 1992, Neural Networks for Vision and Image Processing: Neural substrates for visual attention in the primate brain, MIT Press, Cambridge, MA, USA, pp. 343-364.

Destexhe, A. et al., 1996, In vivo, in vitro and computational analysis of dendritic calcium currents in thalamic reticular neurons, Journal of Neuroscience, 16, 169-185.

Didday, R.L., 1976, A model of visuomotor mechanisms in the frog optic tectum, Mathematical Bioscience, 30, 159-180.

Dougherty, R.C., 1970, A grammar of coördinate conjoined structures: I, Language, 46, 850-898.

Douglas, R. and K. Martin, 1998, The Synaptic Organization of the Brain: Neocortex, Ed. Shepherd, G.M. Oxford University Press, New York, pp. 459-510.

Douglas, R.J. et al., 1995, Recurrent excitation in neocortical circuits, Science, 269, 981-985.

Dowling, J.E. and B.B. Boycott, 1966, Organization of the primate retina: Electron microscopy, Proceedings of the Royal Society of London B, 166, 80-111.

Doya, K., 1999, What are the computations of the cerebellum, the basal ganglia and the cerebral cortex?, Neural Networks, 12, Issue 7-8, 961-974.

Dronkers, N.F., S. Pinker and A. Damasio, 2000, Principles of Neural Science: Language and the aphasias, Eds. Kandel, E.R., J.H. Schwartz and T.M. Jessell Elsevier, New York, pp. 1169-1187.

Duda, R.O., P.E. Hart and D.G. Stork, 2001, Pattern Classification. Wiley, New York.

Dummett, M., 1975, Mind and Language: What is a theory of meaning?, Ed. Guttenplan, S. Oxford University Press, Dummett, M., 1991, The Logical Basis of Metaphysics, Harvard University Press, Dunaevsky, A. et al., 1999, Developmental regulation of spine motility in mammalian CNS, Proceedings of the National Academy of Sciences of the United States of America, 96, 13438-13443.

Earman, J., 1992, Bayes or Bust? A critical examination of Bayesian confirmation theory,

Eccles, J.C., B. Libet and R.R. Young, 1958, The behaviour of chromatolysed motoneurons studied by intracellular recording, Journal of Physiology, 143, 11-40.

Edelman, S. and M.H. Christiansen, 2003, How seriously should we take Minimalist syntax?, Trends in Cognitive Sciences, 7:2, 60-61.

Eijck, J.v., 1984, Varieties of Formal Semantics. Proceedings of the 4th Amsterdam Colloquium, Sept 82.: Discourse Representation, Anaphora and Scopes, Eds. Landman, F. and F. Veltman Foris Publications, Dordrecht, Holland, pp. 103-122.

Eliasmith, C. and C.H. Anderson, 2002, Neural Engineering. Computation, Representation, and Dynamics in Neurobiological Systems. MIT Press, Cambridge, Mass. & London, England.

Eliasmith, C., 1997, Structure without symbols: Providing a distributed account of low-level and high-level cognition,

Engert, F. and T. Bonhoeffer, 1999, Dendritic spine changes associated with hippocampal long-term synaptic plasticity, Nature, 399, 66-70.

Eschenbach, C., 1994, Topological Foundations of Cognitive Science: A mereotopological definition of 'point', Eds. Eschenbach, C., C. Habel and B. Smith Graduiertenkolleg Kognitionswissenschaft, Hamburg, pp. 3-22.

Estes, W.K., 1986, Array models for category learning, Cognitive Psychology, 18, 500-549.

Evans, J.S., J. Clibbens and B. Rood, 1995, Bias in conditional inference: implications for mental models and mental logic, Quarterly Journal of Experimental Psychology A, 48, 644-670.

Evans, J.S., S.E. Newstead and R.M.J. Byrne, 1993, Human reasoning: The psychology of deduction. Erlbaum, Hove, U.K..

Fain, G.L., 1999, Molecular and cellular physiology of neurons. Harvard University Press, Cambridge, Mass.

Farah, M.J., G.W. Humphreys and H.R. Rodman, 1999, Fundamental Neuroscience: Object and face recognition, Eds. Zigmond, M.J. F.E. Bloom, S.C. Landis, J.L. Roberts, L.R. Squire Academic Press, Inc., San Diego, pp. 1339-1361.

Feldman, J. and D.H. Ballard, 1982, Connectionist models and their properties, Cognitive Science, 6, 205-264.

Feldman, J., 1984, Proceedings of the Sixth Annual Conference of the Cognitive Science Society: Computational constraints from biology, Cognitive Science Society, pp. 101.

Felleman, D. and D.C. Van Essen, 1991, Distributed hierarchical processing in primate visual cortex, Cerebral Cortex, 1, 1-47.

Fenstad, J.E., 1998, Discourse, Interaction and Communication. Proceedings of the Fourth International Colloquium on Cognitive Science: Formal Semantics, Geometry, and Mind, Eds. Arrazola, X., K. Korta and F.J. Pelletier Kluwer, Dordrecht, Ferster, D. and K.D. Miller, 2000, Neural mechanisms of orientation selectivity in the visual cortex, Annual Review of Neuroscience, 23, 441-471.

Fiala, J.C. and K.M. Harris, 1999, Dendrites: Dendrite structure, Eds. Stuart, G., N. Spruston and M. Häusser Oxford University Press, Oxford, UK, pp. 1-34.

Field, D.J., 1994, What is the goal of sensory coding?, Neural Computation, 6, 559-601.

Fiengo, R. and H. Lasnik, 1973, The logical structure of reciprocal sentences in English, Foundations of Language, 9, 447-468.

Fillmore, C., 1968, Universals in Linguistic Theory: The case for Case, Eds. Bach, E. and R.T. Harms Holt, Rinehart & Winston, New York, pp. 1-88.

Fillmore, C., 1976, Syntax and Semantics 8: Grammatical Relations: The Case for Case Reopened, Eds. Cole, P. and J. Sadock Academic Press, New York, Fischer, M. et al., 1998, Rapid actin-based plasticity in the dendritic spine, Neuron, 20, 847-854.

FitzHugh, R., 1960, Thresholds and plateaus in the Hodgkin-Huxley nerve equations, Journal of General Physiology, 43, 867-896.

FitzHugh, R., 1961, Impulses and physiological states in theoretical models of nerve membranes, Biophysical Journal, 1, 445-466.

FitzHugh, R., 1969, Biological Engineering: Mathematical models of excitation and propagation in nerve, Ed. Schwan, H.P. McGraw-Hill, New York, pp. 1-86.

Floridi, L., forthcoming, Outline of a Theory of Strongly Semantic Information, Minds and Machines,

Fodor, J., 1983, Modularity of Mind. The MIT Press, Cambridge, Mass..

Fodor, J., 1985, Précis of The modularity of mind, Behavioral and Brain Sciences, 8, 1-6.

Fodor, J., 2000, The mind doesn't work that way : the scope and limits of computational psychology. MIT Press, Cambridge, Mass.

Franks, B., 1995, On Explanation in the Cognitive Sciences: Competence, Idealization and the Failure of the Classical Cascade, The British Journal for the Philosophy of Science, 46, 475-502.

Fried, L.S. and K.J. Holyoak, 1984, Induction of category distributions: A framework for classification learning, Journal of Experimental Psychology: Learning, Memory and Cognition, 10, 234-257.

Friston, K.J., 2002, Beyond Phrenology: What Can Neuroimaging Tell Us About Distributed Circuitry?, Annual Review of Neuroscience, 25; 1, 221-250.

Gallese, V. and A. Goldman, 1998, Mirror neurons and the simulation theory of mind-reading, Trends in Cognitive Sciences, 2, 493-501.

Gallese, V. et al., 1996, Action recognition in the premotor cortex, Brain, 119, 593-609.

Gärdenfors, P., 1990, Induction, conceptual spaces and AI, Philosophy of Science, 57, 78-95.

Gärdenfors, P., 1996, Mental representation, conceptual spaces and metaphors, Synthese, 106, 21-47.

Gärdenfors, P., 2000, Conceptual Spaces: the geometry of thought. MIT Press, Cambridge, Mass. ; London.

Gazdar, G. and G. Pullum, 1976, Papers from the 12th Regional Meeting of the Chicago Linguistic Society: Truth-Functional Connectives in Natural Language, Eds. Mufwene, S., C. Walker and S. Steever Chicago Linguistic Society, Chicago, Illinois, pp. 220-234.

Gazdar, G., 1979, Pragmatics: implicature, presupposition and logical form. Academic Press, London.

Gersho, A. and R.M. Gray, 1992, Vector Quantization and Signal Compression. Kluwer, Norwell, Massachusetts.

Geschwind, N., 1965, Disconnection syndromes in animals and man, Brain, 88, 237-294, 585-644.

Geschwind, N., 1970, The organization of language and the brain, Science, 170, 940-944.

Geschwind, N., 1972, Language and the brain, Scientific American, 226, 76-83.

Gibbons, J.D., 1993, Nonparametric Measures of Association. Sage Publications, Inc., Newbury Park, Calif..

Gibson, J.J., 1972, The Psychology of Knowing: A Theory of Direct Visual Perception, Eds. Royce, J.R. and W. Rozeboom Gordon and Breach, New York, pp. 215-227.

Gibson, J.J., 1977, Perceiving, Acting, and Knowing: The theory of affordances, Eds. Shaw, R.E. and J. Bransford Lawrence Erlbaum Associates, Hillsdale, NJ, Gilbert, C.D. and T.N. Wiesel, 1985, Intrinsic connectivity and receptive field properties, Vision Research, 25, 365-374.

Givón, T., 1970, Notes on the Semantic Structure of English Adjectives, Language, 46, 816-837.

Givón, T., 1978, Syntax and Semantics, vol. 9. Pragmatics: Negation in language: pragmatics, function, ontology, Ed. Cole, P. Academic Press, New York, pp. 69-112.

Gluck, M.A. and C.E. Myers, 2001, Gateway to Memory. An Introduction to Neural Network Modeling of the Hippocampus and Learning. MIT Press, Cambridge, Mass. & London, UK.

Goertzel, B., 1993, The Evolving Mind. Gordon and Breach, Langhorne, Pa..

Goertzel, B., 1993, The Structure of Intelligence. A new mathematical model of mind. Springer-Verlag, Langhorne, Pa..

Goertzel, B., 1994, Chaotic Logic. Language, thought, and reality from the perspective of complex systems science. Plenum Press, New York.

Goertzel, B., 1997, From complexity to creativity. Plenum Press, New York.

Golden, R., 1996, Mathematical Methods for Neural Network Analysis and Design. The MIT Press, Cambridge, USA & London, England.

Goodman, N., 1955, Fact, Fiction, and Forecast. Harvard University Press, Cambridge, Mass..

Gray, E.G., 1959, Electron microscopy of synaptic contacts on dendritic spines of the cerebral cortex, Nature, 183, 1592-1593.

Greenfield, P., 1991, Language, tool and brain: The ontogeny and phylogeny of hierarchically organized sequential behaviour, Behavioral and Brain Sciences, 14, 531-595.

Greenfield, P.M., 1991, Language, tools and brain: The ontogeny and phylogeny of hierarchically organized sequential behaviour, Behavioral and Brain Sciences, 14, 531-595.

Gregg, T.R., 2002, Use of Functional Magnetic Resonance Imaging to Investigate Brain Function,

Grice, H.P., 1975, Syntax and Semantics, vol. 3. Speech Acts: Logic and Conversation, Eds. Cole, P. and J. Morgan Academic Press, New York, NY, pp. 41-58.

Grossberg, S., 1976, Adaptive pattern classification and universal recoding, I: Parallel development and coding of neural feature detectors, Biological Cybernetics, 23, 121-134.

Grossberg, S., 1982, Studies of mind and brain: Neural principles of learning, perception, development, cognition, and motor control. D. Reidel Pub. Co, Dordrecht, Holland & Boston.

Grossberg, S., 1987, Competitive learning: from interactive activation to adaptive resonance, Cognitive Science, 11, 23-63.

Grossman, S., 1989, Algebra and Trigonometry. Saunders College Publishing, Philadelphia.

Grosu, A., 1985, Subcategorization and parallelism, Theoretical Linguistics, 12, 231-240.

Grosu, A., 1987, On an asymmetry in the distribution of island constraints, Lingua, 74, 167-.

Guenther, F.H., 1995, Speech sound acquisition, coarticulation, and rate effects in a neural network model of speech production, Psychological Review, 102, 594-621.

Guenther, F.H., M. Hampson and D. Johnson, 1998, A Theoretical Investigation of Reference Frames for the Planning of Speech Movements, Psychological Review, 105, 611-633.

Gutnick, M.J. and W.E. Crill, 1995, The Cortical Neuron: The cortical neuron as an electrophysiological unit, Eds. Gutnick, M.J. and I. Mody Oxford University Press, New York, USA, pp. 33-51.

Halbasch, K., 1975, An observation on English truth-functions, Analysis, 35, 109-110.

Halmos, P.R., 1950, Measure Theory. Van Nostrand Reinhold Co., New York, USA.

Hartenstein, V. and G. Innocenti, 1981, The arborization of single callosal axons in the mouse cerebral cortex, Neuroscience Letters, 23, 19-24.

Hausman, D.M., 1998, Causal asymmetries. Cambridge University Press, Cambridge, U.K. ; New York.

Haykin, S., 1994, Neural Networks. A Comprehensive Foundation. 1e. Maxwell Macmillan International, New York.

Hebb, D.O., 1949, The Organization of Behavior. John Wiley & Sons, New York.

Helmchen, F., 1999, Dendrites: Dendrites as biochemical compartments, Eds. Stuart, G., N. Spruston and M. Häusser Oxford University Press, Oxford, UK & New York, USA, pp. 161-192.

Helmholtz, H.v., 1925, Treatise on Physiological Optics. Dover, New York.

Hempel, C.G., 1965, Aspects of Scientific Explanation, and Other Essays in the Philosophy of Science. Free Press, New York, N. Y..

Hendry, S.H.C., S.S. Hsiao and M.C. Brown, 1999, Fundamental Neuroscience: Fundamentals of sensory systems, Eds. Zigmond, M.J. F.E. Bloom, S.C. Landis, J.L. Roberts, L.R. Squire Academic Press, Inc., San Diego, pp. 657-670.

Henkel, C.K., 1997, The auditory system. Churchill Livingstone, New York.

Hickok, G. and D. Poeppel, 2000, Towards a functional neuroanatomy of speech perception, Trends in Cognitive Sciences, 4, 131-138.

Hickok, G., 2000, Speech perception, conduction aphasia, and the functional neuroanatomy of

Hickok, G., 2001, Functional Anatomy of Speech Perception and Speech Production: Psycholinguistic Implications, Journal of Psycholinguistic Research, 30, No. 3,

Hille, B., 1992, Ionic channels of excitable membrane. 2e. Sinauer Associates, Sunderland, Massachusetts.

Hinton, G.E. and R.S. Zemel, 1994, Advances in Neural Information Processing Systems 6: Autoencoders, minimum description length and Helmholtz free energy, Eds. Cowan, J.D., G. Tesauro and J. Alspector Morgan Kaufmann Publishers, San Francisco, pp. 3-10.

Hinton, G.E. et al., 1995, The "Wake-Sleep" algorithm for unsupervised neural networks, Science, 268, 1158-1161.

Hinton, G.E., 1989, Connectionist learning procedures, Artificial Intelligence, 40, 185-234.

Hinton, G.E., J.L. McClelland and D.E. Rumelhart, 1986, Parallel distributed processing: Explorations in the microstructure of cognition. Vol. 1:

Foundations: Distributed representations, Eds. Rumelhart, D., J. McClelland and the PDP Research Group. Group MIT Press, Cambridge, Mass., pp. 45-76.

Hintzman, D.L., 1986, 'Schema abstraction' in a multiple-trace memory model, Psychological Review, 93, 411-428.

Hobbs, J., 1979, Coherence and coreference, Cognitive Science, 3, 67-90.

Hobbs, J., 1982, Strategies of Natural Language Processing: Towards an understanding of coherence in discourse, Eds. Lehnert, W. and M.H. Ringle Lawrence Erlbaum, Hillsdale, N.J., pp. 223-244.

Hobbs, J., 1990, Literature and Cognition. Center for the Study of Language and Information, Stanford, Ca..

Hochberg, J., 1998, Handbook of Perception and Cognition. Perception and cognition at century's end: History, philosophy, theory: Gestalt theory and its legacy, Eds. Hochberg, J. and J.E. Cuttings Academic Press, San Diego, pp. 253-306.

Hodgkin, A.L. and A.F. Huxley, 1952, A quantitative description of membrane current and its application to conduction and excitation in nerve, Journal of Physiology, 117, 500-544.

Holmes, W., R. and W. Rall, 1995, The Handbook of Brain Theory and Neural Networks: Dendritic spines, Ed. Arbib, M. The MIT Press, Cambridge, USA & London, England, pp. 289-292.

Hopper, C., 1998, Practicing College Study Skills: Strategies for Success. Houghton Mifflin Company,

Horn, L., 1976, On the Semantic Properties of Logical Operators in English. Indiana University Linguistics Club, Bloomington, Indiana.

Horn, L., 1989, A Natural History of Negation. The University of Chicago Press, Chicago, Illinois.

Horowitz, A.L., 1995, MRI Physics for Radiologists: A Visual Approach. 3e,

Howard, H., 2001, Language and Ideology. Volume 1: Theoretical cognitive approaches: Age/gender morphemes inherit the biases of their underlying dimensions, Eds. Dirven, R., B. Hawkins and E. Sandikcioglu John Benjamins, Amsterdam & Philadelphia, pp. 165-195.

Hubel, D.H. and T.N. Wiesel, 1962, Receptive fields, binocular interaction and functional architecture in the cat's visual cortex, Journal of Physiology (London), 160, 106-154.

Hubel, D.H. and T.N. Wiesel, 1968, Receptive fields and functional architecture of monkey striate cortex., Journal of Physiology, 195, 215-243.

Hume, D., 1955, An Inquiry Concerning Human Understanding. The Liberal Arts Press, New York, NY.

Innocenti, G., 1986, Cerebral Cortex, Vol. 5: General organization of callosal connections in the cerebral cortex, Eds. Jones, E. and A. Peters Plenum, New York, pp. 291-353.

Intrator, N. and L.N. Cooper, 1995, The Handbook of Brain Theory and Neural Networks: Information theory and visual plasticity, Ed. Arbib, M. The MIT Press, Cambridge, USA & London, England, pp. 484-487.

Intrator, N., 1995, The Handbook of Brain Theory and Neural Networks: Competitive learning, Ed. Arbib, M. The MIT Press, Cambridge, USA & London, England, pp. 220-223.

Ivry, R.B. and L.C. Robertson, 1998, The Two Sides of Perception. MIT Press, Cambridge, Mass.

Jackendoff, R., 1972, Semantic Interpretation in Generative Grammar. The MIT Press, Cambridge, Massachusetts.

Jackendoff, R., 1977, X' Syntax: A Study of Phrase Structure. The MIT Press, Cambridge, Massachusetts.

Jackendoff, R., 2002, Foundations of Language: Brain, Meaning, Grammar, Evolution. Oxford University Press, Oxford & New York.

Jacobs, R.A. and S.M. Kosslyn, 1994, Encoding shape and spatial relations: the role of receptive field size in coordinating complementary representations, Cognitive Science, 18, 361-386.

Jakobson, R., 1956, Fundamentals of Language: Two aspects of language and two types of aphasic disturbances, Eds. Jakobson, R. and M. Halle Mouton, The Hague, pp. 55-82.

Jesperson, O., 1917, Negation in English and Other Languages. A. F. Hørst, Copenhagen.

Jesperson, O., 1924, The Philosophy of Grammar. Allen & Unwin, London.

Jiang, Y. et al., 2002, The open pore conformation of potassium channels, Nature, 417, 523-526.

Johnson, M., 1987, The Body in the Mind: The Bodily Basis of Reason and Imagination. University of Chicago Press, Chicago, Illinois.

Johnson-Laird, P.N., R.M.J. Byrne and W. Schaeken, 1992, Propositional reasoning by model, Psychological Review, 99, 418-439.

Jong, F.d., 1984, Linguistics in the Netherlands 1984: Numerals as determiners, Eds. Bennis, H. and W.U.S.v.L. Kloeke Foris, Dordrecht, Holland, pp. 105-114.

Kandel, E.R. and C. Mason, 1995, Essentials of Neural Science and Behavior: Perception of form and motion, Eds. Kandel, E.R., J.H. Schwartz and T.M. Jessell Elsevier, New York, pp. 425-451.

Kandel, E.R. and S. Siegelbaum, 1995, Essentials of Neural Science and Behavior: An introduction to synaptic transmission, Eds. Kandel, E.R., J.H. Schwartz and T.M. Jessell Elsevier, New York, pp. 183-196.

Kanski, Z., 1987, Proceedings of the 1987 Debrecen Symposium on Language and Logic: Logical symmetry and natural language reciprocals, Akadémiai Kiadó, Budapest, pp. 49-68.

Kaplan, D. and L. Glass, 1995, Understanding Nonlinear Dynamics. Springer-Verlag, New York.

Kartalopoulos, S., 1996, Understanding Neural Networks and Fuzzy Logic. Basic concepts and applications. IEEE Press, Piscataway, NJ.

Kastner, S. and L.G. Ungerleider, 2000, Mechanisms of Visual Attention in the Human Cortex, Annual Review of Neuroscience, 23; 1, 315-341.

Katz, B., 1966, Nerve, muscle, and synapse. McGraw-Hill, New York.

Keener, J. and J. Sneyd, 1998, Mathematical Physiology. Springer, New York.

Kehler, A., 2002, Coherence, reference, and the theory of grammar. CSLI Publications, Stanford, Calif.

Keyes, R.W., 1985, What makes a good computational device?, Science, 230, 138-144.

Kiefer, M. et al., 1998, Right Hemisphere Activation during Indirect Semantic Priming: Evidence from Event-Related Potentials, Brain and Language, 64, 377-408.

Kimura, D., 1961, Cerebral dominance and the perception of verbal stimuli, Canadian Journal of Psychology, 15, 166-171.

Kimura, D., 1961, Some effects of temporal lobe damage on auditory perception, Canadian Journal of Psychology, 15, 156-165.

Klauer, K.C. and K. Oberauer, 1995, Testing the mental model theory of propositional reasoning, Quarterly Journal of Experimental PsychologyA, 48A, 671-687.

Klein, R., 1989, Concrete and Abstract Voronoi Diagrams. Springer-Verlag, Berlin, Heidelberg & New York.

Knight, B.W., 1972, Dynamics of encoding a population of neurons, Journal of General Physiology, 59, 734-766.

Knill, D.C., D. Kersten and A. Yuille, 1996, Perception as Bayesian inference: Introduction: A Bayesian formulation of visual perception, Eds. Knill, D.C. and W. Richards Cambridge University Press, New York, pp. 1-22.

Knjazev, J.P., 1998, Typology of Verbal Categories: Papers Presented to Vladimir Nedjalkov on the Occasion of His 70th Birthday: Towards a Typology of Grammatical Polysemy: Reflexive Markers as Markers of Reciprocity, Eds. Kulikov, L. and H. Vater Niemeyer, Tubingen, Germany, pp. 185-193.

Kobatake, E. and K. Tanaka, 1994, Neuronal selectivities to complex object features in the ventral visual pathway of the macaque cerebral cortex, Journal of Neurophysiology, 71, 856-867.

Koch, C., 1999, Biophysics of Computation. Information Processing in Single Neurons. Oxford University Press, New York, USA & Oxford, UK.

Koch, C., T. Poggio and V. Torre, 1983, Nonlinear interaction in a dendritic tree: localization, timing and role of information processing, Proceedings of the National Academy of Sciences of the United States of America, 80, 2799-2802.

Koch, P., 1999, Historical Semantics and Cognition: Cognitive Aspects of Semantic Change and Polysemy: The Semantic Space Have/Be, Eds. Blank, A. and P. Koch Mouton de Gruyter, Berlin, Germany, pp. 279-305.

Koester, J., 1995, Essentials of Neural Science and Behavior: Propagated signaling: The action potential, Eds. Kandel, E.R., J.H. Schwartz and T.M. Jessell Elsevier, New York, pp. 161-178.

Kohonen, T., 1982, Self-organized formation of topologically correct feature maps, Biological Cybernetics, 43, 59-69.

Kohonen, T., 1986, Learning vector quantization for pattern recognition. Technical Report TKK-F-A601. Helsinki University of Technology, Helsinki, Finland.

Kohonen, T., 1989, Self-Organization and Associative Memory. 3rd edition. Springer-Verlag, Berlin, Germany.

Kohonen, T., 1997, Self-Organizing Maps. 2e. Springer-Verlag, Berlin.

Kong, S.-G. and B. Kosko, 1991, Differential competitive learning for phoneme recognition, IEEE Transactions on Neural Networks, 2, 118-124.

Kosko, B., 1991, Stochastic competitive learning, IEEE Transactions on Neural Networks, 2, 522-529.

Kosslyn, S. et al., 1989, Evidence for two types of spatial representations: Hemispheric specialization for categorical and coordinate relations, Journal of Experimental Psychology, 15, 723-735.

Kosslyn, S., 1994, Image and Brain. The Resolution of the Imagery Debate. The MIT Press, Cambridge, Massachusetts, USA & London, England.

Kreiman, J. and D.R. Van Lancker, 1988, Hemispheric specialization for voice recognition: evidence from dichotic listening, Brain and Language, 34, 246-252.

Krifka, M., 1990, Four thousand ships passed through the lock: Object-induced measure functions on events, Linguistics and Philosophy, 13, 487-520.

Kuffler, S.W., 1953, Discharge patterns and functional organization of the mammalian retina, Journal of Neurophysiology, 16, 37-68.

Kuffler, S.W., J.G. Nichols and A.R. Martin, 1984, From Neuron to Brain: A cellular approach to the function of the nervous system, 2e. Sinaur Associates, Sunderland, MA.

Kuruvilla, F.G., P.J. Park and S.L. Schreiber, 2002, Vector algebra in the analysis of genome-wide expression data, Genome Biology, 3, research0011.1-research0011.11.

Labov, W., 1984, Georgetown University Round Table on Language and Linguistics 1984: Intensity, Ed. Schiffrin, D. Georgetown University Press, Washington, pp. 43-70.

Labov, W., 1985, Proceedings of the Eleventh Annual Meeting of the Berkeley Linguistics Society: The Several Logics of Quantification, Eds. Niepokuj, M. M. VanClay, V. Nikiforidou, D. Feder Berkeley Linguistics Society, Berkeley, California, pp. 175-195.

Lakoff, G. and S. Peters, 1969, Modern Studies in English: Readings in Transformational Grammar: Phrasal Conjunction and Symmetric Predicates, Eds. Reibel, D. and S. Schane Prentice-Hall, Englewood Cliffs, N. J., pp. 113-142.

Lakoff, G., 1987, Fire, Woman and Dangerous Things: What Categories Reveal About the Mind. University of Chicago Press, Chicago, Illinois.

Lakoff, G., 1988, Meaning and Mental Representations: Cognitive Semantics, Eds. Eco, U., M. Santambrogio and P. Violi Indiana University Press, Bloomington, pp. 119-154.

Lakoff, R., 1971, Studies in Linguistic Semantics: If's, and's and but's about conjunction, Eds. Fillmore, C.J. and D.T. Langendoen Holt, Rhinehart & Winston, Inc., New York, New York, pp. 114-149.

LaMantia, A.S. and P. Rakic, 1990, Axon overproduction and elimination in the corpus callosum of the developing rhesus monkey, Journal of Neuroscience, 10, 2156-2175.

Langacker, R., 1987, Foundations of Cognitive Grammar, vol. 1: Theoretical prerequisites. Stanford University Press, Stanford, Ca..

Langacker, R., 1991, Foundations of Cognitive Grammar: Volume II, Descriptive Applications. Stanford University Press, Stanford, Ca..

Langendoen, D.T., 1978, The Logic of Reciprocity, Linguistic Inquiry, 9, 177-197.

language, Language and the Brain: Representation and processing, 87-104.

Lapicque, L., 1907, Recherhes quantitatives sur l'excitation électrique des nerfs traité comme une polarisation, J. de Physiol. et de Pathol. Gen., 9, 620-635.

Lasersohn, P., 1995, Plurality, Conjunction and Events. Kluwer Academic, Dordrecht.

Laughlin, S.B. et al., 2000, Information Theory and the Brain: Coding efficiency and the metabolic cost of sensory and neural information, Eds. Baddeley, R., P. Hancock and P. Földiák Cambridge University Press, Cambridge, U.K. & New York, pp. 41-61.

Lea, R.B. et al., 1990, Predicting propositional logic inferences in text comprehension, Journal of Memory & Language, 29, 361-387.

Lea, R.B., 1995, On-line evidence for elaborative logical inferences in text, Journal of Experimental Psychology: Learning, Memory, & Cognition, 21, 1469-1482.

Leech, G., 1970, Towards a Semantic Description of English. Longman, London.

Leech, G., 1981, Possibilities and Limitations of Pragmatics: Pragmatics and Conversational Rhetoric, Eds. Parret, H., H. Sbisa and J. Verschueren Benjamins, Amsterdam, pp. 413-441.

Levine, D.S., 2000, Introduction to Neural and Cognitive Modeling 2e. Lawrence Erlbaum Associates, Publishers, Hillsdale, N.J..

Levinson, S.C., 2000, Presumptive Meanings. The theory of generalized conversational implicature. MIT Press, Cambridge, Mass.

Lewin, K., 1936, Principles of Topological Psychology. McGraw-Hill, New York & London.

Lewis, D., 1973, Counterfactuals. Blackwell, Oxford.

Liberman, A.M. and I.G. Mattingly, 1985, The motor theory of speech perception revised, Cognition, 21, 1-36.

Liberman, A.M. and I.G. Mattingly, 1989, A specialization for speech perception, Science, 243, 489-494.

Liberman, A.M. et al., 1967, Perception of the speech code, Psychological Review, 74, 431-461.

Linebarger, M.C., 1998, Linguistic Levels in Aphasiology: Algorithmic and heuristic processes in agrammatic language comprehension, Eds. Visch-Brink, E. and R. Bastiaanse Singular Publishing Group, San Diego & London, pp. 153-174.

Link, G., 1983, Meaning, Use and Interpretation of Language: The logical analysis of plurals and mass terms: A lattice-theoretical approach, Eds. Bäuerle, R., C. Schwarze and A.v. Stechow Walter de Gruyter, Berlin, pp. 302-323.

Linsker, R., 1988, Self-Organisation in a Perceptual Network, Computer, March,

Löbner, S., 1986, Studies in Discourse Representation Theory and the Theory of Generalized Quantifiers: Quantification as a Major Module in Natural Language Semantics, Eds. Groenendijk, J.A.G., D.d. Jongh and M.B.J. Stokhof Foris Publications, Dordrecht, Holland, pp. 53-86.

Löbner, S., 1987, Generalized Quantifiers: Studies in Linguistics and Philosophy: Natural Language and Generalized Quantifier Theory, Ed. Gärdenfors, P. D. Reidel Publishing Co., Dordrecht, The Netherlands, Maess, B. et al., 2001, Musical Syntax Is Processed in Broca's Area: an MEG Study, Nature Neuroscience, 540-545.

Logothetis, N.K. et al., 2001, Neurophysiological investigation of the basis of the fMRI signal, Nature, 412, 150-157.

Lowe, D., 1995, The Handbook of Brain Theory and Neural Networks: Radial basis function networks, Ed. Arbib, M. The MIT Press, Cambridge, USA & London, England, pp. 779-782.

Lowe, I., 1984, Two Pragmatic Parameters for Logical Connectives, Forum Linguisticum, 8, 140-156.

Lubbe, J.C.A.v.d., 1997, Information Theory. Cambridge University Press, Cambridge, UK.

Luck, S.J. et al., 1997, Neural Mechanisms of Spatial Selective Attention in Areas V1, V2, and V4 of Macaque Visual Cortex, Journal of Neurophysiology, 77 No. 1, 24-42.

Ludlow, P., 1999, Semantics, Tense, and Time. Harvard University Press,

Magee, J.C., 1999, Dendrites: Voltage-gated ion channels in dendrites, Eds. Stuart, G., N. Spruston and M. Häusser Oxford University Press, Oxford, UK, pp. 85-113.

Mainen, Z.F. et al., 1995, A model of spike initiation in neocortical pyramidal neurons, Neuron, 15, 1427-1439.

Maletic-Savatic, M., R. Malinow and K. Svoboda, 1999, Rapid dendritic morphogenesis in CA1 hippocampal dendrites induced by synaptic activity, Science, 283, 1923-1927.

Malt, B.C. and E.E. Smith, 1984, Correlated properties in natural categories, Journal of Verbal Learning and Verbal Behavior, 23, 250-269.

Marin-Padilla, M. et al., 1969, Spine distribution of the layer V pyramidal cell in man: a cortical model, Brain Research, 12, 493-496.

Marr, D., 1971, Simple memory: A theory of archicortex, Proceedings of the Royal Society of London, B176, 841, 23-81.

Marr, D., 1977, Artificial intelligence: a personal view, Artificial Intelligence, 9, 37-48.

Marr, D., 1981, Mind Design. Philosophy, psychology, artificial intelligence. 1e: Artificial intelligence: a personal view, Ed. Haugeland, J. MIT Press, Cambridge, MA, Marr, D., 1982, Vision: A computational investigation into the human representation and processing of visual information. Freeman, New York.

Martin, J.N., 1987, Elements of Formal Semantics. An introduction to logic for students of language. Academic Press, San Diego, California, USA.

Masland, R.H. and E. Raviola, 2000, Confronting complexity: strategies for understanding the microcircuitry of the retina, Annual Review of Neuroscience, 23, 249-284.

Massaro, D., 1988, Some criticisms of connectionist models of human performance, Journal of Memory and Language, 27, 213-234.

May, R., 1985, Logical Form. The MIT Press, Cambridge, MA.

McCawley, J., 1971, Studies in Linguistic Semantics: Tense and Reference in English, Eds. Fillmore, C.J. and D.T. Langendoen Holt, Rhinehart & Winston, Inc., New York, New York, pp. 97-114.

McCawley, J., 1972, Semantics of Natural Language: A program for logic, Eds. Davidson, D. and G. Harman D. Reidel Publishing Company, Dordrecht, Holland, McCawley, J., 1981, Everything that Linguists have Always Wanted to Know about Logic. The University of Chicago Press, Chicago, Illinois.

McClelland, J.L. and D.E. Rumelhart, 1981, An interactive activation model of context effects in letter perception: Part 1. An account of basic findings, Psychological Review, 88, 375-407.

McClelland, J.L. and D.E. Rumelhart, 1989, Explorations in Parallel Distributed Processing: A Handbook of Models, Programs and Exercises. MIT Press, Cambridge, Mass..

McClelland, J.L., 1981, Proceedings of the Third Annual Conference of the Cognitive Science Society: Retrieving general and specific knowledge from stored knowledge of specifics, pp. 170-172.

McClelland, J.L., 1991, Stochastic interactive processes and the effect of context on perception, Cognitive Psychology, 23, 1-44.

McClelland, J.L., B.L. McNaughton and R.C. O'Reilly, 1995, Why there are complementary learning systems in the hippocampus and neocortex: Insights from the successes and failures of connectionist models of learning and memory, Psychological Review, 102, 419-457.

McCormick, D.A., 1998, The Synaptic Organization of the Brain: Membrane properties and neurotransmitter actions, Ed. Shepherd, G.M. Oxford University Press, New York, pp. 37-75.

McCormick, D.A., 1999, Fundamental Neuroscience: Membrane potential and action potential, Eds. Zigmond, M.J. F.E. Bloom, S.C. Landis, J.L. Roberts, L.R. Squire Academic Press, Inc., San Diego, pp. 129-154.

McCulloch, W. and W. Pitts, 1943, A Logical Calculus of the Ideas Immanent in Nervous Activity, Bulletin of Mathematical Biophysics, 5, 115-133.

Medin, D.L. and B.H. Ross, 1989, Advances in the Psychology of Human Intelligence 5: The specific character of abstract thought: categorization, problem-solving, and induction, Ed. Sternberg, R.J. Erlbaum, Hillsdale, NJ, pp. 189-223.

Medin, D.L. and M.M. Schaffer, 1978, Context theory of classification learning, Psychological Review, 85, 207-238.

Medin, D.L. et al., 1982, Correlated symptoms and simulated medical classification, Journal of Experimental Psychology: Human Learning and Memory, 8, 37-50.

Mel, B.W., 1992, Advances in neural information processing systems, vol. 4: The clusteron: toward a simple abstraction for a complex neuron, Eds. Moody, J., S. Hanson and R. Lippmann Morgan Kaufmann, San Mateo, CA, pp. 35-42.

Mel, B.W., 1994, Information Processing in Dendritic Trees, Neural Computation, 6, 1031-1085.

Mel, B.W., 1999, Dendrites: Why have dendrites? A computational perspective, Eds. Stuart, G., N. Spruston and M. Häusser Oxford University Press, Oxford, UK, pp. 271-289.

Mel, B.W., 1999, Think positive to find parts, Nature, 401, 759-760.

Menon, R.S., 2001, Imaging function in the working brain with fMRI, Current Opinion in Neurobiology, 11, 630-636.

Menzies, P., 2001, Counterfactual Theories of Causation, Stanford Encyclopedia of Philosophy,

Mergian, W.H. and J.H.R. Maunsell, 1993, How parallel are the primate visual pathways, Annual Review of Neuroscience, 16, 369-402.

Mervis, C. and E. Rosch, 1981, Categorization of natural objects, Annual Review of Psychology, 32, 89-115.

Miller, G. and P.N. Johnson-Laird, 1976, Language and Perception. Harvard University Press, Cambridge, Mass..

Miller, G., 1956, The magical number 7 ± 2: Some limits on our capacity to store information, Psychological Bulletin, 63, 81-97.

Miller, K.D. and D.J.C. MacKay, 1994, The Role of Constraints in Hebbian Learning, Neural Computation, 6, 100-126.

Milner, B., L. Taylor and R.W. Sperry, 1968, Lateralized suppression of dichotically presented digits after commissural section in man, Science, 161, 184-186.

Moran, J. and R. Desimone, 1985, Selective attention gates visual processing in the extrastriate cortex, Science, 229, 782-784.

Mormann, T., 1993, Natural predicates and topological structures of conceptual spaces, Synthese, 95, 219-240.

Mostowski, A., 1957, On a generalization of quantifiers, Fundamenta Mathematica, 44, 12-36.

Mountcastle, V.B., 1997, The columnar organization of the neocortex, Brain, 120, 701-722.

Moxey, L.M. and A. Sanford, 1993, Prior expectation and the interpretation of natural language quantifiers, European Journal of Cognitive Psychology, 5, 73-91.

Moxey, L.M. and A. Sanford, 2000, Communicating quantities: A review of psycholinguistic evidence of how expressions determine perspectives, Applied Cognitive Psychology, 14, 237-255.

Moxey, L.M. and A.J. Sanford, 1993, Communicating Quantities. A Psychological Perspective. Lawrence Erlbaum Associates, Publishers, Hove, UK & Hillsdale, USA.

Munkres, J., 1975, Topology. A First Course. Prentice-Hall, Englewood Cliffs, NJ.

Murphy, G.L., 2002, The Big Book of Concepts. MIT Press,

Myers, P.S., 1999, Right Hemisphere Damage: Disorders of Communication and Cognition. Singular Publishing Group, San Diego.

Nadeau, S.E., 2000, Aphasia and Language. Theory to practice: Connectionist models and language, Eds. Nadeau, S.E., L.J. Gonzalez-Rothi and B. Crosson Guilford Publications, Inc., New York, NY & London, England, pp. 299-347.

Nagumo, J., S. Arimoto and S. Yoshizawa, 1962, An active pulse transmission line simulating nerve axons, Proceedings of the Institute of Radio Engineers, 50, 2061-2070.

Newell, A., 1990, Unified Theories of Cognition. Harvard University Press, Cambridge, Mass.

Nicholls, J.G. et al., 2001, From Neuron to Brain, 4e. Sinauer Associates, Sunderland, MA.

Nimchinsky, E.A., B.L. Sabatini and K. Svoboda, 2002, Structure and function of dendritic spines, Annual Review of Physiology, 64, 313-353.

Noordhof, P., 1999, Probabilistic Causation, Preemption and Counterfactuals, Mind, 108 , 95-125.

Nosofsky, R.M. and T.J. Palmeri, 1998, A rule-plus-exception model of classification learning, Psychonomic Bulletin and Review, 5, 345-369.

Nosofsky, R.M., 1986, Attention, similarity, and the identification-categorization relationship, Journal of Experimental Psychology: General, 115, 39-57.

Nosofsky, R.M., 1992, From Learning Theory to Connectionism: Essays in Honor of William K. Estes, vol. 1: Exemplars, prototypes, and similarity rules, Eds. Healy, A.F., S.M. Kosslyn and R.M. Schffrin Erlbaum, Hillsdale, NJ, pp. 149-167.

Nusser, Z., 1999, Dendrites: Subcellular distribution of neurotransmitter receptors and voltage-gated ion channels, Eds. Stuart, G., N. Spruston and M. Häusser Oxford University Press, Oxford, UK, pp. 85-113.

Nutt, R., 2002, The History of Positron Emission Tomography, Molecular Imaging and Biology, 4:1, 11-26.

Oaksford, M. and N. Chater, 1994, A rational analysis of the selection task as optimal data selection, Psychological Review, 101, 608-631.

Oaksford, M. and N. Chater, 1996, Rational explanation of the selection task, Psychological Review, 103, 381-391.

Oaksford, M., 2001, Morphological tensions, Trends in Cognitive Sciences, 5:4, 136.

Oaksford, M., L. Roberts and N. Chater, 2002, Relative informativeness of quantifiers used in syllogistic reasoning, Memory and Cognition, 30, 1, 138-149.

Ogawa, S. and T.M. Lee, 1990, Magnetic resonance imaging of blood vessels at high fields: in vivo and in vitro measurements and image simulation, Magnetic Resonance in Medicine, 16, 9-18.

Ogawa, S. et al., 1990, Brain magnetic resonance imaging with contrast dependent on blood oxygenation, Proceedings of the National Academy of Sciences of the United States of America, 87, 9868-9872.

Oirsouw, R.v., 1987, The Syntax of Coordination. Croom Helm, London.

Okabe, A., B. Boots and K. Sugihara, 1992, Spatial Tessellations: Concepts and applications of Voronoi diagrams. John Wiley & Sons, New York, NY.

O'Keefe, J. and L. Nadel, 1978, The Hippocampus as a Cognitive Map. Clarendon Press, Oxford.

Olshausen, B.A. and D.J. Field, 2000, Vision and the Coding of Natural Images, American Scientist, 88, 238-224.

Oram, M. and D. Perrett, 1994, Modeling visual recognition from neurobiological constraints, Neural Networks, 7, 945-972.

O'Reilly, R.C. and J.W. Rudy, in press, Conjunctive Representations, the Hippocampus and Contextual Fear Conditioning, Psychological Review,

O'Reilly, R.C. and Y. Munakata, 2000, Computational Explorations in Cognitive Neuroscience: Understanding the Mind by Simulating the Brain. MIT Press,

O'Reilly, R.C. et al., 1999, Proceedings of the Second International Conference on Cognitive Science: Discrete Representations in Working Memory: A Hypothesis and Computational Investigations, Japanese Cognitive Science Society, Tokyo, Japan, pp. 183-188.

O'Reilly, R.C., K.A. Norman and J.L. McClelland, 1998, Advances in neural information processing systems 10: A hippocampal model of recognition memory, Eds. Jordan, M.I., M.J. Kearns and S.A. Solla MIT Press, Cambridge, MA, pp. 73-79.

Palmer, S. and I. Rock, 1994, Rethinking perceptual organization: The role of uniform connectedness, Psychonomic Bulletin and Review, 1, 29-55.

Pandya, D.N. and B. Seltzer, 1986, Two hemispheres - one brain: The topography of comissural fibers, Eds. Lepore, F., M. Ptito and H. Jasper A. R. Liss, New York, pp. 47-73.

Parry, W. and E. Hacker, 1991, Aristotelian Logic. SUNY Press, Albany.

Parsons, T., 1990, Events in the Semantics of English: A Study in Subatomic Semantics. The MIT Press, Cambridge, Mass.

Parsons, T., 1999, The Traditional Square of Opposition, The Stanford Encyclopedia of Philosophy,

Partee, B.H., 1988, Proceedings of the Fifth Eastern States Conference on Linguistics: Many Quantifiers, Eds. Powers, J. and K.d. Jong The Ohio State University, Columbus, Ohio, pp. 383-402.

Partee, B.H., A.t. Meulen and R.E. Wall, 1990, Mathematical Methods in Linguistics. Kluwer Academic Publishers, Dordrecht, The Netherlands.

Patterson, S., 1998, Discussion. Competence and the classical cascade: a reply to Franks, The British Journal for the Philosophy of Science, 49, 4, 625-636.

Pelletier, F.J., 1977, Or, Theoretical Linguistics, 4, 61-74.

Perceptual Categories, Evolution of Communication, 4, 117-142.

Peters, S. (1966) Coordinate Conjunction in English. Ph.D. thesis, MIT.

Petitto, L.A. et al., 2000, Speech-like cerebral activity in profoundly deaf people while processing signed languages: Implications for the neural basis of human language, Proceedings of the National Academy of Sciences of the United States of America, 97, 13961-13966.

Pfeiffer, P.E., 1995, Basic Probability Topics Using MATLAB. PWS Publishing Company, Boston, USA.

Phillips, W.A. and W. Singer, 1997, In search of common foundations for cortical computation, Behavioral and Brain Sciences, 20, 657-722.

Pick, A., 1913, Clinical studies III: On reduplicative paramnesia, Brain, 26, 260-267.

Pinker, S., 2000, Words and Rules: the ingredients of language. Perennial, New York.

Pinsky, P.F. and J. Rinzel, 1994, Intrinsic and network rhythmogenesis in a reduced Traub model for CA3 neurons, Journal of Computational Neuroscience, 1, 39-60.

Poirazi, P. and B.W. Mel, 2000, The Memory Capacity of Subsampled Quadratic Classifiers: Why Active Dendrites May Remember More, Neural Computation, 12, 1189-205..

Poirazi, P. and B.W. Mel, 2001, Impact of active dendrites and structural plasticity on the memory capacity of neural tissue, Neuron, 29, 779-796.

Popper, K.R., 1959, The Logic of Scientific Discovery. Basic Books, New York.

Posner, M.I. and S.J. Boies, 1971, Components of attention, Psychological Review, 78, 391-408.

Posner, M.I. and S.W. Keele, 1968, On the genesis of abstract ideas, Journal of Experimental Psychology, 77, 353-363.

Posner, M.I. and S.W. Keele, 1970, Retention of abstract ideas, Journal of Experimental Psychology, 83, 304-308.
Posner, M.I., 1978, Chronometric explorations of mind. L. Erlbaum Associates, Hillsdale, N.J..
Pray, L., 2001, Long-Term Potentiation Equals Spinal Growth, The Scientist, 15[24], 28.
Price, C.J., 2001, Functional-imaging studies of the 19th Century neurological model of language, Revue Neurologique, 157, 833-836.
Pulvermüller, F. and H. Preissl, 1994, Explaining aphasias in neuronal terms, Journal of Neurolinguistics, 8, 75-81.
Pulvermüller, F., 1995, Agrammatism, behavioral description and neurobiological explanation, Journal of Cognitive Neuroscience, 7, 165-181.
Pulvermüller, F., 1999, Words in the brain's language, Behavioral and Brain Sciences, 22, 253-336.
Pulvermüller, F., 2002, The Neuroscience of Language. On Brain Circuits of Words and Serial Order. Cambridge University Press,
Pulvermüller, F., B. Mohr and H. Schleichert, 1999, Semantic or lexico-syntactic factors: what determines word-class specific activity in the human brain?, Neuroscience Letters, 275, 81-84.
Purves, D. and e. al., 1997, The Auditory System. Sinauer Associates, Inc., Sunderland, Mass.
Raichle, M.E., 2001, Bold insights, Nature, 412, 128-130.
Rall, W., 1959, Branching dendritic trees and motoneuron membrane resistivity, Experimental Neurology, 1, 491-527.
Rall, W., 1962, Theory of physiological properties of dendrites, Annals of the New York Academy of Sciences, 96, 1971-1092.
Rall, W., 1964, Theoretical significance of dendritic trees for neuronal input-output relations, Ed. Reiss, R. Stanford University Press, Stanford, California, pp. 73-97.
Ramón y Cajal, S., 1888, Estructura de los centros nerviosos de las aves, Rev. Trim. Histol. Norm. Pat., 1, 1-10.
Ramón y Cajal, S., 1891, Significación fisiológica de las expansiones protoplásmicas y nerviosas de la sustancia gris, Congreso Médico Valenciano, June 24,
Ramón y Cajal, S., 1893, Neue darstellung vom histologischen bau des centralnervensystem, Arch. Anat. Entwick., 319-428.
Ramón y Cajal, S., 1911, Histologie du Système Nerveux d l'Homme. Maloine, Paris.
Reed, S.K., 1972, Pattern recognition and categorization, Cognitive Psychology, 3, 382-407.
Regier, T., 1996, The Human Semantic Potential: Spatial Language and Constrained Connectionism. MIT Press, Cambridge, Mass..

Reid, R.C., 1999, Fundamental Neuroscience: Vision, Eds. Zigmond, M.J. F.E. Bloom, S.C. Landis, J.L. Roberts, L.R. Squire Academic Press, Inc., San Diego, pp. 821-851.

Reilly, R.G., 2002, The relationship between object manipulation and language development in Broca's area: A connectionist simulation of Greenfield's hypothesis, Behavioral and Brain Sciences,

Reynolds, J.H., L. Chelazzi and R. Desimone, 1999, Competitive mechanisms subserve attention in macaque areas V2 and V4, Journal of Neuroscience, 19, 1736-1753.

Rinzel, J., 1985, Excitation dynamics: insights from simplified membrane models, Fed. Proc., 44, 2944-2946.

Rips, L., 1994, The Psychology of Proof: Deductive reasoning in human thinking. MIT Press, Cambridge, MA.

Rips, L.J., 1989, Similarity and Analogical Reasoning: Similarity, typicality, and categorization, Eds. Vosniadou, S. and A. Ortony Cambridge University Press, Cambridge [England] & New York, pp. 179-195.

Rizzolatti, G. and M.A. Arbib, 1998, Language within our grasp, Trends in Neurosciences, 21, 188-194.

Rizzolatti, G. et al., 1996, Premotor cortex and the recognition of motor actions, Cognitive Brain Research, 3, 131-141.

Rockland, K.S. and D.N. Pandya, 1979, Laminar origins and terminations of cortical connections of the occipital lobe in the rhesus monkey, Brain Research, 179, 3-20.

Rolls, E.T. and A. Treves, 1998, Neural Networks and Brain Function. Oxford University Press, Oxford & New York.

Rooth, M. and B.H. Partee, 1982, Proceedings of the First West Coast Conference on Formal Linguistics: Conjunction, type ambiguity and wide scope 'or', Eds. Flickenger, D., M. Macken and N. Wiegrand Linguistics Department, Stanford University, Stanford, pp. 353-362.

Rosch, E., 1973, Cognitive Development and the Acquisition of Language: On the internal structure of perceptual and semantic categories, Ed. Moore, T. Academic Press, New York, Rosch, E., 1975, Cognitive representations of semantic categories, Journal of Experimental Psychology: General, 104, 192-233.

Rosch, E., 1978, Cognition and Categorization: Principles of categorization, Eds. Rosch, E. and B.B. Lloyd L. Erlbaum Associates, Hillsdale, N.J., pp. 28-49.

Rosch, E., 1978, New Trends in Cognitive Representation: Prototype classification and logical classification: the two systems, Ed. Scholink, E. Lawrence Erlbaum Associates, Hillsdale, NJ, pp. 73-86.

Rose, R.M. and J.L. Hindmarsh, 1989, The assembly of ionic currents in a thalamic neuron I. The three-dimensional model, Proceedings of the Royal Society of London B, 237, 267-288.

Rosenblatt, F., 1958, The perceptron: a probabilistic model for information storage and organization of the brain, Psychological Review, 65, 386-408.

Rosenblatt, F., 1961, Principles of Neurodynamics. Spartan Press, Washington, D.C..
Ross, B.H. and T.L. Spalding, 1994, Thinking and Problem Solving: Concepts and categories, Ed. Sternberg, R.J. Academic Press, San Diego, pp. 119-148.
Ross, B.H. and V.S. Makin, 1999, The Nature of Cognition: Prototype versus exemplar models in cognition, Ed. Sternberg, R.J. MIT Press, Cambridge, Mass. & London, pp. 205-244.
Roth, E.M. and E.J. Shoben, 1983, The effect of context on the structure of categories, Cognitive Psychology, 15, 346-378.
Rumelhart, D.E. and D. Zipser, 1986, Parallel distributed processing: Explorations in the microstructure of cognition. Vol. 1: Foundations: Feature discovery by competitive learning. §5., Eds. Rumelhart, D., J. McClelland and the PDP Research Group MIT Press, Cambridge, Mass., Russo, E., 2000, Debating the Meaning of fMRI. The degree to which the technology reflects neuronal activity remains unclear, The Scientist, 14[18]:20,
Rumelhart, D.E. and J.L. McClelland, 1981, An interactive activation model of context effects in letter perception: Part 2. The contextual enhancement effect and some tests and extensions of the model, Psychological Review, 88, 375-407.
Rumelhart, D.E. and J.L. McClelland, 1986, On learning the past tense of English verbs. MIT Press, Cambridge, Mass..
Rumelhart, D.E., G.E. Hinton and J.L. McClelland, 1986, A general framework for parallel distributed processing. MIT Press, Cambridge, Mass..
Ryalls, J. and A.R. Lecours, 1996, Broca's first two cases: From bumps on the head to cortical convolutions, Classic Cases in Neuropsychology, 235-242.
Samuels, R., 1998, Evolutionary psychology and the massive modularity hypothesis, The British Journal for the Philosophy of Science, 49, 575-602.
Schachter, P., 1977, Constraints on coördination, Language, 53, 86-103.
Schiller, F., 1992, Paul Broca: Founder of French anthropology, explorer of the brain. Oxford University Press, New York.
Schmerling, S., 1975, Syntax and Semantics, vol. 3. Speech Acts: Asymmetric Conjunction and Rules of Conversation, Eds. Cole, P. and J. Morgan Academic Press, New York, NY, pp. 211-231.
Schütze, C.T., 1996, The empirical base of linguistics. University of Chicago Press, Chicago.
Scoville, W. and B. Milner, 1957, Loss of recent memory after bilateral hippocampal lesions, Journal of Neurology, Neurosurgery and Psychiatry, 20, 11-21.
Segal, M. and P. Andersen, 2000, Dendritic spines shaped by synaptic activity, Current Opinion in Neurobiology, 10, 582-587.
Segal, M., 2001, Rapid plasticity of dendritic spine: hints to possible functions?, Progress in Neurobiology, 63, 61-70.

Segev, I. and M. London, 1999, Dendrites: A theoretical view of passive and active dendrites, Eds. Stuart, G., N. Spruston and M. Häusser Oxford University Press, Oxford, UK & New York, USA, pp. 205-230.

Seidenberg, M.S. and J.H. Hoeffner, 1998, Evaluating Behavioral and Neuroimaging Data on Past Tense Processing, Language, 74, 104-122.

Seidenberg, M.S. and J.L. McClelland, 1989, A distributed, developmental model of word recognition and naming, Psychological Review, 96, 523-568.

Seuren, P., 1984, Operator lowering, Linguistics, 22, 573-627.

Seuren, P.A.M., 2000, Presupposition, negation and trivalence, Journal of Linguistics, 36, 261-297.

Seuren, P.A.M., V. Capretta and H. Geuvers, 2001, The logic and mathematics of occasion sentences, Linguistics and Philosophy, 24, 531-595.

Shannon, C.E. and W. Weaver, 1949, The Mathematical Theory of Communication. University of Illinois Press, Urbana.

Shannon, C.E., 1948, The mathematical theory of communication, Bell System Technical Journal, 27,

Shastri, L., 1990, Connectionism and the computational effectiveness of reasoning, Theoretical Linguistics, 16, 65-87.

Shastri, L., 1991, Advances in connectionist and neural computation theory, vol. 1. High level connectionist models: The relevance of connectionism to AI: A representation and reasoning perspective, Eds. Barnden, J.A. and J.B. Pollack Ablex Pub. Corp, Norwood, N.J., pp. 259-283.

Shastri, L., 2001, From transient patterns to persistent structures: A model of episodic memory formation via cortico-hippocampal interactions, Behavioral and Brain Sciences, under revision,

Shatz, C., 1990, Impulse activity and the patterning of connections during CNS development, Neuron, 5, 745-756.

Shepherd, G.M. and C. Koch, 1998, The Synaptic Organization of the Brain: Introduction to synaptic circuits, Ed. Shepherd, G.M. Oxford University Press, New York, pp. 1-36.

Shepherd, G.M. and R.K. Brayton, 1987, Logic operations are properties of computer-simulated interactions between excitable dendritic spinces, Neuroscience, 21.1, 151-165.

Shepherd, G.M., 1999, Fundamental Neuroscience: Information processing in dendrites, Eds. Zigmond, M.J. F.E. Bloom, S.C. Landis, J.L. Roberts, L.R. Squire Academic Press, Inc., San Diego, pp. 363-388.

Sherman, S.M. and R.W. Guillery, 1998, On the actions that one nerve cell can have on another: distinguishing "drivers" from "modulators", Proceedings of the National Academy of Sciences of the United States of America, 95, 7121-7126.

Sherman, S.M., 2000, A new slant on the development of orientation selectivity, Nature Neuroscience, 3, 524-527.

Siegelmann, H. and E. Sontag, 1995, On the computational power of neural nets, Journal of Computer Systems Science, 50, 132-150.

Singer, W., 2000, The New Cognitive Neurosciences: Response synchronization: A universal coding strategy for the definition of relations, Ed. Gazzaniga, M.S. MIT Press, Cambridge, Mass., pp. 325-338 .

Sloutsky, V.M. and Y. Goldvarg, 1999, Proceedings of the XXI Annual Conference of the Cognitive Science Society: Effects of externalization on representation of indeterminate problems, Eds. Hahn, M. and S. Stones Erlbaum, Mahwah, NJ, pp. 695-700.

Smith, B., 1994, Topological Foundations of Cognitive Science: Topological Foundations of Cognitive Science, Eds. Eschenbach, C., C. Habel and B. Smith Graduiertenkolleg Kognitionswissenschaft, Hamburg, pp. 3-22.

Smith, B., 1996, Mereotopology: A Theory of Parts and Boundaries, Data and Knowledge Engineering, 20, 287-303.

Smith, C., 1969, Modern Studies in English: Readings in Transformational Grammar: Ambiguous Sentences with and, Eds. Reibel, D. and S. Schane Prentice-Hall, Englewood Cliffs, N. J., pp. 75-79.

Smith, E. and D.L. Medin, 1981, Categories and Concepts. Harvard University Press, Cambridge, Mass.

Sommers, F., 1970, The calculus of terms, Mind, 79, 1-39.

Sompolinsky, H. and R. Shapley, 1997, New perspectives on the mechanisms for orientation selectivity, Current Opinion in Neurobiology, 7, 514-522.

Sontag, E., 1995, The Handbook of Brain Theory and Neural Networks: Automata and neural networks, Ed. Arbib, M. The MIT Press, Cambridge, USA & London, England, pp. 119-123.

Sparks, R. and N. Geschwind, 1968, Dichotic listening in man after section of neocortical commissures, Cortex, 4, 3-16.

Spencer, W.A. and E.R. Kandel, 1961, Electrophysiology of hippocampal neurons: IV. Fast potentials, Journal of Neurophysiology, ?, 272-285.

Sperber, D. and D. Wilson, 1986, Relevance: Communication and Cognition. Blackwell, Oxford.

Sperber, D. and D. Wilson, 1995, Relevance: Communication and Cognition, 2e: Postface to the second edition of Relevance: Communication and Cognition, Eds. Sperber, D. and D. Wilson Blackwell, Oxford, pp. 255-279.

Spitzer, M., 1999, The Mind within the Net. Models of learning, thinking, and acting. The MIT Press, Cambridge, Mass.

Spruston, N., G. Stuart and M. Häusser, 1999, Dendrites: Dendritic integration, Eds. Stuart, G., N. Spruston and M. Häusser Oxford University Press, Oxford, UK, pp. 231-270.

Squire, L.R. and B. Knowlton, 1995, The Cognitive Neuorsciences: Memory, hippocampus, and brain systems, Ed. Gazzaniga, M. pp. 825-837.

Squire, L.R. and E.R. Kandel, 1999, Memory. From mind to molecules. Scientific American Library, New York.

Squire, L.R. and S. Zola-Morgan, 1988, Memory: brain systems and behavior, Trends in Neurosciences, 11, 170-175.

Squire, L.R. and S. Zola-Morgan, 1991, The medial temporal memory lobe, Science, 253, 1380-1386.

Stalnaker, R., 1968, Studies in Logical Theory: A theory of conditionals, Ed. Rescher, N. Blackwell, Oxford, pp. 121-136.

Stalnaker, R., 1981, Antiessentialism, Midwest Studies in Philosophy, 4, 343-355.

Stemmer, B., 1999, An On-Line Interview with Noam Chomsky: On the nature of pragmatics and related issues, Brain and Language, 68, 3, 393-401.

Stepanyants, A., P.R. Hof and D.B. Chklovskii, 2002, Geometry and structural plasticity of synaptic connectivity, Neuron, 34, 275-288.

Sternberg, R.J., 1996, Coginitve Psychology, 2e. Harcourt Brace College Publishers, Fort Worth, Texas.

Stoll, R.R., 1979, Set Theory and Logic. Dover Publications, Inc., New York.

Stork, D.G., 1989, Is backpropagation biologically plausible?, Proceedings of the International Joint Conference on Neural Networks, vol. 2., 241-246.

Sugita, Y., 1999, Grouping of image fragments in primary visual cortex, Nature, 401, 269-272.

Talmy, L., 1983, Spatial orientation. Theory, research, and application: How language structures space, Eds. Pick, H.L. and L.P. Acredolo Plenum Press, New York, pp. 225-282.

Tanaka, K., 1996, Inferotemporal cortex and object vision, Annual Review of Neuroscience, 19, 109-139.

Tarski, A., 1935, Der Warheitsbegriff in den formalizierten Sprachen, Studia Philosophica, 1, 261-405.

Tarski, A., 1944, The semantic conception of truth, Philosophy and Phenomenological Research, 4, 341-375.

Thorpe, S., A. Delorme and R. VanRullen, 2001, Spike-based strategies for rapid processing, Neural Networks, 14, 715-725.

Thorpe, S.J., 1995, The Handbook of Brain Theory and Neural Networks: Localized versus distributed representations, Ed. Arbib, M. The MIT Press, Cambridge, USA & London, England, pp. 549-552.

Toni, N. et al., 1999, LTP promotes formation of multiple spine synapses between a single axon terminal and a dendrite, Nature, 402, 421-425.

Touretzky, D.S., 1995, The Handbook of Brain Theory and Neural Networks: Connectionist and symbolic representations, Ed. Arbib, M. The MIT Press, Cambridge, USA & London, England, pp. 243-247.

Toyama, K., M. Kimura and K. Tanaka, 1981, Organization of cat visual cortex as investigated by cross–correlation technique, Journal of Neurophysiology, 46, 202-214.

Trappenberg, T.T., 2002, Fundamentals of Computational Neuroscience. Oxford University Press,

Treisman, A., 1996, The binding problem, Current Opinion in Neurobiology, 6, 171-178.

Troyer, T.W. et al., 1998, Contrast-invariant orientation tuning in cat visual cortex: Feedforward tuning and correlation-based intracortical connectivity, Journal of Neuroscience, 18, 5908-5927.

Trubetzkoy, N., 1969, Principles of Phonology. University of California Press, Berkeley & Los Angeles.

Turing, A.M., 1952, The chemical basis of morphogenesis, Philosophical Transactions of the Royal Society B, 237, 5-12.

Turrigiano, G.G., 1999, Homeostatic plasticity in neuronal networks: the more things change, the more they stay the same, Trends in Neurosciences, 22, 221-227.

Ullman, M.T. et al., 1997, A neural dissociation within language: Evidence that the mental dictionary is part of declarative memory, and that grammatical rules are processed by the procedural system, Journal of Cognitive Neuroscience, 9, 266-276.

Ungerleider, L.G. and M. Mishkin, 1982, Analysis of Visual Behavior: Two cortical visual systems, Eds. Ingle, D., M.A. Goodale and R.J.W. Mansfield Cambridge, pp. MIT Press.

Usher, M. and J.D. Cohen, 1999, Connectionist Models in Cognitive Neuroscience: The Fifth Neural Computation and Psychology Workshop: Short Term Memory and Selection Processes in a Frontal-Lobe Model, Eds. Heinke, D., G.W. Humphreys and A. Olson Springer-Verlag, London, pp. 78-91.

Usrey, W.M., J.B. Reppas and R.C. Reid, 1999, Specificity and strength of retinogeniculate connections, Journal of Neurophysiology, 82, 3527-3540.

Uylings, H.B.M. et al., 1999, The Neurocognition of Language: Broca's language area from a neuroanatomical and developmental perspective, Eds. Brown, C.M. and P. Hagoort Oxford University Press, Oxford, GB, pp. 241-272.

Van Essen, D.C., C.H. Anderson and B.A. Olshausen, 1994, Large-scale neuronal theories of the brain: Dynamic routing strategies in sensory, motor, and cognitive processing, Eds. Koch, C. and J.L. Davis MIT Press, Cambridge, Mass, pp. 1-23.

Van Horn, S.C., A. Erisir and S.M. Sherman, 2000, Relative distribution of synapses in the A-laminae of the lateral geniculate nucleus of the cat, Journal of Comparative Neurology, 416, 509-520.

Van Lancker, D.R., J. Kreiman and J. Cummings, 1989, Voice perception deficits: neuroanatomical correlates of phonagnosia, Journal of Clinical and Experimental Neuropsychology, 11, 665-674.

van Rooy, R., 2001, Proceedings of Tark 2001: Relevance of communicative acts,

Vandeloise, C., 1991, Spatial Prepositions: A Case Study from French. University of Chicago Press, Chicago, Illinois.

Vanier, M. and D. Caplan, 1990, Agrammatic Aphasia: CT-scan correlates of agrammatism, Eds. Menn, L. and L.K. Obler Benjamins, Amsterdam, pp. 97-114.

VanRullen, R. and S.J. Thorpe, 1999, Spatial attention in asynchronous neural networks, NeuroComputing, 26-27, 911-918.

Verkuyl, H., 1981, Formal Methods in the Study of Language, Part II: Numerals and quantifiers in X' Syntax and their semantic interpretation, Eds. Groenendijk, J.A.G., T.M.V. Janssen and M.B.J. Stokhof Matematisch Centrum, Amsterdam, pp. 567-599.

Visch-Brink, E. & Bastiaanse, R. (Eds.) (1998) Singular Publishing Group, San Diego & London.

Voronoi, G.M., 1908, Nouvelles applications des parametres continus la thorie des formes quadratiques, J. Reine Angew. Math., 134, 198-287.

Walker, J.H., 1975, Real world variability, reasonableness judgments, and memory representations for concepts, Journal of Verbal Learning and Verbal Behavior, 14, 241-252.

Weiss, T.F., 1996, Cellular biophysics. Vol. 2: Electrical properties. MIT Press, Cambridge, Massachusetts.

Westerståhl, D., 1989, Handbook of Philosophical Logic, vol 4: Topics in the Philosophy of Language: Quantifiers in Formal and Natural Languages, Eds. Gabbay, D. and F. Guenthner Reidel, Dordrecht, pp. 1-132.

Whittlesea, B.W.A., 1987, Preservation of specific experiences in the representation of general knowledge, Journal of Experimental Psychology: Learning, Memory, and Cognition, 13, 3-17.

Wickens, T.D., 1994, The Geometry of Multivariate Statistics. Lawrence Erlbaum Associates, Inc.,

Wildgen, W., 1994, Process, Image, and Meaning: A Realistic Model of the Meaning of Sentences and Narrative Texts. John Benjamins, Amsterdam & Philadelphia.

Wilson, D. and D. Sperber, 2002, 14: Relevance Theory, Eds. Sperber, D. and D. Wilson pp. 249-.

Wilson, H.R., 1999, Spikes, Decisions, and Actions: The Dynamical Foundations of Neurosciences. Oxford University Press, Oxford, UK & New York, USA.

Yuille, A. and D. Geiger, 1995, The Handbook of Brain Theory and Neural Networks: Winner-Take-All mechanisms, Ed. Arbib, M. The MIT Press, Cambridge, USA & London, England, pp. 1056-1060.

Yuste, R. and A. Majewska, 2001, On the function of dendritic spines, Neuroscientist, 7, 387-395.

Yuste, R. and T. Bonhoeffer, 2001, Morphological changes in dendritic spines associated with long-term synaptic plasticity, Annual Review of Neuroscience, 24, 1071-1089.

Yuste, R. and W. Denk, 1995, Dendritic spines as basic units of synaptic integration, Nature, 375, 682-684.

Yuste, R., A. Majewska and K. Holthoff, 2000, From form to function: calcium compartmentalization in dendritic spines, Nature Neuroscience, 3, 653-659.

Zaidel, E., 1985, The Dual Brain: Language in the right hemisphere, Eds. Benson, D.F. and E. Zaidel Guilford Press, New York, pp. 205-231.

Zeki, S., 2001, Localization and Globalization in Conscious Vision, Annual Review of Neuroscience, 24, 57-86.

Zemel, R.S. (1994) A Minimum Description Length Framework for Unsupervised Learning. Ph.D. thesis, University of Toronto.

Zwarts, J. and Y. Winter, 2000, Vector Space Semantics: a model-theoretic analysis of locative prepositions, Journal of Logic, Language and Information, 9, 171-213.

Zwarts, J., 1997, Vectors as relative positions: a compositional semantics of modified PPs, Journal of Semantics, 14, 57-86.

Index

natural property, 195, 197, 468

CPSIA information can be obtained at www.ICGtesting.com
Printed in the USA
LVOW09*0817100715

445474LV00010B/207/P